METHODS IN MOLECULAR BIOLOGY™

Series Editor
John M. Walker
School of Life Sciences
University of Hertfordshire
Hatfield, Hertfordshire, AL10 9AB, UK

For other titles published in this series, go to
www.springer.com/series/7651

Liposomes

Methods and Protocols
Volume 1: Pharmaceutical Nanocarriers

Edited by

Volkmar Weissig

Department of Pharmaceutical Sciences, Midwestern University College of Pharmacy Glendale, Glendale, AZ, USA

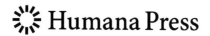

Editor
Volkmar Weissig
Department of Pharmaceutical Sciences
Midwestern University College of Pharmacy Glendale
Glendale, AZ
USA
vweiss@midwestern.edu

ISSN 1064-3745   e-ISSN 1940-6029
ISBN 978-1-60327-359-6   e-ISBN 978-1-60327-360-2
DOI 10.1007/978-1-60327-360-2
Springer New York Dordrecht Heidelberg London

Library of Congress Control Number: 2009933261

© Humana Press, a part of Springer Science+Business Media, LLC 2010
All rights reserved. This work may not be translated or copied in whole or in part without the written permission of the publisher (Humana Press, c/o Springer Science+Business Media, LLC, 233 Spring Street, New York, NY 10013, USA), except for brief excerpts in connection with reviews or scholarly analysis. Use in connection with any form of information storage and retrieval, electronic adaptation, computer software, or by similar or dissimilar methodology now known or hereafter developed is forbidden.
The use in this publication of trade names, trademarks, service marks, and similar terms, even if they are not identified as such, is not to be taken as an expression of opinion as to whether or not they are subject to proprietary rights.
While the advice and information in this book are believed to be true and accurate at the date of going to press, neither the authors nor the editors nor the publisher can accept any legal responsibility for any errors or omissions that may be made. The publisher makes no warranty, express or implied, with respect to the material contained herein.

Cover illustration: Background art is derived from Figure 10 in Chapter 26

Printed on acid-free paper

Humana Press is a part of Springer Science+Business Media (www.springer.com)

Preface

Efforts to describe and model the molecular structure of biological membranes go back to the beginning of the last century. In 1917, Langmuir described membranes as a layer of lipids one molecule thick [1]. Eight years later, Gorter and Grendel concluded from their studies that "the phospholipid molecules that formed the cell membrane were arranged in two layers to form a lipid bilayer" [2]. Danielli and Robertson proposed, in 1935, a model in which the bilayer of lipids is sequestered between two monolayers of unfolded proteins [3], and the currently still accepted fluid mosaic model was proposed by Singer and Nicolson in 1972 [4].

Among those landmarks of biomembrane history, a serendipitous observation made by Alex Bangham during the early 1960s deserves undoubtedly a special place. His finding that exposure of dry phospholipids to an excess of water gives rise to lamellar structures [5] has opened versatile experimental access to studying the biophysics and biochemistry of biological phospholipid membranes.

Although during the following 4 decades biological membrane models have grown in complexity and functionality [6], liposomes are, besides supported bilayers, membrane nanodiscs, and hybrid membranes, still an indisputably important tool for membrane biophysicists and biochemists. In vol. II of this book, the reader will find detailed methods for the use of liposomes in studying a variety of biochemical and biophysical membrane phenomena concomitant with chapters describing a great palette of state-of-the-art analytical technologies.

Moreover, besides providing membrane biophysicists and biochemists with an immeasurably valuable experimental tool, Alex Bangham's discovery has triggered the launch of an entirely new subdiscipline in pharmaceutical science and technology. His observation that the lamellar structures formed by phospholipids exposed to aqueous buffers are able to sequester small molecules has lead to the development of the colloidal drug delivery concept. Following initial studies of enzyme encapsulation in liposomes as an approach towards the treatment of storage diseases [7, 8], a few years later in two New England Journal of Medicine landmark papers, Gregory Gregoriadis outlined the huge carrier potential of liposomes in biology and medicine [9, 10]. The following 2 decades saw immense efforts in academia and in soon-to-be-founded start-up companies to turn Gregoriadis' vision into clinical reality. These 20 years of intense work in liposome laboratories around the world finally culminated with the FDA (USA) approval of the first injectable liposomal drug, Doxil, in February of 1995. Today, liposomes present the prototype of all nanoscale drug delivery vectors currently under development. Lessons learned in the history of over 40 years of Liposome Technology should be heeded by new investigators in the emerging field of pharmaceutical and biomedical nanotechnology. Volume I of this book is dedicated to state-of-the-art aspects of developing liposome-based pharmaceutical nanocarriers.

All chapters were written by leading experts in their particular fields, and I am extremely grateful to them for having spent parts of their valuable time to contribute to this book. It is my hope that together we have succeeded in providing an essential source of practical know-how for every investigator, young and seasoned ones alike, whose research area involves in one way or another phospholipids, glycolipids, and cholesterol.

Last but not least, I would like to thank John Walker, the series editor of "Methods in Molecular Biology," for having invited me to assemble this book and above all for his unlimited guidance and help throughout the whole process.

Glendale, AZ   *Volkmar Weissig*

References

1. Langmuir I (1917) The constitution and structural properties of solids and liquids. II. Liquids. J Am Chem Soc 39:1848–1906
2. Gorter E, Grendel F (1925) On bimolecular layers of lipids on the chromocytes of the blood. J Exp Med 41:439–443
3. Danielli JF, Davson H (1925) A contribution to the theory of permebaility of thin films. J Cell Comp Physiol 5:495–508
4. Singer SJ, Nicolson GL (1972) The fluid mosaic model of the structure of cell membranes. Science 175(23):720–31
5. Bangham AD, Standish MM, Watkins JC (1965) Diffusion of univalent ions across the lamellae of swollen phospholipids. J Mol Biol 13(1):238–52
6. Chan YH, Boxer SG (2007) Model membrane systems and their applications. Curr Opin Chem Biol 11(6):581–7
7. Gregoriadis G, Ryman BE (1971) Liposomes as carriers of enzymes or drugs: a new approach to the treatment of storage diseases. Biochem J 124(5):58P
8. Gregoriadis G, Leathwood PD, Ryman BE (1971) Enzyme entrapment in liposomes. FEBS Lett 14(2):95–99
9. Gregoriadis G (1976) The carrier potential of liposomes in biology and medicine (second of two parts). N Engl J Med 295(14):765–70
10. Gregoriadis G (1976) The carrier potential of liposomes in biology and medicine (first of two parts). N Engl J Med 295(13):704–10

Contents

Preface v
Contributors xi

1. Current Trends in Liposome Research 1
   Tamer A. ElBayoumi and Vladimir P. Torchilin
2. Nanoliposomes: Preparation and Analysis 29
   M.R. Mozafari
3. Preparation of DRV Liposomes 51
   Sophia G. Antimisiaris
4. Elastic Liposomes for Topical and Transdermal Drug Delivery 77
   Heather A.E. Benson
5. Archaebacterial Tetraetherlipid Liposomes 87
   Aybike Ozcetin, Samet Mutlu, and Udo Bakowsky
6. Cationic Magnetoliposomes 97
   Marcel De Cuyper and Stefaan J.H. Soenen
7. Ultrasound-Responsive Liposomes 113
   Shao-Ling Huang
8. Liposome Formulations of Hydrophobic Drugs 129
   Reto A. Schwendener and Herbert Schott
9. Remote Loading of Anthracyclines into Liposomes 139
   Felicitas Lewrick and Regine Peschka Süss
10. Arsonoliposomes: Preparation and Physicochemical Characterization 147
    Sophia G. Antimisiaris and Panayiotis V. Ioannou
11. Liposome-Based Vaccines 163
    Reto A. Schwendener, Burkhard Ludewig, Andreas Cerny, and Olivier Engler
12. Mannosylated Liposomes for Targeted Vaccines Delivery 177
    Suresh Prasad Vyas, Amit K. Goyal, and Kapil Khatri
13. Liposomes for Specific Depletion of Macrophages from Organs and Tissues 189
    Nico van Rooijen and Esther Hendrikx
14. Vesicular Phospholipid Gels 205
    Martin Brandl
15. Environment-Responsive Multifunctional Liposomes 213
    Amit A. Kale and Vladimir P. Torchilin
16. Functional Liposomal Membranes for Triggered Release 243
    Armağan Koçer
17. A "Dock and Lock" Approach to Preparation of Targeted Liposomes 257
    Marina V. Backer and Joseph M. Backer

18	Conjugation of Ligands to the Surface of Preformed Liposomes by Click Chemistry... *Benoît Frisch, Fatouma Saïd Hassane, and Francis Schuber*	267
19	Targeted Magnetic Liposomes Loaded with Doxorubicin.................. *Pallab Pradhan, Rinti Banerjee, Dhirendra Bahadur, Christian Koch, Olga Mykhaylyk, and Christian Plank*	279
20	Liposomes for Drug Delivery to Mitochondria......................... *Sarathi V. Boddapati, Gerard G.M. D'Souza, and Volkmar Weissig*	295
21	Cytoskeletal-Antigen Specific Immunoliposomes: Preservation of Myocardial Viability.. *Vishwesh Patil, Tala Khudairi, and Ban-An Khaw*	305
22	Gadolinium-Loaded Polychelating Polymer-Containing Tumor-Targeted Liposomes.. *Suna Erdogan and Vladimir P. Torchilin*	321
23	Angiogenic Vessel-Targeting DDS by Liposomalized Oligopeptides.......... *Tomohiro Asai and Naoto Oku*	335
24	TAT-Peptide Modified Liposomes: Preparation, Characterization, and Cellular Interaction... *Marjan M. Fretz and Gert Storm*	349
25	ATP-Loaded Liposomes for Targeted Treatment in Models of Myocardial Ischemia.. *Tatyana S. Levchenko, William C. Hartner, Daya D. Verma, Eugene A. Bernstein, and Vladimir P. Torchilin*	361
26	Intracellular ATP Delivery Using Highly Fusogenic Liposomes.............. *Sufan Chien*	377
27	Lipoplex Formation Using Liposomes Prepared by Ethanol Injection.......... *Yoshie Maitani*	393
28	Acid-Labile Liposome/pDNA Complexes.............................. *Michel Bessodes and Daniel Scherman*	405
29	Serum-Resistant Lipoplexes in the Presence of Asialofetuin................. *Conchita Tros de ILarduya*	425
30	Anionic pH Sensitive Lipoplexes..................................... *Nathalie Mignet and Daniel Scherman*	435
31	Liposomal siRNA Delivery... *Jeffrey Hughes, Preeti Yadava, and Ryan Mesaros*	445
32	Complexation of siRNA and pDNA with Cationic Liposomes: The Important Aspects in Lipoplex Preparation....................... *José Mario Barichello, Tatsuhiro Ishida, and Hiroshi Kiwada*	461
33	Effective In Vitro and In Vivo Gene Delivery by the Combination of Liposomal Bubbles (Bubble Liposomes) and Ultrasound Exposure......... *Ryo Suzuki and Kazuo Maruyama*	473
34	Liposomal Magnetofection.. *Olga Mykhaylyk, Yolanda Sánchez Antequera, Dialekti Vlaskou, Edelburga Hammerschmid, Martina Anton, Olivier Zelphati, and Christian Plank*	487

35 Long-Circulating, pH-Sensitive Liposomes..................... 527
    Denitsa Momekova, Stanislav Rangelov, and Nikolay Lambov
36 Serum-Stable, Long-Circulating, pH-Sensitive PEGylated Liposomes.......... 545
    Nicolas Bertrand, Pierre Simard, and Jean-Christophe Leroux

Index *559*

Contributors

YOLANDA SÁNCHEZ ANTEQUERA • *Institute of Experimental Oncology and Therapy Research, Technische Universität München, München, Germany*
SOPHIA G. ANTIMISIARIS • *Laboratory of Pharmaceutical Technology, Department of Pharmacy, University of Patras, and ICE-HT/ FORTH, Patras, Greece*
MARTINA ANTON • *Institute of Experimental Oncology and Therapy Research, Technische Universität München, München, Germany*
TOMOHIRO ASAI • *Department of Medical Biochemistry and Global COE, School of Pharmaceutical Sciences, University of Shizuoka, Shizuoka, Japan*
JOSEPH M. BACKER • *SibTech, Inc., Brookfield, CT, USA*
MARINA V. BACKER • *SibTech, Inc., Brookfield, CT, USA*
DHIRENDRA BAHADUR • *Department of Metallurgical Engineering and Materials Science, IIT Bombay, Mumbai, India*
UDO BAKOWSKY • *Department of Pharmaceutical Technology and Biopharmacy, Philipps-Universität Marburg, Marburg, Germany*
RINTI BANERJEE • *School of Biosciences and Bioengineering, IIT Bombay, Mumbai, India*
JOSÉ MARIO BARICHELLO • *Department of Pharmacokinetics and Biopharmaceutics, Subdivision of Biopharmaceutical Sciences, Institute of Health Biosciences, The University of Tokushima, Tokushima, Japan*
HEATHER A.E. BENSON • *School of Pharmacy, Curtin University of Technology, Perth, WA, Australia*
EUGENE A. BERNSTEIN • *Department of Pharmaceutical Sciences, Bouve College of Health Sciences, Northeastern University, Boston, MA, USA*
NICOLAS BERTRAND • *Faculty of Pharmacy, University of Montreal, Montreal, QC, Canada*
MICHEL BESSODES • *Unité de Pharmacologie Chimique et Génétique, Inserm, U640, Paris, France*
Faculté des Sciences Pharmaceutiques et Biologiques, Univ. Paris Descartes, Paris, France
CNRS, UMR 8151, Paris, France
ENSCP, Paris, France
SARATHI V. BODDAPATI • *Department of Pharmaceutical Sciences, Northeastern University, Boston, MA, USA*
MARTIN BRANDL • *Department of Physics and Chemistry, University of Southern Denmark, Odense M, Denmark*
ANDREAS CERNY • *Ambulatorio di Epatologia, Lugano, Switzerland*
SUFAN CHIEN • *Department of Surgery, University of Louisville, Louisville, KY, USA*

MARCEL DE CUYPER • *Laboratory of BioNanoColloids, Interdisciplinary Research Center, Katholieke Universiteit Leuven, Kortrijk, Belgium*
CONCHITA TROS DE ILARDUYA • *Department of Pharmacy and Pharmaceutical Technology, School of Pharmacy, University of Navarra, Pamplona, Spain*
GERARD G.M. D'SOUZA • *Department of Pharmaceutical Sciences, Massachusetts College of Pharmacy and Health Sciences, Boston, MA, USA*
TAMER A. ELBAYOUMI • *Department of Pharmaceutical Sciences, College of Pharmacy Glendale, Midwestern University, Glendale, AZ, USA*
OLIVIER ENGLER • *Federal Office for Civil Protection (FOCP), Spiez Laboratory, Spiez, Switzerland*
SUNA ERDOGAN • *Department of Radiopharmacy, Faculty of Pharmacy, Hacettepe University, Ankara, Turkey*
MARJAN M. FRETZ • *Department of Pharmaceutics, Utrecht Institute for Pharmaceutical Sciences (UIPS), Utrecht University, Utrecht, The Netherlands*
BENOÎT FRISCH • *Départment de Chimie Bioorganique, Faculté de Pharmacie, Institut Gilbert Lautriat, UMR 7175-LC1 CNRS-Université Louis Pasteur, Strasbourg-Illkirch, France*
AMIT K. GOYAL • *Drug Delivery Research Laboratory, Department of Pharmaceutical Sciences, Sagar, India*
EDELBURGA HAMMERSCHMID • *Institute of Experimental Oncology and Therapy Research, Technische Universität München, München, Germany*
WILLIAM C. HARTNER • *Department of Pharmaceutical Sciences, Bouve College of Health Sciences, Northeastern University, Boston, MA, USA*
FATOUMA SAÏD HASSANE • *Départment de Chimie Bioorganique, Faculté de Pharmacie, Institut Gilbert Lautriat, UMR 7175-LC1 CNRS-Université Louis Pasteur, Strasbourg-Illkirch, France*
ESTHER HENDRIKX • *Department of Molecular Cell Biology and Immunology, VU University Medical Center, Amsterdam, The Netherlands*
SHAO-LING HUANG • *Cardiology Division, Department of Internal Medicine, University of Texas Health Science Center, Houston, TX, USA*
JEFFREY HUGHES • *Department of Pharmaceutics, College of Pharmacy, University of Florida, Gainesville, FL, USA*
PANAYIOTIS V. IOANNOU • *Department of Chemistry, University of Patras, Patras, Greece*
TATSUHIRO ISHIDA • *Department of Pharmacokinetics and Biopharmaceutics, Subdivision of Biopharmaceutical Sciences, Institute of Health Biosciences, The University of Tokushima, Tokushima, Japan*
AMIT A. KALE • *Department of Pharmaceutical Sciences, Center for Pharmaceutical Biotechnology and Nanomedicine, Northeastern University, Boston, MA, USA*
KAPIL KHATRI • *Drug Delivery Research Laboratory, Department of Pharmaceutical Sciences, Sagar, India*
BAN-AN KHAW • *Department of Pharmaceutical Sciences, Center for Cardiovascular Targeting, Bouve College of Health Sciences, Northeastern University, Boston, MA, USA*
TALA KHUDAIRI • *Department of Pharmaceutical Sciences, Center for Cardiovascular Targeting, Bouve College of Health Sciences, Northeastern University, Boston, MA, USA*

HIROSHI KIWADA • *Department of Pharmacokinetics and Biopharmaceutics, Subdivision of Biopharmaceutical Sciences, Institute of Health Biosciences, The University of Tokushima, Tokushima, Japan*

ARMAĞAN KOÇER • *Department of Biochemistry, Groningen Biomolecular Sciences and Biotechnology Institute, Zernike Institute for Advanced Materials, University of Groningen, Groningen, The Netherlands*

CHRISTIAN KOCH • *Institute of Experimental Oncology and Therapy Research, Technische Universität München, München, Germany*

NIKOLAY LAMBOV • *Department of Pharmaceutical Technology and Biopharmaceutics, Faculty of Pharmacy, Medical University-Sofia, Sofia, Bulgaria*

JEAN-CHRISTOPHE LEROUX • *Department of Chemistry and Applied Biosciences, Institute of Pharmaceutical Sciences, ETH Zürich, Zürich, Switzerland*

TATYANA S. LEVCHENKO • *Department of Pharmaceutical Sciences, Bouve College of Health Sciences, Northeastern University, Boston, MA, USA*

FELICITAS LEWRICK • *Department of Pharmaceutical Technology and Biopharmacy, Albert-Ludwigs University, Freiburg, Germany*

BURKHARD LUDEWIG • *Research Department, Kantonal Hospital St. Gallen, St. Gallen, Switzerland*

YOSHIE MAITANI • *Institute of Medicinal Chemistry, Hoshi University, Shinagawa, Tokyo, Japan*

KAZUO MARUYAMA • *Department of Biopharmaceutics, School of Pharmaceutical Sciences, Teikyo University, Kanagawa, Japan*

RYAN MESAROS • *Department of Pharmaceutics, College of Pharmacy, University of Florida, Gainesville, FL, USA*

NATHALIE MIGNET • *Unité de Pharmacologie Chimique et Génétique, Inserm, U640, Paris, France*
Faculté des Sciences Pharmaceutiques et Biologiques, University Paris Descartes, Paris, France
CNRS, UMR8151, Paris, France
ENSCP, Paris, France

DENITSA MOMEKOVA • *Department of Pharmaceutical Technology and Biopharmaceutics, Faculty of Pharmacy, Medical University-Sofia, Sofia, Bulgaria*

M. REZA MOZAFARI • *Phosphagenics R&D Laboratory, Clayton, VIC, Australia*

SAMET MUTLU • *Department of Molecular Biology and Genetics, Istanbul Technical University, Istanbul, Turkey*

OLGA MYKHAYLYK • *Institute of Experimental Oncology and Therapy Research, Technische Universität München, München, Germany*

NAOTO OKU • *Department of Medical Biochemistry and Global COE, School of Pharmaceutical Sciences, University of Shizuoka, Shizuoka, Japan*

AYBIKE OZCETIN • *Department of Pharmaceutical Technology and Biopharmacy, Philipps-Universität Marburg, Marburg, Germany*

VISHWESH PATIL • *Department of Pharmaceutical Sciences, Center for Cardiovascular Targeting, Bouve College of Health Sciences, Northeastern University, Boston, MA, USA*

REGINE PESCHKA-SÜSS • *Department of Pharmaceutical Technology and Biopharmacy, Albert-Ludwigs University, Freiburg, Germany*

CHRISTIAN PLANK • *Institute of Experimental Oncology and Therapy Research, Technische Universität München, München, Germany*

PALLAB PRADHAN • *School of Biosciences and Bioengineering, IIT Bombay, Mumbai, India*
Institute of Experimental Oncology and Therapy Research, Technische Universität München, München, Germany

STANISLAV RANGELOV • *Institute of Polymers, Bulgarian Academy of Sciences, Sofia, Bulgaria*

NICO VAN ROOIJEN • *Department of Molecular Cell Biology and Immunology, VU University Medical Center, Amsterdam, The Netherlands*

DANIEL SCHERMAN • *Unité de Pharmacologie Chimique et Génétique, Inserm, U640, Paris, France; Faculté des Sciences Pharmaceutiques et Biologiques, Univ. Paris Descartes, Paris, France; CNRS, UMR8151, Paris, France; ENSCP, Paris, France*

HERBERT SCHOT • *Institute of Organic Chemistry, Eberhard-Karls University, Tübingen, Germany*

FRANCIS SCHUBER • *Départment de Chimie Bioorganique, Faculté de Pharmacie, Institut Gilbert Lautriat, UMR 7175-LC1 CNRS-Université Louis Pasteur, Strasbourg-Illkirch, France*

RETO A. SCHWENDENER • *Institute of Molecular Cancer Research, University of Zürich, Zurich, Switzerland*

PIERRE SIMARD • *Faculty of Pharmacy, University of Montreal, Montreal, QC, Canada*

STEFAAN J.H. SOENEN • *Laboratory of BioNanoColloids, Interdisciplinary Research Center, Katholieke Universiteit Leuven, Kortrijk, Belgium*

GERT STORM • *Department of Pharmaceutics, Utrecht Institute for Pharmaceutical Sciences (UIPS), Utrecht University, Utrecht, The Netherlands*

RYO SUZUKI • *Department of Biopharmaceutics, School of Pharmaceutical Sciences, Teikyo University, Kanagawa, Japan*

VLADIMIR P. TORCHILIN • *Department of Pharmaceutical Sciences, Center for Pharmaceutical Biotechnology and Nanomedicine, Northeastern University, Boston, MA, USA*

DAYA D. VERMA • *Department of Pharmaceutical Sciences, Bouve College of Health Sciences, Northeastern University, Boston, MA, USA*

DIALEKTI VLASKOU • *Institute of Experimental Oncology and Therapy Research, Technische Universität München, München, Germany*

SURESH PRASAD VYAS • *Drug Delivery Research Laboratory, Department of Pharmaceutical Sciences, Sagar, India*

VOLKMAR WEISSIG • *Department of Pharmaceutical Sciences, Midwestern University College of Pharmacy Glendale, Glendale, AZ, USA*

PREETI YADAVA • *Department of Pharmaceutics, College of Pharmacy, University of Florida, Gainesville, FL, USA*

OLIVIER ZELPHATI • *OZ Biosciences, Marseille, France*

Chapter 1

Current Trends in Liposome Research

Tamer A. ElBayoumi and Vladimir P. Torchilin

Abstract

Among the several drug delivery systems, liposomes – phospholipid nanosized vesicles with a bilayered membrane structure – have drawn a lot of interest as advanced and versatile pharmaceutical carriers for both low and high molecular weight pharmaceuticals. At present, liposomal formulations span multiple areas, from clinical application of the liposomal drugs to the development of various multifunctional liposomal systems to be used in therapy and diagnostics. This chapter provides a brief overview of various liposomal products currently under development at experimental and preclinical level.

Key words: Liposomes, Drug delivery, Drug targeting, Protein and peptide drugs, Gene delivery

1. Introduction

The clinical utility of most conventional therapies is limited either by the inability to deliver therapeutic drug concentrations to the target tissues or by severe and harmful toxic effects on normal organs and tissues. Different approaches have been attempted to overcome these problems by providing "selective" delivery of drugs to the affected area using various pharmaceutical carriers. Among the different types of particulate carriers, liposomes have received the most attention. For more than three decades, liposomes – artificial phospholipid vesicles – obtained by various methods from lipid dispersions in water and capable of encapsulating the active drug, have been recognized as the pharmaceutical carrier of choice for numerous practical applications (1–3). From the biomedical point of view, liposomes are biocompatible, cause very little or no antigenic, pyrogenic, allergic and toxic reactions; they easily undergo biodegradation; they protect the host from any undesirable effects of the encapsulated drug, at the same time protecting the entrapped drugs from the inactivating action of the physiological medium; and, last but not least, liposomes are capable of delivering their content inside many cells (4). Whether the drug is encapsulated in the core or in the bilayer of the liposome is dependent on the characteristics of the drug and the encapsulation process (5). By now, many different methods have been suggested for preparing liposomes of different sizes, structure and size distribution, and a lot of relevant information can be easily found (3, 6–9).

Biodistribution of liposomes is a very important parameter from the clinical point of view. Liposomes can alter both the tissue distribution and the rate of clearance of the drug by making the drug take on the pharmacokinetic characteristics of the carrier (10, 11). The pharmacokinetic variables of the liposomes depend on the physiochemical characteristics of the liposomes, such as size, surface charge, membrane lipid packing, steric stabilization, dose, and route of administration. As with other microparticulate delivery systems, conventional liposomes are vulnerable to elimination from systemic circulation by the cells of the reticuloendothelial system (RES) (12). The primary sites of accumulation of conventional liposomes are the tumor, liver, and spleen compared with non-liposomal formulations (13). Many studies have shown that within the first 15–30 min after intravenous administration of liposomes between 50 and 80% of the dose is adsorbed by the cells of the RES, primarily by the Kupffer cells of the liver (14–16).

2. Recent Examples of Conventional Liposomes in Clinical Therapeutic Applications

On the basis of many studies, involving reducing liposome size and modulating their surface charge, to reduce RES uptake, small (80–200 nm) "conventional" liposomes composed of neutral and/or negatively charged lipids and cholesterol have been prepared. Some of these formulations have already reached the market (Table 1) or are now entering clinical trials. Primary examples are Ambisome® (Gilead Sciences, Foster City, CA, USA) in which the encapsulated drug is the antifungal amphotericin B (Veerareddy and Vobalaboina 2004), Myocet® (Elan Pharmaceuticals Inc., Princeton, NJ, USA) encapsulating the anticancer agent doxorubicin (17), and Daunoxome® (Gilead Sciences), where the entrapped drug is daunorubicin (18). Daunoxome is at present the only pure-lipid MPS-avoiding liposomal formulation; available as a stable ready-to-inject liposomal formulation.

By addition of sphingomyelin and saturated fatty acid chain lipids to the cholesterol-rich liposomes, several formulations have been produced. For example, novel liposomal vincristine,

Table 1
Some liposomal drugs approved for clinical application or under clinical evaluation (in different countries, same drug could be approved for different indications)

Active drug (and product name for liposomal preparation where available)	Indications
Daunorubicin (DaunoXome)	Kaposi's sarcoma
Doxurubicin (Mycet)	Combinational therapy of recurrent breast cancer
Doxorubicin in PEG-liposomes (Doxil, Caelyx)	Refractory Kaposi's sarcoma; ovarian cancer; recurrent breast cancer
Annamycin PEG-liposomes	Doxorubicin-resistant tumors
Amphotericin B (AmBisome)	Fungal infections
Cytarabine (DepoCyt)	Lymphomatous meningitis
Vincristine (Marqibo)	Metastatic malignant uveal melanoma
Lurtotecan (OSI-211)	Ovarian cancer
Irinotecan (LE-SN38)	Advanced cancer
Camptothecin analogue (S-CDK602)	Various tumors
Topotecan (INX-0076)	Advanced cancer
Mitoxantrone (LEM-ETU)	Leukemia, breast, stomach and ovarian cancers
Nystatin (Nyotran)	Topical anti-fungal agent
All-trans retinoic acid (Altragen)	Acute promyelocytic leukemia; non-Hodgkin's lymphoma; renal cell carcinoma; Kaposi's sarcoma
Platinum compounds (Platar)	Solid tumors
Cisplatin	Germ cell cancers, small-cell lung Carcinoma
Cisplatin (SPI-077)	Head and neck cancer
Cisplatin (Lipoplatin)	Various tumors
Paclitaxel (LEP-ETU)	Ovarian, breast and lung cancers
E1A gene-cationic liposome	Various tumors
DNA plasmid encoding HLA-B7	Metastatic melanoma and β2 microglobulin (Allovectin-7)
Liposomes for various drugs and Broad applications	Diagnostic agents (lipoMASC)
BLP 25 vaccine Stimuvax®	Non small cell lung cancer vaccine
HepaXen	Hepatitis B vaccine

(Marqibo®, Hana Biosciences, San Francisco, CA), has received orphan medical product designation in US and Europe and has recently been shown to be clinically effective in the treatment of metastatic malignant uveal melanoma (19). Moreover, INX-0125™ (liposomal vinorelbine) (20) and INX-0076™ (liposomal topotecan) (21), are demonstrating encouraging therapeutic results. Clinical results of OSI-211™ (OSI Pharmaceuticals, Inc., Melville, NY), composed of hydrogenated soy phosphatidylcholine (HSPC) and cholesterol, encapsulating lurtotecan, have demonstrated improved therapeutic advantages, owing mainly to the protection of the active closed-lactone ring form of lurtotecan, consequently improving its anti-tumor toxicity (22, 23) against advanced solid tumors and B-cell lymphoma (24).

Several multi-lamellar liposomal formulations are currently undergoing clinical evaluation. The cardiolipin-based charged liposomal platform has been explored by NeoPharm (NeoPharm Inc., Waukegan, IL) resulting in the development of different formulations including LEM-ETU™ (mitoxantrone-loaded) (25), LEP-ETU™ (paclitaxel – incorporating) (26), and LESN38™ (irinotecan entrapped) (27, 28) liposomes.

Recently, "easy to use" paclitaxel formulation LEP-ETU™, demonstrated bio-equivalence with Taxol® (Bristol-Myers Squibb, New York, NY) and promising activity in Phase I trials. This is mainly due to a superior paclitaxel loading capacity in the liposome bilayer, at a maximum mole percent of about 3.5%. Similarly, paclitaxel has also been encapsulated in other modern formulations of cationic lipid complexes (MBT-0206) that have been shown to be bounded and internalized selectively by angiogenic tumoral endothelial cells after intravenous administration (29).

DepoCyt® (Pacira Pharmaceuticals, San Diego, CA) is a slow release liposome-encapsulated cytarabine formulation, recently approved for intrathecal administration in the treatment of neoplastic meningitis and lymphomatous meningitis (30–32). The Depo-Foam™ platform used in DepoCyt®, is essentially a spherical 20-μm multi-lamellar matrix comprised of phospholipids/lipid mixture, similar to normal human cell membranes (phospholipids, triglycerides and cholesterol) (33).

Annamycin, a semi-synthetic lipophilic doxorubicin analogue, capable of circumventing multidrug-resistance transporters, was incorporated in tween-containing liposomes. (34–36). Currently, this liposomal formulation is investigated in Phase II trials in patients with refractory or relapsed acute lymphocytic leukemia (37).

3. Long-circulating Liposomes

One of the drawbacks of the use of liposomes was the fast elimination from the blood and capture of the liposomal preparations by the cells of the RES, primarily, in the liver. To increase liposomal drug accumulation in the desired areas, the use of targeted liposomes with surface-attached ligands capable of recognizing and binding to cells of interest, and potential induction of the liposimal internalization has been suggested (see Fig. 1). Targeted liposomes offer various advantages over individual drugs targeted by means of polymers or antibodies. One of the most compelling advantages is the dramatic increase in drug payload that can be delivered to the target. Furthermore, the number of ligand molecules exposed on the liposome surface can be increased, improving ligand avidity and degree of the uptake. Immunoliposomes also help provide a "bystander kill" effect, because the drug molecules can diffuse into adjoining tumor cells. Immunoglobulins of the IgG class, and their fragments are the most widely used targeting moieties for liposomes (termed "immunoliposomes" after the modification), which

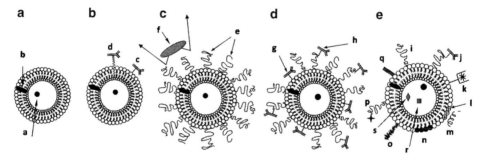

Fig. 1. Evolution of liposomes. (**a**) Early traditional phospholipids "plain" liposomes with water soluble drug (a) entrapped into the aqueous liposome interior, and water-insoluble drug (b) incorporated into the liposomal membrane (these designations are not repeated on other figures). (**b**) Antibody-targeted immunoliposome with antibody covalently coupled (c) to the reactive phospholipids in the membrane, or hydrophobically anchored (d) into the liposomal membrane after preliminary modification with a hydrophobic moiety. (**c**) Long circulating liposome grafted with a protective polymer (e) such as PEG, which shields the liposome surface from the interaction with opsonizing proteins (f). (**d**) Long-circulating immunoliposome simultaneously bearing both protective polymer and antibody, which can be attached to the liposome surface (g) or, preferably, to the distal end of the grafted polymeric chain (h). (**e**) New-generation liposome, the surface of which can be modified (separately or simultaneously) by different ways. Among these modifications are: the attachment of protective polymer (i) or protective polymer and targeting ligand, such as antibody (j); the attachment/incorporation of the diagnostic label (k); the incorporation of positively charged lipids (l) allowing for the complexation with DNA (m); the incorporation of stimuli-sensitive lipids (n); the attachment of stimuli-sensitive polymer (o); the attachment of cell-penetrating peptide (p); the incorporation of viral components (q). In addition to a drug, liposome can loaded with magnetic particles (r) for magnetic targeting and/or with colloidal gold or silver particles (s) for electron microscopy. Reproduced with permission from (2)

could be attached to liposomes without affecting the liposome integrity and antibody properties by covalent binding to the liposome surface or by hydrophobic insertion into the liposomal membrane after modification with hydrophobic residues (38).

Still, despite improvements in the targeting efficacy, the majority of immunoliposomes ended in the liver as a consequence of insufficient time for the interaction between the target and targeted liposome. Better target accumulation can be expected if liposomes can stay in the circulation long enough, which provides more time for targeted liposomes to interact with the target. Prolonged circulation allows also for liposomes to deliver pharmaceutical agents to targets other than the RES. Different methods have been suggested to achieve long circulation of liposomes in vivo, including coating the liposome surface with inert, biocompatible polymers, such as flexible polyethylene glycol (PEG), which form a steric protective layer over the liposome surface and slows down the liposome recognition by opsonins and subsequent clearance (39, 40), see Fig. 1. Moreover, the PEG chains on the liposome surface avoid the vesicle aggregation, improving stability of formulations (41).

It has been shown with a broad variety of examples that, similar to macromolecules, liposomes are capable of accumulating in various pathological areas with affected vasculature (such as tumor, infarcts, and inflammations) via the enhanced permeability and retention (EPR) effect (42, 43), and their longer circulations naturally enhances this way of target accumulation. Doxorubicin, incorporated into long-circulating PEGylated liposomes (Doxil®) demonstrates good activity in EPR-based tumor therapy and strongly diminishes the toxic side effects (cardiotoxicity) of the original drug (44). Evidently, long-circulating liposomes can be easily adapted for the delivery of various pharmaceuticals to tumor and other "leaky" areas. Interestingly, recent evidence showed that PEG-liposomes, previously considered to be biologically inert, still could induce certain side reactions via activation of the complement system (45, 46).

In general, PEGylated liposomes demonstrate dose-independent, non-saturable, log-linear kinetics, and increased bioavailability (16), where incorporation of PEG-lipids causes the liposome to remain in the blood circulation for extended periods of time (i.e., $t_{1/2} > 40$ h) and distribute through an organism relatively evenly with most of the dose remaining in the central compartment (i.e., the blood) and only 10–15% of the dose being delivered to the liver. This is a significant improvement over conventional liposomes where typically 80–90% of the liposome deposits in the liver (16, 47, 48).

Long-circulating liposomes are now investigated in details and widely used in biomedical in vitro and in vivo studies and have also found their way into clinical practice (6, 44). Although these favorable characteristics have extended clinical applications of PEGylated liposomes, recent research calls for some caution, where more investigation towards multiple dose administration or biodistribution in tumor tissues, is warranted. It was recently reported that intravenous injection in rats of PEG-grafted liposomes may significantly alter the pharmacokinetic behavior of a subsequent dose when this dose is administered after an interval of several days (14). This phenomenon, called "accelerated blood clearance" (ABC), appears to be inversely related to the PEG content of liposomes. By the same token, an inverse relationship has been observed between dose and magnitude of the ABC effect (49).

Although, PEG remains the gold standard in liposome steric protection, attempts continue to identify other polymers that could be used to prepare long-circulating liposomes. Earlier studies with various water-soluble flexible polymers have been summarized in (50, 51). More recent papers describe long-circulating liposomes prepared using poly(N-(2-hydroxypropyl)methacrylamide) (52), poly-N-vinylpyrrolidones (53), l-amino acid-based biodegradable polymer-lipid conjugates (54), and polyvinyl alcohol (55).

On the same note, recent research revived the early strategy of liposome surface modification with gangliosides (GM1 and GM3), analogous to erythrocyte membrane, which demonstrated prolonged circulation only in mice and rats. This new application of GM-coated liposomes involved their use for oral administration and delivery to the brain. In particular, Taira et al. (56) suggest that among liposomal formulations used as oral drug carriers, those containing GM1 and GM3 have better possibilities of surviving through the gastrointestinal tract. It was reported that observed higher brain-tracer uptake for GM1 liposomes than for control liposomes in the cortex, basal ganglia, and mesencephalon of both hemispheres; conversely, no significant changes were observed in liposomal liver uptake or blood concentration (57). Another example includes ovalbumin in PEG-coated liposomes induced the best mucosal immune response of all carriers tested (58). To improve protein and peptide bioavailability via the oral route, an oral colon-specific drug delivery system for bee venom peptide was developed that was based on coated alginate gel beads entrapped in liposomes (59).

Subcutaneous administration of PEGylated liposomes has shown unique potential especially for targeting to the lymph nodes, achieving sustained drug release in vivo (60). Earlier research by Allen et al. (61) has demonstrated the feasibility of targeting liposomes to the lymph nodes that was explored for lymphatic delivery of methotrexate (62) and for magnetic resonance imaging (MRI) with Gadolinium-loaded liposomes (63).

Attempts have been done to attach PEG to the liposome surface in a removable fashion to facilitate the liposome capture by the cell after PEG-liposomes accumulate in target site via the EPR effect (42) and PEG coating is detached under the action of local pathological conditions (decreased pH in tumors). New detachable PEG conjugates are described in (64), where the detachment process is based on the mild thiolysis of the dithiobenzylurethane linkage between PEG and amino-containing substrate (such as PE). Low pH-degradable PEG-lipid conjugates based on the hydrazone linkage between PEG and lipid have also been described (65, 66).

4. Clinical Applications of Long-circulating Liposomes

PEGylated liposomal doxorubicin (DOXIL®/Caelyx®) was the first and is still the only stealth liposome formulation to be approved in both USA and Europe for treatment of Kaposi's sarcoma (67) and recurrent ovarian cancer (68, 69). Currently, (DOXIL®/Caelyx®) is undergoing trials for treatment of other malignancies such as multiple myelomas (70), breast cancer (71, 72), and recurrent high-grade glioma (73).

A very similar stealth liposome formulation, but encapsulating cisplatin, SPI-077™ (Alza Corporation, Mountain View, CA, USA), has demonstrated the same evident stealth behavior with an apparent $t_{1/2}$ of approximately 60–100 h. Phase I/II clinical trials of the drug to treat head and neck cancer and lung cancer (74), were showing promising toxicity profile, yet therapeutic efficacy was lacking (75), mainly due to delayed drug release. Hence, another formulation was evaluated, SPI-077 B103 (Alza Corp., Mountain View, CA, USA); they chose B103, where fully hydrogenated soy PC was replaced by unsaturated phospholipids, to decease rigidity of liposomal membrane, aiming for earlier tendency for cisplatin release. However, released drug was not detected in in vitro systems, plasma, or tumor extracellular fluid after administration of either stealth formulation of liposomal cisplatin (76). Similarly, S-CKD602 (Alza Corp., Mountain View, CA, USA), a PEGylated stealth liposomal formulation of CKD-602 – a semisynthetic analog of camptothecin – was submitted for a Phase I trial. After it was demonstrated that the plasma AUC for S-CKD602 was 50-fold that of non-liposomal CKD-602; and showed minimal toxicity and encouraging therapeutic activity (13, 77).

Lipoplatin™ (Regulon Inc. Mountain View, CA, USA) is another pegylated liposomal cisplatin formulation composed of showed plasma half-life is 60–117 h in clinical study, depending on the dose (Boulikas et al. 2005; Stathopoulos et al. 2005). The study also found that Lipoplatin has no nephrotoxicity up to a dose of 125 mg/m² every 14 days without the serious side effects of cisplatin. Clinical evaluation of pegylated liposomal formulation of mitoxantrone (Novantrone®, Wyeth Lederle, Madison, NJ, USA), containing cardiolipin, has displayed promising therapeutic results in acute myeloid leukemia, and prostate cancer (78).

5. Long-circulating Targeted Liposomes

The further development of liposomal carriers involved the attempt to combine the properties of long-circulating liposomes and immunoliposomes in one preparation (79–81). To achieve better selectivity of PEG-coated liposomes, it is advantageous to attach the targeting ligand via a PEG spacer arm, so that the ligand is extended outside the dense PEG brush excluding steric hindrances for its binding to the target. Currently, various advanced technologies are used, and the targeting moiety is usually attached above the protecting polymer layer, by coupling it with the distal water-exposed terminus of activated liposome-grafted polymer molecule (80, 82), see Fig. 1.

One has to note here, that the preparation of modified liposomes with desired properties require chemical conjugation of proteins, peptides, polymers, and other molecules to the liposome surface. In general, the conjugation methodology is based on three main reactions, which are sufficiently efficient and selective: reaction between activated carboxyl groups and amino groups yielding an amide bond; reaction between pyridyldithiols and thiols yielding disulfide bonds; and reaction between maleimide derivatives and thiols yielding thioether bonds. Many lipid derivatives used in these techniques are commercially available (83). Other approaches also exist, for example yielding the carbamate bond via the reaction of *p*-nitrophenylcarbonyl- and amino-group (82, 84, 85).

Although various monoclonal antibodies have been shown to deliver liposomes to many targets, optimization of properties of immunoliposomes still continues. The majority of research relates to cancer targeting, which utilizes a variety of antibodies. Internalizing antibodies are required to achieve a really improved therapeutic efficacy of antibody-targeted liposomal drugs as was shown using B-lymphoma cells and internalizable epitopes (CD19) as an example (86). An interesting concept was developed to target HER2-overexpressing tumors using doxorubicin liposomes, targeted with the Fv fragment of monoclonal antibody anti-HER2 trastuzumab (Herceptin®, Genentech Inc., Vacaville, CA, USA) (87–90). Antibody CC52 against rat colon adenocarcinoma CC531 attached to PEGylated liposomes provided specific accumulation of liposomes in rat model of metastatic CC531 (91).

Nucleosome-specific antibodies capable of recognition of various tumor cells via tumor cell surface-bound nucleosomes improved Doxil® targeting to tumor cells and increased its cytotoxicity (92, 93), and even revealed the potential to improve the Skin side effect of Doxil® (94). Same 2C5 antibody successfully targeted doxorubicin-loaded PEGylated liposomes into human brain U-87 tumor intracranial xenograft in nude mice and significantly enhanced the therapeutic outcome (95). GD2-targeted immunoliposomes with novel antitumoral drug, fenretinide, inducing apoptosis in neuroblastoma and melanoma cell lines, demonstrated strong anti-neuroblastoma activity both in vitro and in vivo in mice (96). EGFR-overexpressing colorectal tumor cells were efficiently targeted with immunoliposomes bearing Fab' from the humanized anti-EGFR monoclonal antibody (97, 98).

Anti-P-selecting-modified liposomes were shown to target the areas of inflammation after an acute myocardial infarction and can deliver pro-angiogenic drugs to this area (99). Combination of immunoliposome and endosome-disruptive peptide improves cytosolic delivery of liposomal drug, increases cytotoxicity, and opens new approach to constructing targeted liposomal systems as shown with diphtheria toxin. A chain incorporated together with pH-dependent fusogenic peptide diINF-7 into liposomes specific towards ovarian carcinoma (100).

Since transferrin (Tf) receptors (TfR) are overexpressed on the surface of many tumor cells, antibodies against TfR as well as Tf itself are among popular ligands for liposome targeting to tumors and inside tumor cells (101). Recent studies involve the coupling of Tf to PEG on PEGylated liposomes in order to combine longevity and targetability for drug delivery into solid tumors (102). Similar approach was applied to deliver into tumors agents for photodynamic therapy including hypericin (103, 104) and for intracellular delivery of cisplatin into gastric cancer (105). Tf-coupled doxorubicin-loaded liposomes demonstrate increased binding and toxicity against C6 glioma (106). Interestingly, the increase in the expression of the TfR was also discovered in post-ischemic cerebral endothelium, which was used to deliver Tf-modified PEG-liposomes to post-ischemic brain in rats (107). Tf (108) as well as anti-TfR antibodies (109, 110) was also used to facilitate gene delivery into cells by cationic liposomes. Tf-mediated liposome delivery was also successfully used for brain targeting. Immunoliposomes with OX26 monoclonal antibody to the rat TfR were found to concentrate on brain microvascular endothelium (111).

Targeting tumors with folate-modified liposomes represents a very popular approach, since folate receptor (FR) expression is frequently overexpressed in many tumor cells. After early studies demonstrated the possibility of delivery of macromolecules (112) and then liposomes (113) into living cells utilizing FR endocytosis, which could bypass multidrug resistance, the interest to folate-targeted drug delivery by liposomes grew fast (see important reviews in refs. (114, 115). Liposomal daunorubicin (116) as well as doxorubicin (117) and 5-fluorouracyl (118) were delivered into various tumor cells both in vitro and in vivo via FR and demonstrated increased cytotoxicity. Recently, the application of folate-modified doxorubicin-loaded liposomes for the treatment of acute myelogenous leukemia was combined with the induction of FR using all-trans retinoic acid (119). Folate-targeted liposomes have been suggested as delivery vehicles for boron neutron capture therapy (120) and used also for targeting tumors with haptens for tumor immunotherapy (121). Within the frame of gene therapy, folate-targeted liposomes were utilized for both gene targeting to tumor cells (122) as well as for targeting tumors with antisense oligonucleotides (123).

In the last few years, antibody-based therapeutics have emerged as important components of therapies for an increasing number of human malignancies(124) and it is expected that several immunoliposomes will be in trials in the near future (77).

The search for new ligands for liposome targeting concentrates around specific receptors overexpressed on target cells (particularly cancer cells) and certain specific components of pathologic cells. Thus, liposome targeting to tumors has been achieved by using vitamin and growth factor receptors (125). Vasoactive intestinal peptide (VIP) was used to target PEG-liposomes with radionuclides to VIP-receptors of the tumor, which resulted in an enhanced breast cancer inhibition in rats (126). PEG-liposomes were targeted by RGD peptides to integrins of tumor vasculature and, being loaded with doxorubicin, demonstrated increased efficiency against C26 colon carcinoma in murine model (127). RGD-peptide was also used for targeting liposomes to integrins on activated platelets and, thus, could be used for specific cardiovascular targeting (128) as well as for selective drug delivery to monocytes/neutrophils in the brain (129). Similar angiogenic homing peptide was used for targeted delivery to vascular endothelium of drug-loaded liposomes in experimental treatment of tumors in mice (130). Epidermal growth factor receptor (EGFR)-targeted immunoliposomes were specifically delivered to variety of tumor cells overexpressing EGFR (131). Mitomycin C in long-circulating hyaluronan-targeted liposomes increases its activity against tumors overexpress hyaluronan receptors (132). The ability of galactosylated liposomes to concentrate in parenchymal cells was applied for gene delivery in these cells, see (133) for review. Cisplatin-loaded liposomes specifically binding chondroitin sulfate overexpressed in many tumor cells were used for successful suppression of tumor growth and metastases in vivo (134). Tumor-selective targeting of PEGylated liposomes was also achieved by grafting these liposomes with basic fibroblast growth factor-binding peptide (135). Intraperitoneal cancer can be successfully targeted by oligomannose-coated liposomes (136).

6. Stimuli-sensitive Liposomes

An interesting example of liposome delivery inside cells involves the use of so-called pH-sensitive liposomes. In this case, the liposome is made of pH-sensitive components and, after being endocytosed in the intact form, it fuses with the endovacuolar membrane under the action of lowered pH inside the endosome, releasing its content into the cytoplasm (see Fig. 2). Studies with pH-sensitive liposomes concentrate around new lipid compositions for imparting pH-sensitivity to liposomes, liposome modification with various pH-sensitive polymers, and combination of the liposomal pH-sensitivity with longevity and ligand-mediated targeting. Thus, long-circulating PEGylated pH-sensitive liposomes, although have a decreased pH-sensitivity, still effectively deliver their contents into cytoplasm (recent review in ref. (137)).

Antisense oligonucleotides are delivered into cells by anionic pH-sensitive PE-containing liposomes stable in the blood,

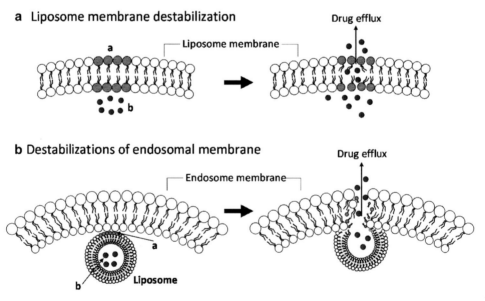

Fig. 2. Fusogenic and stimuli-sensitive liposomes. (**a**) Liposome membrane destabilization. After accumulation in required sites in the body, liposomes containing stimulisensitive components, such as lipids (a) in the membrane and drug (b) inside, and after being subjected to local action of the corresponding stimulus (such as pH or temperature), undergo local membrane destabilization (transfer from *left* to *right* of *panel A*) that allows for drug efflux from the liposome into surroundings. (**b**) Destabilization of endosomal membrane. After being endocytosed by the cell and taken inside the endosome, the liposome containing stimuli (pH)-sensitive components, such as lipids (a) in the membrane and drug (b) inside, can undergo pH dependent membrane destabilization and initiate the destabilization of the lysosomal membrane (transfer from *left* to *right* of *panel B*) that allows for drug efflux into the cell cytoplasm. Reproduced with permission from (2)

however, undergoing phase transition at acidic endosomal pH and facilitating oligo release into cell cytoplasm (review in ref. (138)). New pH-sensitive liposomal additives were recently described including oleyl alcohol (139) and mono-stearoyl derivative of morpholine (140). Serum stable, long-circulating PEGylated pH-sensitive liposomes were also prepared using the combination of PEG and pH-sensitive terminally alkylated copolymer of N-isopropylacrylamide and methacrylic acid (141). The combination of liposome pH-sensitivity and specific ligand targeting for cytosolic drug delivery using decreased endosomal pH values was described for both folate and Tf-targeted liposomes (122, 142). See one of the recent reviews on pH-sensitive liposomes in (143). Liposomes, which can carry on their surface multiple functionalities (for example, targeting ligand and a residue of a cell-penetrating peptide allowing for an effective intracellular delivery of liposomes) and demonstrate different properties depending on the specific conditions of surrounding tissues (for example, lowered pH in tumors) have also been described (65, 66, 144).

7. Virosomes

A new approach in targeted drug delivery has recently emerged, based on the use of certain viral proteins demonstrating a unique ability to penetrate into cells ("protein transduction" phenomenon). It was demonstrated that the trans-activating transcriptional activator (TAT) protein from HIV-1 enters various cells when added to the surrounding media (145). The recent data assume more than one mechanism for cell penetrating peptides and proteins (CPP) and CPP-mediated intracellular delivery of various molecules and particles. TAT-mediated intracellular delivery of large molecules and nanoparticles was proved to proceed via the energy-dependent macropinocytosis with subsequent enhanced escape from endosome into the cell cytoplasm (146) while individual CPPs or CPP-conjugated small molecules penetrate cells via electrostatic interactions and hydrogen bonding and do not seem to depend on the energy (147). Since crossing through cellular membranes represents a major barrier for efficient delivery of macromolecules into cells, CPPs, whatever their mechanism of action is, may serve to transport various drugs and even drug-loaded pharmaceutical carriers into mammalian cells in vitro and in vivo. It was demonstrated that relatively large particles, such as liposomes, could be delivered into various cells by multiple TAT-peptide or other CPP molecules attached to the liposome surface (148–150). Complexes of TAT-peptide-liposomes with a plasmid (plasmid pEGFP-N1 encoding for the Green Fluorescence Protein, GFP) were used for successful in vitro transfection of various tumor and normal cells as well as for in vivo transfection of tumor cells in mice bearing Lewis lung carcinoma (151). TAT-peptide liposomes have been also successfully used for transfection of intracranial tumor cell in mice via intracarotid injection (152).

An interesting example of intracellular targeting of liposomes was described recently, when liposomes containing in their membrane composition mitochonriotropic amphiphilic cation with delocalized positive charge were shown to specifically target mitochondria in inact cells (153). The liposomes were deemed mitochondriotropic due to inclusion of lipidic analogues of triphenylphosphonium cations, which facilitates the efficient subcellular delivery of proapototic ceramide to mitochondria of mammalian cells and improves its activity in vitro and in vivo (154).

Many of the listed functions/properties of liposomes, such as longevity, targetability, stimuli-sensitivity, ability to deliver drugs intracellularly, etc. could, theoretically, be combined in a single preparation yielding a so-called multifunctional liposomal nanocarrier (155) (see Fig. 1).

8. Diagnostic Applications of Liposomes

Liposomes as pharmaceutical nanocarriers find many various applications in addition to already discussed. Thus, the use of liposomes for the delivery of imaging agents for all imaging modalities already has a long history (156). Diagnostic imaging requires that an appropriate intensity of signal from an area of interest is achieved in order to differentiate certain pathologies from surrounding normal tissues, regardless of the modality used. Currently used imaging modalities include gamma-radioscintigraphy; MRI; computed tomography (CT), and ultra-sonography. There exist a variety of different methods to label/load the liposome with a contrast/reporter group: (a) Label could be added to liposomes during the manufacturing process to liposomes (label is incorporated into the aqueous interior of liposome or into the liposome membrane); (b) Label could be adsorbed onto the surface of preformed liposomes; (c) Label could be incorporated into the lipid bilayer of preformed liposomes; (d) Label could be loaded into preformed liposomes using membrane-incorporated transporters or ion channels. In any case, clinically acceptable diagnostic liposomes have to meet certain requirements: (a) The labeling procedure should be simple and efficient; (b) The reporter group should be affordable, stable and safe/easy to handle; (c) Liposomes should be stable in vivo stability with no release of free label; (d) Liposomes need to be stable on storage – within acceptable limits.

The relative efficacy of entrapment of contrast materials into different liposomes as well as advantages and disadvantages of various liposome types were analyzed by Tilcock (157). Liposomal contrast agents have been used for experimental diagnostic imaging of liver, spleen, brain, cardio-vascular system, tumors, inflammations and infections (156, 158).

Gamma-scintigraphy and MR imaging both require a sufficient quantity of radionuclide or paramagnetic metal to be associated with the liposome, to achieve high signal:noise ratio. There are two possible routes to improve the efficacy of liposomes as contrast mediums for gamma-scintigraphy and MRI: to increase the quantity of carrier-associated reporter metal (such as ^{111}In or Gd), and/or enhance the signal intensity. To increase the load of liposomes with reporter metals, amphiphilic chelating polymers, such as N,ε-(DTPA-polylysyl)glutaryl phosphatidyl ethanolamine, were introduced (65, 159, 160). They easily incorporate into the liposomal membrane and sharply increase the number of chelated Gd or ^{111}In atoms attached to a single lipid anchor. In case of MRI, metal atoms chelated into these groups are directly exposed to the water environment, which enhances the signal intensity of the paramagnetic ions and leads to the corresponding enhancement of the vesicle contrast properties. The additional incorporation of amphiphilic PEG into the liposome membrane helps to improve the relaxivity of the contrast ion due to the presence of increased amount of PEG-associated water protons in close vicinity to liposomal membrane. Moreover, coating of liposome surface with PEG polymer will impart STEALTH properties on the formulation, hence, can help in avoiding the contrast agent uptake in the site of injection by resident phagocytic cells. This approach resulted in efficient liposomal contrast agents for the MR imaging of the blood pool (161). MR imaging using pH-responsive contrast liposomes allowed for visualization of pathological areas with decreased pH values (162). Contrast agent-loaded liposomes were also used for in vivo monitoring of tissue pharmacokinetics of liposomal drugs in mice (163). Sterically stabilized superparamagnetic-DTPA liposomes were suggested for MR imaging and cancer therapy(162, 164).

Because of its short half-life and ideal radiation energy, ^{99m}Tc is the most clinically attractive isotope for gamma-scintigraphy. Recently, new methods for labeling preformed glutathione-containing liposomes with various ^{99m}Tc and ^{186}Re complexes were developed (165), which are extremely effective and result in a very stable product.

CT contrast agents (primarily, heavily iodinated organic compounds) were included in the inner water compartment of liposomes or incorporated into the liposome membrane. Thus, Iopromide was incorporated into plain (166) and PEGylated liposomes (167) and demonstrated favorable biodistribution and imaging potential in rats and rabbits. Liposomes for sonography are prepared by incorporating gas bubbles (which are efficient reflectors of sound) into the liposome, or by forming the bubble directly inside the liposome as a result of a chemical reaction, such as bicarbonate hydrolysis yielding carbon dioxide. Gas bubbles stabilized inside the phospholipid membrane demonstrate good performance and low toxicity of these contrast agents in rabbit and porcine models (126).

Furthermore, liquid-filled liposomes have been demonstrated to be echogenic, while the liquid-like composition of the vesicles makes them more resistant to pressure and mechanical stress than encapsulated gas microbubbles. Definity® (Bristol-Myers Squibb Medical Imaging, Inc. New York, NY, USA) is a contrast agent containing perfluoropropane with a phospholipid shell approved in the US for use in cardiology and tumor imaging (168–170). Echogenic liposomes have also been utilized for intravascular ultrasound imaging, targeting the vesicles to the vascular signature associated with arteroma development (171).

9. Liposomes for Vaccination

For vaccination, drug delivery strategies involved more than only using the inactivated microbial antigens as therapeutics. Liposomal vaccine delivery strategy involved harnessing the cell-invading power of viral capsule proteins as well. With this purpose, the liposome surface was modified with fusogenic viral envelope proteins, (172). Initially, virosomes were intended for intracellular delivery of drugs and DNA (173, 174), then became a cornerstone for the development of new vaccines. Delivery of protein antigens to the immune system by fusion-acting virosomes was found to be very effective (175), in particular into dendritic cells (176). As a result, a whole set of virosome-based vaccines have been developed for application in humans and animals.

Special attention was paid to influenza vaccine using virosomes containing the spike proteins of influenza virus (177), since it elicits high titers of influenza-specific antibodies. Trials of virosome influenza vaccine in children showed that it is highly immunogenic and well tolerated (178). A similar approach was used to prepare virosomal hepatitis A vaccine that elicited high antibody titers after primary and booster vaccination of infants and young children (179); the data have been confirmed in healthy adults (180) and in elderly patients (181). Combination of influenza protein-based virosomes with other antigens may be used to prepare other vaccines (182). In general, virosomes can provide an excellent opportunity for efficient delivery of both various antigens and many drugs (nucleic acids, cytotoxic drugs, toxoids)

(182, 183), although they might represent certain problems associated with their stability/leakyness and immunogenicity. Fusion-active virosomes have also been used for cellular of siRNA (184).

Two commercial vaccines based on virosome technology are currently on the market. Epaxal® (Berna Biotech Ltd, Bern, Switzerland), a hepatitis A vaccine, has inactivated hepatitis A virus particles adsorbed on the surface of the immunopotentiating reconstituted influenza virosomes (IRIV). In Inflexal® V (Berna Biotech Ltd) the virosome components themselves are the vaccine protective antigens (185). Recently, in phase I study liposome-encapsulated malaria vaccine (containing monophosphoryl lipid A as adjuvant in the bilayer), the formulation showed induction of higher level of anti-malaria antibody in human volunteers (186). Some liposomal formulations are under investigation in preclinical studies against Yersina pestis, ricin toxin and Ebola Zaire virus (77, 187).

10. Miscellaneous Applications of Liposomal Preparations

During recent years, the topical delivery of liposomes has been applied to different applications and in different disease models (188). Current efforts in this area concentrate around optimization procedures and new compositions. Recently, highly flexible liposomes called transferosomes that follow the trans-epidermal water activity gradient in the skin have been proposed. Diclofenac in transferosomes was effective when tested in mice, rats and pigs (189). The concept of increased deformability of transdermal liposomes is supported by the results of transdermal delivery of pergolide in liposomes, in which elastic vesicles have been shown to be more efficient (190). The combination of liposomes and iontophoresis for transdermal delivery yielded promising results (191, 192).

Photo-dynamic therapy (PDT) is fast developing modality for the treatment of superficial/skin tumors, where photosensitizing agents are used for photochemical eradication of malignant cells In PDT, liposomes are used both as drug carriers and enhancers. Recent review on the use of liposomes in PDT can be found in (193). Targeting as well as the controlled release of photosensitizing agent in tumors may still further increase the outcome of the liposome-mediated PDT. Benzoporphyrin derivative encapsulated in polycation liposomes modified with cetyl-polyethyleneimine was used for antiangiogenic PDT. This drug in such liposomes was better internalized by human umbilical vein endothelial cells and was found in the intranuclear region and associated with mitochondria (194). The commercial liposomal preparation of benzoporphyrin derivative monoacid ring A, known as Visudyne

(Novartis), was active against tumors in sarcoma-bearing mice (195). PDT with liposomal photofrin gives better results in mice with human gastric cancer that with a free drug (196). Another porphyrin derivative (SIM01) in DMPC liposomes also gives better results in PDT, mainly due to better accumulation in the tumor (human adenocarcinoma in nude mice) (197). Liposomal meso-tetrakis-phenylporphyrin was very effective in PDT of human amelanotic melanoma in nudes (198).

An interesting example of a new approach currently in clinical trials is to combine radio-frequency tumor ablation with intravenous liposomal doxorubicin, which resulted in better tumor accumulation of liposomes and increased necrosis in tumors (199, 200).

There is an interest in liposomal forms of "bioenergic" substrates, such as ATP, and some encouraging results with ATP-loaded liposomes in various in vitro and in vivo models have been reported. In a brain ischemia model, the use of the liposomal ATP increased the number of ischemic episodes tolerated before brain electrical silence and death (138). In a hypovolemic shock-reperfusion model in rats, the administration of ATP-liposomes provided effective protection to the liver (201), and also improved the rat liver energy state and metabolism during the cold storage preservation (202). Similar properties were also demonstrated for the liposomal coenzyme Q10 (203). Interestingly, biodistribution studies with the ATP-liposomes demonstrated their significant accumulation in the damaged myocardium (204). Recently, ATP-loaded liposomes were shown to effectively preserve mechanical properties of the heart under ischemic conditions in an isolated rat heart model (205). ATP-loaded immunoliposomes were also prepared possessing specific affinity towards myosin, i.e., capable of specific recognition of hypoxic cells (206) and effectively protected infracted myocardium in vivo (207, 208). Similarly, liposomes loaded with coenzyme Q10 effectively protected the myocardium in infracted rabbits (209).

Active research is in progress in the area of liposomes for the use as vesicular containers, in particular for hemoglobin as blood substitute. Liposome-encapsulated hemoglobin is being developed as an oxygen therapeutic. The spatial isolation of hemoglobin by a lipid bilayer potentially minimizes the cardiovascular/hemodynamic effects associated with other modified forms of hemoglobin. Moreover, the preclinical results showed circulation half-life up to 65 h for this PEGylated liposomal hemoglobin formulation; indicating remarkable physiological longevity where animals tolerated at least 25% of blood exchange without any distress (210–212).

An interesting approach for targeted drug delivery under the action of magnetic field is the use of liposomes loaded with a drug and with a ferromagnetic material. Magnetic liposomes with doxorubicin were intravenously administered to osteosarcoma-bearing hamsters. When the tumor-implanted limb was placed between two poles of 0.4 Tesla magnet, the application of the field for 60 min resulted in fourfold increase in drug concentration in the tumor (213). In the same osteosarcoma model and the magnet implanted into the tumor, magnetic liposomes loaded with adriamycin demonstrated better accumulation in tumor vasculature and tumor growth inhibition (214). Upon intravenous injection in rats, liposomes loaded with ^{99m}Tc-albumin and magnetite demonstrated 25-fold increase in accumulated radioactivity in right kidney, near which a SmCo magnet was implanted, compared to control left kidney (215).

11. Conclusions

The development of "pharmaceutical" liposomes is an ever growing research area with an increasing variety of potential applications, and encouraging results from early clinical applications and clinical trials of different liposomal drugs.

New generation liposomes frequently demonstrate a combination of different attractive properties, such as simultaneous longevity and targetability, longevity and stimuli-sensitivity, targetability and contrast properties etc. These new generation liposomes can also simultaneously entrap more than one therapeutic agent and these liposomal drugs can act/release in a certain coordinated fashion (21, 216, 217).

Thus, liposomes are successfully utilized in all imaginable drug delivery approaches and their use to solve various biomedical problems is steadily increasing. At present, quite a number of various liposomal formulations has received clinical approval, or in advanced clinical trials. We are surely likely to see more liposomal pharmaceuticals on the market in the foreseeable future.

References

1. Lasic DD (1993) Liposomes from physics to applications. Elsevier, Amsterdam
2. Torchilin VP (2005) Recent advances with liposomes as pharmaceutical carriers. Nat Rev Drug Discov 4:145–160
3. Lasic DD, Papahadjopoulos D (eds) (1998) Medical applications of liposomes. Elsevier, New York
4. Connor J, Yatvin MB, Huang L (1984) pH-Sensitive liposomes: Acid-induced liposome fusion. Proc Natl Acad Sci USA 81:1715–1718
5. Lasic DD et al (1992) Gelation of liposome interior. A novel method for drug encapsulation. FEBS Lett 312:255–258
6. Lasic DD, Martin F (eds) (1995) Stealth liposomes. CRC Press, Boca Raton
7. Woodle MC, Storm G (eds) (1998) Long circulating liposomes: Old drugs, new therapeutics. Springer, Berlin
8. Torchilin VP, Weissig V (eds) (2003) Liposomes: A practical approach, Oxford University Press, Oxford, New York

9. Gregoriadis G (ed) (2007) Liposome technology: Liposome preparation and related techniques. Taylor & Francis, London, UK
10. Drummond DC et al (1999) Optimizing liposomes for delivery of chemotherapeutic agents wto solid tumors. Pharmacol Rev 51:691–743
11. Papahadjopoulos D et al (1991) Sterically stabilized liposomes: Improvements in pharmacokinetics and antitumor therapeutic efficacy. Proc Natl Acad Sci USA 88:11460–11464
12. Senior JH (1987) Fate and behavior of liposomes in vivo: A review of controlling factors. Crit Rev Ther Drug Carrier Syst 3:123–193
13. Zamboni WC (2005) Liposomal, nanoparticle, and conjugated formulations of anticancer agents. Clin Cancer Res 11:8230–8234
14. Laverman P et al (2001) Factors affecting the accelerated blood clearance of polyethylene glycol-liposomes upon repeated injection. J Pharmacol Exp Ther 298:607–612
15. Litzinger DC, Buiting AM, van Rooijen N, Huang L (1994) Effect of liposome size on the circulation time and intraorgan distribution of amphipathic poly(ethylene glycol)-containing liposomes. Biochim Biophys Acta 1190:99–107
16. Allen TM, Hansen C (1991) Pharmacokinetics of stealth versus conventional liposomes: Effect of dose. Biochim Biophys Acta 1068:133–141
17. Alberts DS et al (2004) Efficacy and safety of liposomal anthracyclines in phase I/II clinical trials. Semin Oncol 31:53–90
18. Allen TM, Martin FJ (2004) Advantages of liposomal delivery systems for anthracyclines. Semin Oncol 31:5–15
19. Bedikian AY et al (2006) Pharmacokinetics and urinary excretion of vincristine sulfate liposomes injection in metastatic melanoma patients. J Clin Pharmacol 46:727–737
20. Semple SC et al (2005) Optimization and characterization of a sphingomyelin/cholesterol liposome formulation of vinorelbine with promising antitumor activity. J Pharm Sci 94:1024–1038
21. Tardi P et al (2000) Liposomal encapsulation of topotecan enhances anticancer efficacy in murine and human xenograft models. Cancer Res 60:3389–3393
22. Seiden MV et al (2004) A phase II study of liposomal lurtotecan (OSI-211) in patients with topotecan resistant ovarian cancer. Gynecol Oncol 93:229–232
23. Duffaud F et al (2004) Phase II study of OSI-211 (liposomal lurtotecan) in patients with metastatic or loco-regional recurrent squamous cell carcinoma of the head and neck. An EORTC New Drug Development Group study. Eur J Cancer 40:2748–2752
24. Lu C et al (2005) Phase II study of a liposome-entrapped cisplatin analog (L-NDDP) administered intrapleurally and pathologic response rates in patients with malignant pleural mesothelioma. J Clin Oncol 23:3495–3501
25. Ugwu S et al (2005) Preparation, characterization, and stability of liposome-based formulations of mitoxantrone. Drug Dev Ind Pharm 31:223–229
26. Zhang JA et al (2005) Development and characterization of a novel Cremophor EL free liposome-based paclitaxel (LEP-ETU) formulation. Eur J Pharm Biopharm 59:177–187
27. Lei S et al (2004) Enhanced therapeutic efficacy of a novel liposome-based formulation of SN-38 against human tumor models in SCID mice. Anticancer Drugs 15:773–778
28. Pal A et al (2005) Preclinical safety, pharmacokinetics and antitumor efficacy profile of liposome-entrapped SN-38 formulation. Anticancer Res 25:331–341
29. Eichhorn ME et al (2006) Paclitaxel encapsulated in cationic lipid complexes (MBT-0206) impairs functional tumor vascular properties as detected by dynamic contrast enhanced magnetic resonance imaging. Cancer Biol Ther 5:89–96
30. Phuphanich S, Maria B, Braeckman R, Chamberlain M (2007) A pharmacokinetic study of intra-CSF administered encapsulated cytarabine (DepoCyt) for the treatment of neoplastic meningitis in patients with leukemia, lymphoma, or solid tumors as part of a phase III study. J Neurooncol 81:201–208
31. Glantz MJ et al (1999) A randomized controlled trial comparing intrathecal sustained-release cytarabine (DepoCyt) to intrathecal methotrexate in patients with neoplastic meningitis from solid tumors. Clin Cancer Res 5:3394–3402
32. Jaeckle KA et al (2002) An open label trial of sustained-release cytarabine (DepoCyt) for the intrathecal treatment of solid tumor neoplastic meningitis. J Neurooncol 57:231–239
33. Mantripragada S (2002) A lipid based depot (DepoFoam technology) for sustained release drug delivery. Prog Lipid Res 41:392–406
34. Orlandi L et al (2001) Effects of liposome-entrapped annamycin in human breast cancer cells: Interference with cell cycle progression and induction of apoptosis. J Cell Biochem 81:9–22
35. Zou Y, Priebe W, Stephens LC, Perez-Soler R (1995) Preclinical toxicity of liposome-incorporated annamycin: Selective bone marrow

toxicity with lack of cardiotoxicity. Clin Cancer Res 1:1369–1374
36. Zou Y et al (1994) Antitumor activity of free and liposome-entrapped annamycin, a lipophilic anthracycline antibiotic with non-cross-resistance properties. Cancer Res 54:1479–1484
37. Booser DJ et al (2002) Phase II study of liposomal annamycin in the treatment of doxorubicin-resistant breast cancer. Cancer Chemother Pharmacol 50:6–8
38. Torchilin VP (1985) Liposomes as targetable drug carriers. Crit Rev Ther Drug Carrier Syst 2:65–115
39. Klibanov AL, Maruyama K, Torchilin VP, Huang L (1990) Amphipathic polyethyleneglycols effectively prolong the circulation time of liposomes. FEBS Lett 268:235–237
40. Blume G, Cevc G (1993) Molecular mechanism of the lipid vesicle longevity in vivo. Biochim Biophys Acta 1146:157–168
41. Needham D, McIntosh TJ, Lasic DD (1992) Repulsive interactions and mechanical stability of polymer-grafted lipid membranes. Biochim Biophys Acta 1108:40–48
42. Maeda H, Sawa T, Konno T (2001) Mechanism of tumor-targeted delivery of macromolecular drugs, including the EPR effect in solid tumor and clinical overview of the prototype polymeric drug SMANCS. J Control Release 74:47–61
43. Yuan F et al (1994) Microvascular permeability and interstitial penetration of sterically stabilized (stealth) liposomes in a human tumor xenograft. Cancer Res 54:3352–3356
44. Gabizon AA (2001) Pegylated liposomal doxorubicin: Metamorphosis of an old drug into a new form of chemotherapy. Cancer Invest 19:424–436
45. Moghimi SM (2002) Chemical camouflage of nanospheres with a poorly reactive surface: Towards development of stealth and target-specific nanocarriers. Biochim Biophys Acta 1590:131–139
46. Moein Moghimi S et al (2006) Activation of the human complement system by cholesterol-rich and PEGylated liposomes-modulation of cholesterol-rich liposome-mediated complement activation by elevated serum LDL and HDL levels. J Liposome Res 16:167–174
47. Gabizon A, Shmeeda H, Barenholz Y (2003) Pharmacokinetics of pegylated liposomal Doxorubicin: Review of animal and human studies. Clin Pharmacokinet 42:419–436
48. Harris JM, Martin NE, Modi M (2001) Pegylation: A novel process for modifying pharmacokinetics. Clin Pharmacokinet 40:539–551
49. Ishida T et al (2005) Accelerated blood clearance of PEGylated liposomes following preceding liposome injection: Effects of lipid dose and PEG surface-density and chain length of the first-dose liposomes. J Control Release 105:305–317
50. Torchilin VP, Trubetskoy VS (1995) Which polymers can make nanoparticulate drug carriers long-circulating? Adv Drug Deliv Rev 16:141–155
51. Woodle MC (1998) Controlling liposome blood clearance by surface-grafted polymers. Adv Drug Deliv Rev 32:139–152
52. Whiteman KR, Subr V, Ulbrich K, Torchilin VP (2001) Poly(HPMA)-coated liposomes demonstrate prolonged circulation in mice. J Liposome Res 11:153–164
53. Torchilin VP et al (2001) Amphiphilic poly-N-vinylpyrrolidones: Synthesis, properties and liposome surface modification. Biomaterials 22:3035–3044
54. Metselaar JM et al (2003) A novel family of l-amino acid-based biodegradable polymer-lipid conjugates for the development of long-circulating liposomes with effective drug-targeting capacity. Bioconjug Chem 14:1156–1164
55. Takeuchi H, Kojima H, Yamamoto H, Kawashima Y (2001) Evaluation of circulation profiles of liposomes coated with hydrophilic polymers having different molecular weights in rats. J Control Release 75:83–91
56. Taira MC, Chiaramoni NS, Pecuch KM, Alonso-Romanowski S (2004) Stability of liposomal formulations in physiological conditions for oral drug delivery. Drug Deliv 11:123–128
57. Mora M et al (2002) Design and characterization of liposomes containing long-chain N-acylPEs for brain delivery: Penetration of liposomes incorporating GM1 into the rat brain. Pharm Res 19:1430–1438
58. Minato S et al (2003) Application of polyethyleneglycol (PEG)-modified liposomes for oral vaccine: Effect of lipid dose on systemic and mucosal immunity. J Control Release 89:189–197
59. Xing L, Dawei C, Liping X, Rongqing Z (2003) Oral colon-specific drug delivery for bee venom peptide: Development of a coated calcium alginate gel beads-entrapped liposome. J Control Release 93:293–300
60. Phillips WT, Klipper R, Goins B (2000) Novel method of greatly enhanced delivery of liposomes to lymph nodes. J Pharmacol Exp Ther 295:309–313
61. Allen TM, Hansen CB, Guo LS (1993) Subcutaneous administration of liposomes:

A comparison with the intravenous and intraperitoneal routes of injection. Biochim Biophys Acta 1150:9–16
62. Kim CK, Han JH (1995) Lymphatic delivery and pharmacokinetics of methotrexate after intramuscular injection of differently charged liposome-entrapped methotrexate to rats. J Microencapsul 12:437–446
63. Fujimoto Y et al (2000) Magnetic resonance lymphography of profundus lymph nodes with liposomal gadolinium-diethylenetriamine pentaacetic acid. Biol Pharm Bull 23:97–100
64. Zalipsky S et al (1999) New detachable poly(ethylene glycol) conjugates: Cysteine-cleavable lipopolymers regenerating natural phospholipid, diacyl phosphatidylethanolamine. Bioconjug Chem 10:703–707
65. Erdogan S et al (2006) Gadolinium-loaded polychelating polymer-containing cancer cell-specific immunoliposomes. J Liposome Res 16:45–55
66. Kale AA, Torchilin VP (2007) Design, synthesis, and characterization of pH-sensitive PEG-PE conjugates for stimuli-sensitive pharmaceutical nanocarriers: The effect of substitutes at the hydrazone linkage on the ph stability of PEG-PE conjugates. Bioconjug Chem 18:363–370
67. Krown SE, Northfelt DW, Osoba D, Stewart JS (2004) Use of liposomal anthracyclines in Kaposi's sarcoma. Semin Oncol 31:36–52
68. Rose PG (2005) Pegylated liposomal doxorubicin: Optimizing the dosing schedule in ovarian cancer. Oncologist 10:205–214
69. Thigpen JT et al (2005) Role of pegylated liposomal doxorubicin in ovarian cancer. Gynecol Oncol 96:10–18
70. Hussein MA, Anderson KC (2004) Role of liposomal anthracyclines in the treatment of multiple myeloma. Semin Oncol 31:147–160
71. Robert NJ et al (2004) The role of the liposomal anthracyclines and other systemic therapies in the management of advanced breast cancer. Semin Oncol 31:106–146
72. Keller AM et al (2004) Randomized phase III trial of pegylated liposomal doxorubicin versus vinorelbine or mitomycin C plus vinblastine in women with taxane-refractory advanced breast cancer. J Clin Oncol 22:3893–3901
73. Hau P et al (2004) Pegylated liposomal doxorubicin-efficacy in patients with recurrent high-grade glioma. Cancer 100:1199–1207
74. Kim ES et al (2001) A phase II study of STEALTH cisplatin (SPI-77) in patients with advanced non-small cell lung cancer. Lung Cancer 34:427–432
75. Harrington KJ et al (2001) Phase I-II study of pegylated liposomal cisplatin (SPI-077) in patients with inoperable head and neck cancer. Ann Oncol 12:493–496
76. Zamboni WC et al (2004) Systemic and tumor disposition of platinum after administration of cisplatin or STEALTH liposomal-cisplatin formulations (SPI-077 and SPI-077 B103) in a preclinical tumor model of melanoma. Cancer Chemother Pharmacol 53:329–336
77. Immordino ML, Dosio F, Cattel L (2006) Stealth liposomes: Review of the basic science, rationale, and clinical applications, existing and potential. Int J Nanomedicine 1:297–315
78. Adlakha-Hutcheon G, Bally MB, Shew CR, Madden TD (1999) Controlled destabilization of a liposomal drug delivery system enhances mitoxantrone antitumor activity. Nat Biotechnol 17:775–779
79. Torchilin VP et al (1992) Targeted accumulation of polyethylene glycol-coated immunoliposomes in infarcted rabbit myocardium. FASEB J 6:2716–2719
80. Blume G et al (1993) Specific targeting with poly(ethylene glycol)-modified liposomes: Coupling of homing devices to the ends of the polymeric chains combines effective target binding with long circulation times. Biochim Biophys Acta 1149:180–184
81. Abra RM et al (2002) The next generation of liposome delivery systems: Recent experience with tumor-targeted, sterically-stabilized immunoliposomes and active-loading gradients. J Liposome Res 12:1–3
82. Torchilin VP et al (2001) p-Nitrophenylcarbonyl-PEG-PE-liposomes: Fast and simple attachment of specific ligands, including monoclonal antibodies, to distal ends of PEG chains via p-nitrophenylcarbonyl groups. Biochim Biophys Acta 1511:397–411
83. Torchilin V, Klibanov A (1993) Coupling and labeling of phospholipids. In: Cevc G (ed) Phospholipid handbook. Marcel Dekker, New York, pp 293–322
84. Torchilin VP, Weissig V, Martin FJ, Heath TD (2003) Surface modifications of liposomes. In: Torchilin VP, Weissig V (eds) Liposomes: A practical approach, Oxford University Press, Oxford, New York, pp 193–229
85. Klibanov AL, Torchilin VP, Zalipsky S (2003) Long-circulating sterically protected liposomes. In: Torchilin VP, Weissig V (eds) Liposomes: A practical approach, Oxford University Press, Oxford, New York, pp 231–265
86. Sapra P, Allen TM (2002) Internalizing antibodies are necessary for improved therapeutic efficacy of antibody-targeted liposomal drugs. Cancer Res 62:7190–7194

87. Park JW et al (2001) Tumor targeting using anti-her2 immunoliposomes. J Control Release 74:95–113
88. Nellis DF et al (2005) Preclinical manufacture of anti-HER2 liposome-inserting, scFv-PEG-lipid conjugate. 2. Conjugate micelle identity, purity, stability, and potency analysis. Biotechnol Prog 21:221–232
89. Nellis DF et al (2005) Preclinical manufacture of an anti-HER2 scFv-PEG-DSPE, liposome-inserting conjugate. 1. Gram-scale production and purification. Biotechnol Prog 21:205–220
90. Kirpotin DB et al (2006) Antibody targeting of long-circulating lipidic nanoparticles does not increase tumor localization but does increase internalization in animal models. Cancer Res 66:6732–6740
91. Kamps JA et al (2000) Uptake of long-circulating immunoliposomes, directed against colon adenocarcinoma cells, by liver metastases of colon cancer. J Drug Target 8:235–245
92. Lukyanov AN, Elbayoumi TA, Chakilam AR, Torchilin VP (2004) Tumor-targeted liposomes: Doxorubicin-loaded long-circulating liposomes modified with anti-cancer antibody. J Control Release 100:135–144
93. Elbayoumi TA, Torchilin VP (2007) Enhanced cytotoxicity of monoclonal anticancer antibody 2C5-modified doxorubicin-loaded PEGylated liposomes against various tumor cell lines. Eur J Pharm Sci 32:159–168
94. Elbayoumi TA, Torchilin VP (2008) Tumor-specific antibody-mediated targeted delivery of Doxil(R) reduces the manifestation of auricular erythema side effect in mice. Int J Pharm 357:272–279
95. Gupta B, Torchilin VP (2007) Monoclonal antibody 2C5-modified doxorubicin-loaded liposomes with significantly enhanced therapeutic activity against intracranial human brain U-87 MG tumor xenografts in nude mice. Cancer Immunol Immunother 56:1215–1223
96. Raffaghello L et al (2003) Immunoliposomal fenretinide: A novel antitumoral drug for human neuroblastoma. Cancer Lett 197:151–155
97. Mamot C et al (2005) Epidermal growth factor receptor-targeted immunoliposomes significantly enhance the efficacy of multiple anticancer drugs in vivo. Cancer Res 65:11631–11638
98. Mamot C et al (2006) EGFR-targeted immunoliposomes derived from the monoclonal antibody EMD72000 mediate specific and efficient drug delivery to a variety of colorectal cancer cells. J Drug Target 14:215–223
99. Venter JC et al (2001) The sequence of the human genome. Science 291:1304–1351
100. Mastrobattista E et al (2002) Functional characterization of an endosome-disruptive peptide and its application in cytosolic delivery of immunoliposome-entrapped proteins. J Biol Chem 277:27135–27143
101. Hatakeyama H et al (2004) Factors governing the in vivo tissue uptake of transferrin-coupled polyethylene glycol liposomes in vivo. Int J Pharm 281:25–33
102. Ishida O et al (2001) Liposomes bearing polyethyleneglycol-coupled transferrin with intracellular targeting property to the solid tumors in vivo. Pharm Res 18:1042–1048
103. Derycke AS, De Witte PA (2002) Transferrin-mediated targeting of hypericin embedded in sterically stabilized PEG-liposomes. Int J Oncol 20:181–187
104. Gijsens A et al (2002) Targeting of the photocytotoxic compound AlPcS4 to Hela cells by transferrin conjugated PEG-liposomes. Int J Cancer 101:78–85
105. Iinuma H et al (2002) Intracellular targeting therapy of cisplatin-encapsulated transferrin-polyethylene glycol liposome on peritoneal dissemination of gastric cancer. Int J Cancer 99:130–137
106. Eavarone DA, Yu X, Bellamkonda RV (2000) Targeted drug delivery to C6 glioma by transferrin-coupled liposomes. J Biomed Mater Res 51:10–14
107. Omori N et al (2003) Targeting of post-ischemic cerebral endothelium in rat by liposomes bearing polyethylene glycol-coupled transferrin. Neurol Res 25:275–279
108. Joshee N, Bastola DR, Cheng PW (2002) Transferrin-facilitated lipofection gene delivery strategy: Characterization of the transfection complexes and intracellular trafficking. Hum Gene Ther 13:1991–2004
109. Xu L et al (2002) Systemic tumor-targeted gene delivery by anti-transferrin receptor scFv-immunoliposomes. Mol Cancer Ther 1:337–346
110. Tan PH et al (2003) Antibody targeted gene transfer to endothelium. J Gene Med 5:311–323
111. Huwyler J, Wu D, Pardridge WM (1996) Brain drug delivery of small molecules using immunoliposomes. Proc Natl Acad Sci USA 93:14164–14169
112. Leamon CP, Low PS (1991) Delivery of macromolecules into living cells: A method that exploits folate receptor endocytosis. Proc Natl Acad Sci USA 88:5572–5576
113. Lee RJ, Low PS (1994) Delivery of liposomes into cultured KB cells via folate

receptor-mediated endocytosis. J Biol Chem 269:3198–3204
114. Lu Y, Low PS (2002) Folate-mediated delivery of macromolecular anticancer therapeutic agents. Adv Drug Deliv Rev 54:675–693
115. Gabizon A, Shmeeda H, Horowitz AT, Zalipsky S (2004) Tumor cell targeting of liposome-entrapped drugs with phospholipid-anchored folic acid-PEG conjugates. Adv Drug Deliv Rev 56:1177–1192
116. Ni S, Stephenson SM, Lee RJ (2002) Folate receptor targeted delivery of liposomal daunorubicin into tumor cells. Anticancer Res 22:2131–2135
117. Pan XQ, Wang H, Lee RJ (2003) Antitumor activity of folate receptor-targeted liposomal doxorubicin in a KB oral carcinoma murine xenograft model. Pharm Res 20:417–422
118. Gupta Y, Jain A, Jain P, Jain SK (2007) Design and development of folate appended liposomes for enhanced delivery of 5-FU to tumor cells. J Drug Target 15:231–240
119. Torchilin VP et al (2002) Vaccination with nucleosomes results in strong inhibition of tumor growth in various models in mice. Proc Intl Symp Control Rel Bioact Mater 29:1197–1198
120. Stephenson SM et al (2003) Folate receptor-targeted liposomes as possible delivery vehicles for boron neutron capture therapy. Anticancer Res 23:3341–3345
121. Lu Y, Low PS (2002) Folate targeting of haptens to cancer cell surfaces mediates immunotherapy of syngeneic murine tumors. Cancer Immunol Immunother 51:153–162
122. Turk MJ, Reddy JA, Chmielewski JA, Low PS (2002) Characterization of a novel pH-sensitive peptide that enhances drug release from folate-targeted liposomes at endosomal pHs. Biochim Biophys Acta 1559:56–68
123. Leamon CP, Cooper SR, Hardee GE (2003) Folate-liposome-mediated antisense oligodeoxynucleotide targeting to cancer cells: Evaluation in vitro and in vivo. Bioconjug Chem 14:738–747
124. Adams GP, Weiner LM (2005) Monoclonal antibody therapy of cancer. Nat Biotechnol 23:1147–1157
125. Drummond DC et al (2000) Liposome targeting to tumors using vitamin and growth factor receptors. Vitam Horm 60:285–332
126. Dagar S et al (2003) VIP grafted sterically stabilized liposomes for targeted imaging of breast cancer: In vivo studies. J Control Release 91:123–133
127. Schiffelers RM et al (2003) Anti-tumor efficacy of tumor vasculature-targeted liposomal doxorubicin. J Control Release 91:115–122
128. Gupta AS et al (2005) RGD-modified liposomes targeted to activated platelets as a potential vascular drug delivery system. Thromb Haemost 93:106–114
129. Lander ES et al (2001) Initial sequencing and analysis of the human genome. Nature 409:860–921
130. Asai T et al (2002) Anti-neovascular therapy by liposomal DPP-CNDAC targeted to angiogenic vessels. FEBS Lett 520:167–170
131. Mamot C et al (2003) Liposome-based approaches to overcome anticancer drug resistance. Drug Resist Updat 6:271–279
132. Peer D, Margalit R (2004) Loading mitomycin C inside long circulating hyaluronan targeted nano-liposomes increases its antitumor activity in three mice tumor models. Int J Cancer 108:780–789
133. Takakura Y et al (1989) Control of pharmaceutical properties of soybean trypsin inhibitor by conjugation with dextran. II: Biopharmaceutical and pharmacological properties. J Pharm Sci 78:219–222
134. Lee CM et al (2002) Novel chondroitin sulfate-binding cationic liposomes loaded with cisplatin efficiently suppress the local growth and liver metastasis of tumor cells in vivo. Cancer Res 62:4282–4288
135. Terada T et al (2007) Optimization of tumor-selective targeting by basic fibroblast growth factor-binding peptide grafted PEGylated liposomes. J Control Release 119:262–270
136. Ikehara Y, Kojima N (2007) Development of a novel oligomannose-coated liposome-based anticancer drug-delivery system for intraperitoneal cancer. Curr Opin Mol Ther 9:53–61
137. Simoes S et al (2004) On the formulation of pH-sensitive liposomes with long circulation times. Adv Drug Deliv Rev 56:947–965
138. Laham A et al (1988) Intracarotidal administration of liposomally-entrapped ATP: Improved efficiency against experimental brain ischemia. Pharmacol Res Commun 20:699–705
139. Sudimack JJ, Guo W, Tjarks W, Lee RJ (2002) A novel pH-sensitive liposome formulation containing oleyl alcohol. Biochim Biophys Acta 1564:31–37
140. Asokan A, Cho MJ (2003) Cytosolic delivery of macromolecules. II. Mechanistic studies with pH-sensitive morpholine lipids. Biochim Biophys Acta 1611:151–160
141. Roux E et al (2004) Serum-stable and long-circulating, PEGylated, pH-sensitive liposomes. J Control Release 94:447–451

142. Kakudo T et al (2004) Transferrin-modified liposomes equipped with a pH-sensitive fusogenic peptide: An artificial viral-like delivery system. Biochemistry 43:5618–5628
143. Karanth H, Murthy RS (2007) pH-Sensitive liposomes-principle and application in cancer therapy. J Pharm Pharmacol 59:469–483
144. Kale AA, Torchilin VP (2007) Enhanced transfection of tumor cells in vivo using "Smart" pH-sensitive TAT-modified pegylated liposomes. J Drug Target 15:538–545
145. Frankel AD, Pabo CO (1988) Cellular uptake of the tat protein from human immunodeficiency virus. Cell 55:1189–1193
146. Wadia JS, Stan RV, Dowdy SF (2004) Transducible TAT-HA fusogenic peptide enhances escape of TAT-fusion proteins after lipid raft macropinocytosis. Nat Med 10:310–315
147. Rothbard JB et al (2002) Arginine-rich molecular transporters for drug delivery: Role of backbone spacing in cellular uptake. J Med Chem 45:3612–3618
148. Torchilin VP, Rammohan R, Weissig V, Levchenko TS (2001) TAT peptide on the surface of liposomes affords their efficient intracellular delivery even at low temperature and in the presence of metabolic inhibitors. Proc Natl Acad Sci USA 98:8786–8791
149. Tseng YL, Liu JJ, Hong RL (2002) Translocation of liposomes into cancer cells by cell-penetrating peptides penetratin and tat: A kinetic and efficacy study. Mol Pharmacol 62:864–872
150. Gorodetsky R et al (2004) Liposome transduction into cells enhanced by haptotactic peptides (Haptides) homologous to fibrinogen C-termini. J Control Release 95:477–488
151. Torchilin VP et al (2003) Cell transfection in vitro and in vivo with nontoxic TAT peptide-liposome-DNA complexes. Proc Natl Acad Sci USA 100:1972–1977
152. Gupta B, Levchenko T, Torchilin VP (2007) TAT peptide-modified liposomes provide enhanced gene delivery to intracranial human brain tumor xenografts in nude mice. Oncol Res 16:351–359
153. Boddapati SV et al (2005) Mitochondriotropic liposomes. J Liposome Res 15:49–58
154. Boddapati SV et al (2008) Organelle-targeted nanocarriers: Specific delivery of liposomal ceramide to mitochondria enhances its cytotoxicity in vitro and in vivo. Nano Lett 8:2559–2563
155. Torchilin VP (2006) Multifunctional nanocarriers. Adv Drug Deliv Rev 58:1532–1555
156. Torchilin VP (1996) Liposomes as delivery agents for medical imaging. Mol Med Today 2:242–249
157. Tilcock C (1995) Imaging tools: Liposomal agents for nuclear medicine, computed tomography, magnetic resonance, and ultrasound. In: Philippot JR, Schuber F (eds) Liposomes as tools in basic research and industry. CRC Press, Boca Raton, pp 225–240
158. Torchilin VP (1997) Surface-modified liposomes in gamma- and MR-imaging. Adv Drug Deliv Rev 24:301–313
159. Torchilin VP (2000) Polymeric contrast agents for medical imaging. Curr Pharm Biotechnol 1:183–215
160. Erdogan S, Roby A, Torchilin VP (2006) Enhanced tumor visualization by gamma-scintigraphy with 111In-labeled polychelating-polymer-containing immunoliposomes. Mol Pharm 3:525–530
161. Weissig VV, Babich J, Torchilin VV (2000) Long-circulating gadolinium-loaded liposomes: Potential use for magnetic resonance imaging of the blood pool. Colloids Surf B Biointerfaces 18:293–299
162. Lokling KE, Fossheim SL, Klaveness J, Skurtveit R (2004) Biodistribution of pH-responsive liposomes for MRI and a novel approach to improve the pH-responsiveness. J Control Release 98:87–95
163. Viglianti BL et al (2004) In vivo monitoring of tissue pharmacokinetics of liposome/drug using MRI: Illustration of targeted delivery. Magn Reson Med 51:1153–1162
164. Plassat V et al (2007) Sterically stabilized superparamagnetic liposomes for MR imaging and cancer therapy: Pharmacokinetics and biodistribution. Int J Pharm 344:118–127
165. Bao A et al (2003) A novel liposome radiolabeling method using 99mTc-"SNS/S" complexes: In vitro and in vivo evaluation. J Pharm Sci 92:1893–1904
166. Sachse A et al (1993) Preparation and evaluation of lyophilized iopromide-carrying liposomes for liver tumor detection. Invest Radiol 28:838–844
167. Sachse A et al (1997) Biodistribution and computed tomography blood-pool imaging properties of polyethylene glycol-coated iopromide-carrying liposomes. Invest Radiol 32:44–50
168. Xie F, Hankins J, Mahrous HA, Porter TR (2007) Detection of coronary artery disease with a continuous infusion of definity ultrasound contrast during adenosine stress real time perfusion echocardiography. Echocardiography 24:1044–1050
169. Maruyama H et al (2005) Real-time blood-pool images of contrast enhanced ultrasound with definity in the detection of tumour nodules in the liver. Br J Radiol 78:512–518

170. Kitzman DW et al (2000) Efficacy and safety of the novel ultrasound contrast agent perflutren (definity) in patients with suboptimal baseline left ventricular echocardiographic images. Am J Cardiol 86:669–674
171. Morawski AM, Lanza GA, Wickline SA (2005) Targeted contrast agents for magnetic resonance imaging and ultrasound. Curr Opin Biotechnol 16:89–92
172. Mu Y et al (1999) Bioconjugation of laminin peptide YIGSR with poly(styrene co-maleic acid) increases its antimetastatic effect on lung metastasis of B16-BL6 melanoma cells. Biochem Biophys Res Commun 255:75–79
173. Jana SS et al (2002) Targeted cytosolic delivery of hydrogel nanoparticles into HepG2 cells through engineered Sendai viral envelopes. FEBS Lett 515:184–188
174. Cusi MG et al (2004) Efficient delivery of DNA to dendritic cells mediated by influenza virosomes. Vaccine 22:735–739
175. Bungener L, Huckriede A, Wilschut J, Daemen T (2002) Delivery of protein antigens to the immune system by fusion-active virosomes: A comparison with liposomes and ISCOMs. Biosci Rep 22:323–338
176. Bungener L et al (2002) Virosome-mediated delivery of protein antigens to dendritic cells. Vaccine 20:2287–2295
177. Huckriede A, Bungener L, Daemen T, Wilschut J (2003) Influenza virosomes in vaccine development. Meth Enzymol 373:74–91
178. Herzog C, Metcalfe IC, Schaad UB (2002) Virosome influenza vaccine in children. Vaccine 20(Suppl 5):B24–B28
179. Usonis V et al (2003) Antibody titres after primary and booster vaccination of infants and young children with a virosomal hepatitis A vaccine (Epaxal). Vaccine 21:4588–4592
180. Ambrosch F et al (2004) Rapid antibody response after vaccination with a virosomal hepatitis a vaccine. Infection 32:149–152
181. Ruf BR, Colberg K, Frick M, Preusche A (2004) Open, randomized study to compare the immunogenicity and reactogenicity of an influenza split vaccine with an MF59-adjuvanted subunit vaccine and a virosome-based subunit vaccine in elderly. Infection 32:191–198
182. MacGregor RR et al (1998) First human trial of a DNA-based vaccine for treatment of human immunodeficiency virus type 1 infection: Safety and host response. J Infect Dis 178:92–100
183. Moser C, Metcalfe IC, Viret JF (2003) Virosomal adjuvanted antigen delivery systems. Expert Rev Vaccines 2:189–196
184. Huckriede A, De Jonge J, Holtrop M, Wilschut J (2007) Cellular delivery of siRNA mediated by fusion-active virosomes. J Liposome Res 17:39–47
185. Copland MJ, Rades T, Davies NM, Baird MA (2005) Lipid based particulate formulations for the delivery of antigen. Immunol Cell Biol 83:97–105
186. Chen WC, Huang L (2005) Non-viral vector as vaccine carrier. Adv Genet 54:315–337
187. Bramwell VW, Perrie Y (2005) Particulate delivery systems for vaccines. Crit Rev Ther Drug Carrier Syst 22:151–214
188. Cevc G (2004) Lipid vesicles and other colloids as drug carriers on the skin. Adv Drug Deliv Rev 56:675–711
189. Cevc G, Blume G (2001) New, highly efficient formulation of diclofenac for the topical, transdermal administration in ultradeformable drug carriers. Transfersomes Biochim Biophys Acta 1514:191–205
190. Honeywell-Nguyen PL et al (2002) Transdermal delivery of pergolide from surfactant-based elastic and rigid vesicles: Characterization and in vitro transport studies. Pharm Res 19:991–997
191. Vutla NB, Betageri GV, Banga AK (1996) Transdermal iontophoretic delivery of enkephalin formulated in liposomes. J Pharm Sci 85:5–8
192. Han I, Kim M, Kim J (2004) Enhanced transfollicular delivery of adriamycin with a liposome and iontophoresis. Exp Dermatol 13:86–92
193. Derycke AS, de Witte PA (2004) Liposomes for photodynamic therapy. Adv Drug Deliv Rev 56:17–30
194. Takeuchi Y et al (2004) Intracellular target for photosensitization in cancer antiangiogenic photodynamic therapy mediated by polycation liposome. J Control Release 97:231–240
195. Ichikawa K et al (2004) Antiangiogenic photodynamic therapy (PDT) using Visudyne causes effective suppression of tumor growth. Cancer Lett 205:39–48
196. Igarashi A et al (2003) Liposomal photofrin enhances therapeutic efficacy of photodynamic therapy against the human gastric cancer. Toxicol Lett 145:133–141
197. Bourre L et al (2003) In vivo photosensitizing efficiency of a diphenylchlorin sensitizer: Interest of a DMPC liposome formulation. Pharmacol Res 47:253–261
198. Jezek P et al (2003) Experimental photodynamic therapy with MESO-tetrakisphenylporphyrin (TPP) in liposomes leads to disintegration of human amelanotic melanoma

implanted to nude mice. Int J Cancer 103: 693–702
199. Goldberg SN et al (2002) Percutaneous tumor ablation: Increased necrosis with combined radio-frequency ablation and intravenous liposomal doxorubicin in a rat breast tumor model. Radiology 222:797–804
200. Monsky WL et al (2002) Radio-frequency ablation increases intratumoral liposomal doxorubicin accumulation in a rat breast tumor model. Radiology 224:823–829
201. Maeda H, Sawa T, Konno T (2001) Mechanism of tumor-targeted delivery of macromolecular drugs, including the EPR effect in solid tumor and clinical overview of the prototype polymeric drug SMANCS. J Control Release 74:47–61
202. Neveux N, De Bandt JP, Chaumeil JC, Cynober L (2002) Hepatic preservation, liposomally entrapped adenosine triphosphate and nitric oxide production: A study of energy state and protein metabolism in the cold-stored rat liver. Scand J Gastroenterol 37:1057–1063
203. Niibori K et al (1999) Bioenergetic effect of liposomal coenzyme Q10 on myocardial ischemia reperfusion injury. Biofactors 9: 307–313
204. Xu GX et al (1990) Adenosine triphosphate liposomes: Encapsulation and distribution studies. Pharm Res 7:553–557
205. Verma DD, Levchenko T, Bernstein EA, Torchilin V (2004) In: Thirty-first annual meeting of the controlled release society, Controlled Release Society, Honolulu, pp #572
206. Liang W, Levchenko TS, Torchilin VP (2004) Encapsulation of ATP into liposomes by different methods: Optimization of the procedure. J Microencapsul 21:251–261
207. Verma DD et al (2006) ATP-loaded immunoliposomes specific for cardiac myosin provide improved protection of the mechanical functions of myocardium from global ischemia in an isolated rat heart model. J Drug Target 14:273–280
208. Verma DD et al (2005) ATP-loaded liposomes effectively protect the myocardium in rabbits with an acute experimental myocardial infarction. Pharm Res 22:2115–2120
209. Verma DD et al (2007) Protective effect of coenzyme Q10-loaded liposomes on the myocardium in rabbits with an acute experimental myocardial infarction. Pharm Res 24:2131–2137
210. Phillips WT et al (1999) Polyethylene glycol-modified liposome-encapsulated hemoglobin: A long circulating red cell substitute. J Pharmacol Exp Ther 288:665–670
211. Awasthi V et al (2007) Cerebral oxygen delivery by liposome-encapsulated hemoglobin: A positron-emission tomographic evaluation in a rat model of hemorrhagic shock. J Appl Physiol 103:28–38
212. Awasthi VD et al (2004) Kinetics of liposome-encapsulated hemoglobin after 25% hypovolemic exchange transfusion. Int J Pharm 283:53–62
213. Nobuto H et al (2004) Evaluation of systemic chemotherapy with magnetic liposomal doxorubicin and a dipole external electromagnet. Int J Cancer 109:627–635
214. Kubo T et al (2001) Targeted systemic chemotherapy using magnetic liposomes with incorporated adriamycin for osteosarcoma in hamsters. Int J Oncol 18:121–125
215. Babincova M et al (2000) Site-specific in vivo targeting of magnetoliposomes using externally applied magnetic field. Z Naturforsch [C] 55:278–281
216. Minko T et al (2006) New generation of liposomal drugs for cancer. Anticancer Agents Med Chem 6:537–552
217. Al-Jamal WT, Kostarelos K (2007) Construction of nanoscale multicompartment liposomes for combinatory drug delivery. Int J Pharm 331:182–185
218. Veerareddy PR, Vobalaboina V (2004) Lipid-based formulations of amphotericin B. Drugs Today (Barc) 40:133–145
219. Boulikas T, Stathopoulos GP, Volakakis N, Vougiouka M (2005) Systemic Lipoplatin infusion results in preferential tumor uptake in human studies. Anticancer Res 25: 3031–3039
220. Stathopoulos GP et al (2005) Pharmacokinetics and adverse reactions of a new liposomal cisplatin (Lipoplatin): phase I study. Oncol Rep 13:589–595

Chapter 2

Nanoliposomes: Preparation and Analysis

M.R. Mozafari

Abstract

Nanoliposome, or submicron bilayer lipid vesicle, is a new technology for the encapsulation and delivery of bioactive agents. The list of bioactive material that can be incorporated to nanoliposomes is immense, ranging from pharmaceuticals to cosmetics and nutraceuticals. Because of their biocompatibility and biodegradability, along with their nanosize, nanoliposomes have potential applications in a vast range of fields, including nanotherapy (e.g. diagnosis, cancer therapy, gene delivery), cosmetics, food technology and agriculture. Nanoliposomes are able to enhance the performance of bioactive agents by improving their solubility and bioavailability, in vitro and in vivo stability, as well as preventing their unwanted interactions with other molecules. Another advantage of nanoliposomes is cell-specific targeting, which is a prerequisite to attain drug concentrations required for optimum therapeutic efficacy in the target site while minimising adverse effects on healthy cells and tissues. This chapter covers nanoliposomes, particularly with respect to their properties, preparation methods and analysis.

Key words: Cancer therapy, Food nanotechnology, Gene delivery, Mozafari method, Nanoliposomes, Nanotherapy

Abbreviations

Chol	Cholesterol
DCP	Dicetylphosphate
DPPC	Dipalmitoyl phosphatidylcholine
EE	Entrapment efficiency
FFF	Field flow fractionation
HM-liposomes	Lipid vesicles prepared by the heating method
HPLC	High-performance liquid chromatography
OH	Hydroxyl group
LUV	Large unilamellar vesicles
MLV	Multilamellar vesicles
NMR	Nuclear magnetic resonance
PC	Phosphatidylcholine
SPM	Scanning Probe Microscopy
SUV	Small unilamellar vesicles

T_c Phase transition temperature
TLC   Thin layer chromatography
T_m Melting temperature

1. Introduction

Nanoliposomes are nanometric version of liposomes, which are one of the most applied encapsulation and controlled release systems (1). In order to have a better understanding of nanoliposomes, we need to know about the older technology they are developed from, i.e. liposomes. The word liposome derives from two Greek words, lipos (fat) and soma (body or structure), meaning a structure in which a fatty envelope encapsulates internal aqueous compartment(s) (2, 3). Liposomes (also known as bilayer lipid vesicles) are ideal models of cells and biomembranes. Their resemblance to biological membranes makes them an ideal system, not only for the study of contemporary biomembranes, but also in studies investigating the emergence, functioning and evolution of primitive cell membranes (4, 5). Furthermore, they are being used by the food, cosmetic, agricultural and pharmaceutical industries as carrier systems for the protection and delivery of different material including drugs, nutraceuticals, pesticides and genetic material. Liposomes are composed of one or more concentric or non-concentric lipid and/or phospholipid bilayers and can contain other molecules such as proteins in their structure. They can be single or multilamellar, with respect to the number of bilayers they contain, and can accommodate hydrophilic, lipophilic and amphiphilic compounds in their aqueous and/or lipid compartments.

Since the introduction of liposomes to the scientific community, around 40 years ago (6), there have been considerable advances in the optimisation of liposomal formulations and their manufacturing techniques. These include prolonged liposomal half-life in blood circulation, the development of ingenious strategies for tissue and cell targeting and the elimination of harmful solvents used during their preparation (7). The term nanoliposome has recently been introduced to exclusively refer to nano-scale bilayer lipid vesicles since liposome is a general terminology covering many classes of vesicles whose diameters range from tens of nanometers to several micrometers (1). In a broad sense, liposomes and nanoliposomes share the same chemical, structural and thermodynamic properties. However, compared to liposomes, nanoliposomes provide more surface area and have the potential to increase solubility, enhance bioavailability, improve controlled release and enable precision targeting of the encapsulated material to a greater extent (8). The first journal article on nanoliposome technology was published in 2002 (9) and the first book on

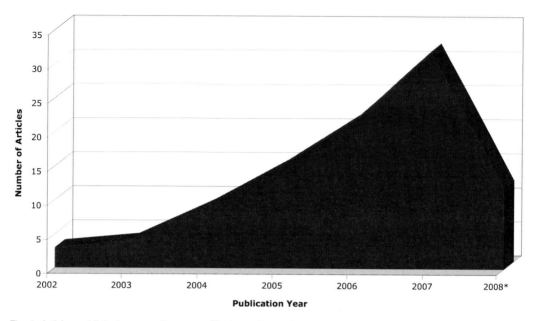

Fig. 1. Articles published on nanoliposomes (*last year incomplete). The database source was Scopus (http://www.scopus.com/scopus/home.url)

nanoliposomes was published more recently in 2005 (1). Figure 1 shows the number of articles and patents published annually since 2002 on nanoliposomes. The sharp increase in the number of these publications is an indication of increasing interest in the field of nanoliposome research.

This entry summarizes the main physicochemical properties of nanoliposomes along with some methods of their preparation and analysis. The methods for the manufacture, isolation, and characterization of nanoliposomes are as diverse as their applications and it is impossible to cover each and every method in a single chapter. Consequently, this chapter describes methods that have been recently developed in our laboratory along with some of the commonly applied procedures for nanoliposome preparation.

1.1. Physicochemical Properties

In order to realize the mechanism of nanoliposome formation and main points in their manufacture, we have to look at their physical and chemical characteristics along with the properties of their constituents.

1.1.1. Chemical Constituents

The main chemical ingredients of nanoliposomes are lipid and/or phospholipid molecules. Lipids are fatty acid derivatives with various head group moieties. When taken orally, lipids are subjected to conversion by gastrointestinal lipases to their constituent fatty acids and head groups. Triglycerides are lipids made from three fatty acids and a glycerol molecule (a three-carbon alcohol with a hydroxyl group [OH] on each carbon atom). Mono- and diglycerides are glyceryl mono- and di-esters of fatty acids. Phospholipids are similar to triglycerides except that the first hydroxyl of the glycerol molecule has a polar phosphate-containing group in place of the fatty acid. Phospholipids are amphiphilic, possessing both hydrophilic (water soluble) and hydrophobic (lipid soluble) groups. The head group of a phospholipid is hydrophilic and its fatty acid tail (acyl chain) is hydrophobic (10). The phosphate moiety of the head group is negatively charged. If the acyl chains only contain single chemical bonds, the lipid is known as '*saturated*' and if there is one or more double bonds in the acyl chains, the lipid is known as '*unsaturated*'. Therefore, saturated lipids have the maximum number of hydrogen atoms.

In addition to lipid and/or phospholipid molecules, nanoliposomes may contain other molecules such as sterols in their structure. Sterols are important components of most natural membranes, and incorporation of sterols into nanoliposome bilayers can bring about major changes in the properties of these vesicles. The most widely used sterol in the manufacture of the lipid vesicles is cholesterol (Chol). Cholesterol does not by itself form bilayer structures, but it can be incorporated into phospholipid membranes in very high concentrations, for example up to 1:1 or even 2:1 molar ratios of cholesterol to a phospholipid such as phosphatidylcholine (PC) (11). Cholesterol is used in nanoliposome structures in order to increase the stability of the vesicles by modulating the fluidity of the lipid bilayer. In general, cholesterol modulates fluidity of phospholipid membranes by preventing crystallization of the acyl chains of phospholipids and providing steric hindrance to their movement. This contributes to the stability of nanoliposomes and reduces the permeability of the lipid membrane to solutes (12).

The amount of cholesterol to be used in the nanoliposomal formulations largely depends on the intended application. For liposomes in the form of multilamellar vesicles (MLV), we have found that both anionic (with dicetylphosphate, DCP, as a negatively charged ingredient) and neutral (without DCP) liposomes containing PC can interact with model membrane systems (in the form of fusion/aggregation) when containing a 10% molar ratio of Chol (13). Anionic vesicles containing 10% Chol were also able to incorporate DNA molecules in the presence of calcium ions. Increasing cholesterol content of these vesicles from 10 to 40%, however, caused them to be unable to interact with model membranes and also unable to incorporate DNA molecules. We concluded that these types of liposomes with 40% or more Chol content couldn't be useful in gene and drug delivery applications. Studies have shown that lipid composition and cholesterol content are among the major parameters in the research and development (R&D) of nanoliposome formulations (13, 14).

1.1.2. Phase Transition Temperature

Physicochemical properties of nanoliposomes depend on several factors including pH, ionic strength and temperature. Generally, lipid vesicles show low permeability to the entrapped material. However, at elevated temperatures, they undergo a phase transition that alters their permeability. Phospholipid ingredients of nanoliposomes have an important thermal characteristic, i.e., they can undergo a phase transition (T_c) at temperatures lower than their final melting point (T_m). Also known as gel to liquid crystalline transition temperature, T_c is a temperature at which the lipidic bilayer loses much of its ordered packing while its fluidity increases. Phase transition temperature of phospholipid compounds and lipid bilayers depends on the following parameters:

- Polar head group;
- Acyl chain length;
- Degree of saturation of the hydrocarbon chains; and
- Nature and ionic strength of the suspension medium.

In general, T_c is lowered by decreased chain length, by unsaturation of the acyl chains, as well as presence of branched chains and bulky head groups (e.g. cyclopropane rings) (15). An understanding of phase transitions and fluidity of phospholipid membranes is essential both in the manufacture and exploitation of liposomes. This is due to the fact that the phase behaviour of liposomes and nanoliposomes determines important properties such as permeability, fusion, aggregation, deformability and protein binding, all of which can significantly affect the stability of lipid vesicles and their behaviour in biological systems (11). Phase transition temperature of the lipid vesicles has been reported to affect the pharmacokinetics of the encapsulated drugs such as doxorubicin (16).

Liposomes made of pure phospholipids will not form at temperatures below T_c of the phospholipid. This temperature requirement is reduced to some extent, but not eliminated, by the addition of cholesterol (17). In some cases, it is recommended that liposome preparation be carried out at temperatures well above T_c of the vesicles. For instance, in the case of vesicles containing dipalmitoyl phosphatidylcholine (DPPC, $T_c=41°C$), it has been suggested that the liposome preparation procedure be carried out at 10°C higher than the T_c at 51°C (18, 19). This is in order to make sure that all the phospholipids are dissolved in the suspension medium homogenously and have sufficient flexibility to align themselves in the structure of lipid vesicles. Following termination of the preparation procedure, usually nanoliposomes are allowed to anneal and stabilize for certain periods of time (e.g. 30–60 min), at a temperature above T_c, before storage.

2. Materials

1. Lecithin and all other phospholipids (Avanti Polar Lipids; Sigma; Lipoid; NOF corporation; Phospholipon GmbH). Alternatively, food-grade or pharmaceutical-grade, lecithin can be used as a substantially cheaper option to the highly purified phospholipids. The nanoliposomal ingredients need to be stored under an oxygen-free atmosphere and checked regularly, e.g. by thin-layer chromatography (TLC) on silica gel plates, for oxidation products and purity evaluation.
2. Cholesterol.
3. Distilled water or Milli-Q reagent grade water.
4. Phosphate buffered saline (PBS pH: 7.4): 137 mM NaCl, 2.7 mM KCl, 4.3 mM Na_2HPO_4, and 1.47 mM KH_2PO_4.
5. Tricine buffer (pH: 7.2): 2 mM tricine, 0.36 M mannitol, 20 mM sodium chloride, 2 mM histidine, 0.1 mM EDTA.
6. Isotonic HEPES buffer (pH: 7.4): 20 mM HEPES and 140 mM NaCl. Adjust the final pH of the buffers to the desired pH value. Autoclave or filter-sterilize and store at room temperature.
7. Chloroform.
8. Methanol.
9. Acetone.
10. Diethyl ether.
11. Ethanol.
12. Liposome hydration or suspension media can comprise any of the above-mentioned buffers or saline solution (0.9% w/v of NaCl in distilled water) or isotonic sucrose solution (9.25% w/v sucrose in distilled water).
13. Mini extrusion device and filters (Avanti Polar Lipids, Inc.; Avestin Inc.)

3. Methods

Hydrated phospholipid molecules arrange themselves in the form of bilayer structures via Van-der Waals and hydrophilic/hydrophobic interactions. In this process, the hydrophilic head groups of the phospholipid molecules face the water phase while the hydrophobic region of each of the monolayers face each other in the middle of the membrane (Fig. 2). It should be noted that formation of liposomes and nanoliposomes is not a spontaneous process and sufficient energy must be put into the system to

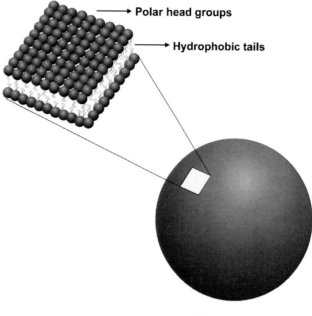

Fig. 2. Enlargement of a section of phospholipid bilayer of a nanoliposome revealing its hydrophilic head groups and hydrophobic section

overcome an energy barrier (for detailed discussion see ref. (20)). In other words, lipid vesicles are formed when phospholipids such as lecithin are placed in water and consequently form bilayer structures, once adequate amount of energy is supplied. Input of energy (e.g. in the form of sonication, homogenisation, heating, etc.) results in the arrangement of the lipid molecules, in the form of bilayer vesicles, to achieve a thermodynamic equilibrium in the aqueous phase. Some of the commonly used methods for the preparation of nanoliposomes are explained in the following section.

3.1. Sonication Technique

Sonication is a simple method for reducing the size of liposomes and manufacture of nanoliposomes (21, 22). The common laboratory method involves treating hydrated vesicles for several minutes with a titanium-tipped probe sonicator in a temperature-controlled environment as explained in the following section.

1. Dissolve a suitable combination of the phospholipid components, with or without cholesterol (see Note 1), in either chloroform or in chloroform–methanol mixture (usually 2:1 v/v).

2. Filter the mixture to remove minor insoluble components or ultrafilter to reduce or eliminate pyrogens.

3. Transfer the solution to a pear-shaped or a round-bottom flask and, employing a rotary evaporator, remove the solvents at temperatures above T_c under negative pressure, leaving a thin layer of dried lipid components in the flask (see Note 2). Other methods of drying the lipid ingredients include lyophilization and spray drying (23).

4. Remove traces of the organic solvents using a vacuum pump, usually overnight at pressures below 0.1 Pa. Alternatively, traces of the organic solvents may be removed by flushing the flask with an inert gas, such as nitrogen or argon.

5. After drying the lipid ingredients, small quantity of glass beads (e.g. with 500 μm diameter) are added to the flask containing the dried lipids following by the addition of a suitable aqueous phase such as distilled water or buffer. Alternative hydration mediums are saline or nonelectrolytes such as a sugar solution. For an in vivo preparation, physiological osmolality (290 mOsmol/kg) is recommended and can be achieved using 0.6% saline, 5% dextrose, or 10% sucrose solution (24). The aqueous medium can contain salts, chelating agents, stabilizers, cryo-protectants (e.g. glycerol) and the drug to be entrapped.

6. The dried lipids can be dispersed into the hydration fluid by hand shaking the flask or vortex mixing for 1–5 min. At this stage, micrometric MLV type liposomes are formed.

7. Transfer the flask containing MLV either to a bath-type sonicator or a probe (tip) sonicator (Fig. 3). For probe sonication, place the tip of the sonicator in the MLV flask and sonicate the sample with 20 s ON, 20 s OFF intervals (to avoid over-heating), for a total period of 10–15 min. At this stage, nanoliposomes are formed, which are predominantly in the form of small unilamellar vesicles (SUV) (see **Note 3**). Alternatively, nanoliposomes can be produced using a bath sonicator as explained in the following section.

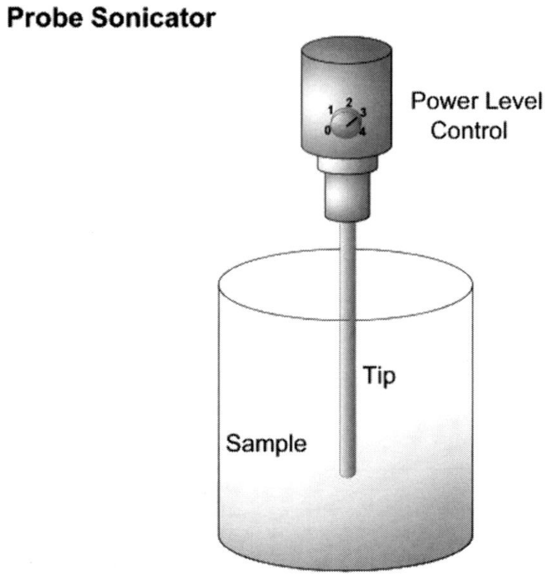

Fig. 3. Schematic representation of a probe-type sonicator

8. Fill the bath sonicator with room temperature water mixed with a couple of drops of liquid detergent. Using a ring stand and test tube clamp, suspend the MLV flask in the bath sonicator. The liquid level inside the flask should be equal to that of outside the flask. Sonicate for a time period of 20–40 min (see Note 4).

9. Store the final product at temperatures above T_c under an inert atmosphere such as nitrogen or argon for 1 h to allow the annealing process to come to completion. Mean size and polydispersity index of vesicles is influenced by lipid composition and concentration, temperature, sonication time and power, sample volume, and sonicator tuning. Since sonication process is difficult to reproduce, size variation between batches produced at different times is not uncommon.

10. Residual large particles remaining in the sample can be removed by centrifugation to yield a clear suspension of nanoliposomes.

3.2. Extrusion Method

Extrusion is a process by which micrometric liposomes (e.g. MLV) are structurally modified to large unilamellar vesicles (LUV) or nanoliposomes depending on the pore-size of the filters used (25–27). Vesicles are physically extruded under pressure through polycarbonate filters of defined pore sizes. A protocol for using a small-sized extruder (Fig. 4) is described in the following section. A mini extruder device (e.g. from Avanti Polar Lipids, Inc., Alabaster, AL, USA; or Avestin Inc., Mannheim, Germany), with 0.5 mL or 1 mL gas-tight syringes can be employed in this procedure.

1. Prepare a liposome sample, such as MLV, as explained earlier.

2. Place one or two-stacked polycarbonate filters into the stainless steel filter-holder of the extruder (Fig. 4).

3. Place the extruder stand/heating block onto a hot plate. Insert a thermometer into the well provided in the heating block. Switch the hot plate on and allow the temperature to reach a temperature above T_c of the lipids (see **Note 5**).

4. In order to reduce the dead volume, pre-wet the extruder parts by passing a syringe full of buffer through the extruder and then discard the buffer (see **Note 6**).

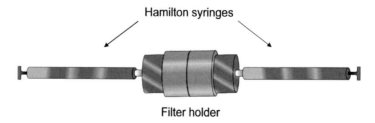

Fig. 4. A small, hand-held, extruder used in the manufacture of nanoliposomes

5. Load the liposome suspension into one of the syringes (donor syringe) of the mini extruder and carefully place the syringe into one end of the extruder by applying a gentle twisting.

6. Place the second syringe (receiver syringe) into the other end of the extruder. Make sure the receiver syringe plunger is set to zero.

7. Insert the fully assembled extruder device into the extruder stand. Insert the stainless-steel hexagonal nut in such a way that any two opposing apexes fall in the vertical plane. Use the swing-arm clips to hold the syringes in good thermal contact with the heating block.

8. Allow the temperature of the liposome suspension to reach the temperature of the heating block (approximately 5–10 min).

9. Gently push the plunger of the filled syringe until the liposome suspension is completely transferred to the empty syringe.

10. Gently push the plunger of the alternate syringe to transfer the suspension back to the original syringe.

11. Repeat the extrusion process for a minimum of seven passes through the filters. In general, the more passes though the filters, the more homogenous the sample becomes. In order to reduce the possibility of sample contamination with larger particles or foreign material, the final extrusion should fill the receiver syringe. Therefore, an odd number of passages through the filters should be performed (see Note 7).

12. Carefully remove the extruder from the heating block. Remove the filled syringe from the extruder and inject the nanoliposome sample into a clean vial.

13. The extruder components can be cleaned by first rinsing with ethanol (or leaving the extruder parts in warm 70% ethanol for few hours) and then rinsing with distilled water.

14. Keep the final product at temperatures above T_c under an inert atmosphere such as nitrogen or argon for 1 h to allow the sample to anneal and stabilize.

3.3. Microfluidization

A method of nanoliposome production without using potentially toxic solvents is the microfluidization technique using a microfluidizer (28–30). Figure 5 shows a schematic representation of a microfluidizer apparatus. This apparatus has been traditionally used in the pharmaceutical industry to make liposomal products (28) and pharmaceutical emulsions (31). More recently, Jafari et al. (30) employed the microfluidizer to produce homogenized milk and flavour emulsions. Microfluidization is based on the principle of dividing a pressure stream into two parts, passing each part through a fine orifice, and directing the flows at each

Microfluidizer

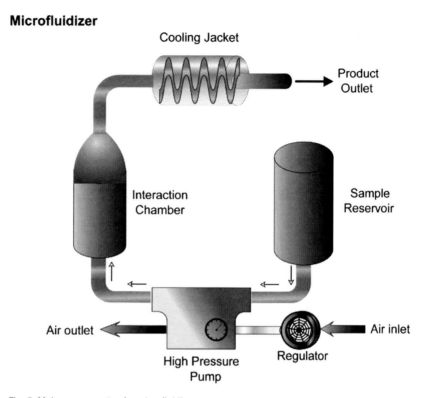

Fig. 5. Main components of a microfluidizer apparatus

other inside the chamber of microfluidizer (30, 32). Within the interaction chamber, cavitation, along with shear and impact, reduces particle sizes of the liposomes. Microfluidizer uses high pressures (up to 10,000 psi) to guide the flow stream through microchannels toward the impingement area (33, 34). The advantages of microfluidization are that: a large volume of liposomes can be formed in a continuous and reproducible manner; the average size of the liposomes can be adjusted; very high capture efficiencies (>75%) can be obtained; and the solutes to be encapsulated are not exposed to sonication, detergents or organic solvents (see Note 8). The process involves only a few steps as exemplified in the following section.

1. Select the ingredients of the nanoliposomes and their suspension medium based on the intended application. The suspension medium is usually an aqueous phase such as deionized/distilled water or buffer. Alternative hydration mediums are saline or nonelectrolytes such as a sugar solution as explained earlier.

2. Prepare a phospholipid dispersion by placing the nanoliposomal ingredients in the suspension medium and then mixing the sample by stirring e.g. employing a blender or by using a homogenizer (see Note 9).

3. Pass the dispersion through the microfluidizer (e.g. an equipment made by Microfluidics International Corp., Newton, MA can be used) by placing the crude suspension of phospholipids in the reservoir and adjusting the air regulator to the selected operating pressure (*see* **Note 10**). With an optimized setting, when the air valve is open, the liquid dispersion flows through a filter into the interaction chamber where it is separated into two streams which interact at extremely high velocities in dimensionally defined microchannels.

4. The suspension can be recycled through the equipment in which case the suspension must be cooled because of the temperature increase in the interaction chamber at high operating pressure (*see* **Notes 11** and **12**).

5. After sample collection, the microfluidizer can be cleaned by recycling 95% ethanol followed by passing distilled water through the system.

6. Leave the nanoliposome suspension at temperatures above T_c under an inert atmosphere such as nitrogen or argon for 1 h to allow the sample to anneal and stabilize.

3.4. Heating Method

Majority of nanoliposome manufacture techniques either involve utilisation of potentially toxic solvents (e.g. chloroform, methanol, diethyl ether and acetone) or high shear force procedures. It has been postulated that residues of these toxic solvents may remain in the final liposome or nanoliposome preparation and contribute to potential toxicity and influence the stability of the lipid vesicles (35–38). Although there are methods to decrease the concentration of the residual solvents in liposomes (e.g. gel filtration, dialysis and vacuum), these are practically difficult and time-consuming procedures. In addition, the level of these solvents in the final formulations must be assessed to ensure the clinical suitability of the products (39). Therefore, it would be much preferable to avoid utilisation of these solvents in nanoliposome manufacture, which will also bring down the time and cost of preparation especially at the industrial scales.

Regarding the utilization of high pressures or high shear forces during nanoliposome manufacture (e.g. as occurs during microfluidization), there are reports on the deleterious effects of these procedures on the structure of the material to be encapsulated (40–45). These hurdles can be overcome by employing alternative preparation methods such as the heating method by which liposomes and nanoliposomes (in addition to some other carrier systems) can be prepared using a single apparatus in the absence of potentially toxic solvents (38, 46–48) as explained in the following section.

1. Hydrate a suitable combination of the phospholipid components, with or without cholesterol (see Note 1) in an aqueous medium for a time period of 1–2 h under an inert atmosphere such as nitrogen or argon. The nanoliposomal ingredients may be hydrated together or separately based on their solubility and T_c (see Note 13).
2. Mix the lipid dispersions along with the material to be encapsulated, in a heat-resistant flask such as a pyrex beaker, and add glycerol to a final volume concentration of 3% (see Note 14). Alternatively a heat-resistant bottle with six baffles can be used for the process as explained and pictured in reference (49).
3. Place the flask or bottle containing the lipids and glycerol on a hot-plate stirrer and mix the sample (e.g. at 800–1,000 rpm) at a temperature above T_c of the lipids for 30 min or until all the lipids dissolved.
4. For the preparation of cholesterol-containing formulations, first dissolve cholesterol in the aqueous phase at elevated temperatures (e.g. 120°C) while stirring (approx. 1,000 rpm) for a period of 15–30 min under nitrogen atmosphere before adding the other components as mentioned earlier.
5. Depending on the nanoliposomal ingredients, sample volume, type of flask used and its number of baffles, as well as type, speed and duration of mixing, nanometric vesicles can be produced without the need to perform filtration or sonication.
6. Leave the nanoliposome suspension at temperatures above T_c under an inert atmosphere such as nitrogen or argon for 1 h to allow the sample to anneal and stabilize.

Liposomes and nanoliposomes prepared by the heating method (HM-liposomes) have been employed successfully as gene transfer vectors (38, 46, 50) as well as drug delivery vehicles (51). Incorporation of plasmid DNA molecules, which are sensitive to high temperatures, to the HM-liposomes was carried out at room temperature by incubation of DNA with the empty, pre-formed, HM-liposomes (38, 46, 50). Another important feature of this method is that it can be easily adapted from small to industrial scales.

Incorporation of drugs into the HM-liposomes can be achieved by several routes including:

1. Adding the drug to the reaction medium along with the liposomal ingredients and glycerol;
2. Adding the drug to the reaction medium when temperature has dropped to a point not lower than the transition temperature (T_c) of the lipids; and
3. Adding the drug to the HM-liposomes after the vesicles are prepared e.g. at room temperature (incorporation of DNA to the HM-liposomes was performed by this route as explained earlier) (7).

Therefore, the heating method has flexibility for the entrapment of various drugs and other bioactives with respect to their temperature sensitivities. Recently, Mozafari and colleagues showed that nanoliposomes prepared by the heating method are completely non-toxic towards cultured cells while nanoliposomes prepared by a conventional method using volatile solvents revealed significant levels of cytotoxicity (38).

3.5. Mozafari Method

A further improved version of the heating method, called Mozafari method, is one of the most recently introduced and one of the most simple techniques for the preparation of liposomes and nanoliposomes (in addition to some other carrier systems). The Mozafari method has recently been employed successfully for the encapsulation and targeted delivery of the food-grade antimicrobial nisin (49). The Mozafari method allows manufacture of carrier systems in one-step, without the need for the pre-hydration of ingredient material, and without employing toxic solvents or detergents from small scales to large, industrial scales. The mentioned method is economical and capable of manufacturing nanoliposomes, with a superior monodispersity and storage stability using a simple protocol and one, single vessel. Encapsulation of nisin (as an example of a substance with low water solubility) in nanoliposomes prepared by the Mozafari method is explained in the following section.

1. Add the liposomal ingredients to a preheated (60°C, 5 min) mixture of nisin (200 μg/ml) and a polyol such as glycerol (final concentration 3%, v/v) in a heat-resistant flask such as a pyrex beaker. Alternatively, a heat-resistant bottle with six baffles can be used as explained and pictured in reference (49).

2. Heat the mixture at 60°C while stirring (approx. 1,000 rpm) on a hotplate stirrer (e.g. RET basic IKAMAG1 Safety Control, IKA, Malaysia) for a period of 45–60 min under nitrogen atmosphere.

3. For the preparation of cholesterol-containing formulations, first dissolve the cholesterol in the aqueous phase at elevated temperatures (c. 120°C) while stirring (approx. 1,000 rpm) for a period of 15–30 min under nitrogen atmosphere before adding the other phospholipid components.

4. Depending on the formulation ingredients, sample volume, type of flask used and its number of baffles, as well as type, speed and duration of mixing, nanometric vesicles can be produced in one step, without the need to perform filtration or sonication.

5. Leave the nanoliposome suspension at temperatures above T_c under an inert atmosphere such as nitrogen or argon for 1 h to allow the sample to anneal and stabilize.

3.6. Characterization and Analysis of Nanoliposomes

Following preparation of nanoliposomes, especially when using a new technique, characterization is required to ensure adequate quality of the product. Methods of characterization have to be meaningful and preferably rapid. Several techniques such as electron microscopy, radiotracers, fluorescence quenching, ultrasonic absorption, electron spin resonance spectroscopy, and nuclear magnetic resonance spectroscopy may be used to characterize nanoliposome formulations. Each technique has characteristic advantages and possible disadvantages. The most important parameters of nanoliposome characterization include visual appearance, size distribution, stability, Zeta potential, lamellarity and entrapment efficiency.

3.6.1. Visualization Techniques

Many imaging techniques are available for visualization of nanoliposomes (52). An optical microscope (phase contrast) can detect particles larger than 300 nm and contamination with larger particles. A polarizing microscope can reveal lamellarity of liposomes. For instance, MLV type liposomes are birefringent and display a Maltese cross (24, 53). The size distribution of nanoliposomes is mainly determined using electron microscopy. Negative staining, freeze-fracture and scanning electron microscopy are the methods most commonly used to characterize nanoliposome structures. A more recently developed microscopic technique with very high resolutions is the Scanning Probe Microscopy (SPM). Two of the most applied SPM techniques are Scanning Tunnelling Microscopy (STM) and Atomic Force Microscopy (AFM). This recent technology gives the possibility to view various biological and non-biological samples under air or water with a resolution up to 3A. By this method, monolayers of various lipids and lipid attached molecules such as antibody fragments can be studied (52).

3.6.2. Size Determination

Size and size distribution (polydispersity) of the formulated nanoliposomes are of particular importance in their characterization. Maintaining a constant size and/or size distribution for a prolonged period of time is an indication of liposome stability. Electron microscopic methods are widely used for establishing the morphology, size and stability of liposomes. With respect to a statistically meaningful analysis of size distribution of the lipid vesicles, methods such as light scattering, which measure the size of large number of vesicles in an aqueous medium, are more appropriate than microscopic techniques. Ideally, these two techniques need to be employed along with other inexpensive and routine laboratory techniques, such as gel permeation chromatography, to provide a comprehensive and reliable characterisation of the nanoliposomal formulations (1, 48).

Each of the currently used particle size determination techniques has its own advantages and disadvantages. Light scattering, for example, provides cumulative average information of the size of a large number of nanoliposomes simultaneously. However, it does not provide information on the shape of the lipidic system (e.g. oval, spherical, cylindrical, etc.) and it assumes any aggregation of more than one vesicle as one single particle.

Electron microscopic techniques, on the other hand, make direct observation possible; hence provide information on the shape of the vesicles as well as presence/absence of any aggregation and/or fusion. The drawback of the microscopic investigations is that the number of particles that can be studied at any certain time is limited. Therefore, the general approach for the determination of size distribution of nanoliposomal formulations should be to use as many different techniques as possible.

3.6.3. Zeta Potential

The other important parameter in liposome characterisation is zeta potential. Zeta potential is the overall charge a lipid vesicle acquires in a particular medium. It is a measure of the magnitude of repulsion or attraction between particles in general and lipid vesicles in particular. Evaluation of the zeta potential of a nanoliposome preparation can help to predict the stability and in vivo fate of liposomes. Any modification of the nanoliposome surface, e.g. surface covering by polymer(s) to extend blood circulation life, can also be monitored by measurement of the zeta potential. Generally, particle size and zeta potential are the two most important properties that determine the fate of intravenously injected liposomes. Knowledge of the zeta potential is also useful in controlling the aggregation, fusion and precipitation of nanoliposomes, which are important factors affecting the stability of nanoliposomal formulations (1).

3.6.4. Lamellarity Determination

The lamellarity of liposomes made from different ingredients or preparation techniques varies widely. This is evidenced by reports showing that the fraction of phospholipid exposed to the external medium has ranged from 5% for large MLV to 70% for SUV (for a review see ref. (54)). Liposome lamellarity determination is often accomplished by ^{31}P NMR. In this technique, the addition of Mn^{2+} quenches the ^{31}P NMR signal from phospholipids on the exterior face of the liposomes and nanoliposomes. Mn^{2+} interacts with the negatively charged phosphate groups of phospholipids and causes a broadening and reduction of the quantifiable signal (22, 24). The degree of lamellarity is determined from the signal ratio before and after Mn^{2+} addition. While frequently used, this technique has recently been found to be quite sensitive to the Mn^{2+} and buffer concentration and the types of liposomes under analysis (22, 24, 54). Other techniques for lamellarity determination include electron microscopy, small angle X-ray scattering (SAXS), and methods that are based on the change in the visible or fluorescence signal of marker lipids upon the addition of reagents.

3.6.5. Encapsulation Efficiency

Encapsulation efficiency is commonly measured by encapsulating a hydrophilic marker (i.e. radioactive sugar, ion, fluorescent dye), sometimes using single-molecule detection. The techniques used for this quantification depend on the nature of the entrapped material and include spectrophotometry, fluorescence spectroscopy, enzyme-based methods, and electrochemical techniques (22, 24, 54).

If a separation technique such as HPLC or FFF (Field Flow Fractionation) is applied, the percent entrapment can be expressed as the ratio of the unencapsulated peak area to that of a reference standard of the same initial concentration. This method can be applied if the nanoliposomes do not undergo any purification (e.g. size exclusion chromatography, dialysis, centrifugation, etc.) following the preparation. Any of the purification technique serves to separate nanoliposome encapsulated materials from those that remain in the suspension medium. Therefore, they can also be used to monitor the storage stability of nanoliposomes in terms of leakage or the effect of various disruptive conditions on the retention of encapsulants. In the latter case, total lysis can be induced by the addition of a surfactant such as Triton X100. Retention and leakage of the encapsulated material depend on the type of the vesicles, their lipid composition and T_c, among other parameters. It has been reported that SUV and MLV type liposomes are less sensitive than LUV liposomes to temperature-induced leakage (Fig. 6). This property of liposomes and nanoliposomes can be used in the formulation of temperature-sensitive vesicles (55).

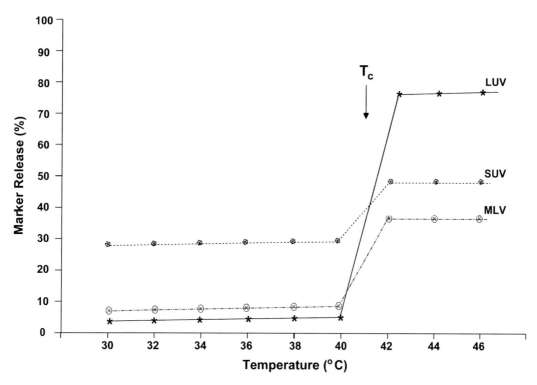

Fig. 6. Temperature-induced release of the water-soluble marker calcein from different types of liposomes composed of DPPC:DCP:Chol (7:2:1) (Mozafari, unpublished data)

Since techniques used to separate nanoliposome-entrapped from free material can potentially cause leakage of contents (e.g. ultracentrifugation) and, in some cases, ambiguity in the extent of separation, research using methods that do not rely on separation are of interest. Reported methods include ^{1}H NMR where free markers exhibit pH sensitive resonance shifts in the external medium versus encapsulated markers; diffusion ordered 2D NMR, which relies on the differences in diffusion coefficients of entrapped and free marker molecules; fluorescence methods where the signal from unencapsulated fluorophores was quenched by substances present in the external solution; and electron spin resonance (ESR) methods which rely on the signal broadening of unencapsulated markers by the addition of a membrane-impermeable agent (54).

4. Notes

1. Total lipid concentrations between 5 and 50 mM can be used in the preparation of nanoliposomes. Previously several of the commercially available lipids were further purified before use. Some of the liposomal ingredients, such as cholesterol, were subjected to recrystallization to remove the oxidation products. However, because of the high quality and availability of phospholipids from commercial sources, at present many researchers do not purify these chemicals any more and directly use them in the preparation of nanoliposomes. For quality control purposes, however, assessment of the purity of lipids prior to liposome preparation is desirable and recommended (24).

2. A suitable size of the flask should be selected during rotary evaporation as to allow for the formation of a thin and homogenous lipid layer inside the flask. Preferably all steps of nanoliposome preparation should be carried out under nitrogen to minimise risk of lipid oxidation. If required, the dried lipid film can be stored in fridge under nitrogen for a couple of weeks, after which should be discarded due to possible deterioration/degradation of the lipids.

3. Probe tip sonicators deliver high-energy input to the lipid suspension but suffer from overheating of the lipid suspension causing degradation. Sonication tips also tend to release titanium particles into the lipid suspension, which must be removed by centrifugation prior to use. For these reasons, bath sonicators are the most widely used instrumentation for preparation of SUV.

4. The bath sonication requires longer process times than probe sonication. However, it has the advantages that it can be carried out in a closed container under nitrogen or argon, and it does not contaminate the lipid with metal from the probe tip. Be sure not to let the bath overheat, and do not drain the bath until it has completely cooled.

5. The extruder apparatus must be fully assembled before inserting it in the heating block, otherwise it will be damaged. The liposome suspension should be kept above the phase transition temperature of the lipids during extrusion.

6. New syringes may have tight fitting parts; to facilitate extrusion, pre-wet syringe barrel and plunger with the suspension medium of the nanoliposomes prior to inserting plunger into barrel.

7. When MLVs are forced through narrow-pore membrane filters under pressure, membrane rupture and resealing occur and encapsulated content leaks out. Therefore, extrusion is performed in the presence of medium containing the final drug concentration, and external solute is removed only after formation is complete (56).

8. It should be noted that the microfluidization process involves a very high shearing force that can potentially damage the structure of material to be encapsulated (40–45). Another disadvantage of the microfluidization method is material loss, contamination and being relatively difficult to scale-up (44).

9. This process may need heating depending on the nanoliposomal ingredients used and their phase transition temperatures (T_c).

10. The conditions of microfluidizer need to be optimized with respect to the pressure, the size of the interaction chamber and the number of passes (usually 3–9 passes is used).

11. The number of sample recirculation depends on the target size of the vesicles, which in turn depends on many parameters such as nanoliposome components. Therefore, the process is repeated until vesicles of a desired size are achieved.

12. The interaction chamber can be packed in ice in order to remove the heat produced during the microfluidization process.

13. In the heating method, the process temperature is based on the properties of the nanoliposomal ingredients (mainly T_c of the lipids), presence or absence of cholesterol, and characteristics of the material to be encapsulated (e.g. melting point and solubility). To avoid subjecting the drug (or other material to be encapsulated) to high temperatures, cholesterol, or

the nanoliposomal ingredient with high T_c (in the absence of Chol), is heated and stirred first.

14. Application of glycerol in the preparation of the HM-liposomes has the following advantages: (a) glycerol is a bioacceptable, non-toxic agent already in use in many pharmaceutical products and can serve as an isotonising agent in the liposomal preparations; (b) unlike the volatile organic solvents employed in the manufacture of conventional liposomes, there is no need for the removal of glycerol from the final preparation; (c) it serves as dispersant and prevents coagulation or sedimentation of the vesicles thereby enhancing the stability of the liposome preparations; (d) it also improves the stability of the liposome preparations against freezing, thawing etc. Therefore, HM-liposomes are also ideal for freeze-drying e.g. in the manufacture of dry powder inhalation products.

References

1. Mozafari MR, Mortazavi SM (eds) (2005) Nanoliposomes: from fundamentals to recent developments. Trafford Publishing Ltd, Oxford, UK
2. Mozafari MR, Flanagan J, Matia-Merino L, Awati A, Omri A, Suntres ZE, Singh H (2006) Review: recent trends in the lipid-based nanoencapsulation of antioxidants and their role in foods. J Sci Food Agric 86:2038–2045
3. Khosravi-Darani K, Pardakhty A, Honarpisheh H, Rao VSM, Mozafari MR (2007) The role of high-resolution imaging in the evaluation of nanosystems for bioactive encapsulation and targeted nanotherapy. Micron 38:804–818
4. Nomura SM, Yoshikawa Y, Yoshikawa K, Dannenmuller O, Chasserot-Golaz S, Ourisson G, Nakatani Y (2001) Towards proto-cells: primitive lipid vesicles encapsulating giant DNA and its histone complex. ChemBioChem 2:457–459
5. Pozzi G, Birault V, Werner B, Dannenmuller O, Nakatani Y, Ourisson G, Terakawa S (1996) Single-chain polyterpenyl phosphates form primitive membranes. Angew Chem Int Ed Engl 35:177–180
6. Bangham AD, Standish MM, Watkins JC (1965) Diffusion of univalent ions across the lamellae of swollen phospholipids. J Mol Biol 13:238–252
7. Mozafari MR (2005) Liposomes: an overview of manufacturing techniques. Cell Mol Biol Lett 10:711–719
8. Mozafari MR (ed) (2006) Nanocarrier technologies: frontiers of nanotherapy. Springer, The Netherlands
9. Zhai GX, Chen GG, Zhao Y, Lou HX, Zhang JS (2002) Study on preparation of low molecular weight heparin nanoliposomes and their oral absorption in rat. J China Pharm Univ 33:200–202
10. Bowtle W (2000) Lipid formulations for oral drug delivery. Pharm Tech Eur 12:20–30
11. New RRC (ed) (1990) Liposomes a practical approach. IRL/Oxford University Press, Oxford, UK
12. Reineccius G (1995) Liposomes for controlled release in the food industry. In: Risch S, Reineccius G (eds) Encapsulation and controlled release of food ingredients. American Chemical Society, Washington DC, pp 113–131
13. Mozafari MR, Hasirci V (1998) Mechanism of calcium ion induced multilamellar vesicle-DNA interaction. J Microencapsul 15:55–65
14. Barenholz Y, Lasic D (eds) (1996) Handbook of nonmedical applications of liposomes, vol III. CRC, Boca Raton, FL
15. Szoka F, Papahadjopoulos D (1980) Comparative properties and methods of preparation of lipid vesicles (liposomes). Annu Rev Biophys Bioeng 9:467–508
16. Gabizon AA, Barenholz Y, Bialer M (1993) Prolongation of the circulation time of doxorubicin encapsulated in liposomes containing a polyethylene glycol-derivatized phospholipid: pharmacokinetic studies in rodents and dogs. Pharm Res 10:703–708
17. Leserman L, Machy P, Zelphati O (1994) Immunoliposome-mediated delivery of nucleic acids: a review of our laboratory's experience. J Liposome Res 4:107–119

18. Jurima-Romet M, Barber RF, Demeester J, Shek PN (1990) Distribution studies of liposome-encapsulated glutathione administered to the lung. Int J Pharm 63:227–235
19. Jurima-Romet M, Barber RF, Shek PN (1992) Liposomes and bronchoalveolar lavage fluid: release of vesicle-entrapped glutathione. Int J Pharm 88:201–210
20. Pegg RB, Shahidi F (1999) Encapsulation and controlled release in food preservation. In: Shafiur Rahman M (ed) Handbook of food preservation. CRC, Boca Raton, FL, pp 611–665
21. Woodbury DJ, Richardson ES, Grigg AW, Welling RD, Knudson BH (2006) Reducing liposome size with ultrasound: bimodal size distributions. J Liposome Res 16:57–80
22. Jesorka A, Orwar O (2008) Liposomes: technologies and analytical applications. Annu Rev Anal Chem 1:801–832
23. Szoka FC, Papahadjopoulos D (1981) Liposomes: preparation and characterization. In: Knight CG (ed) Liposomes: from physical structure to therapeutic application. Elsevier, Amsterdam, pp 51–82
24. Chatterjee S, Banerjee DK (2002) Preparation, isolation, and characterization of liposomes containing natural and synthetic lipids. Methods Mol Biol 199:3–16
25. Hope MJ, Bally MB, Webb G, Cullis PR (1985) Production of large unilamellar vesicles by a rapid extrusion procedure. Characterization of size distribution, trapped volume and ability to maintain a membrane potential. Biochim Biophys Acta 812:55–65
26. MacDonald RC, MacDonald RI, Menco BPM, Takeshita K, Subbarao NK, Hu L (1991) Small volume extrusion apparatus for preparation of large unilamellar vesicles. Biochim Biophys Acta 1061:297–303
27. Berger N, Sachse A, Bender J, Schubert R, Brandl M (2001) Filter extrusion of liposomes using different devices: comparison of liposome size, encapsulation efficiency, and process characteristics. Int J Pharm 223:55–68
28. Vemuri S, Yu CD, Wangsatorntanakun V, Roosdorp N (1990) Large-scale production of liposome by a microfluidizer. Drug Dev Ind Pharm 16:2243–2256
29. Thompson AK, Mozafari MR, Singh H (2007) The properties of liposomes produced from milk fat globule membrane material using different techniques. Lait 87:349–360
30. Jafari SM, He YH, Bhandari B (2006) Nanoemulsion production by sonication and microfluidization – a comparison. Int J Food Prop 9:475–485
31. Silvestri S, Ganguly N, Tabibi E (1992) Predicting the effect of non-ionic surfactants on dispersed droplet radii in submicron oil-in-water emulsions. Pharm Res 9:1347–1350
32. Geciova J, Bury D, Jelen P (2002) Methods for disruption of microbial cells for potential use in the dairy industry – a review. Int Dairy J 12:541–553
33. Kim HY, Baianu IC (1991) Novel liposome microencapsulation techniques for food applications. Trends Food Sci Technol 2:55–61
34. Sorgi FL, Huang L (1994) Large scale production of DC-Chol cationic liposomes by microfluidization. Int J Pharm 144:131–139
35. Kikuchi H, Yamauchi H, Hirota S (1994) A polyol dilution method for mass production of liposomes. J Liposome Res 4:71–91
36. Vemuri S, Rhodes CT (1995) Preparation and characterization of liposomes as therapeutic delivery systems: a review. Pharm Acta Helv 70:95–111
37. Cortesi R, Esposito E, Gambarin S, Telloli P, Menegatti E, Nastruzzi C (1999) Preparation of liposomes by reverse-phase evaporation using alternative organic solvents. J Microencapsul 16:251–256
38. Mozafari MR, Reed CJ, Rostron C (2007) Cytotoxicity evaluation of anionic nanoliposomes and nanolipoplexes prepared by the heating method without employing volatile solvents and detergents. Pharmazie 62: 205–209
39. Dwivedi AM (2002) Residual solvent analysis in pharmaceuticals. Pharm Tech Eur 14:26–28
40. Silvestri S, Gabrielson G, Wu LL (1991) Effect of terminal block on the microfluidization induced degradation of model ABA block copolymer. Int J Pharm 71:65–71
41. CenciaRohan L, Silvestri S (1993) Effect of solvent system on microfluidization-induced mechanical degradation. Int J Pharm 95:23–28
42. Bodmeier R, Chen H (1993) Hydrolysis of cellulose acetate and cellulose acetate butyrate pseudolatexes prepared by a solvent evaporation microfluidization method. Drug Dev Ind Pharm 19:521–530
43. Lagoueyte N, Paquin P (1998) Effects of microfluidization on the functional properties of xanthan Gum. Food Hydrocol 12: 365–371
44. Maa YF, Hsu CC (1999) Performance of sonication and microfluidization for liquid–liquid emulsification. Pharm Dev Technol 4:233–240
45. Kasaai MR, Charlet G, Paquin P, Arul J (2003) Fragmentation of chitosan by

microfluidization process. Innov Food Sci Emerg Tech 4:403–413
46. Mozafari MR, Reed CJ, Rostron C, Kocum C, Piskin E (2002) Construction of stable anionic liposome-plasmid particles using the heating method: a preliminary investigation. Cell Mol Biol Lett 7:923–927
47. Mortazavi SM, Mohammadabadi MR, Khosravi-Darani K, Mozafari MR (2007) Preparation of liposomal gene therapy vectors by a scalable method without using volatile solvents or detergents. J Biotechnol 129:604–613
48. Mozafari MR, Reed CJ, Rostron C (2007) Prospects of anionic nanolipoplexes in nanotherapy: transmission electron microscopy and light scattering studies. Micron 38: 787–795
49. Colas JC, Shi WL, Rao VSNM, Omri A, Mozafari MR, Singh H (2007) Microscopical investigations of nisin-loaded nanoliposomes prepared by Mozafari method and their bacterial targeting. Micron 38:841–847
50. Mozafari MR, Reed CJ, Rostron C, Martin DS (2004) Transfection of human airway epithelial cells using a lipid-based vector prepared by the heating method. J Aerosol Med 17:100
51. Mozafari MR, Reed CJ, Rostron C (2003) 5-fluorouracil encapsulation in colloidal lipid particles: entrapment, release and cytotoxicity evaluation in an airway cell line. Drug Delivery to the Lungs XIV, London, England, pp. 180–183
52. Ozer AY (2007) Applications of light and electron microscopic techniques in liposome research. In: Mozafari MR (ed) Nanomaterials and nanosystems for biomedical applications. Springer, Dordrecht, The Netherlands, pp 145–153
53. Rizwan SB, Dong YD, Boyd BJ, Rades T, Hook S (2007) Characterisation of bicontinuous cubic liquid crystalline systems of phytantriol and water using cryo field emission scanning electron microscopy (cryo FESEM). Micron 38:478–485
54. Edwards KA, Baeumner AJ (2006) Analysis of liposomes. Talanta 68:1432–1441
55. Chandaroy P, Sen A, Hui SW (2001) Temperature-controlled content release from liposomes encapsulating Pluronic F127. J Control Release 76:27–37
56. Mui B, Chow L, Hope MJ (2003) Extrusion technique to generate liposomes of defined size. Methods Enzymol 367:3–14

Chapter 3

Preparation of DRV Liposomes

Sophia G. Antimisiaris

Abstract

Dried reconstituted vesicles (DRV) are liposomes that are formulated under mild conditions and have the capability to entrap substantially high amounts of hydrophilic solutes (compared with other types of liposomes). These characteristics make this liposome type ideal for entrapment of labile substances, as peptide, protein or DNA vaccines and sensitive drugs. In this chapter, we initially introduce all possible types of DRV liposomes (in respect to the encapsulated molecule characteristics and/or their applications in therapeutics) and discuss in detail the preparation methodologies for each type.

Key words: DRV, Protein, Peptide, Hydrophilic drug, Encapsulation yield, Vaccine, DNA, Particulate, Bacteria, Cyclodextrin

Abbreviations

BisHOP	1,2-Bis(hexadecylcycloxy)-3-trimethylaminopropane
CD	Cyclodextrin
CF	5,6-Carboxyfluorescein
CFA	Complete Freund's adjuvant
Chol	Cholesterol
DMPC	1,2-Dimyristoyl-*sn*-glyceroyl-3-phosphocholine
DC-Chol	3b-(*N*,*N*-Dimethylaminoethane)carbamylcholesterol
DOTAP	1,2-Dioleyloxy-3-trimethylammonium propane
DOTMA	*N*-[1-(2,3-Dioleyloxy) propyl]-*N*,*N*,*N*-triethylammonium
DPPC	1,2-Dipalmitoyl-*sn*-glyceroyl-3-phosphocholine
DPPE-PEG2000	1,2-Dipalmitoyl-*sn*-glyceroyl-3-phosphoethanolamine conjugated to polyethylene glycol (MW 2000)
DRV	Dried rehydrated vesicles or Dried Reconstituted vesicles
DSPC	1,2-Distearoyl-*sn*-glyceroyl-3-phosphocholine
HPβ-CD	Hydroxypropyl-beta-CD
H-PC	Hydrogenated PC
Iv	Intravenous

MLV	Multilamellar vesicles
MW	Molecular weight
PA	Phosphatidic acid
PB	Phoshate buffer
PBS	Phosphate buffered saline
PC	Phosphatidylcholine
PCS	Photon correlation spectroscopy
PEG	Polyethylene glycol
PG	Phosphatidylglycerol
PRE	Prednisolone
PS	Phosphatidyl serine
RCM	Radiographic contrast media
rgp63	Recombinant glycoprotein of leishmania
SA	Stearylamine
SM	Sphingomyelin
SUV	Small unilamellar vesicles
Tc	Lipid transition temperature
TO	Triolein

1. Introduction

Dried Reconstituted Vesicles (DRV) (see Note 1), were initially developed in 1984 by Kirby and Gregoriadis (1). They are oligo- or multilamellar liposomes with capability of encapsulating high amounts of aqueous soluble molecules. The fact that the DRV technique involves vesicle formation under mild conditions (e.g., conditions that do not cause decomposition or loss of activity of active substances), makes this technique the method of choice for preparation of liposomal formulations of sensitive active substances as peptides, proteins or enzymes.

High entrapment efficiency is a valuable advantage for any type of liposome, since it results first of all in economy of lipids and active substances. This parameter is very important when functionalized lipids are included in the liposome composition for vesicle targeting to specific receptors, and/or when expensive synthetic drugs are formulated. In addition to economy, high liposome entrapment minimizes the amount of lipid required to deliver a given amount of drug to cells, reducing thus the possibility for (1) saturation of the cells by the lipid and (2) lipid-related toxicity.

The high entrapment capability of DRV's is due to the fact that preformed "empty" small unilamellar vesicles are disrupted during a freeze-drying cycle in the presence of the solute destined for entrapment. Subsequently, during controlled rehydration, which is carried out in the presence of a concentrated solution of the solute (to be encapsulated), the vesicles fuse into large oligolamellar (or multilamellar) vesicles entrapping high amounts of the solute. The size of the initial liposome dispersion used as well as the solute solution characteristics, together with the conditions applying during rehydration of the freeze-dried product, are all important parameters which determine the final size and entrapment efficiency of the DRVs produced. The presence of cryoprotectants during the initial freeze-drying step of DRV preparation results in reduced solute entrapment as previously proven (2), due to the fact that the cryoprotectant preserves the integrity of the "empty" vesicles, preventing disruption and subsequent fusion. It has been reported that by controlling the sugar/lipid mass ratio, different entrapment efficiencies and final vesicle sizes can be achieved (2).

From 1984, when they were first developed, DRV liposomes have been used for liposomal encapsulation of various active substances which may be divided into three main categories: (1) Low MW drug molecules (mainly hydrophilic drugs) (3–20) (2) Proteins or peptides and enzymes (21–26), and (3) DNA or oligonucleotides (26–32). From these categories, the last two are primarily used as liposomal vaccines. Some examples of substances entrapped in DRV liposomes from the last 10 year literature are presented in Table 1.

In addition to hydrophilic drugs, the DRV method has been used for the production of stable liposomal formulations of amphiphilic/lipophilic, or else membrane permeable drugs, after the formation of appropriate aqueous soluble cyclodextrin [CD]-drug complexes which are finally encapsulated in the aqueous vicinity of the DRV vesicles (15, 17, 18). Cyclodextrins are cyclic oligosaccharides (composed of at least six D-(+) glucopyranose units) that form torus-like molecules or truncated cones. CD molecules have a hydrophilic surface and a hydrophobic interior, in which lipophilic drugs can be accommodated to form aqueous soluble CD-drug-complexes. Several applications have been investigated for CD-drug encapsulating DRV liposomes.

1.1. Drug Encapsulating DRVs

Various hydrophilic molecules have been entrapped in DRV liposomes, as radiographic contrast media (RCM) (3), gentamycin (4), thioguanine (7), bupivacaine hydrochloride (8), vancomycin (9), pirarubicin (10), arsenic trioxide (11), aminoglycosides and macrolide antibiotics (12), low-molecular weight heparin (13), etc. In several studies in which the parameters that influence DRV stability and/or encapsulation efficiency are investigated, model molecules are encapsulated in DRVs. In such cases, and especially when the integrity of the liposomes is to be evaluated, it is preferable to use fluorescent dyes, as calcein or 5, 6-carboxyfluorescein (CF), that are entrapped in the vesicles at quenched concentrations, and thus their release can be easily monitored due to the de-quenching of the dye fluorescence intensity upon its release from the vesicles and its dilution in the liposome dispersion media. (1, 2, 14–16)

Table 1
List of compounds that have been entrapped in DRV liposomes for therapeutic applications

Encapsulated substance	Study objective	Reference
Low MW drugs		
Daunorubicion, doxorubicin, riboflavin, sucrose	To control encapsulation yield and size of DRV	(2)
Radiographic contrast agents	Encapsulation ability	(3)
Gentamycin	Formulation optimization	(4)
Carboxyfluorescein (CF)	Coat vesicles with antibodies without inactivating entrapped enzymes	(5)
Hepes buffer	Comparison of integrity of different liposome-types	(6)
Thioguanine	Influence of pH, cholesterol, charge, sonication on EE%	(7)
Bupivacaine hydrochloride	Formulation development	(8)
Vancomycin	Folic acid-coated liposomes for enhancing oral delivery	(9)
Pirarubicin (THP, L-THP)	Evaluation of antitumor effects of pirarubicin DRVs	(10)
Aminoglycoside and macrolide antibiotics	Construct liposomes with high yield entrapment	(11)
Arsenic-Trioxide	Construct stable liposomes with high yield entrapment	(12)
Low-MW- Heparin	Coating stents with heparin-encapsulating liposomes	(13)
Calcein	Study Liposome stability in presence of CD molecules	(14)
Calcein, Prednisolone	Retention of entrapment of amphiphilic drugs/CD complexes	(15)
Carboxyfluorescein (CF)	Study mechanisms involved in liposome-cell interaction	(16)
Dehydroepiandrosterone, Retinol and Retinoic Acid	Method for efficient entrapment of amphiphilic drugs	(17)
Riboflavin	Development of liposomes for photosensitive drugs	(18)
Dexamethasone	Construct stable liposomes with high yield EE%	(19)

Rifampicin (RIF)	Nebulization ability and Stability of liposomal RIF	(20)
Proteins		
Bovine serum albumin	Immunization. Higher responses compared to CFA or plain antigen	(21)
Tetanus toxoid and IgG	Method for attachment of ligands to preformed DRV's	(22)
Glucose oxidase (GO)	Optimization of entrapment of GO in DRV, 24% EE% obtained	(23)
Human gamma-globulin	Preparation and performance evaluation	(24)
t-PA	Optimization of entrapment and freeze-drying of DRV's	(25)
Protein I	Explore protection by Leishmania antigen encapsulated in DRVs	(26)
DNA – Oligonucleotides		
Candida albicans ribosomes	Development of prophylactic anti-Candida vaccine	(26)
Plasmid DNA (encoding hepatitis B surface antigen)	Development of DNA vaccine	(27)
Naked DNA	Compare potency of lipid-based and nonionic surfactant carriers for DNA vaccines	(28)
Plasmid DNA	Feasibility of using High-pressure homogenization-extrusion for DNA-loaded DRV production	(29)
CpG oligodeoxynucleotides (CpG ODN) and (rgp63)	Evaluation of CpG ODN co-encapsulated with rgp63 antigen in cationic liposomes, on immune response	(30)
Plasmid DNA	Feasibility of preserving complexes as dried preparations using a modified DRV method	(31)

As mentioned above, it has been reported that by controlling the sugar/lipid mass ratio, during DRV preparation by the conventional DRV technique (1), the entrapment efficiencies and size distribution of the liposomes produced can be controlled (2).

Ampliphilic or lipophilic drug encapsulating DRV liposomes have also been prepared (15–20). Nevertheless, in most cases amphiphilic or lipophilic drugs (depending on their lipid permeability and their aqueous solubility) rapidly leak out from their liposomal formulations upon dilution of the liposome suspension (the leakage rate being determined by the lipid permeability and aqueous solubility of the drug together with the dilution factor) (15). This is highly likely to occur immediately after i.v. administration of liposomes when the liposome suspension is diluted in the bloodstream. As mentioned earlier, a method to overcome this problem is to entrap an aqueous soluble complex of the drug with cyclodextrin in the aqueous vicinity of DRV liposomes, instead of the drug itself which will probably be located in the lipid bilayers of the liposomes (15, 17, 18). Criticism has been raised about the final result of this approach in terms of its ability to improve the drug retention in the vesicles. In fact, it has been demonstrated for prednisolone (PRE) that the encapsulation of PRE-HP-beta-CD complex does not improve the retention of the drug in the liposomes significantly (compared to liposomes that incorporate the plain drug) but only results in increased drug encapsulation (15). The former failure has been attributed to the rapid displacement of the drug from the CD-complex by other components of the multicomponent drug-in-CD-in-liposome system, as phospholipids and cholesterol that have higher affinity for the CD cavity (compared to the drug). It was recently demonstrated (14) that specific CD molecules have indeed the ability to extract cholesterol molecules and in some cases also phospholipids from liposome bilayers. Thereby, it is important to consider the affinity of the specific drug molecule for the specific cycoldextrin as well as the affinity of the cyclodextrin for the lipid components of the liposome and select the cyclodextrin type and the lipid components of the complex system accordingly, when designing such complex systems, in order to achieve maximum drug retention in the vesicles.

Another application of the drug-in-CD-in-liposome complex system is the construction of liposomal formulations of photosensitive drugs (i.e., drugs that degrade on exposure to light and lose their activity), as riboflavin (18). For this, one or more lipid-soluble UV absorbers, as oil-red-O, oxybenzone or dioxybenzone are entrapped into the lipid phase, while water-soluble ones, as sulisobenzone, are entrapped in the aqueous phase of liposomes together with the drug-CD complex. Study results suggest that liposome-based multicomponent systems could be developed for the protection of photolabile agents in therapeutic and other uses.

1.2. Antigen Encapsulating DRVs

The immunological adjuvant properties of liposomes have been demonstrated more than 30 years ago (33), and extensive research on this subject has shown that liposome adjuvanticity applies to a large variety of protozoan, viral, bacterial, tumor, and other antigens (34, 35). Thereby, the use of liposomes for the encapsulation of protein and DNA vaccines has been extensively studied (27–35). As mentioned before, the DRV technique is the method of choice for the preparation of liposomal formulations of labile compounds that may partly or completely loose their activity when exposed to the conditions applying for the preparation of liposome formulations by other techniques (as contact with organic solvents or sonication). In the case of vaccines, the easy scale-up and high yield entrapment of DRV's are additional assets that make this technique highly advantageous. For these reasons, DRV liposomes have been formulated and tested for their ability to encapsulate and retain a number of different type of vaccines, as peptides, proteins, plasmid DNA, and other macromolecules or even particulates, as attenuated bacteria and spores (36–38). Furthermore the immunological results of such formulations have been studied, and in most of the cases the liposomal vaccines have been demonstrated to perform as good or even better than other adjuvants (39, 40).

In addition to their adjuvanticity, another advantage of antigen-containing DRV suspensions is that they can be freeze-dried in the presence of a cryoprotectant for product shelf life prolongation, without loosing significant amounts of entrapped material upon reconstitution with physiological saline (36–40). However, it is very important to take special care during the initial rehydration of the freeze-dried material: the water added at this stage should be kept to a minimum.

In the following part of this chapter, the methodology for the preparation of DRVs for encapsulation of low molecular weight drugs (mainly hydrophilic), CD-drug complexes, and vaccines (DRVs containing protein, peptide, particulate antigens, or DNA/oligonucleotides [for DNA-vaccines]), are discussed in detail.

2. Materials

2.1. Low MW Drug Encapsulating DRV

1. Egg l-α-phosphatidylcholine [PC] (grade 1) (Lipid Products, Nutfield, UK, or Lipoid, DE), is used in solid state or dissolved (20 mg/ml or 100 mg/ml) in a mixture of $CHCl_3/CH_3OH$ (2:1 v/v), and stored in aliquots at –80°C. The 99% purity of the lipid is verified by thin layer chromatography (see Note 2).

2. Hydrogenated egg phosphatidylcholine [HPC], sphingomyelin [SM], phosphatidic acid [PA], 1,2-dimyristoyl-sn-glyceroyl-3-PC[DMPC],1,2-dipalmitoyl-sn-glyceroyl-3-PC

[DPPC] or 1,2-Distearoyl-*sn*-glyceroyl-3-PC [DSPC] (synthetic-grade 1), (Lipid Products, Nutfield, UK, or Lipoid, DE or Avanti Polar Lipids, USA). Storage conditions and purity tests (see Note 2) are the same as mentioned earlier for PC (with the difference that only 20 mg/ml solutions are made for these lipids).

3. Cholesterol [Chol] (pure) (Sigma-Aldrich, Athens, Greece). Chol is stored desiccated at –20°C. Chol is used for liposome preparation in solid state (powder) or after being dissolved (20 mg/ml or 100 mg/ml) in a mixture of $CHCl_3/CH_3OH$ (2:1 v/v), and stored in aliquots at –80°C.

4. The water used in all solutions is deionized and then distilled [d.d. H_2O].

5. Phosphate buffered saline (PBS), pH 7.4, 10 mM sodium phosphate, 3 mM potassium phosphate, 140 mM NaCl, and 0.2 g sodium azide (to a final concentration of 0.02% w/v; for prevention of bacterial growth). Before adjusting the volume (to 1 L), the pH of the solution is adjusted to 7.40. Sodium azide is not added if the buffer will be used for the preparation of liposomes for in vivo studies.

6. Diluted PBS or Phosphate Buffer pH 7.40, for preparation of CF (or calcein) solution. This buffer is prepared by diluting PBS buffer 10 times with d.d. H_2O (see item 5). This buffer is used for preparation of CF (or calcein) solution, prepared for encapsulation in DRV liposomes.

7. Solution of 5,(6)-carboxyfluorescein [CF] (Eastman Kodak, USA) or calcein (Sigma-Aldrich, Athens, Greece). The solid is dissolved in phosphate buffer pH 7.40 to make a solution of 100 mM, which can be diluted with the same buffer if lower concentration (17 mM) should be used (see Note 3).

8. Triton X-100 (Sigma-Aldrich, Athens, Greece). Triton is used as a 10% v/v solution in the liposome preparation buffer (see below). Usually 1 L solution is prepared, stored at room temperature and used for up to 3 months.

9. β-Cyclodextrin [β-CD] (Sigma-Aldrich, Athens GR).

10. Hydropropyl-β-cyclodextrin (HP β-CD) (Sigma-Aldrich, Athens GR).

11. Prednisolone (PR) (99%) pure, (Sigma-Aldrich, Athens GR).

12. Sephadex G-50 (medium) (Phase Separations, Pharmacia, Sweden). The powder is dispersed in PBS buffer for swelling and the dispersion is subsequently degassed under vacuum. Gel chromatography columns are packed and used for DRV liposome separation from nonencapsulated molecules (as described in detail in the following section).

13. Stewart assay reagent (for determination of phopholipid concentration): For preparation, dissolve 27.03 g of $FeCl_3 \times 6H_2O$ and 30.4 g of NH_4SCN in 1 L of d.d. H_2O. The reagent is stored in dark glass bottles at room temperature and used for up to 1 month.

2.2. Protein (or Peptide) and/or Particulate Encapsulating DRV

1. Egg phosphatidylcholine (PC), phosphatidic acid (PA), phosphatidylglycerol (PG), phosphatidylserine (PS), DSPC, and Chol (for lipid sources and storage see Subheading 2.1 (1–3, and for purity testing, see Note 2).
2. Triolein (TO) (Sigma-Aldrich, Athens, Greece).
3. Stearylamine (SA) (Sigma-Aldrich, Athens, Greece).
4. 1,2-Bis(hexadecylcycloxy)-3-trimethylaminopropane (BisHOP) (Sigma-Aldrich, Athens, Greece).
5. N-[1-(2,3-Dioleyloxy) propyl]-N,N,N-triethylammonium (DOTMA) (Avanti Polar Lipids, USA).
6. 1,2-Dioleyloxy-3-(trimethylammonium propane) (DOTAP) (Avanti Polar Lipids, USA).
7. 3b-(N,N-Dimethylaminoethane)carbamylcholesterol (DC-Chol) (Sigma-Aldrich, Athens, Greece).
8. Sepharose CL-4B (Phase Separations, Pharmacia, Sweden).
9. Polyethylene Glycol 6000 (PEG 6000) (Sigma-Aldrich, Athens, Greece).

2.3. Entrapment of Large Particles, Viruses, or Bacteria into Giant Liposomes

1. Solution 1: PC or DSPC, CHOL, PG, and TO (4:4:2:1 molar ratio, 9 mmol total lipid) in 1.0 ml $CHCl_3$.
2. Solution 2: The same Lipids as in solution 1, but dissolved in 0.5 ml Diethyl Ether.
3. Solutions 3 and 4: Sucrose (Sigma-Aldrich, Athens, Greece): 0.15 M (Sol. 3) and 0.2 M (Sol. 4) in H_2O.
4. Solution 5: Glucose 5% (w/v) in H_2O.
5. Solution 6: Sodium Phosphate buffer (PBS) 0.1 M, pH 7.0, containing 0.9% NaCl.
6. Solution 7: Discontinuous sucrose gradient prepared by the use of two solutions containing 59.7 and 117.0 g of sucrose, respectively, per 100 ml H_2O, in swing-out bucket centrifuge tubes.

3. Methods

In this section of the chapter, the general methodology used to encapsulate any type of material (mostly applying for hydrophilic drugs, peptides or proteins) will be described in detail.

After this, special considerations and methodologies that should be applied for encapsulation of other types of molecules or for special types of applications (liposomal vaccines) will be given in detail.

3.1. General Methodology for DRV Preparation – Preparation of Hydrophilic Compound or Small Molecule Encapsulating DRV Liposomes (1, 2, 15, 17, 18)

For the preparation of DRV liposomes, empty (see Note 4) SUV liposomes dispersed in d.d. H_2O with the appropriate lipid composition and concentration are initially prepared (see Note 5). SUV preparation can be performed by several techniques, depending on the specific lipid composition and concentration required, the most convenient and easiest to use being: (1) Probe sonication in one step (see Note 6), and (2) Size reduction of MLV liposomes (most applied technique).

3.1.1. Preparation of CF-encapsulating DRV's with High Entrapment Yield (1)

3.1.1.1. Thin Film Formation (Step 1)

1. Weigh the required lipid or lipids and Chol (if included in the liposome composition) quantities for the preparation of 1 mL of liposomes and place them in a 50 mL round-bottomed flask; 16.5 μmol of lipid (PC) or 12.5 mg is the amount per each milliliter of aqueous phase that results in the highest encapsulation yield for CF (1).

2. Dissolve the lipids in 1 mL (or more if needed) of a $CHCl_3$/CH_3OH (2:1 v/v) mixture. Alternatively, place the appropriate volumes of pre-formed lipid solutions (see Note 7) in the flask and mix.

3. Connect the flask to a rotor evaporator in order to evaporate the organic solvent under vacuum, until total evaporation and formation of a thin lipid film on the sides of the flask (see Note 8).

4. Flush the thin film with N_2 for at least 10 min and connect the flask (overnight) to a vacuum pump or lyophilizer for total removal of any traces of organic solvents.

5. If desired, the thin films can be sealed with parafilm under N_2, and stored at –20°C for a few days, until being used.

3.1.1.2. Hydration of Thin-Film – MLV Formation (Step 2)

1. Add 1 mL of d.d. H_2O in the flask that contains the thin film.
2. If lipids with high Tc are used as DPPC, DSPC, etc., then the H_2O has to be preheated above the lipid Tc, and the hydration procedure should be performed at that temperature, in a heated water bath.
3. Hydrate the lipid film by repeated vortex agitation.
4. Add glass beads in the flask if needed, in order to facilitate the removal of the lipid from the flask. (In the current case in which plain PC is used the lipid film hydration should be done very easily at room temperature).

3.1.1.3. SUV Formation (Step 3)

1. Place the MLV suspension produced by the method described earlier, in a small test tube for vesicle size reduction by probe sonication.
2. Subject the suspension is to high intensity sonication using a vibra cell Probe sonicator (Sonics and Materials, UK), or other.
3. If a small volume of liposomes are to be prepared (1–3 mL), a tapered micro tip is used, but for larger volumes the conventional tip should be used.
4. Apply sonication (see Note 9) for two 10 min cycles, at least, or until the vesicle dispersion becomes completely transparent.
5. Following sonication, leave the SUV suspensions for 2 h at a temperature higher than the Tc of the lipid used in each case, in order to anneal any structural defects of the vesicles.
6. Remove the titanium fragments (from the probe) and any remaining multilamellar vesicles or liposomal aggregates in the SUV dispersion produced, by centrifugation at 10,000 rpm for 15 min.

3.1.1.4. Preparation of SUV-Solute Mixture (Step 4)

1. Mix the SUV suspension with a solution of the solute that is to be entrapped in the DRVs.
2. Usually 1 mL of SUV suspension is mixed with 1 mL of the solute solution (or 2 mL + 2 mL, etc.).
3. For CF-encapsulating DRV's 1 mL of the empty SUVs are mixed with 1 mL of a 17 mM CF solution in Diluted PBS (see Subheading 2.1).
4. The characteristics of the solute solution are important for the final encapsulation yield; especially the concentration and ionic strength (see Note 10). Some examples of entrapment yields of some substances in DRVs and the conditions applying in each case are presented in Table 2.

3.1.1.5. Freeze-Drying (Step 5) (see Note 11)

1. Place the mixture in a test tube or a 20 mL screw capped bottle, or round bottomed flask for freeze drying.
2. Freeze the empty SUV-solute mixture by swirling the container in the cold liquid containing dewar (taking care not to freeze your fingertips) in order to form a thin layer of frozen liquid on the sides of the container. For freezing, it is advisable to use liquid N_2 or crushed dry ice in acetone (placed in a dewar flask), if available.
3. If the solute is not sensitive to a slow freezing process, the mixture can also be frozen by placing in a freezer for the time period required to achieve complete freezing (depending on the volume and salt content of a sample and the thickness of the layer in the container this may need from a few hours – to overnight freezing [usually most convenient] (see Note 11)).

Table 2
Examples of encapsulation efficiencies of DRV liposomes

Solute entrapped[a]	Lipid composition/lipid concentration[b]	Special consideration	Yield	Reference
		Solute conc.	EE (%)	
CF	PC/Chol (1:1)/16.5	17 mM	40.2	(1)
	PC 16.5	17 mM	54.0	(1)
Glucose	PC/Chol(1:1)/16.5	50 mM	40.0	(1)
Albumin	PC/Chol(1:1)/16.5	10 mg/mL	40.6	(1)
		Solute type/vol.	Amount entrapped[a]	
PRE (plain = PL or as HPβ-CD complex)	PC 6.5	PL (1 mg in lipid phase)	1.3 mg	(15)
	PC/Chol (2:1)/6.5	PL (1 mg in lipid phase)	1.5 mg	(15)
	PC/Chol (2:1)/6.5	HPβ-CD/PRE (6:1)/2 mL	[c]1.3 mg [d]1.7 mg	(15)
	PC 6.5	HPβ-CD/PRE (2:1)/2 mL	[d]3.8 mg	(15)
	PC/Chol (2:1)/6.5	HPβ-CD/PRE (2:1)/2 mL	[d]3.7 mg	(15)
		Sucrose/lipid (mass ratio)	EE (%)/vesicle diameter (nm)	
Penicilin G (5 mg)	PC/Chol (1:1)/16.5	1 3 5	30.5/213 nm 19.5/195 nm 15.5/198 nm	(2)
Riboflavin (1 mg)	PC/Chol (1:1)/16.5	1 3 5	78/591 nm 47.8/168 nm 34.8/145 nm	(2)
		Solute amount	EE (%)	
Tetanus Toxoid	PC/Chol (1:1)/16.5	2 mg	40–82	(36, 38)
BSA		2 mg	40–45%	(38)
Plasmid DNA pGL2	PC/DOPE (1:0.5) 16.5	10–500 μg	44.2	(38)
	PC/DOPE/PS (1:0.5:0.25) 16.5		57.3	(38)
	PC/DOPE/SA (1:0.5:0.25) 16.5		74.8	(38)

[a] or incorporated
[b] μm total lipid/mL (of final liposome suspension)
[c] Diluted to 3 mL before freeze drying
[d] Diluted to 3 mL before freeze drying

4. Connect the sample to a lyophilizer (as a Labconco laboratory lyophilizer) and dry under vacuum (vacuum level below 5 Pa).

3.1.1.6. Rehydration or Reconstitution – DRV Formation (Step 6)

1. Add 100 µL of d.d. H_2O (if applying, this should be preheated above the lipid Tc) and rehydrate the mixture by vortex agitation, taking care to hydrate the full quantity of the lyophilized powder.

2. It is very important and actually determines the encapsulation yield of the DRVs produced to use the smallest possible volume of d.d. H_2O for rehydration of the dried liposome-solute mixture. The typical volume added is 1/10 of the solute (solution) volume, in the current case 100 mL (solutions of 1 mL).

3. After total hydration, leave the mixture to stand for 30 min at room temperature (or higher that the lipid Tc).

4. Replace the remaining volume (900 mL) by adding PBS buffer, and vigorously vortex the mixture. The buffer used should have a tenfold greater osmolarity than the initial solute concentration.

5. In some cases, it is advisable to repeat the first step of rehydration (addition of small volume of water, vortex, annealing) before the final volume adjustment.

6. Keep the suspension above Tc for 30 min.

It is important to understand that when the lyophilized material is hydrated with a tenth of the original solute volume, this results in a tenfold increase in the overall concentration of the solute. And since the liposomes formed are osmotically active, their exposure to hypotonic solutions will result in material loss. This is the reason why a buffer with at least tenfold greater osmolarity of the solute solution should be used (see Note 10).

3.1.1.7. DRV Separation from Nonentrapped Solutes (Step 7)

(In order to use the DRV liposomes, and/or measure the entrapment or encapsulation efficiency (or yield) they should first be separated or purified from not entrapped solute. Depending on the MW of the entrapped solute and on the final size of the DRV liposomes this can be done by centrifugation or size exclusion chromatography).

For separation of the nonencapsulated CF (or calcein) from the DRVs on Sephadex G-50 chromatography columns (see Note 12) eluted with PBS, pH 7.40, the following steps are used:

1. Pre saturate the column with lipids by eluting a liposome sample through the column, in order to finally have lipid recovery that is well over 95%.

2. For lipid recovery calculation measure the lipid concentration in the liposome sample loaded on the column, and the lipid concentration in the liposomal fractions eluted, by a colorimetric assay for phospholipids as the Stewart assay (described

below) (41), and calculate the percent of the loaded lipid which is finally eluted from the column.

3. After assuring that the column has been saturated, load the liposome sample on the column and separate the liposomes from the nonliposome encapsulated molecules.

4. Calculate the amount of CF (or calcein) entrapped in a given volume of vesicles and the amount of lipid in the same volume of the DRV dispersion, as described in the following section.

3.1.1.8. Determination of Entrapped CF (or Calcein)

1. Disrupt the DRVs with a 10% v/v Triton X-100 solution. For this, mix an appropriate volume of the Triton solution in a sample of the DRV dispersion, so that the final concentration of Triton X-100 is 1% v/v.

2. Vigorously mix the dispersion by vortex for at least 2 min (see Note 13).

3. After total disruption of the vesicles, measure the fluorescence intensity (FI) of the sample at 37°C, at EM –470 nm and EX –520 nm and slit band widths 10–10.

4. Finally, calculate the amount of CF (or calcein) entrapped in the DRV's with the help of an appropriate calibration curve of the dye.

3.1.1.9. Determination of Lipid Concentration

The phospholipid content of the liposomes is measured by the Stewart assay (41), a colorimetric technique that is widely used for phospholipid content determination. For this:

1. Mix a sample from the liposome dispersion (20–50 mL, depending on the lipid concentration) with 2 mL Stewart reagent (ammonium ferrothiocyanate 0.1 M) and 2 mL chloroform, in a 2.5 × 10 cm (or higher) test tube.

2. Vortex the mixture vigorously for at least 3 min, in order to extract the complex formed between phospholipid and Stewart reagent in the chloroform phase.

3. Centrifuge the two phase mixture at 5,000 rpm for 5 min, in order to separate the two phases.

4. Remove the top (red) aqueous phase from the test tube by aspiration.

5. Take the chloroform phase out and measure its OD at 485 nm.

6. Finally, calculate the lipid concentration of the sample or samples by comparison of the measured OD-485, with a standard curve (prepared from known concentrations of PC) (see Note 14).

3.1.2. Preparation of Amphiphilic/Lipophilic Drug Incorporating or Encapsulating DRVs (as reported for PRE in (15))

For the preparation of ampliphilic or lipophilic drug incorporating DRV's (this may be needed for comparison between plain drug containing DRVs and CD-Drug complex entrapping DRVs) the general DRV technique (as Subheading 3.1.1), should be modified as follows:

1. In Step 1: If the aqueous solubility of the drug is low, it is advisable to add the drug at this stage, as a concentrated solution in MeOH or $CHCl_3$. In the case of prednisolone (PRE) 100 ul of a 10,000 ppm solution of PRE in MeOH are mixed in the lipid solution.

2. In Step 2: After hydration of the thin film (step 2), the MLV suspension prepared is filtered (Whatmman no. 5) to remove precipitated (nonincorporated in liposomes) drug.

3. In Step 4: 1 ml of the SUV liposomes is mixed with 1 ml of buffer (diluted phosphate buffer) and the mixture is freeze-dried overnight.

4. All following steps are the same with those presented in Subheading 3.1.1. Care has to be taken to use the appropriate buffers before freeze drying and during the initial rehydration step.

3.1.3. Preparation of Drug/CD Complex Encapsulating DRVs (as reported in (15, 18))

As mentioned in the introduction part (Subheading 1.1), amphiphilic or lipophilic drugs can be encapsulated in the aqueous compartments of DRV in the form of soluble inclusion complexes with CD molecules. Thereby, a required initial step of this modified DRV technique is the formation of the CD-Drug inclusion complex.

3.1.3.1. Formation of PRE-CD Inclusion Complexes (42)

1. Mix PRE (2 mg) with 1 mL of d.d. H_2O containing 185 mg β-CD or 50 mg HPβ-CD or 16.7 mg HPβ-CD, and place in a screw capped test tube, or bottle. Higher amount of complexes can be formed by increasing the compound amounts proportionally.

2. Stir the mixtures for 3 days at 20°C (this can be done by placing the screw capped container on a tumbling or circular mixing device). Water soluble inclusion complexes are formed.

3. Centrifuge the milky solution formed with b-CD at 51,000 g for 2 h, or filter the almost clear (or clear) solutions formed with HPb-CD through polycarbonate filters (0.22 mm, Millipore).

4. Calculate the amount of cyclodextrin-solubilized drug spectrophotometrically, by measuring its absorption at 256 nm, and constructing an appropriate calibration curve, from solutions of known drug concentration.

5. Calculate the final cyclodextrin-to-drug molar ratio by taking into account the initial amount of CD added and the final amount of drug measured.

The molar ratio of CD/drug (in this case PRE), plays a significant role in the final vesicle encapsulation efficiency. When an initial HPβ-CD/PRE molar ratio of 2:1 is used for complex preparation, it is calculated that all the drug is in the complex, giving thus a final HP β-CD/PRE complex with a CD/PRE molar ratio of 2:1.

3.1.3.2. Preparation of DRVs

1. Mix 1 mL of empty SUV with 1 mL of the CD-PRE complex (as in **step 4**, of Subheading 3.1.1).
2. Dilute the mixture (by adding an appropriate volume of d.d. H_2O), before freeze drying (see Note 15).
3. Freeze dry the mixture and rehydrate as described in detail in Subheading 3.1.1.
4. In this case, again, care has to be taken to adjust the osmolarity of the mixture before freeze drying and to use the appropriate buffer for dilution of the rehydrated DRVs, making sure that the CD concentration in the solute solution is considered.
5. Separate the nonentrapped inclusion complex and/or PRE from the vesicles by gel-exclusion chromatography.
6. Measure the lipid concentration of the produced liposomes by an appropriate technique (as the Stewart assay presented earlier).
7. Measure the concentration of encapsulated PRE in the DRV's spectrophotometrically, after dissolving DRV samples in MeOH or 2-propanol, and constructing an appropriate calibration curve.

3.1.4. Preparation of Multicomponent DRVs for Photolabile Drugs

The preparation of DRV multicomponent systems of photolabile drugs is achieved by a similar procedure as the one described earlier (Subheading 3.1.3) (17). For this:

1. The lipid soluble light absorbers are added in the lipid phase during thin lipid film preparation (**step 1** of procedure described in Subheading 3.1.1).
2. Aqueous soluble light absorbers are mixed together with the CD-Drug complex and the empty SUV's (**step 4** of procedure described in Subheading 3.1.1).

3.1.5. Preparation of DRV Liposomes with Controlled Entrapment Yield and Vesicle Size (as reported in (2))

Since the addition of cryoprotectants during the freeze drying step (**step 5** in procedure of Subheading 3.1.1) may reduce liposome disruption/fusion and result in decreased encapsulation yield and mean vesicle size of the DRVs produced, it has been demonstrated that by adding specific amounts of sugars in the SUV/solute mixtures prior to freeze drying and by controlling their final concentration therein, the former two DRV characteristics can be controlled accordingly. For this:

1. Add the specified amount of sucrose in the mixture of empty SUV (with mean diameters between 60 and 80 nm) in **step 4** of the general DRV preparation procedure (Subheading 3.1.1), in order to obtain a final sugar-to-lipid mass ratio between 1 and 5 (or else 1–5 mg of sucrose per each mg of phospholipid).
2. Adjust the mixture sugar molarity to a predetermined value by the addition of the appropriate amount of d.d. H_2O (sugar molarity values between 30 and 50 mM were found to give good vesicle encapsulation and considerably small vesicle size, when the sugar/lipid mass ratio was set at 3 (2)).

3. Freeze dry the mixtures overnight. After this, all other steps of the procedure are similar to those described earlier (in Subheading 3.1.1).

After separation of the DRV's from the nonencapsulated solute molecules and measurement of the entrapment it is found that as sugar/lipid mass ratio increases vesicle size decreases together with the encapsulation yield. Some of the solutes studied and the results obtained (2) are presented in Table 2.

3.2. Preparation of Protein (or Peptide) and DNA Vaccines (as reported in (1, 34, 43))

This procedure is similar to the general procedure described in Subheading 3.1.1, for DRV preparation. A few special considerations-suggestions are mentioned in the following section.

3.2.1. Lipid Compositions for Protein and DNA Vaccines

1. Phospholipid (32 mmol) and Chol (32 mmol).
2. Negatively charged liposomes containing 3.2 mmol of PA, PS, or PG.
3. Positively charged liposomes containing 3.2–8 mmol of SA, BisHOP, DOTMA, DOTAP, or DC-CHOL.
4. Depending on the amount of vesicle surface charge required, greater quantities of charged lipids can be added.
5. In all cases, the appropriate quantities of the lipids are dissolved in 2–5 ml of chloroform during step 1 of the general procedure for DRV preparation (see procedure Subheading 3.1.1).

3.2.2. Liposome Preparation Procedure – Vaccine Addition

If the sonication step (**step 3** in Subheading 3.1.1) is not detrimental to the vaccine, the vaccine solution may be added instead of H_2O during the hydration of the thin film (general DRV preparation procedure – **step 2**). For this:

1. Dissolve (up to) 10 mg of the water-soluble vaccine in 2 ml d.d. H_2O or 10 mM sodium phosphate buffer, pH 7.2 [phosphate buffer (PB)].
2. The buffer used can be varied with respect to composition, pH, and molarity, as long as this does not interfere with liposome formation or entrapment yield (see Note 10).
3. The amount of vaccine added can be increased proportionally to the total amount of lipid.
4. Apply sonication and let the produced SUVs to stand for annealing (as described in Subheading 3.1.1).
5. Mix the SUV suspension with water (if the vaccine was added in **step 2** of this procedure) or with the vaccine solution (prepared as described earlier).
6. For the rehydration step (**step 6** in Subheading 3.1.1), it is adviced to use 0.1 ml of H_2O per 32 mmol of phospholipid (but enough H_2O to ensure complete dissolution of the powder)

and keep the sample above Tc for 30 min. Then repeat this step (add another 0.1 mL of water, vortex and keep for 30 min above Tc) before bringing the DRV suspension to volume, with 0.8 ml PB (which is pre warmed above the lipid Tc).

7. The sample is finally allowed to stand for extra 30 min above Tc.

3.2.3. Separation of DRV's from Nonentrapped (Vaccine) Material

This is done by centrifugation:

1. Centrifuged the DRV suspension that contains entrapped and nonentrapped vaccine at 40,000 g for 60 min (4°C).
2. Re-suspend the pellet obtained (vaccine containing DRVs) in H_2O or PB and centrifuge again under the same conditions.
3. Repeat the fore mentioned process once more for total removal of the remaining nonentrapped material.
4. Finally suspend the pellet in 2 ml H_2O or PB.

3.2.4. Vaccine Entrapment Yield Estimation

1. The entrapment yield is calculated by measuring the vaccine in the suspended pellet and the combined supernatants.
2. The easiest way to monitor entrapment is by using a radiolabeled vaccine. If a radiolabel is not available or cannot be used, appropriate quantitative techniques should be employed.
3. To determine the vaccine by such techniques a sample of the liposome suspension is mixed with Triton X-100 (up to 5% final concentration) or 2-propanol (1:1 volume ratio) in order to liberate the entrapped material.
4. If Triton X-100 or the solubilized liposomal lipids interfere with the assay of the material, lipids must be extracted using appropriate techniques.
5. Entrapment yields of this technique range between about 20 and 100%, depending on the amounts of lipid and vaccine used.
6. Highest values are achieved when solutes for entrapment bear a net charge that is opposite to that of the charged lipidic component of liposomes (see Note 16).

3.2.5. Vaccine-containing DRV Size Reduction

If vaccine-containing DRV liposomes must be converted to smaller vesicles (down to about 100-nm z-average diameter) the following procedure is used:

1. Dilute the liposomal suspension obtained in **step 6** (of the general procedure described in Subheading 3.1.1) prior to separation of the entrapped from the nonentrapped vaccine, to 10 ml with H_2O.
2. Pass the dispersion for a specific number of full cycles through a Microfluidizer 110S (Microfluidics), with the pressure gauge set at 60 PSI throughout the procedure to give a flow

rate of 35 ml/min (22). The number of cycles used depends on the vesicle size required and the sensitivity of the entrapped vaccine (e.g., plasmid DNA) (43).

3. The greater the number of cycles the smaller the amount of drug retained by the vesicles (44) (see Note 17).

4. The microfluidized sample volume (approximately 10 ml) may be reduced, if needed, by placing the sample in dialysis tubing and then in a flat container filled with PEG 6000. Removal of excess H_2O from the tubing is relatively rapid (within 30–60 min) and it is, therefore, essential that the sample be inspected regularly.

5. When the required volume has been reached, the sample is treated for the separation of entrapped from nonentrapped vaccine, by molecular sieve chromatography using a Sepharose CL-4B column, in which case vaccine-containing liposomes are eluted at the end of the void volume (see Note 12).

6. Minimum vesicle diameters obtained after 10 cycles of microfluidization are about 100–160 nm, depending on whether microfluidization was carried out in H_2O or PB, using unwashed or washed liposomes.

3.3. Preparation of Giant Liposome DRVs that Entrap Large Particles, Viruses, or Bacteria (as described in (36–38))

3.3.1. Giant Liposome Preparation

1. Mix 1 mL of solution 3 with 1 mL of solution 1 by vortex agitation. Then mix (vortex for 15 s) the resulting water-in-chloroform emulsion with a similar mixture composed of solution 2 and solution 4 (2.5 ml).

2. Place the water-in-oil-in-water double emulsion formed in a 250 ml conical flask.

3. Place the flask under a stream of nitrogen (N_2) and in a shaking incubator (at 37°C) under mild agitation, in order to evaporate the organic solvents. This leads to the generation of (sucrose-containing) giant liposomes.

4. Wash the giant liposomes by centrifugation (in a typical bench centrifuge at 1,000 g) for 5 min over solution 5.

5. Resuspend the liposomal pellet in 1 ml PBS.

6. Mix the suspended pellet with 1 ml of a particulate matter suspension (could be killed or live B. Subtilis spores; killed Bacille Calmette–Guerin (BCG) bacteria, etc.) and freeze-dry the mixture under vacuum, overnight.

7. Rehydrate the freeze-dried material, with 0.1 ml H_2O at 20°C (see Note 18), by vigorous mixing. Let the suspension at peace for 30 min above the lipid Tc.

8. Repeat the previous step with addition of 0.1 ml PBS.

9. Finally, bring the sample to volume by adding 0.8 ml PBS, 30 min later (1 ml total suspension volume).

3.3.2. Giant DRV Liposome Separation from Nonentrapped Particulates

Entrapped particulate material is separated from nonentrapped material (bacteria spores, etc.) by sucrose gradient centrifugation. For this:

1. Place 1 ml of the suspension which contains entrapped and nonentrapped particulates, on top of a sucrose gradient (see solution 7 in Subheading 2.3).
2. Centrifuge for 2 h at 90,000 g, using a swing-out bucket rotor.
3. Following centrifugation, continuously take out 1-ml fractions from the top of the gradient, and assay each fraction for particulate content (it is convenient to use radiolabeled (e.g., ^{125}I-labeled) particulates).
4. The nonentrapped particulates are recovered at the bottom fractions of the gradient, whereas entrapped material is recovered mostly in the top seven fractions of the gradient were liposomes remain (36).
5. Finally, pool the fractions that contain the entrapped spores or bacteria and dialyze them exhaustively against PBS, after placing them in a dialysis tubing (MW-cutoff 10,000), until all sucrose has been eliminated.
6. Centrifuge the dialyzed material and re suspend the liposomal pellet in 1 ml PBS for further use.

Typical values of B. subtilis or BCG entrapment range between 21 and 27% of the material used (36, 37).

4. Notes

1. The abbreviation DRV stands for *Dehydration-Rehydration Vesicles* as initially named by the inventors of this liposome preparation technique (1). However, one will find several other explanations in the relevant literature as: *Dried Rehydrated Vesicles* and *Dried Reconstituted Vesicles*, which are actually the same type of liposomes.
2. The 99% purity of the lipids can be verified by Thin Layer Chromatography on silicic acid precoated plates (Merck, Germany), using a $CHCl_3/CH_3OH/H_2O$ 65:25:4 v/v/v mixture for plate development, and iodine staining for visualization. Pure lipids give single spots.
3. Calcein (as well as CF) is not easily dissolved in buffer with pH 7.40. Therefore, the weighted solid is initially dissolved in NaOH (1 M) which is added dropwise until the full quantity is dissolved and subsequently, the resulting solution is diluted with the appropriate volume of buffer (in order to achieve the required calcein concentration). The pH of the final solution

should be checked and re-adjusted if required, while care has to be taken so that calcein does not precipitate.

4. For the formation of DRV liposomes entrapping solutes that are not sensitive to the conditions used for MLV and/or SUV preparation, it is possible to prepare drug containing liposomes in the initial step of DRV formation. This is particularly important if amphiphilic/lipophilic or in general substances with low aqueous solubility are to be entrapped. However, when there is interest to have a method that can be easily up-scaled for large batch manufacturing, this approach can be problematic.

5. It has been reported that DRV liposomes with comparably high encapsulation efficiency (compared to plain MLV's) can be produced even by using MLV liposomes for the initial drying step (1). Indeed a EE% of 21% for CF was reported when empty MLV liposomes were used (compared to 1.8% when plain MLV were formed using the same CF hydrating concentration), while a 30% EE% was reached when starting from SUV.

6. For "one-step" probe sonication (SUV formation) the lipid or lipids in solid form are placed in an appropriate test tube (with dimension that ensure proper placement on the probe) together with the hydration solution. The mixture is heated above the lipid transition temperature and subsequently probe sonicated. This method may not be applicable in some cases of lipids and when high lipid concentrations are used (>20 mg/mL).

7. The lipids could also be used in the form of solutions in $CHCl_3$/CH_3OH 2:1 v/v that can be initially prepared and stored at –80°C. In this case, the appropriate volume of lipid solution (or each lipid solutions; if several lipids are used in the form of organic solution) is (or are) used. If one step probe sonication is used, the organic solvent is evaporated by a stream of N_2 and the preparation proceeds as described earlier.

8. If the film has irregularities, it is advisable to re-dissolve in $CHCl_3$ (or other easily evaporated organic solvent depending on the solubility of the lipids) and evaporate the organic solvent again, until a nice and even thin film is formed.

9. For probe sonication, the probe tip (or tapered microtip) is immersed into the MLV or lipid dispersion by approximately 1.2–1.5 cm from the surface, taking care so that no part of the tip is in contact with the vial (a mirror is used to be sure). The vial is placed in a ice-cold water tank, to prevent overheating of the liposome dispersion during probe sonication. Alternatively, a glycerol bath which has been pre-heated above the lipid transition temperature can be used in order to prevent overheating of the sample. A simple method for size reduction of MLV liposomes, if a probe socicator is not available, is by extrusion through stacked polycarbonate filters with

appropriate pore dimensions. For this, the filters are placed on a double syringe apparatus (several are commercially available) and the liposome dispersions is passed through the filters from the one syringe to the other several times (at least 10), until the vesicle dispersion size has been appropriately reduced). The system can be immersed in a heated water bath if lipids with high Tc's are used.

10. The solute solution concentration has been demonstrated to influence DRV encapsulation efficiency differently, depending on the solute. As an example, although glucose and CF entrapment values were found to decrease with increasing solute solution concentration, the same was not found true for encapsulation of sodium chloride and potassium chloride (1). For CF, best encapsulation yields in DRVs are demonstrated when a 17 mM solution in a tenfold dilution of an isotonic PBS buffer, is used. The ionic strength of the buffer used to dissolve the solute added at this step, should be at least 10 times less than that of the buffer used for DRV dilution after the hydration step (see below) in order to reduce material losses, due to osmotic activity of liposomes.

11. Although we do not have experience of this in our lab, it has been reported that similar encapsulation yields for DRVs may be obtained by drying down the SUV-solute mixture using other procedures, as drying nonfrozen mixtures at 20°C under vacuum, or under a stream of N_2 (at 20°C or 37°C).

12. A column with dimensions 1×35 cm is sufficient to separate 1 ml of liposome or the DRV liposome dispersion. The column is pre-calibrated and at the same time saturated with a dispersion of empty liposomes mixed with a quantity of the encapsulated material (in each case). The void volume of such columns should be between 7 and 13 mL and the bed volume between 17 and 21 mL.

13. In some cases, especially when rigid liposomes that contain DSPC and Chol are used, the liposomes are difficult to disrupt by using 1% v/v final concentration of Triton X-100 detergent. Then the liposome mixture with Triton can be heated by rapidly immersing in boiling water (in which case care should be taken in order to perform the final measurement after the sample is cooled, in order to avoid mistakes). Another possibility is to use higher final concentration of detergent (in which case the extra dilution of the sample has to be taken into account during calculations).

14. The linear region for a calibration curve by the Stewart assay is between 10 and 100 μg of PC (DPPC, DSPC can also be used at this range).

15. This is a very important step, in order to avoid low encapsulation yields that may be caused by the presence of high CD

concentrations in the sample (which as a oligosacharide will act as a cryoprotectants). For PRE-CD complexes, when the mixtures are diluted up to 10 ml final volume before the freeze drying step, the final encapsulation of PRE is increased by 30% (Table 2) (15).

16. Part of the liposome associated solute may have interacted with the liposomal surface during the entrapment procedure. Thereby, it is essential that actual entrapment of the solute (as opposed to surface-bound solute) is determined. In the case of DNA or proteins, this can be achieved by using deoxyribonuclease (43) and a proteinase (5), respectively, which will degrade the external material.

17. Microfluidization of the sample can also be carried out after removal of nonentrapped vaccine (after **step** 7 of the general procedure in Subheading 1). However, drug retention in this case is reduced. It appears that the presence of nonentrapped drug during micro-fluidization diminishes solute leakage, probably by reducing the osmotic rupture of vesicles and/or the initial concentration gradient across the bilayer membranes (44).

18. It has been observed in this case that rehydration of liposomes which contain "high-melting" DSPC above Tc (50°C) does not have a significant effect on the percentage entrapment of materials (37, 38).

References

1. Kirby C, Gregoriadis G (1984) Dehydration-rehydration vesicles: A simple method for high yield drug entrapment in liposomes. Biotechnology 2:979–984
2. Zadi B, Gregoriadis G (2000) A novel method for high-yield entrapment of solutes into small liposomes. J Liposome Res 10:73–80
3. Seltzer SE, Gregoriadis G, Dick R (1988) Evaluation of the dehydration-rehydration method for production of contrast-carrying liposomes. Invest Radiol 23:131–138
4. Cajal Y, Alsina MA, Busquets MA, Cabanes A, Reig F, Garcia-Anton JM (1992) Gentamicin encapsulation in liposomes: Factors affecting the efficiency. J Liposome Res 2:11–22
5. Senior J, Gregoriadis G (1989) Dehydration-rehydration vesicle methodology facilitates a novel approach to antibody binding to liposomes. Biochim Biophys Acta 1003:58–62
6. Du Plessis J, Ramachandran C, Weiner N, Muller DG (1996) The influence of lipid composition and lamellarity of liposomes on the physical stability of liposomes upon storage. Int J Pharm 127:273–278
7. Casals E, Gallardo M, Estelrich J (1996) Factors influencing the encapsulation of thioguanine in DRV liposomes. Int J Pharm 143:171–177
8. Grant GJ, Barenholz Y, Piskoun B, Bansinath M, Turndorf H, Bolotin EM (2001) DRV liposomal bupivacaine: Preparation, characterization, and in vivo evaluation in mice. Pharm Res 18:336–343
9. Anderson KE, Eliot LA, Stevenson BR, Rogers JA (2001) Formulation and evaluation of a folic acid receptor-targeted oral vancomycin liposomal dosage form. Pharm Res 18:316–322
10. Kawano K, Takayama K, Nagai T, Maitani Y (2003) Preparation and pharmacokinetics of pirarubicin loaded dehydration-rehydration vesicles. Int J Pharm 252:73–79
11. Kallinteri P, Fatouros D, Klepetsanis P, Antimisiaris SG (2004) Arsenic trioxide liposomes: Encapsulation efficiency and in vitro stability. J Liposome Res 14:27–38
12. Mugabe C, Azghani AO, Omri A (2006) Preparation and characterization of dehydration-rehydration vesicles loaded with aminoglycoside

and macrolide antibiotics. Int J Pharm 307: 244–250
13. Koromila G, Michanetzis GPA, Missirlis YF, Antimisiaris SG (2006) Heparin incorporating liposomes as a delivery system of heparin from PET-covered metallic stents: Effect on haemocompatibility. Biomaterials 27:2525–2533
14. Hatzi P, Mourtas SG, Klepetsanis P, Antimisiaris SG (2007) Integrity of liposomes in presence of cyclodextrins: Effect of liposome type and lipid composition. Int J Pharm 333:167–176
15. Fatouros DG, Hatzidimitriou K, Antimisiaris SG (2001) Liposomes encapsulating prednisolone and prednisolone-cyclodextrin complexes: Comparison of membrane integrity and drug release. Eur J Pharm Sci 13: 287–296
16. Manconi M, Isola R, Falchi AM, Sinico C, Fadda AM (2007) Intracellular distribution of fluorescent probes delivered by vesicles of different lipidic composition. Coll Surf B Biointerfaces 57:143–151
17. Loukas YL, Jayasekera P, Gregoriadis G (1995) Novel liposome-based multicomponent systems for the protein of photolabile agents. Int J Pharm 117:85–94
18. McCormack B, Gregoriadis G (1994) Drugs-in-cyclodextrins-in liposomes: A novel concept in drug delivery. Int J Pharm 112: 249–258
19. Kallinteri P, Antimisiaris SG, Karnabatidis D, Kalogeropoulou C, Tsota I, Siablis D (2002) Dexamethasone incorporating liposomes: An in vitro study of their applicability as a slow releasing delivery system of dexamethasone from covered metallic stents. Biomaterials 23:4819–4826
20. Zaru M, Mourtas S, Klepetsanis P, Fadda AM, Antimisiaris SG (2007) Liposomes for drug delivery to the lungs by nebulization. Eur J Pharm Biopharm 67:655–666
21. Gregoriades G, Panagiotidi C (1989) Immunoadjuvant action of liposomes: Comparison with other adjuvants. Immunol Lett 20:237–240
22. Skalko N, Bouwstra J, Spies F, Gregoriadis G (1996) The effect of microfluidization of protein-coated liposomes on protein distribution on the surface of generated small vesicles. Biochim Biophys Acta 1301:249–254
23. Rodriguez-Nogales JM, Pérez-Mateos M, Busto, Ma D (2004) Application of experimental design to the formulation of glucose oxidase encapsulation by liposomes. J Chem Technol Biotechnol 79:700–705
24. Garciia-Santana MA, Duconge J, Sarmiento ME, Lanio-Ruíz ME, Becquer MA, Izquierdo L, Acosta-Dominguez A (2006) Biodistribution of liposome-entrapped human gamma-globulin. Biopharm Drug Disp 27:275–283
25. Ntimenou V, Mourtas S, Christodoulakis EV, Tsilimbaris M, Antimisiaris SG (2006) Stability of protein-encapsulating DRV liposomes after freeze-drying: A study with BSA and t-PA. J Liposome Res 16:403–416
26. Badiee A, Jaafari MR, Khamesipour A (2007) Leishmania major: Immune response in BALB/c mice immunized with stress-inducible protein 1 encapsulated in liposomes. Exp Parasitol 115:127–134
27. Eckstein M, Barenholz Y, Bar LK, Segal E (1997) Liposomes containing Candida albicans ribosomes as a prophylactic vaccine against disseminated candidiasis in mice. Vaccine 15:220–224
28. Perrie Y, Frederik PM, Gregoriadis G (2001) Liposome-mediated DNA vaccination: The effect of vesicle composition. Vaccine 19: 3301–3310
29. Perrie Y, Barralet JE, McNeil S, Vangala A (2004) Surfactant vesicle-mediated delivery of DNA vaccines via the subcutaneous route. Int J Pharm 284:31–41
30. Pupo E, Padrón A, Santana E, Sotolongo J, Quintana D, Dueñas S, Duarte C, Hardy E (2005) Preparation of plasmid DNA-containing liposomes using a high-pressure homogenization-extrusion technique, J Cont Release 104:379–396
31. Jaafari MR, Badiee A, Khamesipour A, Samiei A, Soroush D, Kheiri MT, Barkhordari F, Mahboudi F (2007) The role of CpG ODN in enhancement of immune response and protection in BALB/c mice immunized with recombinant major surface glycoprotein of Leishmania (rgp63) encapsulated in cationic liposome. Vaccine 25:6107–6117
32. Maitani Y, Aso Y, Yamada A, Yoshioka S (2008) Effect of sugars on storage stability of lyophilized liposome/DNA complexes with high transfection efficiency. Int J Pharm 356: 69–75
33. Allison AC, Gregoriadis G (1974) Liposomes as immunological adjuvants. Nature 252:252
34. Gregoriadis G (1990) Immunological adjuvants: A role for liposomes. Immunol Today 11:89–97
35. Gregoriadis G, Saffie R, De Souza JB (1997) Liposome-mediated DNA vaccination. FEBS Lett 402:107–110

36. Antimisiaris SG, Jaysekera P, Gregoriadis G (1993) Liposomes as vaccine carriers. Incorporation of soluble and particulate antigens in giant vesicles. J Immun Meth 166:271–280
37. Gurcel I, Antimisiaris SG, Jaysekera P, Gregoriadis G (1995) In: Shek P (ed) Liposomes in biomedical applications, Harwood Academic, Chur, Switzerland, pp 35–50
38. Gregoriadis G, McCormack B, Obrenovic M, Roghieh S, Zadi B, Perrie Y (1999) Vaccine entrapment in liposomes. Methods 19:156–162
39. Vangala A, Bramwell VW, McNeil S, Christensen D, Agger EM, Perrie Y (2007) Comparison of vesicle based antigen delivery systems for delivery of hepatitis B surface antigen. J Contr Release 119:102–110
40. Lanio ME, Luzardo MC, Alvarez C, Martínez Y, Calderon L, Alonso ME, Zadi B, Disalvo A (2008) Humoral immune response against epidermal growth factor encapsulated in dehydration rehydration vesicles of different phospholipid composition. J Liposome Res 18:1–19
41. Stewart JCM (1980) Colorimetric determination of phospholipids with ammonium ferrothiocyanate. Anal Biochem 104:10–14
42. Loftsson T, Brewster ME (1996) Pharmaceutical applications of cyclodextrins. 1. Drug solubilization and stabilization. J Pharm Sci 85:1017–1025
43. Gregoriadis G, Saffie R, Hart SL (1996) High field incorporation of plasmid DNA within liposomes: Effect on DNA integrity and transfection efficiency. J Drug Targeting 3: 469–475
44. Gregoriadis G, Da Silva H, Florence AT (1990) A procedure for the efficient entrapment of drugs in dehydration-rehydration liposomes (DRVs). Int J Pharm 65:235–242

Chapter 4

Elastic Liposomes for Topical and Transdermal Drug Delivery

Heather A.E. Benson

Abstract

Elastic liposomes have been developed and evaluated as novel topical and transdermal delivery systems. They are similar to conventional liposomes but with the incorporation of an edge activator in the lipid bilayer structure to provide elasticity. Elastic liposomes are applied non-occluded to the skin and have been shown to permeate through the stratum corneum lipid lamellar regions as a result of the hydration or osmotic force in the skin. They have been investigated as drug carriers for a range of small molecules, peptides, proteins and vaccines, both in vitro and in vivo. Following topical application, structural changes in the stratum corneum have been identified and intact elastic liposomes visualised within the stratum corneum lipid lamellar regions, but no intact liposomes have been identified in the deeper viable tissues. The method by which they transport their drug payload into and through the skin has been investigated but remains an area of contention. This chapter provides an overview of the development, characterisation and evaluation of elastic liposomes for delivery into and via the skin.

Key words: Colloids, Elastic liposomes, Liposomes, Transfersomes, Skin penetration enhancement, Drug carriers

1. Introduction

Encapsulation of active ingredients including humectants such as glycerol and urea, sunscreen and tanning agents, enzymes, anti-ageing and acne agents such as retinol, antimicrobials, steroids, hyaluronic acid and natural products, into liposomes is utilised in many cosmetic products. What these preparations have in common is that their target sites are within the skin layers or appendages. Examples of currently available liposome based drug products are less common, although it has been a very active research area with many scientific publications and patent filings in recent years. A variety of colloid systems have been used for encapsulating penetrant molecules, including conventional liposomes, ethosomes, niosomes and elastic liposomes (the initial formulation approach being termed Transfersomes®) (1–3).

Elastic liposomes are claimed to permeate as intact liposomes through the skin layers to the systemic circulation and have been reported to improve in vitro skin delivery (e.g. (4–6)), and in vivo penetration to achieve therapeutic amounts comparable with subcutaneous injection (7). Elastic liposomes were first described by Gregor Cevc (Idea, Munich) who termed them as Transfersomes® (8). The first commercial product based on the Transfersome technology is Diractin, a ketoprofen formulation for management of osteoarthritis (9, 10). Transfersomes are composed of phospholipids, but also contain surfactant such as sodium cholate, deoxycholate, Span, Tween and dipotassium glycyrrhizinate (5, 6, 11). The surfactant acts as an 'edge activator' that destabilises the lipid bilayers and increases deformability of the liposome (12, 13). The formulation may also contain some ethanol (typically up to 10%) and a total lipid concentration of up to 10% in the final aqueous lipid suspension (14, 15). Due to the flexibility conferred on the liposomes by the surfactant molecules, Cevc claimed that Transfersomes squeeze through channels one-tenth the diameter of the Transfersome (15) allowing them to spontaneously penetrate the stratum corneum. Cevc et al. suggested that the driving force for penetration into the skin is the osmotic gradient caused by the difference in water content between the relatively dehydrated skin surface (approximately 20% water) and the aqueous viable epidermis (8, 16). Cevc hypothesised that a Transfersome suspension applied to the skin is subjected to evaporation and to avoid dehydration, Transfersomes must penetrate to deeper tissues. Conventional liposomes remain near the skin surface, dehydrate and fuse with the skin lipids, whilst deformable Transfersomes 'squeeze' through stratum corneum lipid lamellar regions penetrating deeper to follow the osmotic gradient. Consequently, Transfersomes and other elastic liposomes should not be applied under occlusion, as this would decrease the osmotic effect (5, 17, 18).

Cevc originally utilised phosphatidylcholine in combination with the surfactant sodium cholate (15) for his Transfersomes, but many other compositions of elastic liposomes have also been developed and evaluated. In general, phosphatidylcholine (soya, egg or hydrogenated) is used as the lipid. Surfactants used include sodium cholate and deoxycholate, surfactant L-595 (sucrose laurate ester), PEG-8-L (octaoxyethylene laurate ester) (19, 20), dipotassium glycyrrhizinate, (6, 21) Spans and Tweens (22–24). The cationic lipid 1,2-diolyl-3-triethylammonium-propane (DOTAP) in combination with Tween 20 non-ionic surfactant, has been used to form positively charged elastic liposomes that are attracted to the negatively charged skin thereby enhancing skin retention (24, 25). Another approach investigated is replacement of the surfactant with chemicals known to be skin penetration enhancers, such as oleic acid and limonene as the edge activators (26). In addition, many formulations include a low percentage of alcohol (generally ethanol) to reduce the liposome size and zeta potential, and increase drug release and skin flux (11, 27).

2. Materials

2.1. Liposome Preparation and Drug Incorporation

A wide variety of compositions for elastic liposomes have been investigated. The following is based on a typical composition and method of preparation for a lipophilic drug such as estradiol.

1. 1-α-phosphatidylcholine from soybean, 99%, lyophilised powder (Sigma-Aldrich, Sydney, Australia) (see Note 1).
2. Sodium cholate 99% (Sigma-Aldrich, Sydney, Australia).
3. 17-β-Estradiol (Sigma-Aldrich, Sydney, Australia).
4. Buffer salts, ethanol and other chemicals and solvents, analytical grade and used as supplied.
5. Distilled and de-ionised water (Millipore-Q ultra pure water system, Millipore, USA).
6. Polycarbonate membranes, 100 and 200 nm (Milli-Q, Millipore, USA).

2.2. Characterization of Elastic Liposome Formulations

1. Nanosep® centrifugal device (300 kDa MWCO) (Pall Corp., NY, USA).
2. Zetasizer 3000HS (Malvern Instrument, UK).
3. Franz-type diffusion cells (Glass-blowing Service, University of Queensland, Australia).
4. SquameScan 850A and D-Squame discs (CuDerm Corp. Dallas, TX).
5. Agilent 1100 liquid chromatographic system equipped with thermostated autosampler, binary pump, photo diode array detector and Chemstation software (Agilent, USA).
6. Symmetry C_{18} HPLC column (Waters Inc., USA).
7. HPLC solvents (methanol, acetonitrile: Baker, USA).
8. Triton X-100 (Sigma-Aldrich, Sydney, Australia).

3. Methods

3.1. Liposome Preparation and Drug Incorporation

In general, elastic liposomes are prepared using similar methods to conventional liposomes, most commonly using the conventional rotary evaporation sonication method (26, 28–30). A typical preparation method is outlined:

1. Ensure all equipment is clean and dry.
2. Place phosphatidylcholine and sodium cholate (85:15 for final concentration 50 mg/mL lipid) in the round bottom flask of the rotary evaporator that is activated at low speed, at a tilt of approximately 45°.
3. Add 10 mL ethanol (containing estradiol; 1 mg/mL) to dissolve the phospholipid and surfactant (see Note 2).
4. Remove ethanol by rotary evaporation, under a nitrogen stream, at a suitable temperature above the lipid transition temperature. Room temperature is suitable for this formulation. This will leave a film of lipids deposited on the wall of the flask.
5. Final traces of organic solvent can be removed under vacuum for 12 h or overnight.
6. Hydrate the deposited film with water by rotation for 2–4 h without vacuum (see Note 3).
7. Allow the resulting liposome suspension to swell for a further 2–4 h at 4°C temperature.
8. Bath sonicate at 4°C for 30–45 min to reduce liposome size.
9. Extrude the resulting suspension through a sandwich of 200 and 100 nm polycarbonate membranes up to ten times (see Notes 4 and 5).

3.2. Characterization of Elastic Liposome Formulations

3.2.1. Liposome Size, Size Distribution and Zeta Potential Determination

1. Determine liposome size and size distribution by photon correlation spectroscopy (PCS) using a Zetasizer 3000HS.
2. Measure at 25°C with a detection angle of 90°.
3. Correlate the raw data to Z average mean size using a cumulative analysis by the Zetasizer 3000HS software package.
4. Determine the zeta potential of liposomes by Laser Doppler Anemometry, using the Zetasizer 3000HS.
5. All analyses should be performed on samples appropriately diluted with filtered de-ionised water or buffer. For each sample, the mean ± standard deviation of three determinations should be reported.

3.2.2. Vesicular Shape and Surface Morphology

Vesicular shape and surface morphology can be assessed by transmission electron microscopy (TEM) with an accelerating voltage of 100 kV using standard techniques.

3.2.3. Deformability or Elasticity

The deformability of elastic liposomes is determined by assessment of the extrusion of the suspension through a filter membrane of defined pore size (50 nm) under pressure (24, 31). The amount of liposome suspension extruded is measured and the liposome size and shape are determined as previously described. For each liposome suspension, the mean ± standard deviation of three extrusions should be reported. Elasticity is calculated as follows:

$$E = J(r_v/r_p)^2$$

Where E = elasticity, J = amount of suspension extruded in x min (e.g. 5 or 10 min), r_v = liposome size after extrusion and r_p = pore size of filter membrane.

3.2.4. Entrapment Efficiency

The prepared elastic liposome suspension can be separated from the free drug by a suitable separation technique such as filtration with centrifugation (32). The free drug is separated from the liposome suspension using a Nanosep® centrifugal device (300 kDa MWCO) at 16,100 g for 15 min. The liposomes are then lysed by exposure to surfactant solution such as Triton-X (0.5% w/w). The amount of drug entrapped in the liposomes is calculated by the difference between the total amount of drug in the suspension and the amount of non-entrapped drug remaining in the aqueous supernatant. The amount of free estradiol in the supernatant is determined by HPLC analysis using a validated assay procedure (see Note 6). Drug entrapment efficiency is calculated as follows

$$\text{Entrapment Efficiency (\%)} = \left(\frac{\text{Total amount drug} - \text{Amount of free drug}}{\text{Total amount drug}}\right) \times 100$$

3.2.5. Physical Stability Assessment

Physical stability is assessed by placing samples of the elastic liposome suspension into vials that are flushed with nitrogen and sealed. The vials are stored under varied conditions such as light protected or exposed, refrigerated and at room temperature. At different time periods (e.g. 10, 20, 30 days and monthly up to 6 months) the samples are analysed for particle size and residual drug content.

3.3. Drug Release

The experimental setup for the release study comprises a dialysis membrane tube and a glass vessel with diameter just larger than the membrane tube. This is placed in a water bath that is maintained at 37°C.

1. Place phosphate buffered saline (PBS: pH 7.4) solution in the glass vessel and allow to stand in the water bath until equilibrated to 37°C.
2. Tie one end of the dialysis membrane tube tightly to ensure no leakage.
3. Place the liposome suspension in PBS in the dialysis tube and seal the other end.
4. Suspend the dialysis tube within the glass vessel.

5. Remove samples of the buffer solution at time intervals up to 24 h, replacing each sample with fresh pre-warmed PBS.

6. At 24 h, remove the liposome suspension and lyse the liposomes, as previously described, to liberate the estradiol remaining in the liposomes.

7. Measure the estradiol content in all samples by HPLC using a validated assay.

3.4. Skin Permeability and Deposition Measurement

Skin permeability is best measured on excised human skin. Skin can be heat-separated epidermis, dermatomed to a particular thickness (typically 200–500 μm) or whole skin with subcutaneous fat removed. The following protocol outlines a standard skin permeability experiment using human epidermis in static Franz-type diffusion cells:

1. Ethics committee approval is required and skin is generally donated following plastic surgery such as abdominoplasty (see Note 7).

2. Prepare epidermal membranes from full thickness tissue using the heat-separation technique (33). Clean abdominal skin of underlying subcutaneous fat by blunt dissection and immerse in water at 60°C for 1 min. This results in separation at the epidermal–dermal junction. Gently peel epidermis away from the underlying dermis being careful not to damage the epidermis. Place epidermal sheets on aluminium foil, air-dried and stored frozen inside zip-lock bags until required.

3. Insert magnetic flea in receptor chamber and mount epidermal membranes stratum corneum side up in horizontal Franz-type diffusion cells, that are then placed on a magnetic stirrer plate in a water bath (Fig. 1) (see Note 8).

4. Fill receptor chamber with PBS and allow to equilibrate for 8 h in a water bath set at 35–37°C (aim for skin surface temperature of 32°C).

5. Add PBS to the donor chamber and measure resistance across the membrane with a standard multimeter. Any cell with a resistance of less than 20 KΩ should be excluded on the basis that the integrity of the membrane cannot be assured.

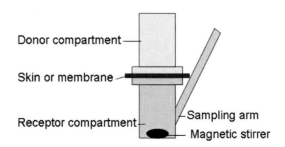

Fig. 1. Franz-type diffusion cell for assessment of skin permeation

6. Remove PBS from the donor and receptor, and refill the receptor chamber with fresh pre-warmed PBS.

7. Add liposome suspension to donor chamber (typically 200–500 μL) that is left unoccluded.

8. Remove samples (typically 200 μL) from the receptor chamber at appropriate time intervals, replacing each sample with an equal volume of pre-warmed PBS. Analyse all samples by HPLC for drug content.

9. Following the final sample, remove and retain any remaining donor solution for analysis (see Note 9).

10. Wash the epidermal surface three times with 500 μL volumes of PBS and retain washings.

11. Remove epidermis from the cell, allow to dry, weigh and extract remaining drug using a solvent extraction technique appropriate to the drug. Analyse extracts to determine drug remaining in skin.

12. Plot cumulative amount of drug in receptor versus time and calculate steady state permeation rate or flux (J_{ss}) and lag time from the slope and x-intercept of the linear portion respectively.

13. Mass balance can be calculated from the amount of drug remaining in the donor, washings, skin and cumulative amount in receptor.

3.5. Pharmacokinetic Evaluation

A suitable animal model can be used for in vivo pharmacokinetic assessment of a topically applied elastic liposome formulation. Rats and mice are most common but tend to provide an over-estimate compared to human skin as the skin is more permeable. Piglets provide a closer approximation to human skin permeability and have been used for some evaluations of elastic liposomes (9, 34). Appropriate animal ethics committee approval is required. A generalised protocol is outlined in the following section but there can be considerable variation depending on the complexity of information sought, i.e., if the determination of distribution into tissues is required in addition to absorption into the circulation:

1. Animals are anaesthetised and liposome suspension applied to a pre-marked test site without occlusion.

2. At a predetermined time, the animal is sacrificed and a blood sample is collected.

3. The test site is tape stripped using D-Squame discs or a suitable alternative adhesive tape. Tape is applied with uniform pressure and removed with a single fluent pull.

4. Subcutaneous and muscle tissue are then dissected and collected separately taking care not to contaminate the deeper tissues during dissection.

5. Extract analytical solutions from all samples using suitable pre-validated solvent extraction procedures. Drug content in all samples should be analysed by a validated HPLC assay or where possible, by scintillation counting if a radiolabelled active compound is available.
6. Suitable control formulations such as plain drug solution and possibly conventional liposome suspension should be applied to additional animals and processed with the same protocol.
7. Drug in plasma verses time post application is plotted and pharmacokinetic parameters of AUC_{0-24h} (area under curve), C_{max} (peak plasma level), t_{max} (time to peak plasma level) and $t_{1/2}$ (plasma half-life) are determined.
8. Drug content in skin (tape strips grouped e.g. 1 and 2 as drug remaining on skin surface, 3–10 and 11–20), subcutaneous tissue and muscle can be compared.

3.6. Summary

It is clear from the body of available literature that elastic liposomes can deliver enhanced amounts of both small and large therapeutic agents into the skin. There is also some evidence for enhanced systemic delivery of small molecules in some cases. Elastic liposomes are relatively easy to formulate and characterise, though a considerable range of liposome forming materials and methods is utilised. The elastic liposome concept can be applied to a variety of compositions, with the potential to optimise the skin deposition and permeability of a range of therapeutic molecules. The exact mechanism of transport of elastic liposomes remains to be elucidated and conclusive evidence for transport of intact liposomes beyond the stratum corneum is lacking. However, increasing applications of enhanced delivery by elastic liposome formulations are being reported, with some Transfersome® products now in advanced clinical trials, such as Diractin (Idea AG, Munich) for enhanced delivery of ketoprofen in the management of osteoarthritis.

4. Notes

1. Phosphatidylcholine of purity at least 90% should be used to form stable liposomes. Lyophilised powder – store in freezer at −20°C.
2. If using a different lipid such as dipalmitoylphosphatidylcholine, or if cholesterol is incorporated in the liposome formulation, the phase transition temperature will be elevated above room temperature. In this case, a suitable temperature just above the phase transition temperature should be used.

3. Ethanol is a suitable solvent for soya phosphatidylcholine. If using other lipids or incorporating cholesterol, ethanol may not dissolve the lipids and alternate solvent systems may be required.

4. If a water-soluble drug is being incorporated, it should be added in the aqueous solution used to hydrate the film and form the liposomes.

5. Liposomes should be freshly prepared.

6. Alternatively radiolabelled estradiol could be used and content analysed by liquid scintillation counting.

7. Some laboratories use cadaver skin. Skin from at least two donors should be used, with skin from each donor used in equal numbers of cells.

8. A variety of other diffusion cells are also available including automated cells with flow-through receptor chambers.

9. It is likely that no suspension will remain on the skin surface as a finite dose was applied without occlusion, therefore, solvent will have evaporated.

References

1. Honeywell-Nguyen PL, Bouwstra JA (2005) Vesicles as a tool for transdermal and dermal delivery. Drug Discov Today Technol 2(1): 67–74
2. Dubey V, Mishra D, Nahar M, Jain NK (2007) Vesicles as tools for the modulation of skin permeability. Expert Opin Drug Deliv 4(6): 579–593
3. Elsayed MM, Abdallah OY, Naggar VF, Khalafallah NM (2007) Lipid vesicles for skin delivery of drugs: reviewing three decades of research. Int J Pharm 332(1–2):1–16
4. Cevc G, Blume G (2001) New, highly efficient formulation of diclofenac for the topical, transdermal administration in ultradeformable drug carriers, Transfersomes. Biochim Biophys Acta 1514(2):191–205
5. El Maghraby GM, Williams AC, Barry BW (1999) Skin delivery of oestradiol from deformable and traditional liposomes: mechanistic studies. J Pharm Pharmacol 51(10): 1123–1134
6. Trotta M, Peira E, Carlotti ME, Gallarate M (2004) Deformable liposomes for dermal administration of methotrexate. Int J Pharm 270(1–2):119–125
7. Cevc G (2003) Transdermal drug delivery of insulin with ultradeformable carriers. Clin Pharmacokinet 42(5):461–474
8. Cevc G, Blume G (1992) Lipid vesicles penetrate into intact skin owing to the transdermal osmotic gradients and hydration force. Biochim Biophys Acta 1104(1):226–232
9. Cevc G, Mazgareanu S, Rother M (2008) Preclinical characterisation of NSAIDs in ultradeformable carriers or conventional topical gels. Int J Pharm 360(1–2):29–39
10. Cevc G, Mazgareanu S, Rother M, Vierl U (2008) Occlusion effect on transcutaneous NSAID delivery from conventional and carrier-based formulations. Int J Pharm 359(1–2): 190–197
11. Boinpally RR, Zhou SL, Poondru S, Devraj G, Jasti BR (2003) Lecithin vesicles for topical delivery of diclofenac. Eur J Pharm Biopharm 56(3):389–392
12. Bouwstra JA, De Graaff A, Groenink W, Honeywell L (2002) Elastic vesicles: interaction with human skin and drug transport. Cell Mol Biol Lett 7(2):222–223
13. Honeywell-Nguyen PL, Gooris GS, Bouwstra JA (2004) Quantitative assessment of the transport of elastic and rigid vesicle components and a model drug from these vesicle formulations into human skin in vivo. J Invest Dermatol 123(5):902–910
14. Cevc G, Schatzlein A, Blume G (1995) Transdermal drug carriers: basic properties,

optimization and transfer efficiency in the case of epicutaneously applied peptides. J Control Rel 36:3–16

15. Cevc G (1996) Transfersomes, liposomes and other lipid suspensions on the skin: permeation enhancement, vesicle penetration, and transdermal drug delivery. Crit Rev Ther Drug Carrier Syst 13(3–4):257–388

16. Cevc G, Blume G, Schatzlein A, Gebauer D, Paul A (1996) The skin: a pathway for systemic treatment with patches and lipid-based agent carriers. Adv Drug Deliv Rev 18(3):349–378

17. Cevc G, Schatzlein A, Richardsen H (2002) Ultradeformable lipid vesicles can penetrate the skin and other semi-permeable barriers unfragmented. Evidence from double label CLSM experiments and direct size measurements. Biochim Biophys Acta 1564(1):21–30

18. Honeywell-Nguyen PL, Wouter Groenink HW, de Graaff AM, Bouwstra JA (2003) The in vivo transport of elastic vesicles into human skin: effects of occlusion, volume and duration of application. J Control Rel 90(2):243–255

19. van den Bergh BA, Vroom J, Gerritsen H, Junginger HE, Bouwstra JA (1999) Interactions of elastic and rigid vesicles with human skin in vitro: electron microscopy and two-photon excitation microscopy. Biochim Biophys Acta 1461(1):155–173

20. van den Bergh BA, Bouwstra JA, Junginger HE, Wertz PW (1999) Elasticity of vesicles affects hairless mouse skin structure and permeability. J Control Rel 62(3):367–379

21. Trotta M, Peira E, Debernardi F, Gallarate M (2002) Elastic liposomes for skin delivery of dipotassium glycyrrhizinate. Int J Pharm 241(2):319–327

22. El Maghraby GM, Williams AC, Barry BW (2000) Skin delivery of oestradiol from lipid vesicles: importance of liposome structure. Int J Pharm 204(1–2):159–169

23. El Maghraby GM, Williams AC, Barry BW (2000) Oestradiol skin delivery from ultradeformable liposomes: refinement of surfactant concentration. Int J Pharm 196(1):63–74

24. Song YK, Kim CK (2006) Topical delivery of low-molecular-weight heparin with surface-charged flexible liposomes. Biomaterials 27(2):271–280

25. Kirjavainen M, Urtti A, Jaaskelainen I, Suhonen TM, Paronen P, Valjakka-Koskela R et al (1996) Interaction of liposomes with human skin in vitro – the influence of lipid composition and structure. Biochim Biophys Acta 1304(3):179–189

26. El Maghraby GM, Williams AC, Barry BW (2004) Interactions of surfactants (edge activators) and skin penetration enhancers with liposomes. Int J Pharm 276(1–2):143–161

27. Jain S, Sapre R, Tiwary AK, Jain NK (2005) Proultraflexible lipid vesicles for effective transdermal delivery of levonorgestrel: development, characterization, and performance evaluation. AAPS PharmSciTech 6(3):E513–E522

28. Cevc G, Blume G, Schatzlein A (1997) Transfersomes-mediated transepidermal delivery improves the regio-specificity and biological activity of corticosteroids in vivo. J Control Rel 45(3):211–226

29. Honeywell-Nguyen PL, Bouwstra JA (2003) The in vitro transport of pergolide from surfactant-based elastic vesicles through human skin: a suggested mechanism of action. J Control Rel 86(1):145–156

30. Elsayed MM, Abdallah OY, Naggar VF, Khalafallah NM (2006) Deformable liposomes and ethosomes: mechanism of enhanced skin delivery. Int J Pharm 322(1–2):60–66

31. Dubey V, Mishra D, Asthana A, Jain NK (2006) Transdermal delivery of a pineal hormone: melatonin via elastic liposomes. Biomaterials 27(18):3491–3496

32. Fry DW, White JC, Goldman ID (1978) Rapid separation of low molecular weight solutes from liposomes without dilution. Anal Biochem 90(2):809–815

33. Kligman A, Christophers E (1963) Preparation of isolated sheets of human stratum corneum. Arch Dermatol 88:70–73

34. Cevc G, Vierl U, Mazgareanu S (2008) Functional characterisation of novel analgesic product based on self-regulating drug carriers. Int J Pharm 360(1–2):18–28

Chapter 5

Archaebacterial Tetraetherlipid Liposomes

Aybike Ozcetin, Samet Mutlu, and Udo Bakowsky

Abstract

Liposomes are widely investigated for their applicability as drug delivery systems. However, the unstable liposomal constitution is one of the greatest limitations, because the liposomes undergo fast elimination after application to the human body. In the presented study, novel archeal lipids were used to prepare liposomal formulations which were tested for their stability at elevated temperatures, at different pH-values and after heat sterilization.

Key words: AFM, GDNT, Tetraether lipids, Liposome, Sulfolobus acidocaldarius, Archae

1. Introduction

Liposomes are bilayered spherical membrane structures which are promising for pharmaceutical and diagnostic use (1). However, there are limitations of the usage of liposomes which is mainly caused by their low stability (2). To reduce this problem, in this study an archael lipid, Glycerol Dialkyl Nonitol Tetraether (GDNT) (see Fig. 1) is chosen to prepare highly stable liposomes. The GDNT is isolated from *Thermoacidophilic Archaebacteria* genus *Sulfolobus acidocaldarius* (3). Archaea are one of the three major domains of life (4). The major difference of these archaea from bacterial and eukaryotic cells is their membrane lipids (5). Archaea do not have any cell wall nevertheless the properties of the cell membrane lipid provide a remarkable long-term stability (6). The archaea are divided into phenotypes of methanogens, halophiles and thermophiles. The lipid used in this study is from a type of thermophilic archaea. *Sulfolobus acidocaldarius* is a well studied extremely thermophilic archae with the optimal growth conditions of 70–80°C and pH 3 (7, 8).

These archael membrane lipids consist of monopolar ether head groups and saturated, branched phytanyl chains which are

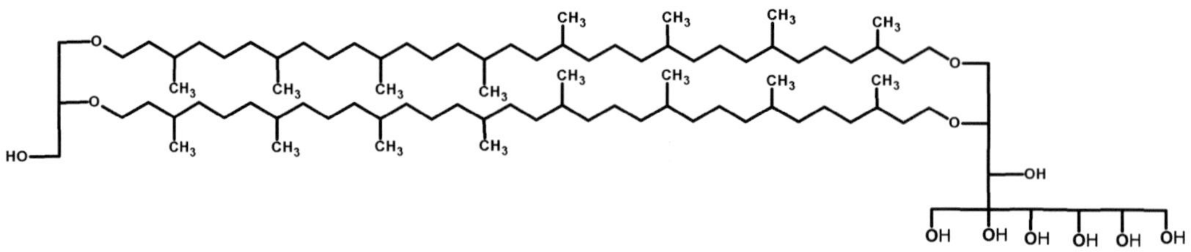

Fig. 1. *GDNT* Glycerol Dialkyl Nonitol Tetraether

mainly attached to the glycerol backbone carbons through ether bonds. This chemical structure provides high stability due to low oxidation capacity and hydrolytic resistance (9).

Liposomes can incorporate hydrophilic and lipophilic drugs and reduce the overall dose and amount of side effects by specific targeting. Also, the relatively stable structure of GDNT liposome offers new range of applications. Liposomal formulations made of GDNT are able to protect the pharmaceuticals from biochemical degradation or metabolism. Therefore, they are interesting for oral applications. Kimura reviewed that a significant quantity of drug entrapped into liposomes can be absorbed by the small-intestinal mucosa (10). However, for oral drug delivery, there are still some questions concerning the stability of liposomes in the acidic milieu of the gastrointestinal tract and their absorption therein (11).

Pulmonary drug delivery is also applicable using liposomes, because they are absorbed through the thin layer of alveolar epithelial cells (vast surface for an adult is 43–102 m^2) (12–14) and transported into the systemic circulation (15–17). However, the inhaled objects can be eliminated by macrophages from the alveoli surface when they are bigger than 260 nm (18).

For the future aspects of tetraetherlipid liposomes, there are still some details to be clarified. This article is concerned with the stability of GDNT liposomes which certainly plays a fundamental role in liposomal delivery.

2. Materials

2.1. Extraction and Hydrolysis of the Tetraether Lipids

1. 25 g Freeze-dried biomass from *Sulfolobus acidocaldarius* (BHP Billiton, Global Technology, Perth Technology Centre, Australia), store at –20°C.
2. Chloroform ($CHCl_3$): methanol (MeOH):5% trichloroacetic acid (TCA) mixture (1:2:1 v/v). Store at RT.
3. MeOH: H_2O mixture (1:1 v/v). Store at RT.
4. 100 ml 1 M methanolic hydrochloric acid. Store at RT.
5. 8 M KOH is prepared to adjust the pH level to pH 14 and to adjust the pH level to pH 3 32% HCl is prepared. Store at RT.

2.2. Separation of the GDNT

1. A chromatography column with a diameter of 4 cm (conventional glass) is filled with 300 g silica gel 60 (Merck, Germany) for separation of lipid fractions.
2. The eluents (1.5 l each) used are $CHCl_3$ followed by $CHCl_3$:diethyl ether (8:2 v/v) and $CHCl_3$: MeOH (8:2 v/v), according to Lo et al. (19).
3. $CHCl_3$: MeOH, (9:1 v:v) is prepared for thin layer chromatography.

2.3. Liposome Formulations

1. Phospholipon 90G (Lipoid, Germany) is dissolved in $CHCl_3$: MeOH (3:1 v/v) to achieve a final concentration of 10 mg/ml solution.
2. For hydrolysis, ultrapure, bidistilled water (pH 5.5) is used.
3. Nylon syringe filters with the pore size of 0.45 μm are supplied from Rotilabo Roth (Carl Roth Karlsruhe Germany).
4. The Extruder is supplied from Avestin Europa GmbH with polycarbonate membranes of 19 mm diameter and 100 nm pore diameter.

2.4. Stability Tests of Liposomes

1. For the pH stability studies pH 2.0 solution (50 ml 0.2 M KCl is adjusted with 0.1 M HCl), pH 4.0 solution, (100 ml 0.1 M potassium hydrogen phthalate is adjusted with 0.1 M HCl) pH 7.4 buffer (129 mM NaCl, 2.5 mM KCl, 7.4 mM Na_2HPO_4, 1.3 mM KH_2PO_4) and pH 9.0 (100 ml 0.025 M NaB_4O_7, 9.2 ml 0.1 M HCl) are prepared.

2.5. Atomic Force Microscopy

1. As substrate for the sample preparation, silicon wafers from Wacker Chemie AG (Munich, Germany) with a natural silicon oxide layer (thickness 3.8 nm) and a surface roughness of 0.3 nm is used. The wafers are split into small pieces of about 1 × 1 cm. The pieces are cleaned in a bath sonicator for 20 min in $CHCl_3$: MeOH (2:1, v:v), then they are dried in a air stream and stored in a dust free atmosphere.
2. NSC 16/Cr-Au cantilevers from Anfatec Instruments AG (Oelsnitz, Germany) with a nominal force constant of 45 N/m, resonant frequency of 170 kHz and a length of 230 μm are used. The sharpness of the tip is less than 10 nm.

3. Methods

3.1. Extraction and Hydrolysis of the GDNT Lipids

1. Different methods for the extraction and purification of the GDNT are established. We used the combined extraction and hydrolysis method according to Bode et al. (20)

2. 25 g freeze-dried biomass from Sulfolobus acidocaldarius grown at 68°C, is transferred into a 1 l flask and into this flask 400 ml of a $CHCl_3$: MeOH: TCA mixture is added.

3. The mixture is heated to a gentle reflux temperature of 60°C for 2 h.

4. After cooling down to room temperature, the mixture is filtered. The filter cake is transferred back to the flask and kept at 60°C in the same solvent mixture as before. This procedure is continued for three more cycles.

5. The collected extractions (upper green layer of precipitate and a clear brown colored solution) are washed with a mixture of MeOH: H_2O (1:1, 800 ml).

6. From the combined chloroform extracts, the solvent is evaporated in vacuum at elevated temperatures.

7. 100 ml 1 M methanolic HCl is added to the residue and this is heated at 80°C for 16 h.

8. Then the reaction mixture is cooled down to room temperature and 100 ml water is added. The pH of the resultant mixture is adjusted to 14 using 8 M KOH.

9. The mixture is subjected to base hydrolysis at 80°C for 1 h, and cooled down to room temperature. Subsequent to this, the mixture is adjusted to pH 3 with 32% HCl.

10. The resultant mixture is extracted with $CHCl_3$ (3 × 200 ml). The chloroform extract is separated, dried with magnesium sulfate and solvent evaporation under vacuum to yield a brown lipid residue.

3.2. Separation of the GDNT

1. The separation of 8 g of the hydrolysed lipid fraction from the total lipid extraction is done via silica gel 60 column chromatography.

2. The samples are collected (50 ml fractions) and their lipid composition is analyzed by thin layer chromatography. The lipids are stained by the use of methanolic sulfuric acid followed by an ashing process. The lipid is visible as dark spots. GDNT shows a R_f value of 0.45 ($CHCl_3$: MeOH, 9:1, v:v) in accordance to literature (0.45 (19) and 0.35 (20)).

3. The fractions containing GDNT are collected and the solvents are evaporated at elevated temperature. The purified lipid is stored at −20°C before use.

3.3. Preparation of Liposomes

1. Preparation of liposomes conducted with different molar ratios of GDNT and Phospholipon. As a stock solution, Phospholipon is dissolved in chloform:methanol (2:1, v/v) and 10 mg/ml solution is prepared.

2. Different liposome compositions are prepared (see Fig. 2).

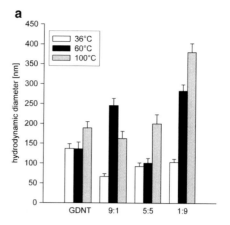

Formulation (molar ratio)	Mean Diameter [nm] (PDI)	Zeta-Potential
1:9 GDNT:Phospholipon	96.6 (0.197)	−14.6±0.285
5:5 GDNT: Phospholipon	88.2 (0.224)	−11.2±0.127
9:1 GDNT: Phospholipon	69.0 (0.223)	−5.28±0.079
10:0 GDNT:Phospholipon	137.5 (0.295)	−15.3±0.60

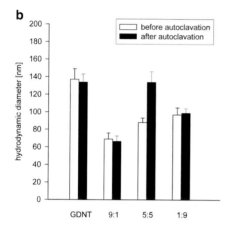

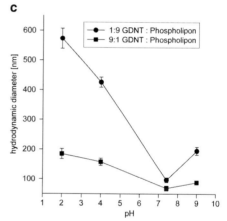

Fig. 2. Different liposomal formulations containing the tetraether lipid GDNT are characterized regarding their stability. The initial diameter ranged from 69 nm for the mixture 9:1 GDNT: Phospholipon to 137.5 nm for the pure GDNT. The negative zeta potential is representative for all liposomal formulations. (**a**) The size stability of the liposomes is tested at different temperatures. The measurements are done after incubating 40 μl liposome solutions in 240 μl bidestilled water for 4 h. The results show the stabilizing effect of GDNT content in the liposome structure. (**b**) The diagram represents the stability of particles after autoclavation. The samples are incubated 15 min at 121°C and 29 psi for sterilization. (**c**) Both of the samples are investigated after incubating 4 days in different pH solutions. Standard deviations are calculated from three independent measurements

3. The liposome preparation procedure is based on film formation and hydration. After transferring the mixture of Phospholipon and GDNT solutions into 10 ml round bottom flasks, the chloroform-methanol solution is removed by evaporation at 300 mbar and 45°C to obtain a lipid film.

4. For hydration, bidistilled water is used to prepare the liposomes in the concentration of 10 mg/ml. To form the lipid vesicles, a bath sonicator at 45°C is used. After obtaining a dispersion of the lipid in water, sonication is continued with a probe type sonicator to increase the energy input. For the following processes, the sample is transferred into a 50 ml plastic tube.

5. Sonication is continued for 8 min (30 s sonication followed by 30 s rest). The power was set to level 6. (see Notes 1 and 2). If a clear dispersion is not achieved, the sonication can be continued.

6. After sonication the samples are filtered by syringe filters with the pore size of 0.45 µm to separate large vesicles that may blockade the extruder membrane (see Note 3).

7. For the preparation of ~100 nm liposomes, the formulations are extruded through a 0.1 µm polycarbonate membrane. Preheating of the extruder above the main phase transition of the lipid mixture (40°C) is essential to provide an effective liposome extrusion (see Note 4). The extrusion is conducted 21 times to each sample. The achieved liposomes have a diameter of 100–150 nm. If the initial size measurements verify larger diameters, the extrusion can be repeated 11 times more.

8. All the size measurements in this study are performed on a NanoZS (Malvern instruments GmbH, Germany). The diluted liposomal formulations (10:60, v:v, bidistilled water) are measured in a micro cuvette (Malvern instruments GmbH, Germany).

9. Zeta potential measurements are also performed on the NanoZS. All samples are diluted 1:10 (v:v) with water. A folded electrophoresis cell is used (Malvern instruments GmbH, Germany) for Laser Doppler Anemometry (LDS) measurements.

10. The cuvette or electrophoresis cell is placed in the NanoZS and allowed to equilibrate to the pre-set temperature (25°C).

11. The manufacturer's software automatically adjusts the Laser attenuation and measurement position. For each size and zeta potential determination, 3 measurements consisting of 6 sub runs with a duration of 10 s are averaged.

12. In Fig. 2, the initial diameter and zeta potential values of liposome formulations are presented.

3.4. Stability Testing

3.4.1. Thermostability Testing

1. After measuring the initial diameters and the zeta potentials of liposomes, from each composition, 100 µl aliquots are mixed with 600 µl bidistilled water in a glass tube.

2. The samples are transferred into a metal tube holder and incubated at 36°C, 60°C, 100°C for 4 h in a cabinet heater.

3. After incubation, when the samples are at room temperature, the diameter values are determined with NanoZS.

4. The results of thermostability tests are presented in Fig. 2.

3.4.2. Stability During Autoclavation

1. The autoclavation is performed on a standard autoclave 3850 ELC (Systec GmbH, Germany).

2. Due to the properties of high thermostability the autoclavation of liposomes were considered to be applicable, which brings large sterilization possibilities along with.

3. Before autoclavation, the liposome samples were diluted with bidestilled water in the ratio of 10:60 which is the same dilution value to be used in zeta-size measurements.

4. The samples were loaded into the autoclave and treated with a standard autoclave procedure for solutions (15 min at 121°C, 29 psi pressure, saturated steam).

5. The size measurements in Fig. 2. It could be shown that the liposomal formulations containing GDNT can be autoclaved. The diameter is relatively constant.

3.4.3. pH Stability

As a first test for the stability of the formulations in the gastrointestinal tract, liposomes are incubated in solutions of different pH-values mimicking the gastrointestinal environment.

1. pH stability tests are performed by incubating the liposomes in different solutions. From each liposomal formulation, 100 μl aliquots are incubated in 600 μl of pH 2.0, pH 4.0, pH 7.4, pH 9.0 buffer solutions.

2. The initial diameters and zeta potentials are measured with a NanoZS (Malvern Instruments, Germany, HeNe Laser 633 nm, 173°C scattering angle, 25°C) right after mixing the liposomes with buffer solutions.

3. After 4 days of incubation, hydrodynamic diameters are determined as it is explained in Chap. 3.3.8.

4. With the size measurements, the effect of GDNT stabilizing effect is shown. The results are presented in Fig. 2. With the increase of GDNT molar ratio, the liposome size is much more stable compared to the low molar content of GDNT.

3.5. Atomic Force Microscopy

1. Atomic force microscopy (AFM) is performed on a Digital Nanoscope IV Bioscope (Veeco Instruments, Santa Barbara, CA). The AFM is vibration and acoustically damped (21). All measurements are performed in tapping mode. The applied force to the sample surface is adjusted to a minimum to avoid the damage of the sample. The sample is investigated and scanned under constant force. The scan speed is proportional to the scan size and the scan frequency is between 0.5 and 1.5 Hz. Images are obtained by displaying the amplitude signal of the cantilever in the trace direction, and the height signal in the retrace direction, both signals being simultaneously recorded. The results are visualized either in height (the real height of the sample in a resolution on 0.3 nm) or in amplitude mode (the damping of the frequency signal).

2. Various methods are available for sample preparation. A very convenient procedure for the preparation of liposomes is the self-assembly technique (22). Small pieces of silicon wafers (about 1 × 1 cm) as substrate material are placed into the sample dispersion for 20 min at room temperature and the liposomes are allowed to adsorb to the surface under equilibrium conditions. The silicon substrates are removed from the dispersions. The samples are dried at room temperature and investigated within 2 h.

3. The liposomes containing GDNT are stable against the substrate surface, while conventional liposomes tend to spread to the surface and form a supported lipid bilayer as shown in Fig. 3.

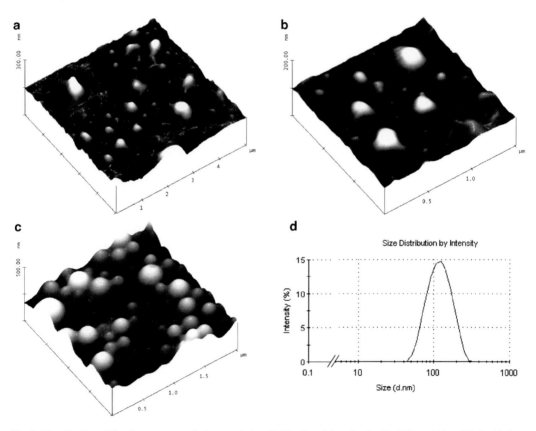

Fig. 3. Visualization of the liposome morphology and size distribution determined with AFM and NanoZS. (**a, b**) Pure phospholipon liposomes adhered on silicon wafer as substrate. The liposomes have diameters between 80 and 250 nm with an average diameter of 178 ± 12 nm. The liposomes tend to spread to the surface, because of the low membrane stability. (**c**) Pure GDNT liposomes with an average diameter of 137 ± 8 nm (PDI 0.295 ± 0.017) and a zeta potential of −15.3 ± 0.60 mV. The liposomes are stable and show a spherically, round shape. (**d**) Size distribution of the pure GDNT liposomes measured with NanoZS

4. Notes

1. The container should be small enough so that the sonicator probe can immerse deeply (1–2 cm) in the sample but large enough so that the probe does not touch the sides or bottom of the container. 50 ml plastic tubes with round bottoms are used.
2. Probe type sonication heats the solution very quickly; to avoid damage of lipids and liposomes, an ice bath in a beaker is prepared and placed securely on a ring stand in the sound proof box of the sonicator.
3. For the filtration of liposome solutions the whole sample is transferred through the filter into plastic tubes.
4. When working with an extruder, it is important to assure that the retainer nuts are tight to avoid the leaking of the liposome solutions.

References

1. Torchilin VP (2005) Recent advances with liposomes as pharmaceutical carriers. Nat Rev Drug Discov 4:145–160
2. Sharma A, Sharma US (1997) Liposomes in drug delivery: progress and limitations. Int J Pharm 154:123–140
3. Brock T, Brock K, Belly R, Weiss R (1972) Sulfolobus: a new genus of sulfur-oxidizing bacteria living at low pH and high temperature. Arch Mikrobiol 84:54–68
4. Woese CR, Kandler O, Wheelis LL (1990) Towards a natural system of organisms: proposal for the domains archaea, bacteria and eucarya. Proc Natl Acad Sci 87:4576–4579
5. De Rosa M, Gambacorta A (1988) The lipids of archaebacteria. Prog Lipid Res 27:153
6. Gambacorta A, Gliozzi A, DeRosa M (1995) Archaeal lipids and their biotechnological application. World J Microbiol Biotechnol 11:115–131
7. Brock TD, Brock KM, Belly RT, Weiss RL (1972) *Sulfolobus*: A new genus of sulfur-oxidizing bacteria living at low pH and high temperature. Arch Microbiol 84:54–68
8. Grogan DW (1989) Phenotypic characterization of the archeabacterial genus *Sulfolobus*: comparison of five wild-type strains. J Bacteriol 171:6710–6719
9. Kates M (1978) The phytanyl ether-linked polar lipids and isoprenoid neutral lipids of extremely halophilic bacteria. Prog Chem Fats Other Lipids 15:301–342
10. Kimura T (1988) Transmucosal passage to liposomal drugs. In: Gregoriadis G (ed) Liposomes as drug carriers. Wiley, Chichester, pp 635–647
11. Gilligan CA, Li Wan Po A (1991) Oral vaccines: design and delivery. Int J Pharma 75:1–24
12. Weibel ER (1973) Morphological basis of alveolar–capillary gas exchange. Physiol Rev 53:419–495
13. Gehr P, Bachofen M, Weibel ER (1978) The normal human lung: ultrastructure and morphometric estimation of diffusion capacity. Respir Physiol 32:121–140
14. Stone KC, Mercer PR, Gehr P (1992) Allometric relationships of cell numbers and size in the mammalian lung. Am J Respir Cell Mol Biol 6:235–243
15. Kellaway IW, Farr SJ (1990) Liposomes as drug delivery systems to the lung. Adv Drug Del Rev 5:149–161
16. Patton JS, Fishburn CS, Weers JG (2004) The lung as a portal of entry for systemic drug delivery. Proc Am Thorac Soc 1:338–344
17. Patton JS, Byron PR (2007) Inhaling medicines: delivering drugs to the body through the lungs. Nat Rev Drug Discov 6:67–74

18. Niven RW (1995) Delivery of biotherapeutics by inhalation aerosol. Crit Rev Ther Drug Carrier Syst 12:151–231
19. Lo SL, Montague CE, Chang EL (1989) Purification of glycerol dialkyl nonitol tetraether from Sulfolobus acidocaldarius. J Lipid Res 6:944–949
20. Bode ML, Buddoo SR, Minnaar SH, du Plessis CA (2008) Extraction, isolation and NMR data of the tetraether lipid calditoglycerocaldarchaeol (GDNT) from Sulfolobus metallicus harvested from a bioleaching reactor. Chem Phys Lipids 154(2): 94–104
21. Oberle V, Bakowsky U, Zuhorn IS, Hoekstra D (2000) Lipoplex formation under equilibrium conditions reveals a three-step mechanism. Biophys J 79:1447–1454
22. Kneuer C, Ehrhardt C, Radomski MW, Bakowsky U (2006) Selectins-potential pharmacological targets? Drug Discov Today 11(21–22):1034–1040

Chapter 6

Cationic Magnetoliposomes

Marcel De Cuyper and Stefaan J.H. Soenen

Abstract

Magnetoliposomes (MLs) consist of nanosized, magnetisable iron oxide cores (magnetite, Fe_3O_4) which are individually enveloped by a bilayer of phospholipid molecules. To generate these structures, the so-called water-compatible magnetic fluid is first synthesized by co-precipitation of Fe^{2+} and Fe^{3+} salts with ammonia and the resulting cores are subsequently stabilized with lauric acid molecules. Incubation and dialysis of this suspension with an excess of sonicated, small unilamellar vesicles, ultimately, results in phospholipid-Fe_3O_4 complexes which can be readily captured from the solution by high-gradient magnetophoresis (HGM), reaching very high yields. Examination of the architecture of the phospholipid coat reveals the presence of a typical bilayered phospholipid arrangement. Cationic MLs are then produced by confronting MLs built up of zwitterionic phospholipids with vesicles containing the relevant cationic lipid, followed by fractionation of the mixture in a second HGM separation cycle. Data, published earlier by our group (Soenen et al., ChemBioChem 8:2067-2077, 2007) prove that these constructs are unequivocal biocompatible imaging agents resulting in a highly efficiënt labeling of biological cells.

Key words: Immobilized phospholipid membranes, Magnetite, Magnetisable biocolloids, Magnetisable liposomes, Magnetoliposomes (cationic), Magnetophoresis (high-gradient), Ultrasmall superparamagnetic iron oxides, Phospholipid vesicles, USPIOs

1. Introduction

A variety of (magnetic) nanoparticles have been used in the past 10 years to label biological cells (1, 2). Within this research topic, a crucial goal that has to be met by the nanocolloids under consideration are an efficient internalization by the cell without evoking toxic effects. In addition, a profound knowledge of the intracellular processing route, ultimately, may lead to the creation of a powerful imaging signal that remains stable over a long time period. These requirements are fulfilled when using cationic magnetoliposomes (MLs) (3, 4).

The first reports describing the production of well characterized magnetoliposomes date from the late 1980s (5, 6). The structures are built up of a nm-sized magnetite (Fe_3O_4) core, in which the original stabilizing lauric acid coat is first replaced by a phospholipid bilayer during incubation and dialysis of the fatty acid coated particle with preformed small unilamellar phospholipid vesicles (see Note 1). Mechanistically, it has been proven that the process of ML formation is controlled by the spontaneous transfer of phospholipids according to the so-called aqueous transfer model (7) (see Note 2). As MLs can be categorized as 'ultrasmall superparamagnetic iron oxides' (commonly designated as USPIOs), a high-gradient magnetic field (HGM) is needed to withdraw the MLs from the incubation mixture (8, 9). On the basis of a detailed structural analysis of the coat, we found earlier that the inner phospholipid leaflet is strongly chemisorbed on the iron oxide surface, whereas the outer one is more loosely bound by physisorption (7), allowing flexibility in modifying its characteristics (10–12). This feature is exploited in the generation of cationic MLs, where intermolecular transfer of lipids can occur in a mixture of neutral MLs and vesicles bearing, for instance, the cationic 1,2-diacyl-3-trimethylammonium-propane (13). Indeed, after fractionation of the mixture in a second HGM cycle, it is found that cationic lipid molecules, originally residing in the outer leaflet of the vesicles show up in the ML population.

2. Materials

All chemicals used have a pro analysis grade.
Note: Concentrated solutions of HCl, HNO_3, $HClO_4$, H_2SO_4, chromic acid and NH_4OH are either corrosive, explosive, highly toxic and/or hazardous; take the necessary precautions in handling these substances!

2.1. Magnetic Fluid

1. Concentrated solutions of $FeCl_2 \cdot 4aq$ and $FeCl_3 \cdot 6aq$ (Merck, Darmstadt, Germany) are strongly acidic. Be careful. Lauric acid (Sigma, Bornem, Belgium) is used as a solid product.
2. A permanent magnet, for instance, disassembled from a loudspeaker is most satisfactory.

2.2. Phospholipid Vesicles

1. Throughout the experiments, a 5 mM TES [2-((Tris (hydroxymethyl)-methyl)amino)ethanesulfonic acid; Sigma] buffer, pH 7.0 is used.
2. (Phospho)lipids : dimyristoylphosphatidylcholine (DMPC; MW 678), dimyristoylphosphatidylglycerol (DMPG; MW 689)

and 1,2-distearoyl-3-trimethylammonium-propane (DSTAP; MW 703) are from Avanti Polar Lipids (Birmingham, AL).

3. A probe-type ultrasonic disintegrator (MSE, 150 W), equipped with either an exponential (for 3 mL volumes) or a 3/8″ probe (for volumes up to 50 mL) is used to prepare phospholipid vesicles.

2.3. Magnetoliposomes

1. Magnetic fluid stock solution and vesicles (see Subheadings 3.1 and 3.2, respectively).

2. Dialysis membranes (MW cut-off 12.000; SpectraPor no. 2 dialysis tubing; Spectrum Laboratories, Medicell Laboratories, London, UK).

3. A water-cooled Bruker electromagnet (Type BE15) (Karlsruhe, Germany), equipped with pole pieces set at a distance of 3 mm from one another, is used. For ML fractionations (see Subheading 3.4) the instrument operates at 80 V and 30 A and produces under these conditions a magnetic field intensity of approximately 1.5 Tesla (see Note 3). The diameter of the pole caps is 15 cm, allowing more than 20 samples to be fractionated simultaneously.

4. The magnetic filter device consists of pieces of tubing (inner/outer diameter : 0.078/0.125 in.; Silastic Medical-grade Tubing, Midland, Michigan, USA) plugged with the stainless steel fibers (Type 430; Bekaert Steel Cooperation, Belgium) (see Note 4). It is of crucial importance to LOOSELY pack these fibers within the tubing so that during ML fractionation the shear force of the buffer stream passing through the filter is not too high; otherwise, capturing the MLs on the iron wires may be hampered – see Subheading 3.4). Excellent results are obtained with 60 mg loosely packed fibers put in the earlier-mentioned tubing of about 7 cm length (see Note 5).

To construct the magnetic filtration set up, the magnetic filter is installed in a conduit system through which buffer is pumped by means of a peristaltic pump (Fig. 1).

2.4. Phosphate and Iron Determination

Phosphate and iron dosages are done spectrophotometrically in the visible wavelength zone.

The chemicals for phosphate determination are sodium molybdate ($Na_2MoO_4 \cdot 2aq$) and hydrazinium chloride (NH_2-$NH_2 \cdot 2HCl$), both from Merck. The stock solution used to construct the calibration curve contains 250.70 mg $Na_2HPO_4 \cdot 12aq$/L (Merck), corresponding to a concentration of 0.700 μmol phosphate/mL.

The indicator substance for iron determination is Tiron [4,5-dihydroxy-1,3-benzenedisulfonic acid, disodium salt] (Acros Organics, Geel, Belgium). The stock solution used to prepare the calibration samples contains 1g Fe/L (solution of $Fe(NO_3)_3 \cdot 4aq$ in HNO_3 0.5N) (Panreac Quimica, Barcelona, SP).

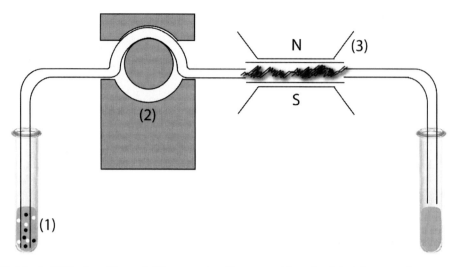

Fig. 1. Sketch of a HGM set-up. The crude ML preparation (1), recovered from the dialysis bag (see Subheading 3.3) is pumped by means of a peristaltic pump (2) through the magnetic filter device (3). The consecutive steps in the fractionation process are illustrated in Fig. 2

2.5 Electron micrographs are taken on a Zeiss EM10C apparatus. Copper grids (3.05 mm diameter; 200 mesh; Type G200-Cu) were from EMS (Aurion, Wageningen, The Netherlands). Uranyl acetate (UCB, Belgium) is used as a negative stain.

3. Methods

3.1. Preparation Recipe of the Magnetic Fluid (see Note 6)

The protocol, largely based on Khalafalla and Reimers's work (14), goes as follows:

1. In separate beakers 12 g $FeCl_2 \cdot 4aq$ and 24 g $FeCl_3 \cdot 6aq$ are each solubilized in 50 mL distilled water. (CAUTION: This solution is very acidic!) and then added together in a beaker of 250 mL (see Note 7).

2. Slowly add 50 mL of concentrated NH_4OH 56% (=28% NH_3), meanwhile rapidly stirring the solution with a corrosion resistant mechanical paddle. The Fe_3O_4 cores produced in this way are not stabilized and sink as a heavy, black precipitate.

3. Start the first magnetic decantation step by putting the beaker on a (permanent) magnet for about 15 min, and then pour off the clear supernatant while keeping the recipient positioned on the magnet.

4. Wash the precipitate with 100 mL containing 5 mL NH_4OH_{conc} and 95 mL water, and decant again after about 15 min. Repeat this step four times.

5. Heat the gel-like slurry in a boiling water bath to 80–90°C (temperature of precipitate, not the water bath!) and add 6 g of the lauric acid surfactant in solid form, meanwhile stirring with a glass rod. As a result of particle peptization the slurry becomes gradually liquefied and slightly foams. Heating is continued until foam formation ceases (after approximately 7 min).

6. Add 50 mL of water and centrifuge the solution at 500 g for 7 min to remove large clusters which might have formed. The supernatant is collected and stored.

Using the above-described experimental conditions, a stock solution of about 114 mg Fe_3O_4/mL is obtained. If desired, a further dilution with water can be executed without inducing precipitation. This is why a solution of the water-adapted, lauric acid-coated iron oxide cores is called a 'dilution-insensitive' magnetic fluid (14) (see Note 8).

3.2. Phospholipid Vesicles

1. Powdered phospholipids (e.g., DMPC, DMPG or mixtures thereof – see Notes 9 and 10) are weighted directly in a thermostable sonication vial and solubilised either in a minimum volume of chloroform or a chloroform/CH_3OH mixture. Otherwise, in case of phospholipid stock solutions in organic solvents, the desired amounts are pipetted. To improve solubilisation the vial can be set in a water bath at higher temperature (see Note 11).

2. The organic solvent is removed by evaporation in a stream of N_2, thereby depositing a thin lipid film on the glass wall. Care must be taken that the solvent is completely removed, which can be a hard job for the last residual solvent molecules. Routinely, the sonication vial is set overnight in an exsiccator under high vacuum.

3. The lipid film is then hydrated and solubilised in 5 mM TES Buffer, pH 7.0. Depending on experimental requirements, the final phospholipid concentration is usually taken between 0.1 and 20 mg/mL (see Note 12).

4. The transducer tip of the sonicator is immersed in the suspension at least 1 cm under the surface. Avoid to touch the glass walls during operation.

5. Sonication is done at 18 μm peak-to-peak power intensity (corresponding to about 75% of full power) at a temperature above the phase transition temperature (see Note 13).

6. Titanium micro debris lost from the tip during sonication is removed by centrifugation at 500 g for 10 min above the transition temperature (see Note 14).

7. If desired, the precise phospholipid concentration can be calculated by means of a phosphate determination (see Subheading 3.5.1).

3.3. Magnetoliposome Preparation

1. 200 mg DMPC (or DMPG or DMPC-DMPG mixtures) is sonicated in 15 mL 5 mM TES buffer, pH 7.0 (see Subheading 3.2 and Notes 9 and 10).
2. 0.35 ml of the laurate stabilized magnetic fluid is added (see Note 15). The solution is mixed and transferred into dialysis tubes.
3. Dialysis against 5 mM TES buffer is done for at least 3 days with very frequent buffer changes (at least 15 times) above the transition temperature of the phospholipids used. During this step, the laurate molecules are slowly removed from the iron oxide surface and concomitantly replaced by phospholipid molecules (6, 7). At the end, a clear solution without any precipitate should be obtained.

3.4. High-Gradient Magnetophoresis (Fig. 2)

1. Position the magnetic filter device(s) in the 3 mm-gap between the two pole faces of the electromagnet.

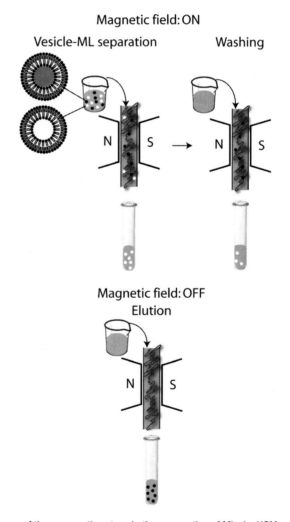

Fig. 2. Scheme of the consecutive steps in the preparation of MLs by HGM

2. Through each magnetic filter 0.75 mL of the sample is pumped at a rate of 12 mL/h.

3. A further washing of the retentate with 0.75 mL of TES buffer is necessary to remove iron oxide-free vesicles, which otherwise remain between the filter wires by capillarity.

4. The peristaltic pump and the magnetic field are switched off (in this sequence!).

5. The retentate is flushed out of the filter by the shear force of a buffer stream at high speed (0.5 L/h) and collected. Gentle squeezing of the pieces of tubing containing the magnetic filter may facilitate the release of the MLs from the filter. In general, the recovery expressed in terms of Fe_3O_4 is 95–99% (see Note 16). The stability of the ML preparation significantly improves by keeping the MLs above the gel-to-liquid crystalline phase transition temperature of the constituting phospholipid(s).

3.5. Magnetoliposome Characterization

In the following section, a few means are presented allowing to check the quality of the ML coat. Other ones (construction of adsorption isotherms, detergent extractions, …) can be found in the literature (6).

3.5.1. Phospholipid/Fe_3O_4 Ratio

For a 14 nm-diameter iron oxide core covered with an intact phospholipid bilayer, typically, a phospholipid /Fe_3O_4 (mmol/g) ratio between 0.7 and 0.8 is calculated (see Note 17). The recipies to measure phosphate and iron content of the ML samples are described in the following section:

1. Phospholipid determination is done by the phosphomolybdenum blue method (15). All measurements (calibration + unknown samples) are done in triplicate (see Note 18). The following protocol is followed:

    (a) First, a series of calibration solutions is prepared starting from a stock solution containing 0.700 μmol phosphate/mL, which is then diluted 2.5, 5, 7.5 and 10 times.

    (b) Then, to 100 μL of both the calibration solutions and the ML samples (if need be diluted to keep the phospholipid concentration below 0.700 μmol/mL when working with highly concentrated samples), 100 μL $HClO_4$ is added and the mixture is heated for 45 min at 180–200°C (see Note 19). During this chemical digestion process white fumes are circling around in the tubes.

    (c) After cooling to ambient temperature, 1 mL of a *working* reagent is added, which is prepared based on a sodium molybdate-containing stock solution. The following sequence is followed to prepare the latter solution. (i) 400 mg $NH_2NH_2 \cdot 2HCl$ dissolved in 14 mL HCl 4N is mixed with a solution containing 10 g Na_2MoO_4 in 60 mL HCl 4N (see Note 20), (ii) the mixture is heated for 20 min in a bath at 60°C, and (iii) after cooling, 14 mL of concentrated sulphuric acid is added slowly, meanwhile further cooling and vigorously mixing, (iv) the final volume is adjusted to 100 mL with water. To prepare the working reagent, 5.5 mL of the above-described stock solution is mixed with 26 mL H_2SO_4 1N and further diluted with distilled water to reach a final volume of 100 mL (see Note 21).

(d) Reaction of the *working* solution (1 mL) with the digested (calibration + unknown) samples occurs in a boiling water bath for 15 min. After cooling 1.5 mL H_2SO_4 1N is added to each tube and the absorbance is measured at 820 nm (see Note 22). In case the absorption of the sample solution(s) is too high, it is diluted with H_2SO_4 1N.

2. Iron determination

The method used is based on complex formation of Fe^{3+} with Tiron producing a red color which can be measured spectrophotometrically at 480 nm (16). Common plastic 3 mL-tubes are used. All determinations are done in triplicate.

(a) First, a calibration curve is constructed starting from a stock solution, containing 1 mg Fe^{3+}/mL, which is diluted with HCl_{conc} (37%) and HNO_{3conc} (65%) and distilled water, according to the scheme given in the following section (Table 1). The final concentration of the different dilutions equals 0; 10; 20; 30; 50; 70, 100 and 150 μg Fe/mL.

(b) The dilution scheme for measurement of the iron content of MLs is given in Table 2 (see Note 23).

(c) To 0.5 mL of the calibration and (diluted) sample solutions, add 0.6 mL of a mixture composed of 100 μL Tiron

Table 1
Dilution of the iron stock solution to prepare the calibration curve samples

Tube Nr.	μL Fe stock[a]	mL HCl (37%)	mL HNO_3 (65%)	mL H_2O_{dist}	Final Fe concentration
0	0	0.6	0.2	4.20	0
1	50	0.6	0.2	4.15	10
2	100	0.6	0.2	4.10	20
3	150	0.6	0.2	4.05	30
4	250	0.6	0.2	3.95	50
5	350	0.6	0.2	3.85	70
6	500	0.6	0.2	3.70	100
7	750	0.6	0.2	3.45	150

[a]Fe concentration of stock solution equals 1 mg/mL

Table 2
Possible dilutions of the ML sample in order to get absorbencies which are located within the range of Fe-concentrations of the calibration curve (see Table 1)

Dilution factor	mL sample	mL HCl	mL HNO$_3$	mL H$_2$O$_{dist}$
2	1.000	0.24	0.08	0.68
5	0.400	0.24	0.08	1.28
10	0.200	0.24	0.08	1.48
15	0.133	0.24	0.08	1.55
20	0.100	0.24	0.08	1.58
25	0.080	0.24	0.08	1.60
30	0.133	0.48	0.16	3.23
40	0.100	0.48	0.16	3.26
50	0.080	0.48	0.16	3.28
75	0.053	0.48	0.16	3.31
100	0.100	1.20	0.40	8.30

0.25M (see Note 24) and 0.5 mL KOH 4N and, subsequently, 1 mL phosphate buffer, 0.2M, pH 9.5.
The red color ($A^{480\,nm}$) develops immediately and reaches a maximum after about 15 min. The color remains stable for a few hours.

(d) By multiplying the iron concentration of the sample, as deduced from the calibration curve, (expressed in µg Fe/mL) by 1.38 the Fe$_3$O$_4$ concentration (µg/mL) in the ML samples is calculated.

3.5.2. Transmission Electron Microscopy

(a) A droplet of the ML suspension, adjusted to about 5 mM Fe, is deposited on a Formvar-coated grid.

(b) After 5 min, excess fluid is drained away with a piece of filter paper and the sample is stained with a drop of 0.5% uranylacetate (UCB, Belgium) in Milli Q water (see Note 25).

(c) After dehydration, the grid samples are examined in the electron microscope. A representative sample of an electron micrograph, clearly showing the translucent ML envelope, is shown in Fig. 3.

3.5.3. Organic Solvent Extraction

(a) A sample of the MLs is brought in a Teflon disk sealed screw cap vial and a mixture of CHCl$_3$ and CH$_3$OH (molar ratio 3/1) is added. The final organic solvent/water volume ratio equals 5/1 (see Note 26).

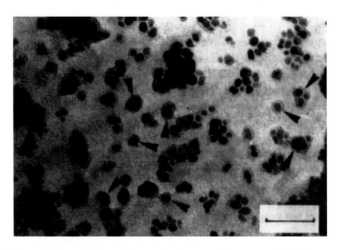

Fig. 3. Transmission electron micrograph of magnetoliposomes. Note the covering around the magnetite particles (indicated by the *arrows*) (Scale bar = 40 nm) From ref. (6), Copyright Springer-Verlag

(b) The vial is then attached to the stirring shaft of a motor driven mixer, positioned at an angle of about 45°C, and vigorously rotated (60 rpm) for 1 day at room temperature.

(c) The dark-brown magnetite-containing phase is aspirated and collected for iron and phosphate determinations (see the forementioned discussion). The data show that the original phospholipid/Fe_3O_4 ratio is dropped by about 2/3 indicating that the outer lipid layer is removed.

3.6. Cationic Magnetoliposomes (Fig. 4)

Production of cationic MLs containing, for instance, 3.33% DSTAP (see Note 27), proceeds according to the following consecutive steps (3, 4, 13):

1. First, neutral DMPC-MLs are prepared as outlined in Subheading 3.3. The data for the preparation are as follows: DMPC concentration 10.22 μmol/mL; 12.07 mg Fe_3O_4/mL; mmol DMPC/g Fe_3O_4 ratio of 0.84; volume 15 mL.

2. 15 mL of the zwitterionic DMPC-ML suspension is incubated at 37°C with 15 mL of intramembraneously mixed DMPC/DSTAP (90/10 molar ratio) vesicles. The latter were made by weighing 93.54 mg DMPC and 10.78 mg DSTAP in the sonication vial and vesicles (15 mL) were made thereof as described in Subheading 3.2 (see Note 28).

3. After 1 day, the mixture is subjected to a second magnetophoresis cycle and the cationic MLs are recovered. A value of 0.75 is found for the mmol *phospho*lipid/g magnetite ratio.

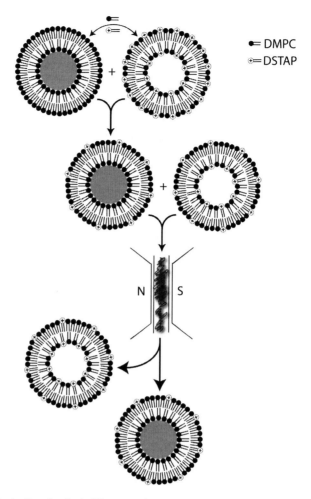

Fig. 4. Production of cationic MLs occurs in a two step procedure. Neutral MLs are first incubated with DMPC-DSTAP sonicated vesicles allowing exchange of the lipid molecules residing in the outer shell of the bilayer of both particle populations. Then, the cationic MLs are recovered from the mixture by HGM

4. Notes

1. The term "Magnetoliposome" is also used to designate rather undefined phospholipid-iron oxide complexes, or extruded large liposomes (diameter >100 nm) containing several magnetic cores (17). In reading papers on magnetoliposomes, it is strongly advisable to carefully control the physical appearance of the nanocolloids under consideration.

2. The capacity of phospholipid molecules to spontaneously jump out of a phospholipid bilayer is of crucial importance. Features of the individual phospholipid molecules (e.g., charge of the polar headgroup, fatty acyl chain length and presence/absence of unsaturated bonds), bilayers (e.g., curvature, 'melting

behavior') and external conditions (e.g., temperature, pH and ionic strength of the medium) play a key role (11, 18, 19).

3. For magnetic attraction of superparamagnetic particles such as MLs, a most critical parameter is the magnetic field gradient and to a lesser extent the magnetic field strength (8). Consequently, high-gradient magnetophoretic separations can also be successfully executed with a moderately strong permanent magnet.

4. In selecting the steel wool, two criteria have to be fulfilled: (i) it should be corrosion resistant to not interfere with the iron determination of the ML preparations, and (ii) it should be highly magnetisable which is not evident for common stainless steel materials.

5. To construct the magnetic filter the steel fibers, oriented in the same direction, are smoothly curled up along their long axis (similar to rolling a cigarette) and then the resulting plug is carefully turned into the tubing by applying a rotating motion.

6. To limit a person's exposure to hazardous and unpleasant fumes, all manipulations are done in a fume cupboard.

7. To avoid rapid oxidation by air oxygen, it is important to use the Fe(II) solution immediately after solubilisation. In the mixture, the Fe(II)/Fe(III) ratio equals 2:3 while their ideal ratio in magnetite is 1:2. As outlined by Khalafalla and Reimers (14) and as experimentally verified (10), due to air oxidation during preparation, some Fe(II) ions are oxidized to Fe(III) ultimately resulting close to a 1:2 ratio.

8. In case storage occurs under an air atmosphere the dark black color gradually converts to dark brown. This reaction is well known in magnetic colloid chemistry. It concerns an oxidative conversion of magnetite (Fe_3O_4) to maghemite (γ-Fe_2O_3). As the magnetic saturation of both iron oxides does not largely differ, this process will not influence the magnetic features (20, 21). In case oxidation has to be avoided, the stock solution of the magnetic fluid can be stored under an inert atmosphere.

9. In this work, we will mainly focus on DMPC and DMPG lipids but some other phospholipids work as well. However, if the geometry of a phospholipid molecule (i.e., the contribution of the polar headgroup versus the apolar part – see ref. (22, 23)) does not allow the formation of small unilamellar vesicles it will be difficult to generate stable MLs. For instance, in selected experimental conditions, pure phosphatidylethanolamine membranes are known to be destabilized (24).

10. Experimentally, we observed that the stability of DMPC vesicles increases by adding a small amount (>1 mol% is enough) of negatively charged phospholipids such as DMPG.

11. Incubation occurs by preference above the gel-to-liquid crystalline phase transition temperature which, for instance, equals

23°C for both DMPC and DMPG. For the corresponding dipalmitoyl-analog it is 41°C and for the dioleoyl-analog it is well below 0°C (25).

12. At this step, it is highly advisable to control the pH of the solution as phospholipid suppliers do not always clearly indicate the ionization state of the lipids.

13. In practice, power setting of the sonicator is fine-tuned so that no frothing and only minimal disturbance of the solution surface occurs. The turbidity of the final dispersion not only depends on the concentration but also to a large extent on the type of phospholipid used. For DMPC and DMPG the sonication period lasts for about 15 min, after which no further decrease in turbidity can be visually observed, resulting in a slightly white suspension when using DMPC and a completely clear solution with the use of DMPG.

14. A classical desktop centrifuge will be OK for this purpose. At this stage, carefully control the bottom of the tube as the presence of a white precipitate points to removal of phospholipid aggregates; this may occur with phospholipids which, for geometrical reasons, are not apt to form small unilamellar vesicles (22, 23).

15. Experimentally, we found that for the dimyristoylphospholipids under consideration a phospholipid/magnetite weight ratio of 5 is sufficient to construct an intact outer leaflet of the bilayer coat. In case phospholipid types with longer, more hydrophobic chains are used, this value can be decreased. If one wants to construct monolayered magnetoliposomes, a phospholipid/magnetite weight ratio of 0.3 should be taken. These MLs, however, are far less stable and, driven by the hydrophobic effect, tend to cluster and precipitate in just a few hours (6, 7).

16. In case multiple filters were used at the same time to prepare a larger batch of MLs, the pieces of filter tubing can be connected in tandem and then eluted with buffer. If only a small elution volume is used highly concentrated ML solutions can be obtained.

17. The calculations take into account the density of magnetite, the surface of a 14 nm-diameter particle, the cross-sectional area of the polar headgroup of a phospholipid molecule and the fact that a phospholipid *bi*layer is surrounding the solid core. Within the highly curved lipid bilayer, almost 2/3 of the lipids are located in the outer layer and 1/3 in the inner one (6).

18. In our hands, it is absolutely necessary to thoroughly clean the glass tubes in chromic acid. As the explosive $HClO_4$ is used to convert organic into inorganic phosphate, the presence of large amounts of oxidizable organic compounds in the solution(s) should be avoided as this may cause a burst of flame!

19. When putting on a heating block, keep in mind that the tube holes at the border of the block have a lower temperature.
20. Solubilisation of both chemicals is facilitated by stirring but even then, it may ask quite a long time. The combined solution is stored in the dark and remains stable for at least 2 years.
21. During the latter step, the color changes from blue to yellow. The solution is stable at room temperature for at least 1 week.
22. Hint: To minimize errors due to contamination of residual droplets within the cuvette, the different tubes are put in increasing intensity of blue color and measured in that sequence.
23. Due to the presence of a phospholipid bilayer coat around the iron oxide cores, the latter are quite reluctant to the solubilisation process. Therefore, it is advisable to add only the acids (37% HCl – 65% HNO_3; v/v 3/1), seal the tubes with a stopper to circumvent evaporation, heat till about 60°C until a clear yellow – not brown! – color appears and then add the required amount of water.
24. If stored in the dark at ambient temperature, this solution remains stable for about 3 months.
25. It is recommended to filter the solution through a 0.22 μm pore size Millipore filter just before use.
26. For other valuable solvent compositions: see ref. (26).
27. Often, the dioleoyl-analog (DOTAP) is used in lipid formulations. However, DOTAP-containing MLs are found to be more toxic to biological cells (3).
28. Note that in this experimental set-up, the molar amount of lipids in the vesicle population equals the molar amount of DMPC in the MLs. During the incubation step DMPC and DSTAP, spontaneously percolate between both colloidal particles. However, for thermodynamical reasons, i.e., slow transmembraneous flip-flop movements, only the outer leaflet of the ML coat and the outer shell of the vesicle membrane are involved in the exchange process (3, 13). As 2/3 of the total lipid contents is present in the outer layer of the vesicles and MLs, an equilibrium will be reached when 1/3 of the lipids has transferred. Thus, if the starting vesicles contain 10% DSTAP, ultimately, 3.33% arrives in the ML population.

Acknowledgments

S.J.H.S. is a recipient of a research grant from the Institute for the Promotion of Innovation through Science and Technology in Flanders (IWT-Vlaanderen).

References

1. Slotkin JR, Cahill KS, Tharin SA, Shapiro EM (2007) Cellular magnetic resonance imaging: nanometer and micrometer size particles for noninvasive cell localization. NeuroTherapeutics 4:428–433
2. Wilhelm C, Gazeau F (2008) Universal cell labelling with anionic magnetic nanoparticles. Biomaterials 29:3161–3174
3. Soenen SJH, Baert J, De Cuyper M (2007) Optimal conditions for labelling 3T3 fibroblasts with magnetoliposomes without affecting cellular viabilities. Chembiochem 8:2067–2077
4. Soenen SJH, Vercauteren D, Braeckmans K, Noppe W, De Smedt S, De Cuyper M (2009) Stable long-term intracellular labelling with fluorescently-tagged cationic magnetoliposomes. Chembiochem 10:257–267
5. De Cuyper M, Joniau M (1987) Biomagnetic particles: a synthetic approach. Arch Internat Physiol Biochim 95:B15
6. De Cuyper M, Joniau M (1988) Magnetoliposomes – formation and structural characterization. Eur Biophys J 15:311–319
7. De Cuyper M, Joniau M (1991) Mechanistic aspects of the adsorption of phospholipids onto lauric acid stabilized Fe_3O_4 nanocolloids. Langmuir 7:647–652
8. Whitesides GM, Kazlauskas RJ, Josephson L (1983) Magnetic separations in biotechnology. Trends Biotechnol 1:144–148
9. Bucak S, Jones DA, Laibnis PE, Hatton TA (2003) Protein separations using colloidal magnetic nanoparticles. Biotechnol Prog 19:477–484
10. De Cuyper M, Müller P, Lueken H, Hodenius M (2003) Synthesis of magnetic Fe_3O_4 particles covered with a modifiable phospholipid coat. J Phys: Condens Matter 15:S1425–S1436
11. De Cuyper M, Crabbe A, Cocquyt J, Van der Meeren P, Martins F, Santana MHA (2004) PEGylation of phospholipids improves their intermembrane exchange rate. Phys Chem Chem Phys 6:1487–1492
12. De Cuyper M, Soenen SJH, Coenegrachts K, Ter Beek L (2007) Surface functionalization of magnetoliposomes in view of improving iron oxide-based magnetic resonance imaging contrast agents: anchoring of gadolinium ions to a lipophilic chelate. Anal Biochem 367:266–273
13. De Cuyper M, Caluwier D, Baert J, Cocquyt J, Van der Meeren P (2006) A successful strategy for the production of cationic magnetoliposomes. Z Phys Chem 220:133–141
14. Khalafalla SE, Reimers GW (1980) Preparation of dilution-stable aqueous magnetic fluids. IEEE Trans Magn 16:178–183
15. Vaskovsky VE, Kostetsky EY, Vasendin IM (1975) A universal reagent for phospholipid analysis. J Chromatogr 114:129–141
16. Yoe J, Jones A (1944) Colorimetric determination of iron with disodium-1,2-dihydroxybenzene-3,5-disulfonate. Ind Eng Chem Anal Ed 16:111–115
17. Martina M-S, Fortin J-P, Ménager C, Clément O, Barrat G, Grabielle-Madelmont C, Gazeau F, Cabuil V, Lesieur S (2005) Generation of superparamagnetic liposomes revealed as highly efficient MRI contrast agents for in vivo imaging. J Am Chem Soc 127:10676–10685
18. De Cuyper M, Joniau M, Dangreau H (1983) Intervesicular phospholipid transfer – a free-flow electrophoresis study. Biochemistry 22:415–420
19. De Cuyper M, Joniau M, Engberts JBFN, Südholter EJR (1984) Exchangeability of phospholipids between anionic, zwitterionic and cationic membranes. Colloids Surf 10:313–319
20. Lawaczeck R, Menzel M, Pietsch H (2004) Superparamagnetic iron oxide particles: contrast media for magnetic resonance imaging. Appl Organometal Chem 18:506–513
21. Sun Y-K, Ma M, Zhang Y, Gu N (2004) Synthesis of nanometer-size maghemite particles from magnetite. Colloids Surf A 245:15–19
22. Nagle JF, Tristram-Nagle S (2000) Structure of lipid bilayers. Biochim Biophys Acta 1469:159–195
23. Förster G, Meister A, Blume A (2001) Chain packing modes in crystalline surfactant and lipid bilayers. Curr Opin Coll Interface Sci 6:294–302
24. Boggs JM (1987) Lipid intermolecular hydrogen bonding: influence on structural organization and membrane function. Biochim Biophys Acta 906:353–404
25. Marsh D (1990) Handbook of lipid bilayers. CRC Press, Boca Raton, Fl, USA
26. De Cuyper M, Noppe W (1996) Extractability of the phospholipid envelope of magnetoliposomes by organic solvents. J Colloid Interface Sci 182:478–482

Chapter 7

Ultrasound-Responsive Liposomes

Shao-Ling Huang

Abstract

Ultrasound-responsive liposomes are drug-loaded liposomes that contain a small amount of gas (often air). Co-encapsulation of a pharmaceutic along with this gas renders the liposomes acoustically active, allowing for ultrasound imaging as well as controlled release of the contents through ultrasound stimulation. Methods for the facile production of gas-containing liposomes with simultaneous drug encapsulation are available. Conventional procedures are used to prepare liposomes composed of phospholipid and cholesterol, namely, hydration of the lipid film followed by sonication. After sonication, the gas is introduced by one of two methods. The first method involves freezing and lyophilizing the sonicated liposomes in the presence of mannitol, the relevant property of which appears to be that it accentuates freezing damage to the lipid membranes. The other technique employs freezing of liposomes under elevated pressure of the desired gas. The concept of ultrasound-mediated drug delivery has many potential applications to specific clinical conditions such as cancer, thrombus, arterial restenosis, myocardial infarction, and angiogenesis because of its ability to localize the delivery of therapeutic agents that would cause side effects if given in large amounts systemically.

Key words: Phospholipids, ultrasound, Mannitol, Freeze-thawing, Controlled release, Drug encapsulation, Pressure-freezing, gases, air

1. Introduction

Liposomes are characterized by a lipid bilayer structure with clearly separated hydrophilic and hydrophobic regions. Hydrophilic portions of bilayer lipids are directed towards the internal and external aqueous phases, whereas hydrophobic portions of both lipid layers are directed towards each another, forming the internal core of the membrane. A useful feature of liposomes used for drug delivery is that they allow for localization and encapsulation both water-soluble and water-insoluble substances, either together or separately. Water-soluble materials are entrapped in

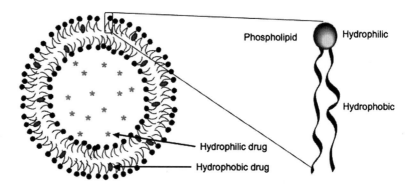

Fig.1. Schematic diagram of hydrophilic drug and hydrophobic drug incorporated into a liposome

the aqueous core, while water-insoluble and oil-soluble hydrophobic drugs or other agents reside within the bilayer (1) (Fig. 1). Recently, liposomes have been used to entrap both gas and hydrophilic drug simultaneously (2, 3). When a gas is encapsulated, for energetic reasons, it can be presumed to reside between the two monolayers of the liposome bilayer or perhaps as a monolayer-covered air bubble *within* the aqueous compartment of liposomes (3).

When gas-carrying liposomes contain hydrophilic drugs within the aqueous compartment, they respond to application of ultrasound by releasing their contents (ultrasound-responsive liposomes). This is likely due to the rarefaction phase of ultrasound leading to expansion of the air (or gas) pocket and stressing the liposome membrane. If the pressure drop and the resultant air pocket expansion is large enough, the stress will exceed the elastic limit and the liposome ruptures. When the integrity of the vesicle is compromised, some or all of the contents (depending upon how long resealing takes) will be released.

Characteristics of ultrasound-responsive liposomes that make them particularly suitable for drug and gene delivery are as follows:

1. Ultrasound responsive liposomes have high drug and gene loading properties, comparable to those of conventional liposomes.

2. Echogenic liposomes containing entrapped therapeutic agents can be conjugated with antibodies and targeted to specific disease sites, allowing high local concentrations and low systemic toxicity.

3. The discharge of entrapped contents can be controlled to produce bolus release with a single high amplitude ultrasonic pulse, or sustained release through a series of low amplitude pulses. These options could be particularly valuable in those cases where it is important to raise the local concentration to a therapeutic level and maintain it for some time.

4. Cavitation caused by ultrasound-triggered destruction of gas pockets in ultrasound-responsive liposomes bubbles increases the permeability of cells and tissues, thus facilitating the transport of the drug or gene into the targeted cells and tissues.
5. The ultrasound reflectivity of entrapped gas in echogenic liposomes allows real time image-guided drug and gene delivery.

The freeze-drying method for making ultrasound-responsive liposomes involves liposomal freezing in the presence of mannitol followed by lyophilization (3) (Fig. 2). In this method, the presence of mannitol in freezing and lyophilization is critical for generation of the ultrasound-responsiveness. Unlike trehalose, which is used as a cryoprotectant to preserve liposome structure upon freezing, mannitol lacks cryoprotection activity (4–6), yet has been used effectively in the production of ultrasound-responsive liposome dispersions (7). Evidently, the important function of mannitol (and, probably also other solutes that do not interfere with crystallization during freezing) is to cause membrane disruption and fusion (8), maximizing the exposure of disrupted lipid hydrophobic bilayers to air. Freeze-thawing in the presence of mannitol leads to an increase in liposomal size from 90 to 800 nm, which indicates a fusion of lipid bilayers during freezing. In addition, a fluffy cake is generated after lyophilization in the presence of mannitol, which increases the area of liposomes exposed to air.

The procedure of freeze-drying in the presence of mannitol has proven to be a superior method for simultaneous encapsulation

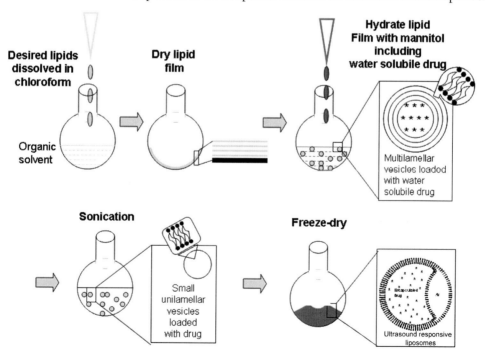

Fig. 2. Schematic diagram of ultrasound-sensitive liposome preparation procedure.

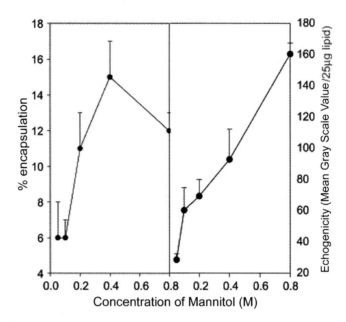

Fig. 3. Encapsulation efficiency of calcein (left) and echogenicity of acoustically active liposomes (EggPC:DPPE:DPPG:CH 69:8:8:15) (right) made by freeze-drying in the presence of different concentrations of mannitol. Mean ± SD. $n=6$. Reproduced with permission from (3)

of gas as well as hydrophilic solutions (Fig. 3). One freeze-thaw cycle in the presence of mannitol was found to capture appreciately about 15% of an external calcein (calcein is a convenient marker for determining trapped volumes) solution within the echogenic liposomes. Encapsulation efficiency can be markerly improved (to 20%) without diminishing echogenicity by simply increasing the number of freeze-thaw cycles to three or four, resulting in liposomes that contain both air and hydrophilic drugs.

Another unique, simple and highly efficient method for gas and hydrophilic solute co-encapsulation into liposomes is the pressure-freeze method. Compared to the lyophilization method, the pressure-freeze method has the advantage of facile encapsulation of the gas phase (2) (Table 1). This method includes three additional steps after sonication. The first step is to increase the gas solubility in lipid solution by increasing the gas pressure. According to Henry's Law, the solubility of a gas in a liquid is directly proportional to the pressure of that gas over the liquid. Thus, if the pressure is increased, the gas molecule concentration in solution is increased. This is important as gas uptake by the liposomes should be proportional to the amount of gas in solution. Step two is freezing, which serves two purposes: increasing the local concentration of dissolved gas and nucleating formation of small pockets of gas. Gasses, like other solutes, are more soluble in liquid water than in solid ice. Thus, as the ice crystals grow, dissolved gas is progressively displaced from ice to unfrozen solution,

Table 1
Comparsion of freeze-lyophilization method and pressured-freezing method for ultrasound-responsive liposome preparation

Freeze-lyophilization method	Pressure-freezing method
Lipid film hydration	Lipid film hydration
Sonication	Sonication
	Introduction of gas to hydrated lipids under pressure.
Freezing	Freezing in the presence of mannitol
	Pressure release
Thawing	Thawing

with the result that the dissolved gas becomes increasingly concentrated in the ever-diminishing volume of liquid solution. When the dissolved gas concentration becomes sufficiently high, a gas bubble may nucleate and grow (9). Indeed, it has long been known that during freezing, air is released and often trapped as bubbles in the resultant ice. This phenomenon has important biological consequences in that bubble formation contributes to freezing damage in long-term preservation of cells and tissues (10, 11). During the freezing, mannitol crystallizes out of solution and, along with ice crystals, damages the liposome's bilayer, providing sites for gas bubble nucleation (12). In the final step, the sample has to be decompressed before thawing because the gas concentration in the solution is high upon initially melting as it contains most of the air that was dissolved in the suspension upon pressurization. On the other hand, the ice contained within the liposomes will melt first and immediately expose the lipid to ambient (1 atm) pressure. This initial melting phase is not only highly supersaturated with gas, but it also contains gas pockets that will grow when exposed to ambient pressure. Hence, gas will come out of solution, expanding the gas nuclei that formed during freezing. The result is the formation of gas pockets that are stabilized by a monolayer of lipid.

The procedures developed lead to both gas and aqueous phase encapsulation. Given the high encapsulation of ultrasound-sensitive liposomes, it is clear in the latter method, like conventional liposomes the solute is captured by each particle in rough proportion to the size of the particle. It does not follow, however, that each liposome contains gas. Although there is a difficult on determining the distribution of gas among single particle, it does seem clear that almost all of the liposomes do contain some air. This conclusion is based on the observation that 95% of FITC-labeled liposomes floated to the top of a mixture of the echogenic liposomes and with 0.32 mol/L mannitol. Since the liposomes were prepared in the same concentration of mannitol, their density would be similar to that of the solution, and only those containing air would float. Accordingly, it appears that 95% of the liposomes in the preparation contain air and hence, that a similar proportion contains both air and solute.

2. Materials

1. Composition of ultrasound-responsive liposome (see **Note 1**).

Lipid		Molecular weight	Transition temperature
Egg phosphatidylcholine (Avanti Polar Lipids; Alabaster, AL) (see Note 2)	Egg PC	760.08	N/A
Dipalmitoyphosphatidylethanolamine (see Note 3)	DPPE	691.97	41.6°C
Dipalmitoylphosphatidylglycerol (see Note 4)	DPPG	744.96	41°C
Dihexanoylphosphatidylcholine (see Note 5)	DHPC	481.57	−3°C
Cholesterol (see Note 6)	CH	386.654	N/A

2. Lipid solution preparation: Typically, lipids are dissolved in a chloroform:methanol (9:1) mixture at a concentration of 20 mg lipid/mL to obtain a clear lipid solution. Higher concentrations may be used if lipid solubility permits. Lipid solutions are stored at −20°C.

3. Calcein solution: Calcein (2′,7′-[(bis[carboxymethyl]-amino) methyl]-fluorescein)(Sigma- Sigma Chemical Co. (St. Louis, MO, USA) (see **Note 7**). Stock solutions of calcein (80 mM) are made by quickly dissolving solid calcein at pH 9.0 and then adjusting the pH to 7.5.

4. Calcein fluorescence quench solution: Cobalt chloride ($CoCl_2$) (Sigma-Aldrich, St. Louis, MO) (see **Note 8**) is dissolved in 50 nM 3-morpholinepropanesulphonic acid (MOPS) buffer (pH 7.5).

5. Permeabilization solution: 10% Triton X-100 solution.

6. Sonitron1000 ultrasound device (RichMar, Inola, OK).

7. Gases: Argon (Airgas Inc., Chicago, IL) and octafluorocyclobutane (Specialty Gases of America, Inc., Toledo, OH).

3. Methods

3.1. Ultrasound-Responsive Liposomal Preparation by Freeze-Drying Method

1. Remove all lipids used in preparation from refrigerator and warm up to room temperature with a hair dryer. Some lipids such as DPPE and DPPG stock solutions may precipitate upon storage in the refrigerator. To disperse such a preparation, it may be heated in an oven at 42°C and then cooled to room temperature with unheated air from a hair dryer.
2. Mix lipids (as showing on Table 2 for preparing 30 mg of appropriate sample) in a 250 mL round bottom flask according to the desired lipid molar ratio. For liposomal conjugation with an antibody or similar targeting agent, MPB-PE or PEG-PE is to be added.
3. After mixing lipids, add approximately about 1 mL of chloroform and gently shake the flask in order to obtain uniform organic solution.
4. Evaporate solvent under a gentle argon stream in the chemical fume hood until there is no chloroform smell. Slowly rotate the flask in a 50°C water bath to prevent cholesterol crystallization. Take precautions to prevent water splashing into the flask.
5. Place flask in a desiccator under vacuum (<100 milliTorr) for 2–4 h for complete removal of solvent.
6. Remove flask from desiccator.
7. Hydrate the dried lipid film with 0.1 mM calcein (*see* **Note 9**) in 0.32 M mannitol (*see* **Note 10**). The resultant liposomes have a10mg/mL final concentration. Swirl the flask until all of the lipid is off the walls (*see* **Note 11**).

Table 2
Working table for making a 30 mg of ultrasound-sensitive liposomes correspond to 69:8:8:15 molar percent of Egg PC, DPPE:DPPG:CH. For conjugation, DPPE is substituted by MPB-PE or PEG2000-PE, otherwise composition remains the same

Lipids	Acoustic liposomes	Acoustic liposomes for antibody conjugation
EPC	22.56 mg	21.8 mg
DPPE	2.38 mg	–
MPB-PE	–	3.35 mg
DPPG	2.56 mg	2.48 mg
CH	2.50 mg	2.41 mg

8. Sonicate the liposomal dispersion for 5 min in bath sonicator.
9. Measure the absorbance of the resulting solution at 440 nm. If A_{440} is more than 0.7 for a 0.5 mm optical path length, sonicate the sample additionally until the absorbance is less than 0.7 (see Note 12).
10. Freeze the liposomal solution at −70°C for at least ½ h followed by thawing the frozen liposomes to room temperature. Repeat three times (see Note 13).
11. Freeze the liposome to −70°C again.
12. Top the solution with argon and freeze on dry ice. Add ethanol to the bottom of the dry ice container to obtain a good thermal contact between the dry ice and the vial. Don't let the ethanol splash into the vial. Freezing takes about 30 min.
13. Cover the sample using caps with holes and nylon filters. Put the sample in a VirTis Model 6 bench top freeze-dry apparatus (VirTis Corp., Gardiner, NY). The temperature in the lyophilizer should be less than −50°C, the pressure less than 50 mTorr. Lyophilize for 24 h (see Note 14).
14. The lyophilized dry cake should be stored in 4°C until use and re-suspended at a concentration of 10 mg/mL in water immediately before using.

3.2. Ultrasound-Responsive Liposomal Preparation by Freeze Under Pressure Method

1. Remove all lipids used in preparation from refrigerator and warm up to room temperature with a hair dryer. Some lipids such as DPPE and DPPG stock solutions may precipitate upon storage in the refrigerator. To disperse such a preparation, it may be heated in an oven, at 42°C and then cooled to room temperature with unheated air from a hair dryer.
2. Mix lipids (as showing on Table 2 for preparing 30 mg of appropriate sample) in a 250 mL round bottom flask according to the desired lipid molar ratio. For liposomal conjugation with an antibody or similar targeting agent, MPB-PE or PEG-PE is to be added.
3. After mixing lipids, add approximately about 1 mL of chloroform and gently shake the flask in order to obtain uniform organic solution.
4. Evaporate solvent under a gentle argon stream in the chemical fume hood until there is no chloroform smell. Slowly rotate the flask in a 50°C water bath to prevent cholesterol crystallization. Take precautions to prevent water splashing into the flask.
5. Place flask in a desiccator under vacuum (<100 milliTorr) for 2–4 h for complete removal of solvent.
6. Remove flask from desiccator.
7. Hydrate the dried lipid film with 0.1 mM calcein (see Note 9) in 0.32 M mannitol (see Note 10). The resultant liposomes

have a 10 mg/mL final concentration. Swirl the flask until all of the lipid is off the walls (see Note 11).

8. Sonicate the liposomal dispersion for 5 min in bath sonicator.
9. Measure the absorbance of the resulting solution at 440 nm. If A_{440} is more than 0.7 for a 0.5 mm optical path length, sonicate the sample additionally until the absorbance is less than 0.7 (see Note 12).
10. Transfer a total of 500 µl of the sonicated liposomal dispersion to a 1.8 mL screw-cap borosilicate glass vial (12 × 32 mm) capped with open screw caps containing Teflon-covered silicon rubber septa.
11. Introduce the desired gas into the vial through the septum with a 10 mL syringe fitted with a 27Gx1/2″ needle (see Note 15). Incubate the pressurized-gas/liposome dispersion for a ½ h at room temperature. Freeze the sample by cooling to −78°C on dry ice for at least ½ h (see Note 14).
12. Release the pressure by unscrewing the caps immediately after their removal from dry ice (see Note 16).
13. Thaw the depressurized frozen liposomes by exposure to room temperature, allowing the temperature of the dispersion to change from −78°C to 24°C within 10 min.

3.3. Determination of Encapsulation Efficiency

1. Add 25 µL of resuspended liposomes (10 mg lipid/mL) to 500 µL of 50 mM MOPS buffer containing 110 mM NaCl.
2. Measure the fluorescence (F_b) at 490 nm Ex, 520 nm Em.
3. Add 5 µL 10 mM $CoCl_2$ to quench the fluorescence of external calcein so that the residual fluorescence represents the entrapped calcein.
4. Measure the fluorescence (F_a) after quenching.
5. Add 10 µL 10% triton X-100 to lyse the liposomes, allowing the release of entrapped calcein. Measure the background fluoresence at zero encapsulated volume. Measure the fluorescence (F_{totq}) after lysing liposomes.
6. Calculate the encapsulation efficiency. The fluorescence is measured before and after the addition of 5 µL of 10 mM of $CoCl_2$, and after the addition of 10 µL. 10% triton X-100 at 490 nm Ex, 520 nm Em. The % encapsulation is calculated as follows.

$$\%\text{Encapsulation} = \left(\frac{F_a - F_{totq}}{F_b - F_{totq}}\right) \times 100 \qquad (1)$$

3.4. Measurement of Echogenicity and Videodensitometric Analysis

1. Add 1 mL of pure water to 10 mg lipid dry cake and gently swirl flask to re-suspend liposomes.
2. Take 12.5 µL of the re-suspended liposomal dispersion and dilute to 5 mL in 12 × 16 mm glass vials.

3. Image the diluted liposomes with a IVUS 20-MHz high-frequency intravascular ultrasound (IVUS) imaging catheter (Boston Scientific Inc., Sunnyvale, CA) or comparable instrument. Instrument settings for gain, zoom, compression, and rejection levels are constant for all samples. Record images in real time for subsequent playback and image analysis.

4. Playback images acquired in digital format or on videotape through the IVUS unit.

5. Digitize images to a 640 × 480 pixel spatial resolution (approximately 0.045 mm/pixel) and 8-bit (256 gray levels) amplitude resolution using Snappy Video Snapshot software (Play Inc., Rancho Cordova, CA) and custom image-acquisition software (Image-Pro Plus software V4.1; Media Cybernetics, Silver Spring, MD).

6. Calculate the mean gray scale value (MGSV) for each digitized image. The annular area between the vial wall and the imaging catheter has to be manually outlined (excluding the area of the strut artifact of the IVUS imaging catheter). Designate this as the area of interest (AOI). This method provides an estimate of the total ultrasound reflectivity of the whole sample (Fig. 4).

7. Compute the mean gray scale value of all pixels in the AOI using the Image Histogram function in Image-Pro Plus software.

8. Digitalize and analyze the mean gray scale value (MGSV) of the entire image.

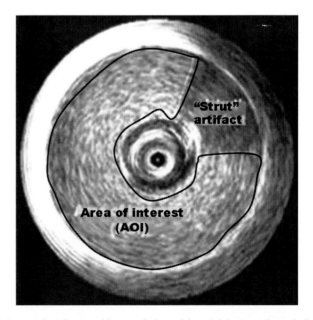

Fig. 4. Intravascular ultrasound image of glass vial containing an echogenic liposomal dispersion. Manually selected area of interest and the excluded strut artifact are displayed. Mean gray scale values were obtained via videodensitometric analysis of the area of interest

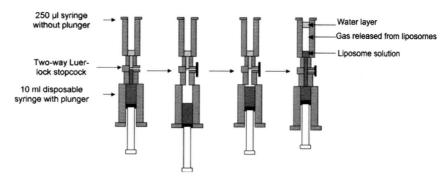

Fig. 5. Apparatus for quantifying air content of ELIP, consisting of a 10-mL disposable syringe attached to a 250-µL syringe without a plunger through a two-way Luer-lock stopcock

3.5. Measurement of the Amount of Gas in the Lipid Dispersions

1. Draw 2 to 3 mL of a 10 mg/mL lipid dispersion into a 10-mL disposable syringe.
2. Connect the 10-mL disposable syringe to a laboratory-made gas volume-measuring device (Fig. 5) through a two-way Luer-lock stopcock.
3. Open the stopcock to allow the pressure in the whole system to reach atmospheric pressure.
4. Hold the device vertically. Inject approximately 20-µL solution from the 10 mL disposable syringe into the 250-µL syringe to create a water level and to expel all air from the Lauer luck in between the two syringes.
5. Close the stopcock and withdraw the plunger of the large syringe to generate a vacuum of approximately 0.03 to 0.06 atm for 30 s. Repeat the process 3 times.
6. Open the stopcock. Due to gas release from the liposomal dispersion into solution, the air pocket will rise to the top of the lipid dispersion. Inject the air pocket followed by a small volume of lipid dispersion into the barrel of the microliter syringe by depressing the plunger of the large syringe.
7. The volume of released gas should display between two columns of liquid, water at the top and lipid dispersion at the bottom. Record the volume from the microliter syringe scale.

3.6. Ultrasound-Triggered Release

1. Ultrasound-triggered release experiments are performed in a chamber as illustrated in Fig. 6. This chamber consists of a Costar transwell insert with a 0.4-µm pore polyester membrane resting above a sheet of Rho-c rubber in a water bath containing Phosphate Buffered Saline (PBS) solution. The open top of the transwell insert allows for introduction of the liposomal dispersion (400 µL) and placement of the ultrasound probe. The polyester membrane allows for a 100% transmis-

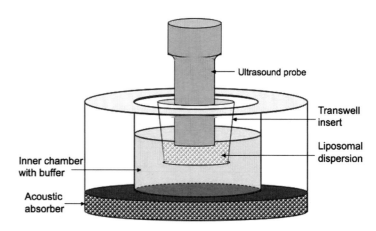

Fig. 6. Experimental apparatus for ultrasound-triggered release (see text for details)

sion of ultrasound and the Rho-c rubber eliminates the reflection of ultrasound wave.

2. Dilute 100 µL of the 10 mg lipid/mL calcein-containing, acoustically active liposomes, to 500 µL with 50 mM 3-morpholinepropanesulphonic acid (MOPS) buffer containing 110 mM NaCl (to maintain isosmolality with the liposomal contents of 320 mM mannitol and 0.1 mM calcein) into the transwell chamber.
3. Add 5 µL of 40 mM $CoCl_2$ into the chamber to quench the external fluorescence.
4. Measure the fluorescent intensity (F_{in}) of the suspension.
5. Apply ultrasound (1 MHz, 2 W/cm², 100% duty cycle; probe size 1.2 cm diameter) for 10s using a Sonitron (*see* **Note 17**).
6. Measure the fluorescence intensity ($F_{ultrasound}$).
7. Add 25 µL of 10% Triton X-100.
8. Measure fluorescent intensity (F_{totq}).
9. Calculate the amount of released calcein upon ultrasound application.

$$\% \text{release} = \left(\frac{F_{in} - F_{ultrasound}}{F_{in} - F_{totq}} \right) \times 100 \qquad (2)$$

4. Plot the Release Curve (see Note 18) (Fig. 7)

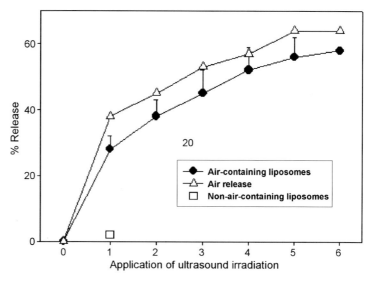

Fig. 7. Ultrasound-triggered release of calcein and air from acoustically active liposomes composed of EggPC:DPPE:DPPG:CH at molar ratio of 69:8:8:15 including 4% DHPC. Mean ± SD, $n = 6$. Internal control: "non-acoustically active liposomes" were evaluated for calcein release. Reproduced with permission from (3)

5. Notes

1. The composition of ultrasound responsive liposomes can be varied (cationic, anionic, neutral lipid) according to the specific drug encapsulated. The same preparation methods can be used for all compositions).

2. More highly saturated lipids can be used to replace Egg PC. It is well known that increased saturation of the constituent lipids increases the rigidity of lipid monolayers and bilayers and reduces their permeability to small molecules, and increases their ability to resist the compressive effects of surface tension (13, 14).

3. MPB-PE or PEG-PE be used to replace DPPE when antibody conjugation is needed.

4. PG has a negative charge, which facilitates bilayer hydration, encourages the formation of liposomes, and hinders aggregation of liposomes once formed. However, high concentration of negatively charged lipids should be avoided if the liposomes are to be used for systemic liposome delivery, since high concentrations of PG stimulate macrophage clearance.

5. DHPC increases the sensitivity of liposomes to ultrasound application. DHPC is a short chain lipid, which can stabilize broken edges of bilayers (15, 16) allowing the openings created by membrane stretch to remain open longer, allowing more contents to be released.

6. CH has well-known effects on lipid bilayer rigidity and stability (17). Bilayer stability effects are reflected by the effect of CH concentration on phase transition temperature (18), saturated lipid segregation, bilayer fusion temperature and bilayer fluidity (19).

7. Calcein is a stable, hydrophilic, highly fluorescent dye to which lipid membranes are essentially impermeable. It has been used as traceable "stand-in" for hydrophilic drugs in a variety liposomal drug delivery experiments.

8. Co^{2+} combines with calcein to form a chelate that does not have fluoresence in the neutral pH range. However, in the presence of EDTA, a stronger Co^{2+} chelator, calcein is released and can be readily detected at very low concentrations. Therefore, the inclusion of EDTA or other strong chelators in these measurements will cause interfere to the quenching effect of Co^{2+}.

9. For convenience in the assay, calcein was used as a hydrophilic drug stand-in. It registers the fraction of the aqueous entrapped phase and hence provides a good measure of how much of a hydrophilic drug would be encapsulated by echogenic liposomes.

10. Hydration should be with a solution of physiological osmolality.

11. The temperature of the hydration step should be above the gel-liquid crystal transition temperature of the lipid to allow the lipid to hydrate in its fluid phase with adequate agitation. Hydration time may differ slightly among lipid species and structures. We believe that good hydration prior to sonication makes the sizing process easier and improves the homogeneity of the preparation.

12. The mean diameter of the liposomes after the initial sonication step and before freezing is 90 ± 30 nm (first column) and increases to 711 ± 105 nm after freezing and thawing in the presence of 0.32 M mannitol.

13. One freeze-thaw cycle in the presence of mannitol allows approximately 15% of the external calcein solution to be captured by the liposomes. The encapsulation efficiency can be markedly improved (to 20%) without diminishing echogenicity by simply increasing the number of freeze-thaw cycles to three or four.

14. The process of using a laboratory scale lyophilizer involves significant pressure differences and the potential for glassware

implosion, along with the possibility of frostbite or tissue damage associated with exposure to cryogenic materials. Operators should wear appropriate eye and hand protection at all times when working with a lyophilizer.

15. The Teflon-covered silicon rubber septum should not release a significant volume of gas for at least 24 h.

16. It is important to release the pressure before thawing the liposomal solution. Only a small amount of gas (10 µL) is taken up by the liposomal suspension (5 mg lipids) when the pressure is released before the freezing step or after thawing step. The osmoticant mannitol solution without liposomes takes up about 5–10 µL of gas. In contrast, releasing pressure after freezing and before thawing results a much higher gas incorporation (40 µL/5 mg lipid). This indicates that air is entrapped when liposomes are frozen under applied pressure and thawed at ambient pressure.

17. The parameters (frequency, intensity, duration, pulse interval, etc.) for the ultrasound application have a major effect on ultrasound triggered release (20, 21). Ultrasound parameters should hence be optimized for each liposome preparation. Furthermore, there is evidence that clinical doppler ultrasound can induce more extensive release of drugs from liposomes (20) than Sonitron.

18. Echogenic liposomes seem to constitute one of the most sensitive ultrasound-controlled release systems yet described, with ultrasound having been used to trigger release from several different echogenic liposomal drug delivery preparations such as tissue plasminogen activator (tPA) and papaverin (22, 23).

References

1. Nii T, Ishii F (2005) Encapsulation efficiency of water-soluble and insoluble drugs in liposomes prepared by the microencapsulation vesicle method. Int J Pharm 298: 198–205
2. Huang SL, McPherson DD, Macdonald RC (2008) A method to co-encapsulate gas and drugs in liposomes for ultrasound-controlled drug delivery. Ultrasound Med Biol 34(8): 1272–1280
3. Huang SL, MacDonald RC (2004) Acoustically active liposomes for drug encapsulation and ultrasound-triggered release. Biochim Biophys Acta 1665:134–141
4. Crowe JH, Crowe LM, Carpenter JF, Wistrom CA (1987) Stabilization of dry phospholipid bilayers and proteins by sugars. Biochem J 242:1–10
5. Crowe JH, Whittam MA, Chapman D, Crowe LM (1984) Interactions of phospholipid monolayers with carbohydrates. Biochim Biophys Acta 769:151–159
6. Tsvetkova NM, Phillips BL, Crowe LM, Crowe JH, Risbud SH (1998) Effect of sugars on headgroup mobility in freeze-dried dipalmitoylphosphatidylcholine bilayers: solid-state 31P NMR and FTIR studies. Biophys J 75:2947–2955
7. Demos SM, Alkan-Onyuksel H, Kane BJ, Ramani K, Nagaraj A, Greene R, Klegerman M, McPherson DD (1999) In vivo targeting of acoustically reflective liposomes for intravascular and transvascular ultrasonic enhancement. J Am Coll Cardiol 33:867–875
8. Castile JD, Taylor KM (1999) Factors affecting the size distribution of liposomes produced by

freeze-thaw extrusion. Int J Pharm 188: 87–95

9. Zhang XH, Maeda N, Craig VS (2006) Physical properties of nanobubbles on hydrophobic surfaces in water and aqueous solutions. Langmuir 22:5025–5035

10. Carte AE (1961) Air Bubbles in Ice. Proc Phys Soc 77:12

11. King CJ (1974) Pressurized freezing for retarding shrinkage after drying: a quantitative interpretation. Cryobiology 11:121–126

12. Craig VSJ (1996) Formation of micronuclei responsible for decompression sickness. J Colloid Interface Sci 183:260–268

13. Bloemen PG, Henricks PA, van Bloois L, van den Tweel MC, Bloem AC, Nijkamp FP, Crommelin DJ, Storm G (1995) Adhesion molecules: a new target for immunoliposome-mediated drug delivery. FEBS Lett 357:140–144

14. Buchanan KD, Huang S, Kim H, Macdonald RC, McPherson DD (2008) Echogenic liposome compositions for increased retention of ultrasound reflectivity at physiologic temperature. J Pharm Sci 97:2242–2249

15. Booth PJ, Riley ML, Flitsch SL, Templer RH, Farooq A, Curran AR, Chadborn N, Wright P (1997) Evidence that bilayer bending rigidity affects membrane protein folding. Biochemistry 36:197–203

16. Glover KJ, Whiles JA, Wu G, Yu N, Deems R, Struppe JO, Stark RE, Komives EA, Vold RR (2001) Structural evaluation of phospholipid bicelles for solution-state studies of membrane-associated biomolecules. Biophys J 81:2163–2171

17. Halstenberg S, Heimburg T, Hianik T, Kaatze U, Krivanek R (1998) Cholesterol-induced variations in the volume and enthalpy fluctuations of lipid bilayers. Biophys J 75:264–271

18. Chaudhury MK, Ohki S (1981) Correlation between membrane expansion and temperature-induced membrane fusion. Biochim Biophys Acta 642:365–374

19. Toner M, Cravalho EG, Karel M (1990) Thermodynamics and kinetics of intracellular ice formation during freezing of biological cells. J Appl Phys 67:1582–1593

20. Smith DA, Porter TM, Martinez J, Huang S, MacDonald RC, McPherson DD, Holland CK (2007) Destruction thresholds of echogenic liposomes with clinical diagnostic ultrasound. Ultrasound Med Biol 33:797–809

21. Coussios CC, Holland CK, Jakubowska L, Huang SL, MacDonald RC, Nagaraj A, McPherson DD (2004) In vitro characterization of liposomes and Optison by acoustic scattering at 3.5 MHz. Ultrasound Med Biol 30:181–190

22. Tiukinhoy-Laing SD, Huang S, Klegerman M, Holland CK, McPherson DD (2007) Ultrasound-facilitated thrombolysis using tissue-plasminogen activator-loaded echogenic liposomes. Thromb Res 119:777–784

23. McPherson DD, Huang S, Holland CK (2008) Echogenic liposomes for molecular targeted therapeutic delivery. J Acoust Soc Am 123:3216

Chapter 8

Liposome Formulations of Hydrophobic Drugs

Reto A. Schwendener and Herbert Schott

Abstract

Here, we report methods of preparation for liposome formulations containing lipophilic drugs. In contrast to the encapsulation of water-soluble compounds into the entrapped aqueous volume of a liposome, drugs with lipophilic properties are incorporated into the phospholipid bilayer membrane. Water-soluble molecules, for example, cytotoxic or antiviral nucleosides can be transformed into lipophilic compounds by attachment of long alkyl chains, allowing their stable incorporation into liposome membranes and taking advantage of the high loading capacity lipid bilayers provide for lipophilic molecules. We created a new class of cytotoxic drugs by chemical transformation of the hydrophilic drugs cytosine-arabinoside (ara-C), 5-fluoro-deoxyuridine (5-FdU) and ethinylcytidine (ETC) into lipophilic compounds and their formulation in liposomes.

The concept of chemical modification of water-soluble molecules by attachment of long alkyl chains and their stable incorporation into liposome bilayer membranes represent a very promising method for the development of new drugs not only for the treatment of tumors or infections, but also for many other diseases.

Key words: Liposomes, Lipophilic drugs, Lipophilic ara-C drugs, NOAC, Duplex drugs

1. Introduction

Liposomes are predominantly used as carriers for hydrophilic molecules that are encapsulated within the aqueous inner volume which is confined by the lipid bilayer. These molecules generally do not interact with the lipid moiety of the vesicle. Long circulating liposomes modified with poly(ethylene glycol) (PEG) and other formulations carrying encapsulated cytotoxic drugs such as doxorubicine, paclitaxel, vincristine, lurtotecan and others are clinically approved chemotherapeutic liposome formulations (1–5).

In contrast, many lipophilic drugs or prodrugs can only be applied therapeutically by use of potentially toxic solubilizing agents such as detergents or polymers or by development of complex pharmaceutical formulations (6–8). Therefore, in view of such potential disadvantages, many hydrophobic drugs are not further developed into clinically used medicines. Such technical drawbacks can be resolved by incorporation of lipophilic drugs into the bilayer matrix of phospholipid liposomes. We and others chose the approach of the chemical transformation of water-soluble molecules of known cytotoxic properties into lipophilic drugs or prodrugs. Some recent examples of modifications of antitumor drugs and their formulation in liposomes are gemcitabine, paclitaxel, methotrexate, 5-iodo-2′-deoxyuridine and cytosine arabinoside (ara-C) (9–15).

We selected ara-C as a compound of well known cytotoxic properties and transformed the nucleoside into lipophilic derivatives. Due to the insolubility of the resulting compounds, we developed formulations in which the lipophilic moieties of the molecules serve as anchor for a stable incorporation into the lipid bilayer membranes of small unilamellar liposomes, taking advantage of the high loading capacity of the phospholipid bilayers. To introduce lipophilic anchors ara-C was modified with long acyl and alkyl chains, preferably of similar chain lengths as the phospholipids, allowing optimal alignment within the lipid bilayer matrix. Out of a series of N^4-alkyl derivatives of ara-C, the most effective compound, N^4-octadecyl-ara-C (NOAC) was extensively studied by us (Fig. 1) (16–18). In contrast to ara-C, NOAC is a highly lipophilic drug with an extreme resistance towards deamination. Liposome formulations of NOAC showed excellent anti-tumor activities after oral and parenteral therapy in several tumor models. From a large number of studies, we conclude that the mechanisms of action of the N^4-alkyl-ara-C derivatives are distinct from ara-C and that such lipophilic derivatives represent a new class of cytotoxic nucleoside drugs (19–22). Most hydrophobic drugs interact with lipoproteins which are the major transport vehicles for lipids and cholesterol throughout the aqueous environment of the blood and lymph circulatory systems (23, 24). We could show that liposome-incorporated NOAC is transferred to lipoproteins, mainly to the low- and high-density lipoproteins (LDL and HDL), respectively (25–27). Thus, the strong affinity of NOAC to lipoproteins, and of lipophilic drugs in general,

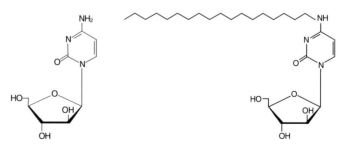

Fig. 1. Chemical structures of the hydrophilic parent compound 1-β-D-arabinofuranosylcytosine (ara-C, mol. wt. 243.2) and its lipophilic N^4-alkyl derivative N^4-octadecyl-1-β-D-arabinofuranosylcytosine (NOAC, mol. wt. 495.7).

might be exploited for an enhanced drug uptake in tumor cells that express high numbers of LDL receptor molecules.

Recently, we further modified NOAC by the synthesis of new duplex drugs by combination of the clinically used cancer drugs ara-C, 5-fluorodeoxyuridine (5-FdU) and the highly active new compound ethynylcytidine (1-(3-C-ethynyl-β-D-ribopentofuranosyl)-cytosine, ETC) with NOAC, yielding the heterodinucleoside phosphates arabinocytidylyl-N^4-octadecyl-1-β-D-arabinofuranosyl-cytosine (ara-C-NOAC), 2′-deoxy-5-fluorouridylyl-N^4-octadecyl-1-β-D-arabinofuranosylcytosine (5-FdU-NOAC) and ETC-NOAC (3′-C-ethynylcytidylyl-(5′ → 5′)-N^4-octadecyl-1-β-D-arabinofuranosylcytosine) as shown in Fig. 2 (28–32).

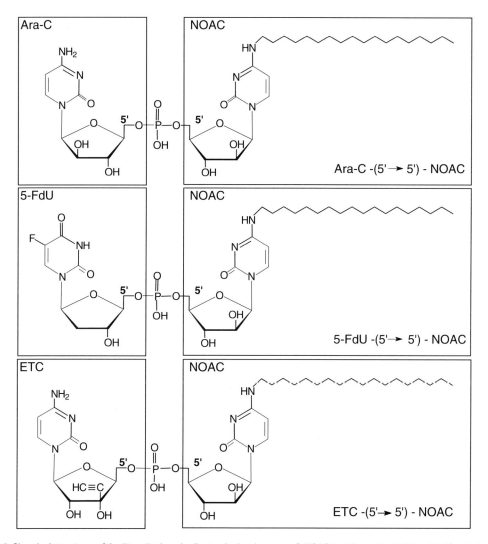

Fig. 2. Chemical structures of the 5′ → 5′ phosphodiester duplex drugs ara-C-NOAC (arabinocytidylyl-(5′ → 5′)-N^4-octadecyl-1-β-D-arabinofuranosylcytosine, mol. wt. 801 g/mol), 5-FdU-NOAC (2′-deoxy-5-fluorouridylyl-(5′ → 5′)-N^4-octadecyl-1-β-D-arabinofuranosylcytosine, mol. wt. 804 g/mol) and ETC-NOAC (3′-C-ethynylcytidylyl-(5′ → 5′)-N^4-octadecyl-1-β-D-arabinofuranosylcytosine, mol. wt. 825 g/mol)

The cytotoxic activity of such duplex drugs is expected to be more effective as compared to the monomeric nucleosides. Due to the combination of the effects of both active molecules that can be released in the cells as monomers or as the corresponding monophosphates, it can be anticipated that mono-phosphorylated nucleosides are directly formed in the cytoplasm after enzymatic cleavage of the duplex drugs. Thus, mono-phosphorylated molecules would not have to pass the first phosphorlyation step, which is known to be rate limiting.

In previous studies performed with similar heterodinucleoside phosphate dimers composed of the antivirally active nucleosides azidothymidine, dideoxycytidine and dideoxyinosine and formulated in liposomes we found significantly different pharmacokinetic properties and superior antiviral effects in comparison to the parent hydrophilic nucleosides (33, 34). Thus, the chemical modification of cytotoxic nucleosides and their formulation in liposomes render these new hetero-dinucleoside compounds interesting candidates for further developments.

Here, we present the methods of preparation of liposomes as carriers for lipophilic nucleosides and heterodinucleoside drugs. We do not describe in details the methods used to evaluate the cytotoxic properties of the lipophilic drug formulations. For comprehensive information, we refer to our publications and to the related literature.

2. Materials

2.1. Liposome Preparation (Extrusion Method)

1. Soy phosphatidylcholine (SPC) (L. Meyer GmbH, Hamburg, Germany), store at −20°C, prepare a stock solution, e.g. of 20–100 mg/mL by dissolving SPC in methanol/methylene chloride (1:1, v/v).
2. Cholesterol (see **Note 1**).
3. D,L-α-Tocopherol, store at −20°C, make a stock solution, e.g. of 10 mg/mL by dissolving D,L-α-tocopherol in methanol/methylene chloride (1:1, v/v).
4. Phosphate buffer, PB: 13 mM KH_2PO_4, 54 mM $NaHPO_4$, pH 7.4 (see **Note 3**).
5. Round bottom flasks (20–100 mL).
6. Rotatory evaporator, e.g. Rotavap (Büchi AG, Flawil, Switzerland).
7. Lipex™ high pressure extruder (35) (Northern Lipids Inc., 8855 Northbrook Court, Burnaby, BC, Canada, Website: http://www.northernlipids.com).

8. Nuclepore membranes of defined pore sizes: 400, 200, 100 nm (Sterlitech Corp., Kent, WA, USA or Sterico AG, Wangen, Switzerland).

9. Sterile filters 0.45- or 0.2-µm and plastic syringes, various suppliers.

3. Methods

The methods described in the following section outline (1) the preparation of liposomes by filter extrusion and (2), detergent dialysis. In the past decades, a large number of methods of liposome preparation have been developed and refined. For comprehensive information, we refer to corresponding chapters of this book volume and the literature (36, 37). We favor the use of the two methods described in the following section that are recommendable because of their ease, versatility and high quality of liposomes they produce.

3.1. Preparation of NOAC, ara-C-NOAC, 5-FdU-NOAC and ETC-NOAC Liposomes by High Pressure Filter Extrusion

1. Liposomes are prepared by sequential filter extrusion of the lipid/drug mixtures. The basic composition for the preparation of 5.0 mL liposomes is 1.0 g soy phophatidylcholine (SPC, L. Meyer GmBH, Hamburg, Germany), 125 mg cholesterol (Fluka, Buchs, Switzerland) (see **Note 1**), 6 mg D,L-α-tocopherol (Merck, Darmstadt, Germany) and the lipophilic drug at concentrations of 1–10 mg/mL.

2. The solid lipids and the lipophilic drugs (see Figs. 1 and 2), either as powder or stock solutions are dissolved in 5–10 mL methanol/methylene chloride (1:1, v/v) in a round bottom flask (see **Note 2**). PEG-modified liposomes are obtained by addition of PEG(2000)-DPPE (28 mg/mL) to the basic lipid mixtures (see **Note 3**).

3. After removal of the organic solvents by rotary evaporation (40–45°C, 60 min) the dry lipid mixture is solubilized with phosphate buffer PB (67 mM, pH 7.4) by vigorous agitation (see **Note 4**).

4. The mixture is then subjected to repetitive extrusion through Nuclepore polycarbonate (Sterlitech Corp., Kent, WA, USA or Sterico AG, Wangen, Switzerland) filters (400-, 200- and 100-nm pore size) using a Lipex™ Extruder (Northern Lipids, Inc.) (see **Note 5**).

5. Finally, the liposomes are sterilized by filtration (0.45- or 0.2-µm sterile filters). Mean hydrodynamic diameters of vesicles (liposomes, nanospheres, nanobeads) can be determined with dynamic laser light scattering instruments, e.g. the NICOMP

380 particle sizer, Particle Sizing Systems (Sta. Barbara, CA, USA). Incorporation of the lipophilic drugs is estimated to range between 95 and 100% according to previous determinations (38) (see **Note 5**).

3.2. Preparation of NOAC, ara-C-NOAC, 5-FdU-NOAC and ETC-NOAC Liposomes by Dialysis

1. Small unilamellar vesicles (SUV) of 50–200 nm mean size can also be prepared using detergent dialysis methods as described (39, 40). The same lipid/drug compositions as given in Subheading 3.1, **step 1** are used with the only difference that the detergent is also added to the organic solution. Controlled removal of detergent from mixed lipid/detergent/drug micelles yields liposomes of high size homogeneity and stability.
2. The detergent sodium cholate (see **Note 6**) is added at a ratio of total lipids to detergent of 0.6 mol, including the lipophilic drug.
3. The dry lipid/detergent/drug film is dispersed in PB (see **Note 4**) and left 1–2 h or over night at room temperature for equilibration.
4. Detergent is removed by controlled dialysis of the mixed micelles against 3–5 L of PB or PB-Man (volume ratio = 1 to 1000) for 12–15 h at room temperature, e.g. using a Mini-Lipoprep instrument (Harvard Apparatus, Holliston, Massachusetts, Website: http://www.harvardapparatus.com) (see **Notes 6** and **7**).

4. Conclusion

With the chemical transformation of water-soluble nucleosides into lipophilic compounds, followed by their incorporation into lipid bilayer membranes of liposomes, a new class of cytotoxic drug formulations is obtained that can be applied for the treatment of tumors by parenteral and oral routes. Lipophilic ara-C derivatives, particularly the extensively studied drug NOAC, and the novel duplex drugs composed of NOAC and the nucleosides ara-C, 5-FdU and ETC represent very promising new anticancer drugs of high cytotoxic activity, ability to circumvent resistance mechanisms, and strong apoptosis inducing capability.

We conclude that the chemical modification of water-soluble molecules by attachment of long alkyl chains and their stable incorporation into the bilayer membranes of small unilamellar liposomes represent a very promising example of taking advantage of the high loading capacity lipid bilayers offer for lipophilic drugs. The combination of chemical modifications of water soluble drugs of known pharmacological activities with their formulation in liposomes represents a valuable method for the development of novel pharmaceutical preparations, not only for the treatment of tumors or infectious diseases, but also for many other disorders.

5. Notes

1. Cholesterol (e.g. from Fluka, purum quality, >95%) should be recrystallized from methanol. Cholesterol of minor quality or purity should be avoided, since liposome membrane stability can be reduced.

2. Detachment of the lipid mixtures from the glass walls of the round bottom flasks can be accelerated by addition of small glass beads (2–3 mm diameter) and vigorous shaking. Preferably, the glass beads are added to the organic lipid solution before evaporation of the solvents. This will facilitate detachment and dispersion of the lipid film.

3. Other lipid compositions with synthetic lipids, hydrogenated SPC (HSPC) and PEG-modified phospholipids are often used, especially for liposome formulations intended for parenteral applications use (long circulating or "stealth" liposomes) (41). Several analytical methods to follow loss of lipids during the preparation steps are available. Radioactively labeled lipids (^{3}H-DPPC, ^{14}C-DPPC) or cholesterol (^{3}H-cholesterol) or ^{3}H-cholesteryl hexadecyl ether (NEN Life Science Products, Boston, MA, USA) or lipophilic fluorescence dyes (e.g. lipophilic BODIPY derivatives, Molecular Probes) are added at appropriate amounts to the initial lipid mixtures.

4. If the liposome preparations are intended to be stored for longer time periods, they may be frozen or lyophilized, provided that they are prepared in a phosphate buffer that contains a cryoprotectant. We use an iso-osmolar phosphate-mannitol buffer of the following composition: 20 mM phosphate buffer (0.53 g/L KH_2PO_4 plus 2.87 g/L $Na_2HPO_4.2H_2O$) plus 230 mM mannitol (42.0 g/L mannitol), (PB-Man)

5. The concentration of the lipophilic drugs NOAC, ara-C-NOAC, 5-FdU-NOAC or ETC-NOAC in the liposomes can be varied from 1 mg/mL to about 10 mg/mL, depending on the concentration required for biological activity (e.g. based on corresponding IC_{50}-values), the phospholipid concentration, the lipid composition and the method of liposome preparation. The concentrations of incorporated drugs can be determined by reverse phase HPLC (38).

6. The preparation of liposomes from mixed detergent/lipid micelles can also be done with other detergents, such as n-alkyl-glucosides (n=6–9), octyl-thioglucoside or N-octanoyl-N-

methylglucamin (MEGA-8, Fluka). Interestingly, the choice of detergent influences the size of the resulting liposomes. Thus, liposomes prepared from n-octyl-glucoside/phospholipid/cholesterol mixed micelles have an average size of 180 nm, whereas those made with n-hexyl-glucoside are 60 nm in diameter (39). Detergent removal by conventional dialysis using semipermeable dialysis tubes (e.g. Spectrapor, mol. wt. cut off 12,000–14,000 Da) is not recommended because, due to a concentration gradient which is formed within the dialysis tube, heterogenous and unstable liposomes will be produced.

7. When synthetic lipids are used, detergent removal has to be performed above the corresponding transition temperature T_c of the lipid. Hence, when for example dipalmitoylphosphatidyl choline (DPPC) is used as main liposome forming lipid, a temperature above its T_c of 41°C has to be chosen. Additional membrane forming components (cholesterol, lipophilic drugs, etc.) depress the T_c by several degrees.

References

1. Hofheinz RD, Gnad-Vogt SU, Beyer U, Hochhaus A (2005) Liposomal encapsulated anti-cancer drugs. Anticancer Drugs 16:691–707
2. Perez-Lopez ME, Curiel T, Gomez JG, Jorge M (2007) Role of pegylated liposomal doxorubicin (Caelyx) in the treatment of relapsing ovarian cancer. Anticancer Drugs 18: 611–617
3. Porter CA, Rifkin RM (2007) Clinical benefits and economic analysis of pegylated liposomal doxorubicin/vincristine/dexamethasone versus doxorubicin/vincristine/dexamethasone in patients with newly diagnosed multiple myeloma. Clin Lymphoma Myeloma S4: S150–S155
4. Thomas DA, Sarris AH, Cortes J, Faderl S, O'Brien S, Giles FJ, Garcia-Manero G, Rodriguez MA, Cabanillas F, Kantarjian H (2006) Phase II study of sphingosomal vincristine in patients with recurrent or refractory adult acute lymphocytic leukemia. Cancer 106:120–127
5. Hennenfent KL, Govindan R (2006) Novel formulations of taxanes: a review. Old wine in a new bottle? Ann Oncol 17:735–749
6. Strickley RG (2004) Solubilizing excipients in oral and injectable formulations. Pharm Res 21:201–230
7. Fahr A, Liu X (2007) Drug delivery strategies for poorly water-soluble drugs. Expert Opin Drug Deliv 4:403–416
8. ten Tije AJ, Verweij J, Loos WJ, Sparreboom A (2003) Pharmacological effects of formulation vehicles: implications for cancer chemotherapy. Clin Pharmacokinet 42:665–685
9. Brusa P, Immordino ML, Rocco F, Cattel L (2007) Antitumor activity and pharmacokinetics of liposomes containing lipophilic gemcitabine prodrugs. Anticancer Res 27: 195–199
10. Bergman AM, Kuiper CM, Noordhuis P, Smid K, Voorn DA, Comijn EM, Myhren F, Sandvold ML, Hendriks HR, Fodstad O, Breistol K, Peters GJ (2004) Antiproliferative activity and mechanism of action of fatty acid derivatives of gemcitabine in leukemia and solid tumor cell lines and in human xenografts. Nucleosides Nucleotides Nucleic Acids 23:1329–1333
11. Stevens PJ, Sekido M, Lee RJ (2004) A folate receptor-targeted lipid nanoparticle formulation for a lipophilic paclitaxel prodrug. Pharm Res 21:2153–2157
12. Pignatello R, Puleo A, Puglisi G, Vicari L, Messina A (2003) Effect of liposomal delivery on in vitro antitumor activity of lipophilic conjugates of methotrexate with lipoamino acids. Drug Deliv 10:95–100
13. Zerouga M, Stillwell W, Jenski LJ (2002) Synthesis of a novel phosphatidylcholine conjugated to docosahexaenoic acid and methotrexate that inhibits cell proliferation. Anticancer Drugs 13:301–311

14. Harrington KJ, Syrigos KN, Uster PS, Zetter A, Lewanski CR, Gullick WJ, Vile RG, Stewart JS (2004) Targeted radiosensitisation by pegylated liposome-encapsulated 3′,5′-O-dipalmitoyl 5-iodo-2′-deoxyuridine in a head and neck cancer xenograft model. Br J Cancer 91:366–373
15. Hamada A, Kawaguchi T, Nakano M (2002) Clinical pharmacokinetics of cytarabine formulations. Clin Pharmacokinet 41:705–718
16. Rubas W, Supersaxo A, Weder HG, Hartmann HR, Hengartner H, Schott H, Schwendener RA (1986) Treatment of murine L1210 leukemia and melanoma B16 with lipophilic cytosine arabinoside prodrugs incorporated into unilamellar liposomes. Int J Cancer 37:149–154
17. Schwendener RA, Schott H (1992) Treatment of L1210 murine leukemia with liposome – incorporated N^4-hexadecyl-1-β-D-arabinofuranosyl-cytosine. Int J Cancer 51:466–469
18. Schwendener RA, Schott H (2005) Lipophilic arabinofuranosyl cytosine derivatives in liposomes. Methods Enzymol 391:58–70
19. Horber DHC, Schott H, Schwendener RA (1995) Cellular pharmacology of a liposomal preparation of N^4-hexadecyl-1-β-D-arabinofuranosylcytosine, a lipophilic derivative of 1-β-D-arabinofuranosylcytosine. Br J Cancer 71:957–962
20. Horber DH, von Ballmoos P, Schott H, Schwendener RA (1995) Cell cycle dependent cytotoxicity and induction of apoptosis by N^4-hexadecyl-1-β-D-arabinofuranosylcytosine, a new lipophilic derivative of 1-β-D-arabinofuranosylcytosine. Br J Cancer 72:1067–1073
21. Horber DH, Schott H, Schwendener RA (1995) Cellular pharmacology of N^4-hexadecyl-1-β-D-arabinofuranosylcytosine (NHAC) in the human leukemic cell lines K-562 and U-937. Cancer Chemother Pharmacol 36:483–492
22. Schwendener RA, Friedl K, Depenbrock H, Schott H, Hanauske AR (2001) In vitro activity of liposomal N^4-octadecyl-1-β-D-arabinofuranosylcytosine (NOAC), a new lipophilic derivative of 1-β-D-arabino-furanocylcytosine on biopsized clonogenic human tumor cells and hematopoietic precursor cells. Invest New Drugs 19:203–210
23. Wasan KM, Dion R, Brocks DR, Lee SD, Sachs-Barrable K, Thornton SJ (2008) Impact of lipoproteins on the biological activity and disposition of hydrophobic drugs: implications for drug discovery. Nat Rev Drug Discov 7:84–99
24. Rensen PC, de Vrueh RL, Kuiper J, Bijsterbosch MK, Biessen EA, van Berkel TJ (2001) Recombinant lipoproteins: lipoprotein-like lipid particles for drug targeting. Adv Drug Deliv Rev 47:251–276
25. Koller-Lucae SKM, Schott H, Schwendener RA (1997) Pharmacokinetic properties in mice and interactions with human blood in vitro of liposomal N^4-octadecyl-1-β-D-arabinofuranosylcytosine (NOAC), a new anticancer drug. J Pharmacol Exp Ther 282:1572–1580
26. Koller-Lucae SKM, Suter MJ, Rentsch KM, Schott H, Schwendener RA (1999) Metabolism of the new liposomal anticancer drug N^4-octadecyl-1-β-D-arabinofuranosylcytosine (NOAC) in mice. Drug Metab Dispos 27:342–350
27. Koller-Lucae SM, Schott H, Schwendener RA (1999) Low density lipoprotein and liposome mediated uptake and cytotoxic effect of N^4-octadecyl-1-β-D-arabinofuranosylcytosine (NOAC) in Daudi lymphoma cells. Br J Cancer 80:1542–1549
28. Horber DH, Cattaneo-Pangrazzi RM, von Ballmoos P, Schott H, Ludwig PS, Eriksson S, Fichtner I, Schwendener RA (2000) Cytotoxicity, cell cycle perturbations and apoptosis in human tumor cells by lipophilic N^4-alkyl-1-β-D-arabinofuranosylcytosine derivatives and the new heteronucleoside phosphate dimer arabinocytidylyl-(5′→5′)-N^4-octadecyl-1-β-D-ara-C. J Cancer Res Clin Oncol 126:311–319
29. Cattaneo-Pangrazzi RM, Schott H, Wunderli-Allenspach H, Derighetti M, Schwendener RA (2000) Induction of cell cycle-dependent cytotoxicity and apoptosis by new heterodinucleoside phosphate dimers of 5-fluorodeoxyuridine in PC-3 human prostate cancer cells. Biochem Pharmacol 60:1887–1896
30. Cattaneo-Pangrazzi RM, Schott H, Schwendener RA (2000) The novel heterodinucleoside dimer 5-FdU-NOAC is a potent cytotoxic drug and a p53-independent inducer of apoptosis in the androgen-independent human prostate cancer cell lines PC-3 and DU-145. Prostate 45:8–18
31. Marty C, Ballmer-Hofer K, Neri D, Klemenz R, Schott H, Schwendener RA (2002) Cytotoxic targeting of F9 teratocarcinoma tumours with anti-ED-B fibronectin scFv antibody modified liposomes. Br J Cancer 87:106–112
32. Takatori S, Kanda H, Takenaka K, Wataya Y, Matsuda A, Fukushima M, Shimamoto Y, Tanaka M, Sasaki T (1999) Antitumor mechanisms and metabolism of the novel antitumor nucleoside analogues, 1-(3-C-ethynyl-β-D-ribo-pentofuranosyl)cytosine and 1-(3-C-ethynyl-β-D-ribo-pentofuranosyl)uracil. Cancer Chemother Pharmacol 44:97–104

33. Schwendener RA, Gowland P, Horber DH, Zahner R, Schertler A, Schott H (1994) New lipophilic acyl/alkyl dinucleoside phosphates as derivatives of 3′-azido-3′-deoxythymidine: inhibition of HIV-1 replication in vitro and antiviral activity against Rauscher leukemia virus infected mice with delayed treatment regimens. Antiviral Res 24:79–93

34. Peghini PA, Zahner R, Kuster H, Schott H, Schwendener RA (1998) *In vitro* inhibition of hepatitis B virus replication and pharmacokinetic properties of new lipophilic dinucleoside phosphate derivatives. Antivir Chem Chemother 9:117–126

35. Mayer LD, Hope MJ, Cullis PR (1986) Vesicles of variable sizes produced by a rapid extrusion procedure. Biochim Biophys Acta 858:161–168

36. Gregoriadis G (ed) (2007) Liposome technology, 3rd edn, vol I–III. Informa Healthcare, Inc., New York

37. Torchilin VP (2003) Weissig V (eds) Liposomes: a practical approach, 2nd edn. The Practical Approach Series, 264. Oxford University Press, Oxford

38. Rentsch KM, Schwendener RA, Schott H, Hänseler E (1997) Pharmacokinetics of N^4-octadecyl-1-β-D-arabinofuranosylcytosine (NOAC) in plasma and whole blood after intravenous and oral application in mice. J Pharm Pharmacol 49:1076–1081

39. Schwendener RA, Asanger M, Weder HG (1981) The preparation of large bilayer liposomes: controlled removal of *n*-alkyl-glucoside detergents from lipid/detergent micelles. Biochem Biophys Res Commun 100: 1055–1062

40. Schwendener RA (1986) The preparation of large volumes of homogeneous, sterile liposomes containing various lipophilic cytostatic drugs by the use of a capillary dialyzer. Cancer Drug Deliv 3:123–129

41. Allen TM (1994) Long-circulating (sterically stabilized) liposomes for targeted drug delivery. Trends Pharmacol Sci 15: 215–220

Chapter 9

Remote Loading of Anthracyclines into Liposomes

Felicitas Lewrick and Regine Süss

Abstract

The following chapter introduces a remote loading procedure for anthracyclines focussing on the well-established drug doxorubicin.

The key advantage of remote loading is that it leads to higher drug to lipid ratios and encapsulation efficiencies compared to conventional passive trapping techniques like hydration of dried lipid films with aqueous drug solutions.

The method presented is appropriate to produce sterile liposomal doxorubicin formulations with a final concentration of 2 mg/mL doxorubicin, which can be applied not only in vitro but also in vivo.

Key words: Aseptic remote loading, Doxorubicin, Ammonium sulfate, Tangential flow filtration

1. Introduction

Various agents from the substance class of anthracyclines act as effective anticancer agents. The mostly used drug of this class is doxorubicin which is applied in the treatment of a broad range of solid tumors and tumors of hematological origin.

However, the administration of free doxorubicin often leads to dose-limiting side effects such as cardiotoxicity and myelosupression. This toxicity can be reduced by liposomal encapsulation of doxorubicin because of the modified biodistribution of the drug (1). Additionally, the efficiency of the drug is improved due to the passive targeting effect of liposomes.

An optimal loading procedure for doxorubicin into liposomes aims at a high drug to lipid ratio and an encapsulation efficiency of almost 100% to render the separation of unencapsulated drug unnecessary.

For weak bases like doxorubicin, this can be achieved by remote loading, which is based on the generation of a transmembrane pH-gradient as driving force for an accumulation of the drug inside the liposome.

Several direct and indirect methods to produce such a gradient across the liposomal bilayer are described in the literature (2). All of these concepts follow a common principle: The interior pH of the liposome is acidified, whereas the exterior pH-value is adjusted to physiological conditions. The uncharged DXR which is incubated with these liposomes diffuses into the vesicles. It becomes protonated intravesiculary which inhibits membrane repermeation and results in an accumulation inside the liposomes (3). Due to the doxorubicin precipitation either through drug self association or precipitation with salts present in the interior liposomal buffer, it is possible to attain doxorubicin levels within the liposome that exceed the solubility of the drug (2).

This chapter introduces loading via an ammonium sulfate gradient as shown in Fig. 1. It is one of the most common remote loading procedures which is also employed for the commercially available liposomal doxorubicin formulation Caelyx®.

The required pH-gradient is achieved indirectly. After encapsulation of an ammonium sulfate buffer the extraliposomal buffer is exchanged by an ammonium ion-free buffer. Given the higher permeation coefficient of ammonia as compared to the permeation coefficient of protons the ammonium ion gradient leads to a pH-gradient resting stable over the entire loading period (2–5).

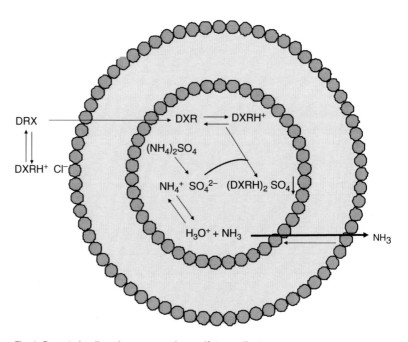

Fig. 1. Remote loading via an ammonium sulfate gradient

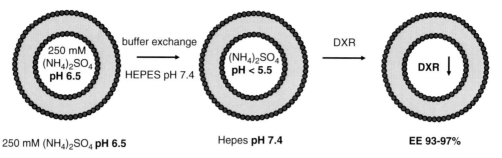

Fig. 2. Three steps of the remote loading procedure

The remote loading procedure consists of three steps: preparation of the liposomes, establishment of an ion-gradient and loading of the drug into the preformed liposomes (Fig. 2).

For in vivo experiments, it is necessary to achieve sufficient drug concentrations in small volumes (2 mg/mL) and the preparations should be sterile. Especially, the required concentration of 2 mg/mL requires a system for buffer exchange without diluting the liposomes. Furthermore, liposomes cannot be sterilized after preparation and must thus be produced under aseptic conditions. These points were taken into account for the method described here.

2. Materials

1. Doxorubicin hydrochloride (Ph. Eur. Reference standard) from the European Directorate for the Quality of Medicine and Healthcare (Council of Europe), Strasbourg, France. Store at –20 °C.
2. Glass vials for lyophilisation as well as corresponding rubber stopper with 3 mL fill volume (Schott, Jena, Germany). These are autoclaved before use.
3. Soy phosphatidylcholine (SPC, >98% purity) (Lipoid, Ludwigshafen, Germany). Store at –20 °C. Unfreeze in an exsiccator before us.
4. Cholesterol (Chol). Cholesterol was once recrystallized from ethanol (2 g/100 mL) before use. Store at –20 °C. Unfreeze in an exsiccator before us.
5. HEPES buffered saline pH 7.4: 140 mM NaCl and 10 mM N-[2-Hydroxyethyl] piperazin-N'-[2-ethanesulfonic acid] (HEPES, Carl Roth, Karlsruhe, Germany) are dissolved in purified water, adjusted to pH 7.4 and autoclaved.
6. Ammonium sulfate buffer: 250 mM ammonium sulfate are dissolved in purified water, adjusted to pH 6.5 and autoclaved.
7. Lipex Basic Extruder, Northern Lipids Inc., Burnaby, BC, Canada. The extrusion device is autoclaved before use.

8. Polycarbonate membranes 200 and 80 nm (Nuclepore, Pleasanton, USA).

9. Minimate™ Tangential Flow Filtration Capsule 50 kDa cut-off and Ultrasette™ Lab Tangential Flow Device 10 kDa cut-off/screen channel (Pall Filtron, Dreieich, Germany). Both devices are stored in sterile 0.1N sodium hydroxide solution and sanitized before use (see Subheading 3.3).

10. Microkros® Hollow Fiber Modules (Spectrum Labs, Eindhoven, Netherlands) for single use.

11. Zetamaster S (Malvern, Herrenberg, Germany) with auto analysis option, Malvern Software Version 1.4.1 for determination of the liposomal hydrodynamic diameter by photon correlation spectroscopy (PCS).

12. Cholesterol FS® (DiaSys Diagnostic Systems, Holzheim, Germany) for rapid cholesterol quantification (analysis within 5 min).

3. Methods

3.1. Aseptic Preparation of Doxorubicin Aliquots

The solid doxorubicin HCl is dissolved in sterile purified water at a concentration of 1 mg/mL. Two milliliters of this solution are filled as single use aliquot into each autoclaved lyophilisation vial under the laminar flow bench with light protection. Immediately afterwards, the lyophilisation process is started (freezing: 1 h at −50 °C, main drying: 42 h at −20 °C, secondary drying: 6 h at 30 °C). Afterward, the aliquots can be stored at 4 °C for up to 3 months.

3.2. Preparation of Liposomes

1. The basic liposome composition used in this protocol consists of SPC/Chol 7:3 (molar ratio). Lipids are dissolved in dichloromethane in an autoclaved round bottom flask and the solvent is dried by rotary evaporation followed by high vacuum for 1 h. This lipid film can be stored at −20 °C for several weeks.

2. The following steps take place under the laminar flow bench using sterile vials, syringes and needles.

3. The lipid film is hydrated using the appropriate volume of sterile ammonium sulfate buffer 250 mM pH 6.5 yielding final total lipid concentrations of 20 mM (see Note 1).

4. The resulting dispersion of multilamellar large vesicles is homogenized by extrusion using the Lipex Basic Extruder (fill volume: 10 mL) driven by nitrogen pressure up to 12 bars. For this purpose, the dispersion was extruded seven times through polycarbonate membranes with 200 nm pores and 11 times through 80 nm pores.

5. The particle size was determined by photon correlation spectroscopy (PCS) to be within a range of 105 ± 10 nm with a polydipersity index below 0.08. Liposomes can be stored up to 2 weeks at 2–8 °C.

3.3. Establishment of a Gradient (see Note 2)

1. Sanitization of tangential flow devices Minimate™ and Ultrasette™: Prior to use, the tangential flow devices are sanitized with sterile 0.1N NaOH solution at 35 °C. Thus, 200 mL of the sanitization fluid are added to a reservoir, tempered and circulated for about 1 h at moderate pump rates (50–100 mL/min).

2. Flushing of tangential flow devices Minimate™ and Ultrasette™: After sanitization, the tangential flow devices are flushed with sterile water at room temperature. At least 500 mL should be pumped through the Minimate™ and at least 1 L should be pumped through the Ultrasette™. For the Minimate™ devices the flow through the filtrate tubing is increased by producing a backpressure by tightening the retentate screw clamps. The clamps should be tightened carefully until the filtrate flow rate is approximately equal to the retentate flow rate. This procedure has to be repeated with sterile HEPES buffered saline pH 7.4 which shall be exchanged against the exterior buffer of the liposomes (ammonium sulfate buffer) in the next step.

3. The buffer exchange is performed with the Ultrasette™ (see Note 3). Therefore, initially 30 mL of the 20 mM liposomes in ammonium sulfate buffer are diluted with 30 mL of HEPES buffered saline pH 7.4 in the reservoir. Tangential filtration is started immediately stirring constantly with a pump rate of 500 mL/min. The amount of filtrate is replaced continuously with HEPES buffered saline pH 7.4 to keep the volume in the reservoir at about 60 mL until further 270 mL HEPES buffered saline are added. Thereafter, liposomes are concentrated as high as possible (approximately to a volume of 50 mL). The following steps are performed to optimize the recovery of the product. The filtrate tubing is closed during further circulation of the liposomes for 2 min. The end of the retentate tubing is then placed into a collection vessel and the liposomes are pumped out. When the reservoir volume reaches the bottom, HEPES buffer saline pH 7.4 is added up to the volume of the hold up volume of the system (approximately 40 mL). The buffer is pumped into the system once again until the reservoir is almost empty to replace the volume containing liposomes. At the end of this procedure, the volume of the liposomal preparation is about 100 mL.

4. Two Minimate™ devices connected in parallel are used to regain a concentration of liposomes of 20 mM. Tangential filtration is continued immediately with this device with a flow rate of 100 mL/min until the volume in the reservoir is about 15 mL.

Then, filtrate tubings are again closed during further circulation of the liposomes for 2 min to optimize product recovery before the liposomes are pumped out (see **Note 4**).

5. Liposomes are sterile filtered through 0.2-μm pores.

6. The concentration of cholesterol in the resulting liposomal preparation is determined using a quantification kit (cholesterol FS®). The advantage of this assay is the quick performance. The samples can be measured within 5 min at 37 °C. The concentration of cholesterol allows calculating the total lipid concentration of the liposomes (which should be at least 20 mM).

3.4. Loading

1. The required amount of doxorubicin aliquots is equilibrated to room temperature.

2. One milliliter of the at least 20 mM liposomal preparation is pipetted to each 2 mg doxorubicin HCl lyophilisate under the laminar flow bench. The vial is closed and vortexed until the doxorubicin HCl is dissolved completely in the exterior Hepes buffer saline pH 7.4 (see **Notes 2** and **5**).

3. The solution is stored at 7 °C over night (see **Note 6**). During this period of time, loading takes place.

4. The resulting preparation is stable for 12 days with respect to encapsulation efficiency and size if stored at 2–8 °C.

3.5. Quantification

1. The encapsulation efficiency is quantified after the loading procedure based on fluorescence dequenching of self-associated doxorubicin in liposomes and own fluorescence of diluted doxorubicin outside the liposomes. Measurements are performed at λ_{exc} 480 nm and λ_{em} 590 nm. Ten microliters of the sample are added to 3 mL HEPES buffered saline and measured immediately. Afterwards, the 100% value of dequenched doxorubicin of the measured sample is determined by addition of 10 μL Triton X-100 (10% V/V) (5). This method yields encapsulation efficiencies of >95%.

2. The doxorubicin concentration is determined at 495 nm using UV/vis measurement after lysis of the liposomes with Triton X-100 (final concentration of 0.5% V/V) (5).

4. Notes

1. For the method described earlier, at least 30 mL of a 20 mM liposomal preparation are necessary to result in a final concentration of 20 mM liposomes after buffer exchange due to the retention volumes of the tangential flow devices (Ultrasette: 40 mL, 2 Minimate connected in parallel: 15 mL).

2. The ammonium gradient and hence the pH gradient resulting from buffer exchange is only stable during the loading period. At the beginning the interior pH value declines to a pH less than 5.5, whereas the exterior buffer is at pH 7.4. For SPC/Chol liposomes (using the fluorescent probe HPTS) the interior pH was found to reach pH 6.8 after 2 h changing to approximately pH 7.4 after 12 h. Thus, buffer exchange, determination of the total lipid concentration (measured by cholesterol determination) and loading must be performed in a single working process without interruption.

3. Buffer exchange must be as efficient as possible because ammonium sulfate residues in the external buffer lead to a considerable decrease of encapsulation efficiencies.

4. MicroKros® devices can be used as an alternative method for buffer exchange. These devices are hollow fibre cartridges for single use offered for different process volumes from 2 to 50 mL. They allow for a quick buffer exchange without diluting and can be used by manual operation or in connection with a peristaltic pump. For efficient buffer exchange, the sample is diluted 1:1 (V/V) with HEPES buffered saline and concentrated to the original volume. This step is repeated eight times.

5. The ratio of total lipid to doxorubicin should be at least 3:1 (mol/mol) for successful remote loading. The preparation described in this chapter (1 mL 20 mM liposomes/2 mg doxorubicin) corresponds to a ratio of approximately 6:1 (mol/mol).

6. For liposomal loading and storage of the liposomes, temperatures less than 2 °C must be avoided to prevent lower encapsulation efficiencies and leakage.

References

1. Gabizon A, Dagan A, Goren D, Barenholz Y, Fuks Z (1982) Liposomes as in vivo carriers of adriamycin: reduced cardiac uptake and preserved antitumor activity in mice. Cancer Res 42:4734–4739
2. Abraham, S. A., Waterhouse, D. N., Mayer, L. D., Cullis, P. R., Madden, T. D., and Bally, M. B (2005) The liposomal formulation of doxorubicin. Methods Enzymol 391:71–97
3. Haran G, Cohen R, Bar LK, Barenholz Y (1993) Transmembrane ammonium sulfate gradients in liposomes produce efficient and stable entrapment of amphipathic weak bases. Biochim Biophys Acta 1151:201–215
4. Barenholz Y (2001) Liposome application: problems and prospects. Curr Opin Colloid Interface Sci 2001:66–77
5. Fritze A, Hens F, Kimpfler A, Schubert R, Peschka-Süss R (2006) Remote loading of doxorubicin into liposomes driven by a transmembrane phosphate gradient. Biochim Biophys Acta 1758:1633–1640

Chapter 10

Arsonoliposomes: Preparation and Physicochemical Characterization

Sophia G. Antimisiaris and Panayiotis V. Ioannou

Abstract

Arsonoliposomes (ARSL) which are liposomes that contain arsonolipids in their membranes have shown interesting anticancer and antiparasitic activity in vitro. Their lipid composition (the specific arsonolipids and/or phospholipids used for their preparation, and the relative amounts of each lipid type) highly influences their physicochemical properties as well as their in vivo kinetics and antiparasitic activity; however, their cytotoxicity towards cancer cells is minimally – if at all – modified. ARSL are prepared by a modification of the "one step" method followed or not by sonication (for formation of sonicated or non-sonicated ARSL, respectively). Arsonoliposomes may be composed only of arsonolipids (containing or not cholesterol) [plain ARSL], or they may contain mixtures of arsonolipids with phospholipids (with or without Chol) [mixed ARSL]. Herein, we describe in detail the preparation and physicochemical characterization of ARSL.

Key words: Arsenic, Arsenolipid, Arsonolipid, Liposomes, Arsonoliposomes, Anti-cancer, Anti-parasitic, Drug delivery, Biodistribution

Abbreviations

Ars	Arsonolipid
ARSL	Arsonoliposome
As	Arsenic
C6	Brain Glioma cell line (Rat)
CF	5,6-Carboxyfluorescein
Chol	Cholesterol
Cryo-EM	Cryo electron microscopy
DMSO	Dimethyl sulfoxide
DPPC	1,2-Dipalmitoyl-*sn*-glyceroyl-3-phosphocholine
DPPE-PEG2000	1,2-Dipalmitoyl-*sn*-glyceroyl-3-phosphoethanolamine conjugated to polyethylene glycol (MW 2000)
DSPC	1,2-Distearoyl-*sn*-glyceroyl-3-phosphocholine
DSPE-PEG2000	1,2-Distearoyl-*sn*-glyceroyl-3-phosphoethanolamine conjugated to polyethylene glycol (MW 2000)

EDTA	Ethylenediamine tetraacetic acid
EM	Electron microscopy
FCS	Fetal calf serum
GH3	Pituitary tumor cells
HL-60	Human leukemia cell line
HUVEC	Human umbilical vein endothelial cell
IC50	50% Growth inhibition concentration
MEC	Minimum effective concentration
NB4	Human leukemia cell line
P	Phosphorus
PBS	Phosphate buffered saline
PC	Phosphatidylcholine
PC3	Prostate cancer cell line
PCS	Photon correlation spectroscopy
PEG	Polyethylene glycol
SUV	Small unilamellar vesicles
TBS	Tris-buffered saline

1. Introduction

Arsenolipids are naturally occurring arsenic-containing lipids that have been discovered in natural sources (1). The arsonolipids or 2,3-diacyloxypropylarsonic acids, are analogs of phosphonolipids (Fig. 1 – compound 1), in which P has been replaced by As (Fig. 1 – compound 2) and they have not been discovered in nature. The synthesis of arsonolipids [Ars] has been explored and a simple one-pot method with moderate yield (2) is currently available for the preparation of racemic or optically active arsonolipids. Recently, this method was reinvestigated (3). After being synthesized and characterized, it was anticipated that Ars may express selective anticancer activity, by being incorporated into the membranes of cancerous cells which may result in the modification of their

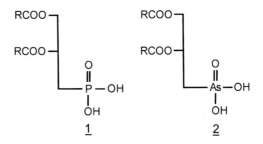

Fig. 1. Chemical structure of phosphonolipids (1) and arsonolipids (2). Arsonolipids (Ars) with R = lauric acid (C12); myristic acid (C14); palmitic acid (C16), and stearic acid (C18) were used for ARSL construction.

organization (due to differences between the polar head groups of arsonolipids and phospholipids). Another mode of action can be the reduction of As(V) to As(III) by biological thiols (4, 5) giving 2,3-diacyloxypropyldithioarsonites [R-As(SR′)$_2$]. This can be especially true in the case of certain cancer cell types that are known to have increased thiol levels (in comparison to normal cells) (6). When arsonolipids (dispersed in DMSO) were found to be inactive against various cancer cell lines in NIH (US-National Institute of Health) screening tests (7), it was hypothesized that perhaps, if these lipids were incorporated in vesicular structures they may interact differently with cancer cells and this may result in increased cytotoxicity.

The formation of liposomes [or better arsonoliposomes (ARSL)], composed solely of arsonolipids (Ars with R = lauric acid (C12); myristic acid (C14); palmitic acid (C16) and stearic acid (C18) (Fig. 1) have been used for ARSL construction), mixed or not with cholesterol (Chol) (plain ARSL), or composed of mixtures of Ars and phospholipids (as phosphatidylcholine [PC] or 1,2-distearoyl-*sn*-glyceroyl-PC [DSPC]) and containing or not Chol (mixed ARSL), was not an easy task. Several liposome preparation techniques (thin-film hydration, sonication, reversed phase evaporation, etc.) were initially tested, but were not successful to form vesicles. Thereby a modification of the so called "one step" or "bubble" technique (8), in which the lipids (in powder form) are mixed at high temperature with the aqueous medium, for an extended period of time, was developed. This technique was successful for the preparation of arsonoliposomes (plain and mixed) (9). If followed by probe sonication, smaller vesicles (compared to those formed without any sonication [nonsonicated]) could be formed [sonicated ARSL] (9). Additionally, sonicated PEGylated ARSL (ARSL that contain polyethyleneglycol [PEG]-conjugated phospholipids in their lipid bilayers) were prepared by the same modified one-step technique followed by sonication (10).

By applying the modified one-step liposome-preparation technique (that is described in detail in Subheading 3), several types of plain or mixed ARSL, composed of different lipids and at different ratios (Table 1), some also PEGylated, were prepared and characterized physicochemically (9–12). Vesicle mean diameters and size distributions, vesicle zeta-potential and efficiency to encapsulate hydrophilic substances (e.g. calcein) have been measured, and the physical stability of ARSL was evaluated by measuring their size distribution during extend periods of storage.

In order to evaluate the membrane integrity of the different ARSL types, the release of vesicle-encapsulated 5,6-carboxyfluorescein (9) or calcein (10, 11), has been measured.

As the in vivo absorption of certain ARSL types was very low (13), as demonstrated after intraperitoneal injection of ARSL in

Table 1
Arsonoliposomes that have been constructed and corresponding physicochemical properties. Values are from refs. 9–12

Ars	ARSL lipid comp.	ARSL type[a]	Mean diam. (nm)	Z-potential (mV)	Stability[b] MI-B/MI-FCS/PS	Ca^{2+}-induced aggregation
C12	Plain Ars	1/son	251/116	−46.4/−50.8	VP/−/	−
	Ars/Chol 1:1	1	247	−	VG/−/−	−
	Ars/DSPC 1:1	1	346	−	P/−//	−
	Ars/Chol 2:1	son	103	−63.5	−/−/VG	VP
	PC/Ars/Chol 12:8:10	son	73	−42.0	−/−/VG	VP
	PC/Ars/Chol 17:3:10	son	63	−23.9	−/−/VG	−
C14	Plain Ars	1/son	329/118	−44.1/−51.3	P/−/−	−
	Ars/Chol 1:1	1	384	−	G/−/−	−
	Ars/DSPC 1:1	1	−	−	P/−/−	−
	Ars/Chol 2:1	son	108	−65.4	G/−/VG	−
	PC/Ars/Chol 12:8:10	son	87	−43.0	G/G/VG	−
	PC/Ars/Chol 17:3:10	son	76	−28.2	−/−/VG	−
C16	Plain Ars	1/son	273/130	−57.7/−57.2	G/−/	−
	Ars/Chol 1:1	1	362	−	VG/−/	−
	Ars/DSPC 1:1	1	247	−	P/−/	−
	Ars/Chol 2:1	son	111	−69.5	G/VP/VG	VP
	PC/Ars/Chol 12:8:10	son	91	−50.3/−32.2 (in EDTA)[c]	G/NG/G	VP
	PC/Ars/Chol 17:3:10	son	78		G/NG/G	−
	DSPC/Ars/Chol 12:8:10/+PEG+	son	80/103	−42.1/−14.9 (in EDTA)[c]	VG/VG/VG	VG
	DSPC/Ars/Chol 17:3:10/+PEG+	son	100/92	−26.8/−2.9 (in EDTA)[c] −17.4/−4.0 (in EDTA)[c]	VG/NG/VG	VG
C18	Plain Ars	1/son	265/−	−38.5/−40.1	VP/−/VP	VP
	Ars/Chol 1:1	1	427	−	VG/−/VP	VP
	Ars/DSPC 1:1	1	290	−	P/−/VP	VP
	Ars/Chol 2:1	son	121	−59.2	G/−/G	VP
	PC/Ars/Chol 12:8:10	son	93	−48.4	−/−/VG	VP
	PC/Ars/Chol 17:3:10	son	75	−32.1	−/−/G	−

[a]ARSL preparation method (*1* one step; *son* sonicated)
[b]*MI-B* membrane integrity in buffer; *MI-FCS* membrane integrity in FCS; *PS* physical stability (mean size stability)
[c]Measured in presence of 1 mM EDTA
VP very poor; *P* poor; *NG* not good; *G* good; *VG* very good

mice; it was hypothesized that this was due to (blood) calcium-induced vesicle aggregation (a logical assumption if one considers the negative charge of arsonolipids, and highly negative zeta-potential values of ARSL (Table 1)). Indeed, after evaluation of the calcium-induced aggregation of ARSL by turbidity and size distribution measurements, it was shown that certain ARSL compositions were highly aggregated and subsequently fused into very large vesicles, in the presence of physiologically relevant calcium concentrations (12). Nevertheless, some types of ARSL and especially PEGylated ARSL were found to have high membrane integrity (10–12) and also retain their size after prolonged incubation in divalent cation-containing media, as presented in Table 1 (10).

The physicochemical differences between the various types of ARSL which were prepared and studied, were found to influence their in vivo kinetics (14) and their antiparasitic/antitrypanocidal activity (15, 16), but not their in vitro anticancer activity, at least not to a substantial level (17) (Table 2). In a series of in vitro studies (17–19), on several types of cancer cell lines (HL-60, C6, PC3, NB4, GH3), it has been demonstrated that all the types of ARSL studies (Table 2) have increased cytotoxicity towards cancer cells (as measured after 24 or 48 h of co-incubation) in comparison with normal cell types (as HUVEC cells) evaluated under the same experimental conditions.

ARSL types which have been found to have selective anticancer activity and substantial (for in vivo applications) stability, are currently being evaluated as carriers of anticancer drugs (encapsulated in their aqueous compartments and/or incorporated in their membrane) that may be delivered preferentially to cancer cells with the help of specific ligands.

Herein, we discuss in detail the preparation, physicochemical characterization, and evaluation of in vitro integrity and physical stability of arsonoliposomes.

2. Materials

2.1. Arsonoliposome Preparation

1. *rac*-Arsonolipids [Ars] (with acyl chains: lauryl (C12), myristyl (C14), palmityl (C16), and stearyl (C18)), are synthesized as described previously (2, 3). They are stored desiccated in aliquots (−80°C).

2. Egg l-α-phosphatidylcholine [PC] (grade 1) (Lipid Products, Nutfield, UK, or Lipoid, DE), is used in solid state or dissolved (20 or 100 mg/mL) in a mixture of $CHCl_3/CH_3OH$ (2:1 v/v), and stored in aliquots at −80°C. The 99% purity of all lipids is verified by thin layer chromatography (see Note 1).

3. 1,2-Distearoyl-*sn*-glyceroyl-3-phosphocholine [DSPC] (synthetic – grade 1), and 1,2-dipalmitoyl-*sn*-glyceroyl-3-phosphocholine [DPPC] (Lipid Products, Nutfield, UK, or Lipoid, DE or Avanti Polar Lipids, USA). Storage conditions and purity tests (see Note 1) are the same as mentioned earlier for PC (with the difference that only 20 mg/mL solutions are made for these lipids).

Table 2
Arsonoliposomes that have been evaluated for anticancer and antiparasitic activity and in vivo distribution. Values are from refs. 13–19

Ars	ARSL lipid comp.	Anti-parasitic activity (in vitro)		Anti cancer activity[a] IC50 (×10⁻⁵ M) (in vitro)							In vivo distribution %Dose/tissue in blood (1 h post-injection [IP])
		IC50 (μM)[b] L-donov	MEC[c] (μM) T-bb	HL60	C6	GH3	PC3	NB4	HUVEC		
C12	Ars/Chol 2:1	1.70	0.58	4.8	4.8	1.6	–	–	70	–	
	PC/Ars/Chol 12:8:10	0.43	0.87	4.2	4.8	1.2	–	–	85	–	
	PC/Ars/Chol 17:3:10	–	–	11.3	–	1.9	–	–	–	–	
C14	Ars/Chol 2:1	–	–	1.10	6.7	9.0	–	–	161	–	
	PC/Ars/Chol 12:8:10	–	–	0.71	5.2	13.6	–	–	–	–	
	PC/Ars/Chol 17:3:10	–	–	0.71	–	–	–	–	–	–	
C16	PC/Ars/Chol 12:8:10	0.97	0.24	0.85	3.7	3.1	1.2	2.4	253	Poor (0.15%)	
	PC/Ars/Chol 17:3:10	0.21	0.20	2.9	14.6	5.5	17.2	22	–	–	
	DSPC/Ars/Chol 12:8:10	–	5	6.1	–	–	2.6	1.5	117	Good (5.2%)	
	DSPC/Ars/Chol 17:3:10	–	–	–	–	–	0.54	–	–	–	
	DSPC/Ars/Chol 12:8:10/+PEG	–	5	–	–	–	1.3	8.2	80.6	Very good (>6%)	
	DSPC/Ars/Chol 17:3:10/+PEG	–	–	–	–	–	18	12.6	–	–	
C18	Ars/Chol 2:1	–	–	<0.75	2.1	6.9	–	–	195	–	
	PC/Ars/Chol 12:8:10	–	–	10.6	6.2	6.4	–	–	–	–	
	PC/Ars/Chol 17:3:10	–	–	9.7	5.5	–	–	–	–	–	

[a]Anticancer activity – IC50 is expressed as 10^{-5} M arsonolipid concentration needed to be incubated for 24 h with 100,000 cells
[b]This is the miltefosine resistant strain of Leishmania donovani L82 parasite; after 72 h incubation with ARSL
[c]MEC after 24 h incubation of Trypanosoma brucei-brucei CMP acute strain with ARSL

4. 1,2-Distearoyl-*sn*-glyceroyl-3-phosphoethanolamine, conjugated to polyethylene glycol (MW 2000) [DSPE-PEG2000] and 1,2-dipalmitoyl-*sn*-glyceroyl-3-phosphoethanolamine conjugated to polyethylene glycol (MW 2000) [DPPE-PEG2000] (Avanti Polar Lipids, USA). Storage conditions are the same as mentioned earlier for PC (with the difference that only 5 mg/mL solutions are made for these lipids). The purity of these lipids was not checked.

5. Cholesterol [Chol] (pure) (Sigma–Aldrich, Athens, Greece). Chol is stored desiccated at –20°C. Chol is used for ARSL preparation in solid state (powder) or after being dissolved (20 or 100 mg/mL) in a mixture of $CHCl_3/CH_3OH$ (2:1 v/v), and stored in aliquots at –80°C.

6. Triton X-100 . Triton is used as a 10% v/v solution in the liposome preparation buffer (see below). Usually 1 L solution is prepared, stored at room temperature and used for up to 3 months.

7. The water used in all solutions is deionized and then distilled [d.d. H_2O].

8. PBS (phosphate buffered saline), pH 7.40 for liposome preparation (for cases in which empty liposomes are prepared). This buffer is also used as elution buffer, for cleaning ARSL (from non-encapsulated molecules) by gel filtration. In each liter, this buffer contains: sodium phosphate (Sigma) 0.05 M, NaCl 150 mM and sodium azide (Sigma) 0.2 g (to a final concentration of 0.02% w/v; for prevention of bacterial growth). Before adjusting the volume (to 1 L), the pH of the solution is adjusted to 7.40.

9. Solution of calcein or 5,(6)-carboxyfluorescein [CF] (Eastman Kodak, USA). The solid is dissolved in buffer pH 7.4 (see in the following section) to make a solution of 100 mM (see Note 2).

10. Phosphate buffer pH 7.40, for preparation of calcein (or CF) solution. In each liter, this buffer contains: sodium phosphate 0.05 M, NaCl (Merck, DE) 20 mM and sodium azide 0.2 g (to a final concentration of 0.02% w/v; for prevention of bacterial growth). Before adjusting the volume (to 1 L), the pH of the solution is adjusted to 7.40. This buffer is used for preparation of calcein (or CF) solution, so that the final solution is iso-osmolar to PBS, and thus the ARSL prepared are osmotically stable when diluted with PBS (20).

2.2. ARSL Entrapment Efficiency and Membrane Integrity Evaluation

1. Sephadex G-50 (medium) (Phase Separations, Pharmacia, Sweden). The solid is dispersed in PBS buffer for swelling and the dispersion is subsequently degassed under vaccum. Gel chromatography columns are packed and used for ARSL separation from non-encapsulated molecules (as described in detail in the following section).

2. Stewart assay reagent: For preparation, dissolve 27.03 g of $FeCl_3·6H_2O$ and 30.4 g of NH_4SCN in 1 L of d.d. H_2O. The reagent is stored in dark glass bottles at room temperature and used for up to 1 month.

3. Reagents for measurement of arsenic content of arsonoliposomes: Fuming nitric acid (Merck, DE), nitric acid (0.2 % v/v in d.d. H_2O), cold (4°C) hydrogen peroxide 30% (v/v), nickel nitrate solution (5% m/V) prepared by dissolving an appropriate amount of the corresponding high purity salt (+99.999%, Aldrich) in water.

4. Pyrolytic graphite-coated tubes (Perkin-Elmer).

5. Arsenic standard solution of 1,000 mg/L (Merck, DE)

2.3. Vesicle Mean Diameter and zeta-Potential Measurement

1. Polycarbonate filters with 0.22-mm pore size (Millipore, UK).

2. Filtered PBS (see the aforementioned instructions). PBS is prepared as described earlier and then filtered through polycarbonate filters. The filtered buffer is used for dilution of ARSL dispersions before measuring their size distribution and zeta-potential

3. Zeta sizer cuvettes (Malvern, UK).

2.4. ARSL Physical Stability and Calcium Induced Aggregation

1. Tris buffered saline (TBS) pH 7.40. One liter of this buffer contains: trizma base 0.05 M, NaCl 150 mM and sodium azide 0.2 g (to a final concentration of 0.02% w/v; for prevention of bacterial growth). Before adjusting the volume (to 1 L) the pH of the solution is adjusted to 7.40, with concentrated HCl.

2. Solutions of calcium chloride: 10 mM solution is prepared in d.d. H_2O (or in TBS buffer), and used the same day.

3. Solution of ethylene diamine tetra acetic acid (EDTA) (Sigma-Aldrich, Athens, Greece). 10 mM solution is prepared in d.d. H_2O, and used the same day.

3. Methods

3.1. Arsonoliposome Preparation

A modification of the "one step method" (8), has been developed (9) for the preparation of arsonoliposomes. For this, the appropriate amount of lipid or lipids as powders (see Note 3) are weighted, placed in a 20-mL screw cap bottle (Fig. 2) and mixed with the lipid hydrating solution (d.d. H_2O or phosphate-buffered saline (PBS) pH 7.40, or a solution containing the molecules intended to be encapsulated in the vesicles), which has been previously heated (at 50–80°C, depending on the transition temperature of the arsonolipids used, see Note 4). A small magnetic bar is placed in the screw-capped bottle and the lipid

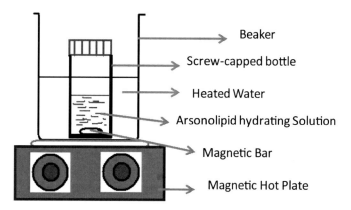

Fig. 2. Experimental setup used for ARSL preparation

mixture is subsequently magnetically stirred vigorously on a (magnetic) hot plate for 6–12 h, taking care to preserve the temperature of the liquid in the bottle for the whole period of time(see Note 5). In some cases, depending on the initial temperature of the aqueous phase and the stirring intensity, vesicles are formed significantly faster (in a few hours). When arsonoliposomes that include PC in their lipid composition are prepared, a much lower temperature is required (around 50–60°C), because of the low transition temperature of PC. After formation of liposomes, the samples are left to anneal for at least 1 h at the liposome preparation temperature used in each case.

For reduction of arsonoliposome size, the suspension produced by the method described earlier is subjected to high intensity sonication using a vibra cell Probe sonicator (Sonics and Materials, UK). A tappered microtip is used when a small volume of liposomes are to be prepared (1–3 mL), but for larger volumes, the conventional tip should be used. Sonication (see Note 6) is applied for two 10 min cycles, at least, or until the vesicle dispersion becomes completely transparent. Following sonication, the ARSL suspensions are left to stand for 2 h at a temperature higher than the transition temperature of the lipids used in each case (2, 7), in order to anneal any structural defects of the vesicles. The titanium fragments (from the probe) and any remaining multilamellar vesicles or liposomal aggregates in the small vesicle dispersion produced, are subsequently removed by centrifugation at 10,000 rpm (= $7800 \times g$) for 15 min.

Final and definite proof that indeed vesicles are formed after the "one step method" and that they still exist after sonication can be established by observing the morphology of the arsonoliposome dispersions using different types of electron microscopy (EM), as discussed elsewhere (9–12). Additionally, the ability of the vesicles to encapsulate aqueous soluble markers as carboxyfluorescein or calcein (see below) serves as proof that vesicular structures are present in most of the arsonoliposome dispersions prepared.

3.2. ARSL Entrapment Efficiency and Membrane Integrity Evaluation

3.2.1. ARSL Entrapment Efficiency

For calculation of entrapment efficiency of ARSL, the concentration of encapsulated material and liposomal lipid are measured. Calcein and 5,(6) carboxyfluorescein (CF) have been used as encapsulated materials. Initially, the non-encapsulated calcein or CF is separated from ARSL dispersions on Sephadex G-50 chromatography columns (see Note 7) eluted with PBS, pH 7.40, that renders the liposomes osmotically stable (20). The column is presaturated with lipids and therefore the lipid recovery in all cases should be well over 95% (this can be calculated by measuring the lipid concentration in the liposome sample loaded on the column, and the lipid concentration in the liposomal fractions eluted, by a colorimetric assay for phospholipids (21) as described in the following section).

After separating liposomes from non-liposome encapsulated molecules, the amount of calcein (or CF) entrapped in a given volume of vesicles is determined as well as the amount of lipid in the same volume of the ARSL dispersion, as described in the following section.

A. *Determination of ARSL entrapped calcein (or CF)*: For this, the ARSL are disrupted by a 10% v/v Triton X-100 solution. A specific volume of this solution is mixed in a sample of the ARSL dispersion, so that the final concentration of Triton X-100 is 1% v/v. Subsequently, the sample is vigorously mixed by vortex for at least 2 min (see Note 8). After total disruption of the vesicles the fluorescence intensity of the sample is measured at 37°C, with EM at 490 nm and EX at 520 nm and 10–10 slit band widths. Finally, the amount of calcein (or CF) entrapped in the ARSL is calculated with the help of an appropriate calibration curve of the dye, which is constructed for this purpose.

B. *Determination of Lipid concentration in ARSL*: The arsonolipid content of arsonoliposomes is determined by atomic absorption spectrophotometry after digestion with concentrated nitric acid, as previously reported (22). In brief, 20 µL from each (ARSL) suspension are digested with 2 mL of nitric acid, in a 25 mL conical flask. The flask is heated on a hot plate placed under a hood, by slowly increasing the temperature to 90–100°C. The solution is allowed to evaporate to dryness (but not charred), and the residue is taken up with 3 mL HNO_3 and 3 mL cold (4°C) 30% H_2O_2. A reaction is then initiated by slowly heating the mixture and the rate of decomposition of H_2O_2 is controlled by frequently removing the flask from the hot plate. The solution is then brought to a brief boil, cooled, and diluted to 50.0 mL with d.d. H_2O.

The total arsenic in the samples obtained as described earlier, is determined by graphite furnace atomic absorption spectroscopy technique (GFAAS). A computer-controlled atomic absorption spectrometer (AAnalyst 300, Perkin-Elmer) equipped with a graphite furnace (HGA-800, Perkin-Elmer) was used in our studies (13). The absorption is measured at a wavelength of 193.7- and a 0.70-nm slit bandwidth. Deuterium lamp continuous background correction is used throughout the measurements to eliminate spectral interferences. Pyrolytic graphite-coated tubes (Perkin-Elmer) are used and the atomization process is done at the tube wall. Argon at a 250 mL/min flow rate is used as a purge gas. The addition of matrix modifier converts the As to a less volatile compound and thus the char temperature may be increased to 1,400°C. An aqueous arsenic standard solution of 1,000 mg/L (Merck) is used for the preparation of aqueous calibration standards of lower concentrations (20–300 ppb). These standards are prepared daily, acidified with nitric acid and stored in polyethylene containers. The final nitric acid concentration is 0.2% (w/v).

In the case of mixed ARSL, the phospholipid content of the liposomes is measured by the Stewart assay (21), a colorimetric technique that is widely used for phospholipid content determination. For this, a sample of the liposome dispersion (20–50 µL) is mixed with 2 mL Stewart reagent (ammonium ferrothiocyanate 0.1 M) and 2 mL chloroform. The mixture is then vortexed vigorously for at least 3 min, in order to extract the complex formed between phospholipid and Stewart reagent in the chloroform phase. After this, the samples are centrifuged at 5,000 rpm (= 1950 × g) for 5 min, and the OD of the chloroform phase is measured at 485 nm. Finally, the lipid concentration of the samples is calculated by comparison with a standard curve (prepared from known concentrations of PC) (see Note 9).

3.2.2. ARSL Membrane Integrity Evaluation

The leakage of small water-soluble dyes encapsulated in the aqueous interior of liposomes during their preparation is often used as a method to study their membrane integrity during incubation under various conditions (temperature, pH, presence of serum proteins, etc.). In the case of ARSL's, the release of CF or calcein has been used as a measure of the vesicle membrane integrity, during incubation of ARSL in buffer or in presence of serum proteins [80% FCS] at 37°C under mild agitation. Calcein (or CF) is encapsulated in the vesicles in a quenched concentration (100 mM), and, therefore, its release from the membrane can be calculated without separation of free and liposomal dye, as reported before (23). In brief, 20 µL of the incubated ARSL dispersion are drawn out from each incubation tube and diluted with 4 mL of PBS, pH 7.40. The fluorescence intensity of the samples is then measured (EM 490 nm, EX 520 nm, slit–slit: 10–10), before and after the addition of Triton X-100 at a final concentration of 1% v/v. The percent of calcein (or CF) latency (% latency) is determined, by the following equation:

$$\% \text{Latency} = \frac{F_T - F_I}{F_T} \times 100,$$

where F_I and F_T are the fluorescence intensity values of the sample in the absence and presence of 1% Triton X-100 (final concentration; see Note 8), respectively (values obtained after mixing the samples with Triton are corrected for dilution).

3.3. Vesicle Size Distribution and zeta-Potential Measurement

A total of 50 µL of the ARSL dispersions are diluted with 20 mL of filtered buffer (0.22-mm pore size, polycarbonate filters, Millipore, UK) and sized immediately by photon correlation spectroscopy (Model 4700C, Malvern Instruments, UK), which enables the mass distribution of particle size to be obtained, according to the manufacturer. Size distribution measurements are made at 25°C with a fixed angle of 90° and the sizes quoted are the z average mean (dz) for the ARSL hydrodynamic diameter.

In some cases, the size of ARSL dispersions are re-measured after incubation in appropriate media and temperature for 24 and/or 48 h, in order to confirm aggregation (10–12).

For surface charge determination, ARSL dispersions are diluted with filtered PBS pH 7.40 and their electrophoretic mobility is measured at 25°C by Photon Correlation Spectroscopy [PCS] (Zetasizer 5000, Malvern Instruments, UK). Finally, the zeta potential values of the dispersions are calculated by the instrument from their electrophoretic mobility, by application of the Smolowkovski equation.

3.4. ARSL Physical Stability and Calcium Induced Aggregation

3.4.1. ARSL Physical Stability (Vesicle Self-Aggregation)

It is well known that liposomes have a tendency to aggregate and subsequently, –in some cases – fuse into larger particles during storage. The physical stability of ARSL, in terms of their size preservation during incubation in the media they have been prepared in, may be evaluated by measuring the vesicle mean diameter and size distribution (as described earlier) or by measuring the turbidity of the ARSL vesicle dispersions, at various time points during their incubation (immediately after preparation, as well as after 2, 4 and 24 or 48 h). For this, the ARSL are diluted with PBS or d.d. H_2O (if they have been prepared in d.d. H_2O) in order to have a final lipid concentration of 0.065 mM, and at various time points the turbidity of the dispersions is measured at a wavelength of 500 nm, by a spectrofluorometer (as Shimatzu RF-1501) with both emission and excitation wavelengths set at 500 nm, and slits at 10–10 (see Note 10).

3.4.2. Calcium-Induced Aggregation of ARSL

Because of the negative surface charge of ARSL (Table 1), divalent cations are expected to induce their aggregation, after removal of

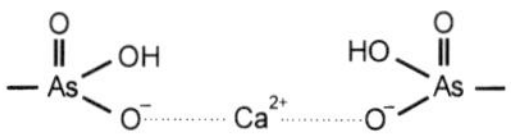

Fig. 3. Mechanism of ARSL divalent cation-induced aggregation

H_2O molecules, by acting as bridges (Fig. 3) as suggested previously for phosphatidylserine-containing liposomes (24).

Calcium-induced vesicle aggregation of ARSL may be studied by measuring the turbidity of the vesicle dispersions (as described in Subheading 3.4.1), for physical stability evaluation. ARSL, which are prepared in d.d. H_2O (or TBS, see Note 11), are diluted with solutions of calcium chloride in order to have final calcium concentrations in the dispersions between 0.43 and 1.8 mM (see Note 12), and at the same time final lipid concentration of 0.065 mM.

Calcium-induced ARSL aggregation is evaluated by measuring the turbidity of liposome dispersions, as described in Subheading 3.4.1, immediately after mixing with $CaCl_2$, as well as after 2, 4, and 24 or 48 h. The initial turbidity of each ARSL dispersions in H_2O (or buffer) at the beginning of each experiment (time 0) is taken as starting point.

The effect of EDTA on the calcium-induced turbidity change of the ARSL dispersions may be evaluated by remeasuring the turbidity after adding a tenfold amount of EDTA (compared with the final calcium concentration of the sample), and correcting the measured turbidity value by the dilution factor. The decrease in turbidity due to sample dilution is accounted for by performing blank experiments (diluting some samples with H_2O). This control experiment would indicate if the vesicles are fusing or perhaps changing morphology, or if only loose aggregates – that can be easily disassembled upon $CaCl_2$ removal – are formed.

4. Notes

1. The 99% purity of the lipids can be verified by Thin Layer Chromatography on silicic acid precoated plates (Merck, Germany), using a $CHCl_3/CH_3OH/H_2O$ 65:25:4 v/v/v mixture for plate development, and iodine staining for visualization. Pure lipids give single spots.
2. Calcein (as well as CF) is not easily dissolved in buffer with pH 7.40. Therefore, the weighted solid is initially dissolved in NaOH (1 M) which is added dropwise until the full quantity is dissolved and subsequently, the resulting solution is diluted with the appropriate volume of buffer (in order to achieve the

required calcein concentration). The pH of the final solution should be checked and re-adjusted if required, while care has to be taken so that calcein does not precipitate.

3. The lipids could also be used in the form of solutions in $CHCl_3/CH_3OH$ 2:1 v/v that can be initially prepared and stored at −80°C. In this case, the appropriate volume of lipid solution (or each lipid solutions; if several lipids are used in the form of organic solution) is (or are) placed in the 20-mL screw-capped bottle, and evaporated under a gentle stream of nitrogen and mild heating. After all the organic solvent has been evaporated, the container is flushed for extra 10 min with N_2 and then the heated hydration solvent is placed in the bottle and ARSL preparation proceeds as described earlier.

4. Transition temperatures of arsonolipids (Ars) range between 72 and 80°C (for *rac*, R and S, C12 Ars), 50 and 60°C (for *rac*, R and S, C14 Ars), and 65 and 72°C (for *rac*, R and S, C16 Ars), as published earlier for arsonolipid dispersion at pH 8.0 (7).

5. In order to preserve the high temperature of the lipid hydrating solution in the screw-capped bottle, a 50-mL beaker with water is placed on the hot (magnetic) plate and the bottle is placed in the beaker (see Fig. 2). The water in the beaker is occasionally gently stirred and re-filled, to compensate for evaporation.

6. For probe sonication, the probe tip (or tapered microtip) is immersed into the ARSL dispersion by approximately 1.2–1.5 cm from the surface, taking care so that no part of the tip is in contact with the vial (a mirror is used to be sure). The vial is placed in a ice-cold water tank, to prevent overheating of the liposome dispersion during probe sonication.

7. A column with dimensions 1 × 35 cm is sufficient to separate 1 mL of liposome or ARSL dispersion. The column is pre-calibrated and at the same time saturated with a dispersion of empty ARSL mixed with a quantity of the encapsulated material (in each case). The void volume of such columns should be between 7 and 13 mL and the bed volume between 17 and 21 mL.

8. In some cases, especially when rigid ARSL that contain DSPC and Chol are used, the ARSL are difficult to disrupt by using 1% v/v final concentration of Triton X-100 detergent. Then the ARSL mixture with Triton can be heated by rapidly immersing in boiling water (in which case, care should be taken in order to perform the final measurement after the sample is cooled, in order to avoid mistakes). Another possibility is to use higher final concentration of detergent (in which case the extra dilution of the sample has to be taken into account during calcein (or CF) % latency calculation).

9. The linear region for a calibration curve by the Stewart assay is between 10 and 100 μg of PC (DPPC, DSPC, or mixtures with Ars (see below) can also be used at this range). The Stewart assay was found to detect arsonolipids (when present in high concentrations in the ARSL dispersions). Thereby, known concentrations of phospholipids mixed with arsonolipids (at the specific analogy used in the ARSL samples measured, in each case), are initially measured and their values are used for calibration curve construction.

10. When evaluating liposome aggregation (or physical stability) by turbidity measurements, it is advisable to use spectrofluorometers that are equipped with thermostated sample holders with magnetic stirring capability, in order to avoid flaws in measurement due to precipitation of the large vesicles.

11. In calcium induced vesicle aggregation studies, it should be avoided to use phosphate buffers in order to avoid experimental mistakes due to formation and precipitation of calcium phosphates, which will make the vesicle dispersions highly turbid.

12. Higher calcium concentrations (calcium concentration in plasma is somewhat higher than the highest concentration used in this study: >2.0 mM compared with 1.8 mM, respectively), were not used, because these experiments are conducted in water or buffer and not in the presence of serum proteins, and thus all $CaCl_2$ ions are available for interaction with the vesicles, which is not the case in serum, where some of the $CaCl_2$ ions may be bound to proteins.

13. For more information about studies performed with ARSL, see review articles (25, 26).

Acknowledgments

Part of the work presented in this review was partly supported by the General Secretariat of Research and Technology, Athens Greece (PENED 95).

The authors would like to thank all the scientists that have contributed in this on-going research, and especially, Prof D. Fatouros, Prof. P. Loiseau, Prof. P. Frederik and Prof. P. Klepetsanis.

References

1. Dembitsky VM, Levitsky DO (2004) Arsonolipids. Prog Lipid Res 43:403–448
2. Serves SV, Sotiropoulos DN, Ioannou PV, Jain MK (1993) One pot synthesis of arsonolipid via thioarsenite precursors. Phosphorous Sulfur Silicon 81:181–190
3. Tsivgoulis GM, Ioannou PV (2008) A reinvestigation of the synthesis of arsonolipids (2, 3-diacyloxypropylarsonic acids). Chem Phys Lipids 152:113–121
4. Serves SV, Charalambidis YC, Sotiropoulos DN, Ioannou PV (1995) Reaction of arsenic

(III) oxide, arsenous and arsenic acids with thiols. Phosphorous Sulfur Silicon 105:109–116
5. Timotheatou D, Ioannou PV, Scozzafava A, Briganti F, Supuran CT (1996) Carnonic anhydrase interaction with lipothioasrenites: a novel class of isozymes I and II inhibitors. Met Based Drugs 3:263–268
6. Kigawa J, Minagawa Y, Kanamori Y, Itamochi H, Cheng X, Okada M, Oishi T, Terakawa N (1998) Glutathione concentration may be a useful predictor of response to second-line chemotherapy in patients with ovarian cancer. Cancer 82:697–702
7. Serves SV, Tsivgoulis GM, Sotiropoulos DN, Ioannou PV, Jain MK (1992) Synthesis of (R)- and (S)-1, 2-diacyloxypropyl-3-arsonic acids: optically active arsonolipids. Phosphorous Sulfur Silicon 71:99–105
8. Talsma H, Van Steenbergen M, Borchert J, Crommelin D (1994) A novel technique for the one-step preparation of liposomes and non-ionic surfactant vesicles in a continuous gas stream: the 'bubble method'. J Pharm Sci 83:276–280
9. Fatouros D, Gortzi O, Klepetsanis P, Antimisiaris SG, Stuart MCA, Brisson A, Ioannou PV (2001) Preparation and properties of arsonolipid containing liposomes. Chem Phys Lipids 109:75–89
10. Piperoudi S, Fatouros D, Ioannou PV, Frederik P, Antimisiaris SG (2006) Incorporation of PEG-lipids in arsonoliposomes can produce highly stable arsenic-containing vesicles of specific lipid composition. Chem Phys Lipids 139:96–106
11. Piperoudi S, Ioannou PV, Frederik P, Antimisiaris SG (2005) Arsonoliposomes: effect of lipid composition on their stability. J Liposome Res 15:187–197
12. Fatouros DG, Piperoudi S, Gortzi O, Ioannou PV, Frederik P, Antimisiaris SG (2005) Physical stability of sonicated arsonoliposomes: effect of calcium ions. J Pharm Sci 94:46–55
13. Antimisiaris SG, Klepetsanis P, Zachariou V, Giannopoulou E, Ioannou PV (2005) In vivo distribution of arsenic after i.p. injection of arsonoliposomes in balb-c mice. Int J Pharm 289:151–158
14. Zagana P, Haikou M, Klepetsanis P, Giannopoulou E, Ioannou PV, Antimisiaris SG (2008) In vivo distribution of arsonoliposomes: effect of vesicle lipid composition. Int J Pharm 347:86–92
15. Antimisiaris SG, Ioannou PV, Loiseau PM (2003) In vitro antileishmanial and trypanocidal activities of arsonoliposomes and preliminary in vivo distribution. J Pharm Pharmacol 55:647–652
16. Zagana P, Klepetsanis P, Ioannou PV, Loiseau PM, Antimisiaris SG (2007) Trypanocidal activity of arsonoliposomes: effect of vesicle lipid composition. Biomed Pharmacother 61:499–504
17. Zagana P, Haikou M, Giannopoulou E, Ioannou PV, Antimisiaris SG (2009) Arsonoliposome interaction with cells in culture. Effect of pegylation and lipid composition. J Mol Nutr Food Res 53:592–599
18. Gortzi O, Papadimitriou E, Kontoyannis C, Antimisiaris SG, Ioannou PV (2002) Arsonoliposomes, a Novel class of arsenic-containing liposomes: effect of palmitoyl-arsonolipid-containing liposomes on the viability of cancer and normal cells in culture. Pharm Res 19:79–86
19. Gortzi O, Papadimitriou E, Antimisiaris SG, Ioannou PV (2003) Cytotoxicity of arsono-lipid containing liposomes towards cancer and normal cells in culture: effect of arsono-lipid acyl chain length. Eur J Pharm Sci 18:175–183
20. Cleland M, Allen T (1981) Serum induced leakage of liposome entrapped contents. Biochim Biophys Acta 597:418–426
21. Stewart JCM (1980) Colorimetric determination of phospholipids with ammonium ferrothiocyanate. Anal Biochem 104:10–14
22. Desaulniers JAH, Sturgeon RE, Berman SS (1985) Atomic absorption determination of trace metals in marine sediments and biological tissues using a stabilized temperature platform furnace. At Spectrosc 6:125–127
23. Senior J, Gregoriadis G (1982) Stability of small unilamellar liposomes in serum and clearance from the circulation: the effect of the phospholipids and cholesterol components. Life Sci 30:2123–2136
24. Wilschut J, Duzgunes N, Hong K, Hoekstra D, Papahadjopoulos D (1983) Retention of aqueous contents during divalent cation-induced fusion of phospholipid vesicles. Biophys Biochim Acta 734:309–318
25. Fatouros D, Ioannou PV, Antimisiaris SG (2006) Novel nanosized arsenic containing vesicles for drug delivery: arsonoliposomes. J Nanosci Nanotechnol 6:2618–2687
26. Antimisiaris SG (2007) Arsonoliposomes for drug delivery. J Drug Deliv Sci Technol 17:377–388

Chapter 11

Liposome-Based Vaccines

Reto A. Schwendener, Burkhard Ludewig, Andreas Cerny, and Olivier Engler

Abstract

Here, we report methods of preparation of liposome vaccine formulations for the entrapment of antigenic peptides and antigen encoding plasmid DNAs. Two examples of liposomal vaccine formulations producing highly effective immune responses are given. Firstly, a formulation with encapsulated antigenic peptides derived from the hepatitis C virus NS4 and the core proteins, and secondly, the encapsulation of a plasmid DNA encoding the gp33 glycoprotein of the lymphocytic choriomeningitis virus (LCMV). Vaccination with liposomal HCV peptides in HLA-A2 transgenic mice by subcutaneous injections induced strong cytotoxic T cell responses as shown by lysis of human target cells expressing HCV proteins. The immunogenicity of the liposomal peptide vaccines was further enhanced by incorporation of immunostimulatory CpG oligonucleotide sequences, shown by a strong increase of the frequency of IFN-γ secreting cells that persisted at high levels for long periods of time. With the LCMV model, we could show that upon intradermal injection, plasmid–DNA liposomes formed LCMV gp33 antigen depots facilitating long-lasting in vivo antigen loading of dendritic cells (DC), followed by a strong immune response. Our data show that liposomal formulations of peptide or plasmid–DNA vaccines are highly effective at direct in vivo antigen loading and activation of DC leading to protective antiviral and anti-tumor immune responses.

Key words: Liposomes, Peptides, DNA, Immunostimulatory oligonucleotides, CpG, Immunization, Dendritic cells, Adjuvants, Vaccines, HCV, LCMV

1. Introduction

With the availability of well-characterized antigens, in particular with highly purified proteins or synthetic peptides, more effective and safer vaccines can be developed. However, this approach may be hampered by the fact that many antigens are often poorly immunogenic when administered alone, necessitating the development of suitable adjuvants that have the ability to potentiate the immunogenic effect of a given antigen, preferably with little or no side effects. Adjuvants can be divided into two groups, based on their principal mechanisms of action: (1) vaccine delivery systems (1–5) and (2) immunostimulatory adjuvants (6, 7). Vaccine delivery systems are generally composed of particles of comparable dimensions to pathogens as bacteria and viruses (e.g. liposomes, microemulsions, immunostimulatory complexes and other nano- or microparticle systems) (8–11).

These systems function mainly to target associated antigens to antigen-presenting cells (APC). Currently, complex formulations are being developed in which carrier systems are exploited both for the delivery of antigens and of co-administered immunostimulatory adjuvants either to isolated APCs (macrophages, dendritic cells) or by direct in vivo applications (12–14). Such approaches are used to ensure that both antigen and adjuvant are delivered to the same population of APCs. Additionally, particulate delivery systems can specifically target the adjuvant effect to the key cells of the immune system, reducing systemic distribution and minimizing induction of adverse reactions. Small unilamellar liposomes have a significant potential as delivery systems for the co-administration of antigens (peptides, lipopeptides) and of immunostimulatory adjuvants, including CpG oligonucleotides or DNA encoding antigens and/or immunostimulatory sequences (15). Additionally, the efficacy of liposome-based vaccines can be improved by targeting them more effectively and specifically to the APCs by exploiting various scavenger and other receptors as their targets (16–18) or by enhancing their cell uptake properties by modification with cell penetrating peptides (19).

Here, we present methods of preparation of liposomal vaccines and results obtained in our laboratories with small unilamellar liposomes as carriers of antigen peptides and peptide encoding DNA plasmids, demonstrating their high potential as therapeutic vaccine formulations against infectious diseases and cancers.

2. Materials

2.1. Liposome Preparation

1. Soy phosphatidylcholine (SPC) (L. Meyer GmbH, Hamburg, Germany), store at −20°C, prepare a stock solution, e.g. of 20–100 mg/mL by dissolving SPC in methanol/methylene chloride (1:1, v/v).

2. Cholesterol (see Note 1).

3. D,L-α-Tocopherol (Merck, Darmstadt, Germany), store at −20°C, make a stock solution, e.g. of 10 mg/mL by dissolving D,L-α-tocopherol in methanol/methylene chloride (1:1, v/v).

4. 1-Palmitoyl-2-oleoyl-sn-glycero-3-phosphocholine (POPC) (Avanti Polar Lipids, Alabaster, AL). Store at −20°C. Make a stock solution as described for SPC.

5. Didodecyldimethyl ammoniumbromide (DDAB) (Fluka, Buchs, Switzerland). Store at 4°C.

6. Phosphate buffer, PB: 13 mM KH_2PO_4, 54 mM $NaHPO_4$, pH 7.4 (see Note 2).

7. Round bottom flasks (20–100 mL).

8. Rotatory evaporator, e.g. Rotavap (Buechi AG, Flawil, Switzerland).

9. Lipex™ high pression extruder (Northern Lipids Inc., 8855 Northbrook Court, Burnaby, BC, Canada).

10. Nuclepore membranes of defined pore sizes: 400; 200; 100 nm (Sterlitech Corp., Kent, WA, USA or Sterico AG, Wangen, Switzerland).

11. BSS, balanced salt solution.

12. TE buffer: 10 mM Tris–HCl, pH 8, 10 mM EDTA.

13. Sterile filters, 0.45 or 0.2 μm and plastic syringes, various suppliers.

2.2. Peptides, Plasmids and Adjuvants

1. The HLA-A2 restricted $CD8^+$ T cell epitope peptide from the Hepatitis C virus core protein c132 (aa 132–140; DLMGYIPLV; >95% purity) and the peptide NS1851 from the NS4 protein (aa 1851–1859; ILAGYGAGV; >95% purity) were from Neosystems (Strasbourg, France) and stored at 4°C.

2. The immunostimulatory oligonucleotide ODN1668 (5´-TCCATGACG-TTCCTGATGCT-3´), referred to as CpG was synthesized by Microsynth (Balgach, Switzerland) and stored at −20°C.

3. The plasmid pEGFPL-33A was kindly provided by Stefan Oehen (20). This vector is composed of a DNA insert coding for a FLAG Tag (DYKDDDDK) and the gp33 epitope (KAVYNFATM) flanked N-terminally by three leucines and C-terminally by four alanines inserted in the pEGFP-N3 expression vector (Clontech, Palo Alto, CA). The gp33 peptide is the immunodominant epitope of the lymphocytic choriomeningitis virus glycoprotein.

2.3. Enzymatic Digestion of Non-entrapped Plasmid DNA and Determination of Encapsulated Plasmid DNA

1. DNase I (RNase free) (Roche Diagnostics, Rotkreuz, Switzerland).

2. Magnesium chloride ($MgCl_2$) and EDTA.

3. Biogel A-15 (10 × 1 cm) column (Bio-Rad Laboratories) or corresponding product.

4. Phosphorus-32 (Amersham Biosciences, Amersham, UK).

3. Methods

The methods described in the following section outline (1) the preparation of liposomes by high pressure filter extrusion, (2) the encapsulation of peptide antigens and immunostimulatory CpG oligonucleotides and (3) the preparation of plasmid DNA liposomes encoding the LCMV GP33 peptide antigen.

We do not describe in details the immunological methods (immunization, ^{51}Cr release assay, ELISPOT, ELISA and flow cytometry) used for the analysis of the immune responses induced by the liposome vaccines (Subheading 3.2). For comprehensive information, we refer to our publications (21–23) and to the related literature.

3.1. Liposome Preparation

Liposomes belong to the most studied particulate carrier systems. In the past decades, a vast number of liposome preparation methods for the encapsulation of a large variety of molecules have been developed and refined. We refer to the corresponding literature and to our publications for more information. We recommend the high pressure filter extrusion method for the preparation of peptide or DNA containing liposomes because of its ease, versatility, up-scaling options and high quality of the liposomes produced.

3.1.1. Liposome Compositions

Liposomes can be composed of a large selection of phospholipids and additional lipophilic compounds like cholesterol, poly(ethylene glycol) lipids (PEG), glycolipids, and antioxydants. Depending on the intended application, different lipid compositions have to be selected. The "state-of-the-art" liposomes used for intravenous applications, e.g. liposomes carrying cytotoxic antitumor drugs, are those composed of lipids containing hydrophilic carbohydrates or polymers, mainly poly(ethylene glycol) modified phospholipids. Such PEG- or "stealth" liposomes evade fast absorption in the mononuclear phagocyte system and have long blood circulating times (24). Liposome formulations carrying antigens intended as vaccines are administered by subcutaneous or intradermal injection and usually do not require further modifications, since the targets are phagocytosing cells such as macrophages and dendritic cells that are mainly localized at the site of injection. Their composition can be kept quite simple by choosing phospholipids (SPC, synthetic phospholipids) and cholesterol as main components. Nevertheless, they may be modified by specific molecules (e.g. mannosylated lipids) that recognize and bind to receptors expressed on APCs (17). Another important feature facilitating broad applications and up-scaling of the preparations is that antigen and DNA containing liposomes can be formulated as stable lyophilized products by addition of appropriate cryoprotectants (25).

3.1.2. Liposome Preparation by High Pressure Filter Extrusion

Peptide liposome vaccines were prepared by freeze–thawing of the lipid/peptide mixtures followed by sequential filter extrusion.

Preparation of Liposomes Containing Antigenic HCV Peptides and Immunostimulatory CpG Oligonucleotides (21)

1. The basic composition for the preparation of 5.0 mL liposomes was 1.0 g soy phosphatidylcholine (SPC, L. Meyer GmbH, Hamburg, Germany), 125 mg cholesterol (Fluka, Buchs, Switzerland) (see **Note 1**) and 6 mg D,L-α-tocopherol (Merck, Darmstadt, Germany) as antioxidant.

2. The solid lipids were dissolved in methanol/methylene chloride (1:1, v/v) in a round bottom flask or corresponding amounts of stock solutions were added.

3. After removal of the organic solvents by rotary evaporation (40–45°C, 30–60 min), the dry lipid mixture was solubilized with the HCV c132 or NS4 1851 peptide (4 mg/mL) dissolved in phosphate buffer PB (67 mM, pH 7.4) by vigorous agitation (see **Notes 2** and **3**). In addition to the peptide antigens, some formulations contained immunostimulatory CpG oligonucleotides (250 nmol/mL in Tris/EDTA buffer) which were added likewise to the lipids.

4. The mixture was then subjected to 3–5 freeze–thaw cycles (liquid nitrogen–water 40°C; see **Note 4**), followed by repetitive extrusion through Nuclepore (Sterlitech Corp., Kent, WA, USA or Sterico AG, Wangen, Switzerland) filters (800, 400 and 200 nm pore size) using a Lipex™ Extruder (Northern Lipids Inc.) (see **Note 5**).

5. Liposomes were filter sterilized (0.45- or 0.2-μm sterile filters) and diluted in BSS (balanced salt solution). Peptide encapsulation was estimated to range between 80 and 90% according to previous determinations (22). Non-entrapped peptides and CpG oligonucleotides can either be kept in the preparation or removed by dialysis.

Preparation of Liposomes Containing a Peptide-Encoding Plasmid

For example, the preparation of 5 mL liposomes containing 1.5 mg/mL pEGFPL-33A plasmid encoding the LCMV gp33 peptide and the green fluorescent protein EGFP is given.

1. The lipids (456 mg 1-palmitoyl-2-oleyl-sn-glycero-3-phosphocholine, POPC and 5.6 mg didodecyldimethyl-ammoniumbromide, DDAB) in a 20-mL round bottom flask were dissolved in an appropriate amount of methylene chloride/methanol (1:1, v/v, ~20 mL).

2. The organic solvent was removed by rotary evaporation (40–45°C, 30–60 min). The plasmid DNA solution (1.5 mg/mL in Tris/EDTA buffer) was added to the dry lipid film and the lipids solubilized by vigorous agitation (see Note 3).

3. Then the mixture was subjected to 3–5 freeze–thaw cycles followed by filter extrusion as described in Subheading 3.1.2, step 1.

4. Finally, the liposomes were filter sterilized (0.45- or 0.2-μm sterile filters) and diluted in BSS.

Enzymatic Digestion of Non-entrapped Plasmid DNA and Determination of Encapsulated Plasmid DNA

Non-entrapped plasmid–DNA was removed by enzymatic digestion, followed by separation of liposomes and digested plasmid–DNA by column chromatography.

1. DNase I (80 U/μg) and magnesium chloride (5 mM) were added to 1 mL of the DNA-liposomes.
2. This mixture was incubated for 3 h at 37°C and the reaction stopped with EDTA (7 mM).
3. The digested non-encapsulated DNA was separated from the liposomes by gel chromatography on a Biogel A-15 column (10 × 1 cm) equilibrated with Tris buffer (50 mM). Separation was achieved by elution with Tris buffer (10 mM) and collection of 0.5–1.0 mL fractions. The plasmid–DNA liposomes were eluted in fractions 2–5, whereas digested plasmid DNA was retained in the column.
4. To determine the fraction of encapsulated plasmid–DNA, the plasmid was radioactively labeled with ^{32}P (see **Note 6**). Plasmid labeling was done according to conventional methods in trace amounts before encapsulation into liposomes. The percentage of encapsulated plasmid–DNA was determined after separation from digested non-encapsulated DNA. As alternative method, DNA-encapsulation can be monitored on agarose gels (0.8%) by application of untreated and detergent solubilized liposomes (e.g. Triton X-100 or octyl-glucoside). Results show that the superhelical conformation of the plasmid–DNA is preferentially encapsulated in the liposomes (not shown).

3.2. Immunization with Liposomal Antigen Formulations

3.2.1. Efficacy of CD8 T Cell Induction by Liposomes Containing HCV Epitopes and Influence of an Immunostimulatory CpG Oligonucleotide (21)

Six- to eight-week-old HDD mice, transgenic (tg) for HLA-A2.1 (A0201) major histocompatibility complex (MHC) class I and deficient for both H-2D^b and murine β$_2$-microglobulin (β$_2$m) (26) were immunized with the HLA-A2 restricted CD8$^+$ T cell epitope from the Hepatitis C virus NS4 protein NS1851 (aa 1851–1859; ILAGYGAGV) or from the core protein c132 (aa 132–140; DLMGYIPLV).

1. The HLA-2.1 tg mice were injected subcutaneously (s.c.) at the base of the tail with 50 μL of the liposome formulation (~130 μg peptide) with or without immunostimulatory CpG molecules and as control with a saline solution containing the peptide (~130 μg peptide). Mice received three injections at a 2-week interval and the response was analyzed 2 weeks after the last injection.
2. Spleen cells (4 × 10^6) were isolated to analyze the CTL response and restimulated with peptide-pulsed and irradiated spleen cells (2 × 10^6 cells in 2 mL medium). On day 3, IL-2 (2.5 U/mL) was

added. The specific lysis of peptide pulsed HLA-A2 transfected target cells (EL-4S3-Rob HDD) was analyzed in a standard 4 h ^{51}Cr release assay. Spontaneous and maximal release was determined from wells containing medium alone or after lysis with 1N HCl, respectively. Lysis was calculated by the formula:

% Lysis = (release in assay − spontaneous release)/(maximum release − spontaneous release)×100.

Peptide-specific lysis was determined as the percentage of lysis obtained in presence or in absence of the peptide.

3. For the IFN-γ ELISPOT assay splenocytes were re-stimulated over night with peptide NS1851 as described for the CTL assay. Then, 10^5 and 10^4 cells were transferred to precoated ELISPOT plates (U-CyTech, Utrecht, Netherlands) and incubated for 5 h. Spot formation was analyzed as described in the manufacturer's protocol.

4. To perform intracellular cytokine staining spleen cells (8×10^6 cells in 2 mL medium) were re-stimulated over night with 10 mg/mL peptide (c132 or a control peptide). Cells were, subsequently treated for 2.5 h with Brefeldin A (Golgy Stop, BD Bioscience) and permeabilization/fixation was performed with Cytoperm/Cytofix (BD Bioscience) according to the manufacturer's protocol. Surface CD8 was stained with anti-mouse CD8α-FITC antibodies (BD Bioscience) and IFN-γ staining was done with anti-mouse IFN-γ-PE antibodies or isotype control antibodies (BD Bioscience). Fluorescence was analyzed on a Coulter Epics XL-MCL flow cytometer (Coulter Corp., Hialeah, FL, USA). The CTL responses and IFN-γ production are shown in Fig. 1. Encapsulation of the peptide NS1851 into liposomes was sufficient to induce a specific and strong CTL response against peptide-pulsed target cells. No specific cytotoxicity was detected in the negative control formulations consisting of peptides solubilized in 0.9% NaCl (open circles) or empty liposomes (data not shown). Co-administration of immunostimulatory CpG molecules resulted in augmented target cell lysis (Fig. 1a).

IFN-γ is a potent immunostimulatory and anti-viral cytokine. The frequency of specific IFN-γ-secreting cells stimulated by liposomal formulations exclusively containing the CTL epitope NS1851 or combinations with CpG, respectively, was evaluated by ELISPOT assay. Two weeks after three immunizations, high numbers of specific IFN-γ-secreting cells in mice immunized with liposomes containing NS1851 (~0.2% of total spleen cells) were detected which further increased in mice immunized with liposomal formulations containing CpG (~0.7% of total spleen cells) (Fig. 1b).

The frequency of IFN-γ-producing cells was further analyzed by intracellular cytokine staining. The results shown in Fig. 2

Cytotoxic T cell response and IFN-γ production after subcutaneous immunization with NS1851 peptide

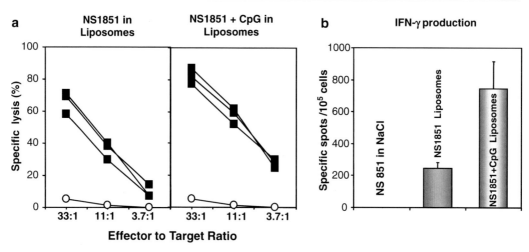

Fig. 1. CTL responses (**a**) and IFN-γ production (**b**) after s.c. injection of NS1851 liposomes (~130 μg peptide; three times every 2 weeks) with or without immunostimulatory CpG (250 nmol) or as controls the peptide in saline solution. The CTL response was analyzed 2 weeks after the last injection and measured in a standard ^{51}Cr release assay after 5 days of re-stimulation in vitro. *Black squares* represent the response induced by liposomal formulations and *open circles* the response of the control peptide. IFN-γ production was measured by ELISPOT assay after stimulation with peptide pulsed cells

Intracellular IFN-γ production after c132 peptide immunization

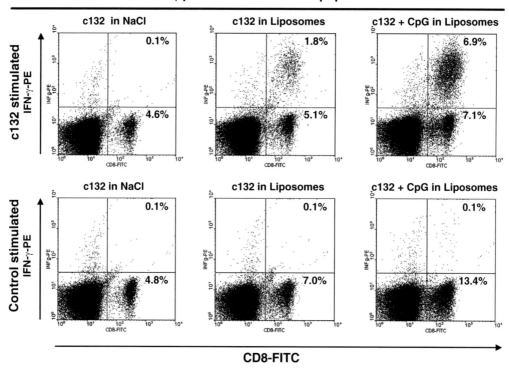

Fig. 2. Analysis of IFN-γ-producing splenocytes by intracellular cytokine staining. Mice were immunized with c132 liposomes with or without CpG or with the peptide in 0.9% NaCl. Intracellular IFN-γ production was analyzed 14 days after the second immunization upon stimulation with the specific peptide c132 or a control peptide. Indicated is a representative example of IFN-γ producing CD8$^+$ T cells in each vaccine group. IFN-γ production of CD8$^+$ T cells derived from control immunized mice was always at background levels (0.1%)

indicate that in mice immunized with liposomes containing only the c132 peptide, approximately 1.8% of total splenocytes (corresponding to 26% of CD8+ T cells) produced IFN-γ in response to peptide stimulation, while in mice immunized with liposomes containing c132 plus CpG approximately 6.9% of splenocytes (corresponding to 49% of CD8+ T cells) produced IFN-γ upon specific stimulation. IFN-γ production of CD8+ T cells determined from control immunized mice was always at background levels (0.1%).

3.2.2. Induction of a LCMV Antigen Specific Immune Response with pEGFPL-33A Liposomes

The induction of the LCMV gp33 specific CTL response after intradermal immunization with a liposome formulation of pEGFPL-33A DNA is shown by specific lysis of re-stimulated spleen cells 9 days after the first immunization in Fig. 3.

1. Three C57BL/6 mice per group were injected intradermally with 50 µg of liposomal plasmid DNA (A, C) or 50 µg free plasmid DNA (B, D) followed by a second treatment after 48 h. Three control mice received 50 µg of an antiviral peptide antigen (gp33) in incomplete Freund's adjuvant (IFA) intradermally. Nine days after the first immunization, spleen cells were re-stimulated for 5 days and analyzed in a ^{51}Cr-release assay on peptide labeled and unlabeled EL-4 cells. The spontaneous ^{51}Cr-release was <14% (22).

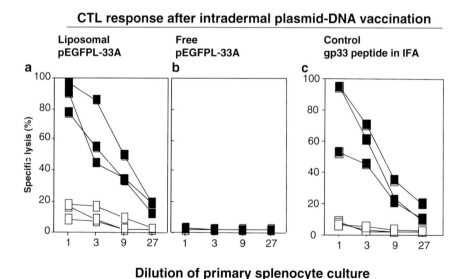

Fig. 3. CTL response after intradermal plasmid vaccination. Three C57BL/6 mice per group were injected intradermally with 50 µg of liposomal plasmid DNA (**a, c**) or 50 µg free plasmid DNA (**b**). After 48 h the treatment was repeated. As control three mice received 50 µg of the antiviral peptide antigen gp33 in IFA. Nine days after the first immunization isolated spleen cells were re-stimulated for 5 days and analyzed in a ^{51}Cr-release assay on peptide labeled (*filled squares*) and unlabeled (*open squares*) EL-4 cells. The spontaneous ^{51}Cr-release was <14%

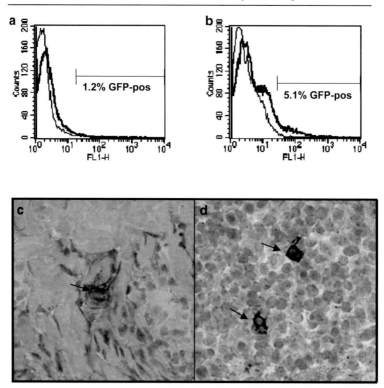

Fig. 4. Efficiency of in vivo transfection of dendritic cells after intradermal immunization with a liposomal plasmid–DNA vaccine. Mice were treated intradermally with 80 μg liposomal plasmid–DNA and after 36 h a sample of the skin at the injection site and regional lymph nodes were isolated (Lnn. inguinalis, axillaris, brachialis) and analyzed. (**a**) Flow cytometric analysis of a lymph node cell suspension revealed specific fluorescence in 1.2% of the cells produced by the green fluorescent marker protein (EGFP; *thick line* in **a**). (**b**) Dendritic cells were enriched by separation from lymph node cells by gradient centrifugation. Specific EGFP-fluorescence in the DC population was 5.1% (*thick line* in **b**). (**c**) Immunohistochemical detection of EGFP-positive cells of typical DC morphology (*arrow*) in the cutaneous tissue at the injection site. EGFP expression was visualized by incubation with an anti-EGFP antibody, followed by alkaline phosphatase staining. (**d**) Immunohistochemical detection of EGFP-positive cells in the T-cell area of a regional lymph node (Ln. inguinalis; *arrows*)

3.2.3. Detection of EGFP Expressing Antigen Presenting Cells After Intradermal Vaccination with pEGFPL-33A DNA Liposomes

The efficiency of the in vivo transfection of APCs after intradermal application of liposomal pEGFPL-33A DNA (80 μg) to the left flank of C57BL/6 mice was analyzed by flow cytometry.

1. Regional lymph nodes were removed 36 h after injection. The flow cytometric analysis of the crude lymph node cell suspension showed that 1.2% of the cells expressed the enhanced green fluorescent protein (EGFP) in comparison to cells from untreated controls (Fig. 4a).

2. Dendritic cells were enriched by gradient centrifugation up to 50–60% purity. The specific EGFP fluorescence of this cell population increased to 5.1% compared to control DCs (Fig. 4b).

3. The transfection of APCs with plasmid-liposomes was further analyzed by immunohistochemistry. In Fig. 4c, d, expression of EGFP in the skin at the injection site and in a regional lymph node is shown.

4. Conclusion

The examples of the vaccination experiments performed with liposome formulations of antigen peptides and/or antigen encoding DNA demonstrate that strong antigen-specific immune responses are obtained. Small unilamellar liposomes of the basic composition phosphatidylcholine and cholesterol have a significant potential as delivery systems for the co-administration of peptide antigens and of immunostimulatory adjuvants, including CpG oligonucleotides, whereas the lipid composition POPC and DDAB was effective for the encapsulation of DNA encoding antigens and/or immunostimulatory sequences. The liposome formulations have the advantage of ease of production at large scale, low costs of the components and proven safety. The results presented here clearly indicate that liposomal antigen delivery in vivo is a promising approach to induce efficient antiviral and anti-tumor immune responses with relevance for human applications. The process of RNA interference by short interfering RNA sequences (siRNA) represents a new class of molecules with a high potential for medical applications. Thus, the delivery of liposome encapsulated siRNA opens new opportunities for the development of novel antiviral and antitumor treatment modalities for prophylaxis and treatment of virus infections and cancer in humans (27, 28). Furthermore, it appears likely that liposome based delivery systems modified by specific targeting functions to antigen presenting cells are well suited to further amplify the immune responses that mediate protection against viral infections or rapidly growing tumors.

5. Notes

1. Cholesterol (e.g. from Fluka, purum quality, >95%) should be recrystallized from methanol. Cholesterol of minor quality should be avoided, since liposome membrane stability can be reduced. Store at 4°C, make a stock solution of 10 mg/mL by dissolving cholesterol in methanol.
2. If the liposome vaccines are intended to be stored for longer periods of time, they may be frozen or lyophilized, provided

that they are prepared in a phosphate buffer that contains a cryoprotectant. We use an iso-osmolar phosphate-mannitol buffer of the following composition: 20 mM phosphate buffer (0.53 g/L KH_2PO_4 plus 2.87 g/L $Na_2HPO_4 \cdot 2H_2O$) plus 230 mM mannitol (42.0 g/L mannitol).

3. The detachment of the lipid mixtures from the glass walls of the round bottom flasks can be accelerated by addition of small glass beads (2–3 mm diameter), followed by vigorous shaking. Preferably, the glass beads are added to the organic lipid solution before evaporation of the solvents. The encapsulation efficiency of hydrophilic molecules into the trapped volume of the liposomes is significantly increased by using the freeze–thaw method as described (29). Encapsulation efficiency is further improved by performing the freeze–thaw cycles at high lipid per volume concentrations (e.g. 200 mg or more lipid/mL) as used in our studies. If synthetic lipids as liposome forming components are used, temperatures above the corresponding lipid transition temperature T_c have to be applied for the preparation process.

4. If liquid nitrogen is not available, other freezing methods can be used, e.g. freezing the lipid suspension in a –80°C freezer or using other refrigerants like dry ice.

5. Mean hydrodynamic diameters of vesicles (liposomes, nanospheres, nanobeads) can be determined by dynamic laser light scattering, e.g. the NICOMP 380 particle sizing instrument, Particle Sizing Systems (Sta. Barbara, CA, USA).

6. Plasmid–DNA is trace labeled with ^{32}P-DNA at a ratio of 1,000 to 1 w/w. As an alternative, fluorescence labeled DNA can be used as trace label.

Acknowledgments

The authors thank Federica Barchiesi and Bernhard Odermatt for their valuable contributions.

References

1. Singh M, Chakrapani A, O'Hagan D (2007) Nanoparticles and microparticles as vaccine-delivery systems. Expert Rev Vaccines 6:797–808
2. Almeida AJ, Souto E (2007) Solid lipid nanoparticles as a drug delivery system for peptides and proteins. Adv Drug Deliv Rev 59:478–490
3. Kersten G, Hirschberg H (2004) Antigen delivery systems. Expert Rev Vaccines 3:453–462
4. Azad N, Rojanasakul Y (2006) Vaccine delivery – current trends and future. Curr Drug Deliv 3:137–146
5. Peek LJ, Middaugh CR, Berkland C (2008) Nanotechnology in vaccine delivery. Adv Drug Deliv Rev 60:915–928
6. O'Hagan DT, Valiante NM (2003) Recent advances in the discovery and delivery of vaccine adjuvants. Nat Rev Drug Discov 2:727–735

7. Liang MT, Davies NM, Blanchfield JT, Toth I (2006) Particulate systems as adjuvants and carriers for peptide and protein antigens. Curr Drug Deliv 3:379–388
8. Altin JG, Parish CR (2006) Liposomal vaccines – targeting the delivery of antigen. Methods 40:39–52
9. de Jong S, Chikh G, Sekirov L, Raney S, Semple S, Klimuk S, Yuan N, Hope M, Cullis P, Tam Y (2007) Encapsulation in liposomal nanoparticles enhances the immunostimulatory, adjuvant and anti-tumor activity of subcutaneously administered CpG ODN. Cancer Immunol Immunother 56:1251–1264
10. Torchilin VP (2006) Multifunctional nanocarriers. Adv Drug Deliv Rev 58:1532–1555
11. Fenske DB, Cullis PR (2008) Liposomal nanomedicines. Expert Opin Drug Deliv 5:25–44
12. Zhou Y, Bosch ML, Salgaller ML (2002) Current methods for loading dendritic cells with tumor antigen for the induction of antitumor immunity. J Immunother 25:289–303
13. Tacken PJ, Torensma R, Figdor CG (2006) Targeting antigens to dendritic cells in vivo. Immunobiology 211:599–608
14. Reddy ST, Swartz MA, Hubbell JA (2006) Targeting dendritic cells with biomaterials: developing the next generation of vaccines. Trends Immunol 27:573–579
15. Chen W, Huang L (2008) Induction of cytotoxic T-lymphocytes and antitumor activity by a liposomal lipopeptide vaccine. Mol Pharm 5:464–471
16. Foged C, Arigita C, Sundblad A, Jiskoot W, Storm G, Frokjaer S (2004) Interaction of dendritic cells with antigen-containing liposomes: effect of bilayer composition. Vaccine 22:1903–1913
17. Sprott GD, Dicaire CJ, Gurnani K, Sad S, Krishnan L (2004) Activation of dendritic cells by liposomes prepared from phosphatidylinositol mannosides from *Mycobacterium bovis* bacillus Calmette-Guérin and adjuvant activity in vivo. Infect Immun 72:5235–5246
18. van Broekhoven CL, Parish CR, Demangel C, Britton WJ, Altin JG (2004) Targeting dendritic cells with antigen-containing liposomes: a highly effective procedure for induction of antitumor immunity and for tumor immunotherapy. Cancer Res 64:4357–4365
19. Marty C, Meylan C, Schott H, Ballmer-Hofer K, Schwendener RA (2004) Enhanced heparin sulfate proteoglycan-mediated uptake of cell-penetrating peptide-modified liposomes. Cell Mol Life Sci 61:1785–1794
20. Oehen S, Junt T, Lopez-Macias C, Kramps TAN (2000) Antiviral protection after DNA vaccination is short lived and not enhanced by CpG DNA. Immunology 99:163–169
21. Engler OB, Schwendener RA, Dai WJ, Wölk B, Pichler W, Moradpour D, Brunner T, Cerny A (2004) A liposomal peptide vaccine inducing CD8[+] T cells in HLA-A2.1 transgenic mice, which recognise human cells encoding hepatitis C virus (HCV) proteins. Vaccine 23:58–68
22. Ludewig B, Barchiesi F, Pericin M, Zinkernagel RM, Hengartner H, Schwendener RA (2000) *In vivo* antigen loading and activation of dendritic cells via a liposomal peptide vaccine mediated protective antiviral and antitumor immunity. Vaccine 19:23–32
23. Marty C, Schwendener RA (2004) Cytotoxic tumor targeting with scFv antibody-modified liposomes. In: Ludewig B, Hoffmann MW (eds) Adoptive immunotherapy: methods and protocols. Methods in Molecular Medicine, vol 109. Humana, Totowa, NJ, pp 389–401
24. Allen TM (2002) Ligand-targeted therpeutics in anticancer therapy. Nat Rev Cancer 2:750–762
25. van Winden EC (2003) Freeze-drying of liposomes: theory and practice. Methods Enzymol 367:99–110
26. Ureta-Vidal A, Firat H, Perarnau B, Lemonnier FA (1999) Phenotypical and functional characterization of the CD8[+] T cell repertoire of HLA-A2.1 transgenic, H-2K[b] null D[b] null double knockout mice. J Immunol 163:2555–2560
27. Hügle T, Cerny A (2003) Current therapy and new molecular approaches to antiviral treatment and prevention of hepatitis C. Rev Med Virol 13:361–371
28. Ge Q, Filip L, Bai A, Nguyen T, Eisen HN, Chen J (2004) Inhibition of influenza virus production in virus-infected mice by RNA interference. Proc Natl Acad Sci U S A 101:8676–8681
29. Mayer LD, Hope MJ, Cullis PR (1986) Vesicles of variable sizes produced by a rapid extrusion procedure. Biochim Biophys Acta 858:161–168

Chapter 12

Mannosylated Liposomes for Targeted Vaccines Delivery

Suresh Prasad Vyas, Amit K. Goyal, and Kapil Khatri

Abstract

Mannosylated liposomes appear to be a promising and potential carrier system for delivery of proteins, peptides, or nucleic acids. The present chapter describes novel mannosylated liposomes, which increase the intracellular targeting of immunogen to dendritic cells and macrophages possessing the specific receptors. The liposomes used in the present investigation were prepared by hand-shaken method and characterized for size, shape, surface charge, encapsulation efficiency, ligand binding, and specificity and uptake studies. The immune-stimulating activity of the liposomes was studied by measuring antigen-specific antibody titer following subcutaneous administration of different liposomal formulations in BALB/c mice. It was found that O-palmitoyl mannan (OPM)-coated liposomes showed better uptake efficiency. In vivo studies revealed that the OPM-coated liposomes exhibited significant higher serum antibody response and stronger TH1/TH2-based cellular responses. In conclusion, novel vesicular constructs are useful nanosized carriers having superior surface characteristics – for active interaction with the antigen-presenting cells and subsequent processing and presentation of antigen.

Key words: Liposomes, Mannan, Dendritic cells, Vaccines, Mannosylated liposomes

1. Introduction

Liposomes as vaccine delivery vehicles have demonstrated substantial advantages as they are less toxic, targetable, can maintain the antigen integrity, and are easy to prepare too. Antigen delivery with liposomes as carrier systems provides options and opportunities for designing bio-stable- and/or site-specific immunization. Liposomes can augment both humoral and cell-mediated immunity against a wide variety of antigens, including various protein/peptide and DNA derived from bacterial and viral sources, ovalbumin, bovine serum albumin, and influenza subunit vaccine. The adjuvanticity of liposomes appears to be dependent on structural characteristics such as vesicle size, surface charge, lipid to antigen or plasmid DNA ratio, the number of lamellae, and the rigidity of the bilayer (1, 2). Besides, change in the composition coating of liposomes with specific ligand(s) or polymer may also enhance the adjuvanticity of liposomes. Polysaccharide/polypeptide/polymer-coated liposomes were developed by several groups, which showed selectivity, specificity, mucoadhesivity, and adjuvanticity (3–6). The attachment of a ligand that can be recognized by a specific mechanism would endow a carrier with the ability to target a specific population of cells. In the search for ligand-directed delivery systems, several ligands including asialoglycoproteins, galactose, mannose, transferrin, and antibodies have been used to improve the delivery of biomolecules to the target cells. Therefore, the incorporation of such ligands into liposomes would improve the target-cell specificity and immune response (3, 6–8). Recently, the emphasis has been laid upon the carbohydrate (C-type lectin receptors)-mediated liposomal interactions with the target cells. Among the various carbohydrate ligands such as glycoproteins, glycolipids, viral proteins, polysaccharides, lipopolysaccharides, and other oligosaccharides, mannan-anchored liposomal systems have shown a tremendous potential in drug delivery, targeting, as well as in immunization. The activity of mannose receptors has been demonstrated in dendritic cells, macrophages such as kupffer cells, peritoneal and pulmonary macrophages, and hepatic, sinusoidal endothelial cells (8–12). Therefore, mannosylated ligands can be used as carriers for these types of cells. Mannan coating will not only stabilize the vesicles but also acts as a ligand for mannose receptors expressed on macrophages and dendritic cells (4, 6, 8–12). Here we describe novel mannosylated liposomes consisting of a palmitoilated mannan, which increase the intracellular targeting efficiency to immuno-competent antigen-presenting cells possessing the corresponding receptors, such as dendritic cells and macrophages.

2. Materials

2.1. Preparation of Palmitoylated Liposomes

1. Distearyl phosphatidylcholine (DSPC).
2. Dioleoyl phosphatidylethanolamine (DOPE).
3. Cholesterol.
4. Palmitoyl chloride.
5. Mannan.
6. Sephadex G-50.
7. Concanavalin A.
8. Mica sheet.

9. Dimethylformamide.
10. Dry pyridine.
11. Dry diethylether.
12. Absolute ethanol.
13. Zetasizer nano ZS-90 (Malvern Pvt Ltd., UK).
14. Nanoscope a AFM system (Digital Instruments, Santa Barbara, CA).
15. BCA protein assay or Hoechst 33258 dye binding assay.
16. Phosphate-buffered saline (PBS) (pH 7.4) containing 5 mM calcium chloride and 5 mM magnesium chloride.

2.2. Culture and Uptake Analysis of Dendritic Cells

1. Roswell Park Memorial Institute-1640 (RPMI-1640) medium supplemented with 10% fetal calf serum (FCS)
2. Granulocyte monocyte colony stimulating factor (GM-CSF) and Interleukin-4 (IL-4)
3. Sample staining buffer: (0.1% (w/v) NaN_3 and 1.0% (w/v) bovine serum albumin in Hanks balanced salt solution)
4. Labeled or unlabeled antibody
5. Propidium iodide (PI) solution (50 μg/mL propidium iodide in 0.1% (w/v) sodium citrate)
6. 4% *para*-formaldehyde

2.3. Immunological Studies

1. Microtiter plate (Nunc-Immune Plate® Fb96 Maxisorb, Nunc, USA)
2. Pentobarbital injection
3. Coating buffer: 0.04-M carbonate buffer, pH 9.6 (sodium carbonate 1.59 g/L, sodium hydrogen carbonate 2.93 g/L in distilled water)
4. Washing buffer: 0.1% (v/v) Tween 20 in PBS, pH 7.4
5. Horseradish, peroxidase-labeled goat, anti-mouse antibodies (Sigma-Aldrich Co., USA)
6. H_2O_2
7. *O*-Phenylenediamine dihydrochloride in 0.05 M phosphate–citrate buffer, pH 5.0
8. 1N H_2SO_4

3. Methods

The surface-modified purpose-specific versions of liposomes (mannosylated liposomes) offer potentials of exquisite levels of specificity and targetability. The affinity and selectivity of the anchored mannan toward its complementary mannose receptor (type-I C-type lectin) is a desirable prerequisite that enhances antigen uptake and subsequent presentation via both the MHC class I and class II pathway (4, 5, 7, 10). Over the years, various strategies have been developed for the coating of the liposomal surface with natural or hydrophobized mannan which include use of mannose-lipid conjugates as raw material for the preparation of liposomes; direct binding of mannose derivatives by chemical reaction to plain liposomes; simple adsorption of their palmitoyl or cholesteroyl derivatives or natural mannan onto the surface of liposomes (12–14). Coating of liposomes with mannan can be performed by coincubation of aqueous solutions of mannan derivatives with preformed liposomal dispersion or covalent coupling of mannan derivative with preformed liposomes or mixing of mannan derivatives with liposomal lipid constituents during the formation of liposomes (Fig. 1). This architects a polysaccharide (mannan)-based artificial cell wall on the outermost surface of the liposomes. Recently, palmitoyl conjugates of mannan have been employed by our groups to coat the liposomes. When added to liposomes, the hydrophobic anchors interact with the outer half of the bilayer orienting and projecting hydrophilic portion toward the aqueous bulk.

3.1. Preparation of Liposomes

1. Take DSPC, DOPE, and cholesterol in the molar ratio of 2:1:1 in a 50-mL round bottom flask and dissolve in 10 mL of chloroform: methanol (1:1). Flush the flask with nitrogen and maintain the controlled vacuum (see **Note 1**).

2. Rotate the flask to evaporate the solvent leaving a stack of thin layers on the wall of round bottom of flask (see **Note 2**).

3. Keep the flask for 6 h to ensure the complete removal of the solvent system.

4. Add 10 mL of phosphate buffer saline in which protein/antigen is previously dissolved (see **Notes 3** and **4**).

5. Seal the flask and hydrate the lipid film in rotary evaporator or using a manual shaker for 72 h.

6. Purify the prepared liposomes from excess protein/antigen by size exclusion chromatography using a column of sephadex G-50 or by dialysis.

7. Store the immunogenic vesicles at $4 \pm 1°C$ under nitrogen and protect from light until use.

3.2. Synthesis of OPM

1. Dissolve the mannan (100 mg) to be palmitoylated in 1 mL of dry dimethyl formamide (DMF).

2. Dissolve palmitoyl chloride (10 mg) in DMF in the presence of dry pyridine (1.0 mL).

3. Immediately add 1 mL palmitoyl chloride solution to the 1 mL of mannan solution. Mix to dissolve.

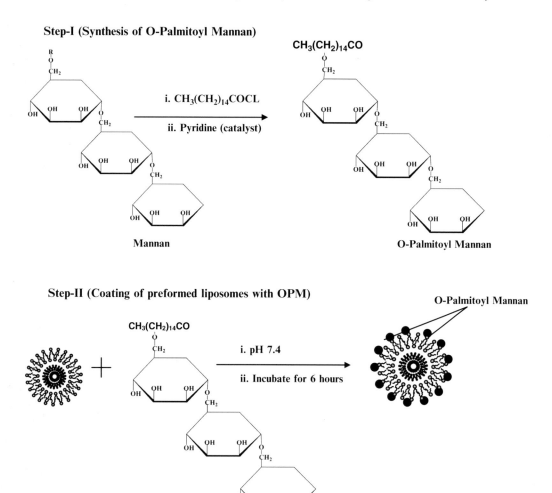

Fig. 1. Preparation of O-palmitoyl mannan-anchored vesicles. Liposomes were prepared by hand-shaking method and coating of liposomes with these hydrophobized polysaccharides (OPM) was performed by incubation of aqueous solutions of polysaccharide derivatives with preformed liposomal dispersion

4. React the mixture at 60°C and stir for 6 h.
5. Slowly pour the resultant mixture into absolute ethanol (100 mL) under vigorous stirring and collect the precipitate by centrifugation.
6. Wash the precipitate with 50 mL of absolute ethanol and 25.0 mL of dry diethyl ether, and dry in vacuum at $50 \pm 1°C$ for 1 h (see **Note 5**).

3.3. Coating of OPM on the Surface of Liposomes

1. Prepare liposomal vesicles using any established methods or by as describe in above procedure.
2. Dissolve the different concentration of O-palmitoyl mannan (OPM) (as synthesized above) in double distilled water.

3. Incubate the prepared liposomal vesicles with OPM at room temperature in the presence of an inert gas atmosphere to prevent lipid oxidation.

4. Optimum ratio of lipid to synthesized polymer and incubation time is determined by changing the concentration of polymer (OPM) used for coating of the surface of liposomes incubation for different time intervals (see **Note 6**).

5. Purify the surface-modified liposomes from excessive unbound polysaccharide by centrifugation ($25,000 \times g$ for 25 min).

6. Store the surface-modified, immunogenic liposomal vesicles at $4 \pm 1 °C$ under nitrogen and protect from light until use.

3.4. Particle Size, Polydispersity Index, and Zeta Potential Measurements of Mannosylated Liposomes

Particle size, polydispersity index, and zeta potential measurements of mannosylated liposomes are determined by Zetasizer nano ZS-90.

1. Deposit 20 mL of liposomal formulation onto freshly cleaved mica and keep the formulation in place for 90 s.

2. Wash the unadsorbed material with distilled water (0.2 μm filtered).

3.5. AFM Analysis of Mannosylated Liposomes

3. Observe the AFM image in tapping mode in air with single-crystal silicon cantilevers on a Nanoscope, a AFM system.

4. Optimize the scanning speed on the basis of liposomal size.

5. Capture the AFM image by both height and amplitude modes (An example of AFM image of liposomes are shown in Fig. 2).

3.6. Determination of Encapsulation Efficiency

The concentration of immunogen in liposomal formulation is determined by the ultracentrifugation of the prepared liposomal formulation.

1. Ultracentrifuge the liposomal formulation at $40,000 \times g$ for 40 min.

2. Remove the supernatant and extrude the pellet through 0.2-μm membrane filter.

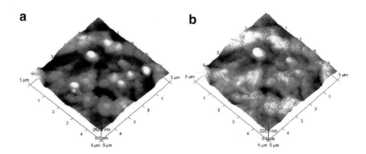

Fig. 2. AFM image of OPM-coated and-uncoated liposomes: (**a**) Plain liposomes, and (**b**) OPM-coated liposomes. Liposomes were prepared by hand-shaking method and a drop of liposomal formulation was transferred to a mica sheet. Each image shows the spherical shape determined from the AFM observation. AFM image of the liposomes indicated that plain liposomes have lower size compared to OPM-coated liposomes

3. Quantify the concentration of antigen (protein/pDNA) by a micro-BCA protein assay or Hoechst-33258 dye-binding assay.

4. Calculate the concentration of antigen from the respective absorbance/fluorescence.

3.7. Ligand-Binding Specificity of Liposome-Anchored Ligands

In vitro, ligand-specific affinity of mannosylated liposomes toward exogenously provided lectin Concanavalin A (Con-A) can be used as a measure of activity for OPM-anchored liposomes toward mannose receptor.

1. Add 200 µl of plain OPM (standard) or OPM-coated liposomes formulation (mannosylated or neutral) to 1-mL Con A (1 mg/mL) in PBS (pH 7.4).

2. Observe the increase in turbidity at 550 nm at variable time interval using UV spectrophotometer or measure the size of Con A-aggregated liposomal formulations by particle size analyzer (An example of in vitro ligand-binding specificity of liposomes is shown in Fig. 3).

3.8. Uptake Analysis of Mannosylated Liposomes Via Dendritic Cells

3.8.1. Isolation and Culture of Dendritic Cells

1. Isolate murine bone marrow cells from bone (femurs and tibias).

2. Generate dendritic cells (DCs) from murine bone marrow cells by washing out the bone marrow two to three times with RPMI-1640 medium (see **Note 7**).

3. Plate the DCs in six-well tissue culture plates at 2×10^6 cells per well in RPMI-1640 medium supplemented with 10% fetal calf serum (FCS) (see **Note 8**).

4. Incubate for 1 h at 37°C/5% CO_2 (see **Note 7**).

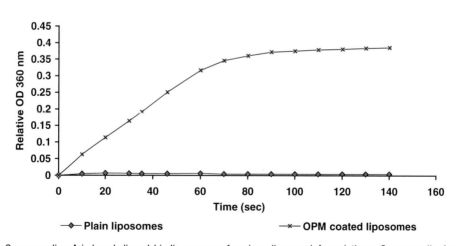

Fig. 3. Concanavaline-A-induced, ligand-binding assay of various liposomal formulations. Concanavalin A solution (100 µg/mL) was added to 0.2 mL of diluted liposomes formulations and absorbance at 550 nm was measured as a function of time. The data were subtracted from the blank experiment conducted without an addition of concanavalin A. Binding of the terminal mannose residues of the mannan to the ConA causes agglutination of the complex in solution resulting in an increase in turbidity. However, plain liposomes did prevent the binding of Con A as evidenced by the lower turbidity. The accessibility of the mannan to the liposomes is critical for in vivo cell binding

5. Remove the nonadherent cells by gentle washing and culture the adherent cells with RPMI-10 FCS enriched with 500 U/mL GM-CSF and 8 ng/mL IL-4 to generate DCs.

6. Replace the culture medium every third day.

7. Count the yield of viable dendritic cells by flow cytometry or by simply counting in a hemacytometer.

3.8.2. Flow Cytometric Analysis for Evaluation of Kinetics of Uptake

1. Add 100 mL of each liposomal formulation to the cultured DCs (on the seven of culture) wells.

2. Remove excess of formulations by washing with ice-cold phosphate-buffered saline (PBS).

3. Harvest the cells.

4. After stimulating the cells, keep in the dark on an ice bath and wash three times in ice-cold, phosphate-buffered saline (PBS) or twice in ice-cold, ethanoic acid buffer followed by twice in cold PBS.

5. Resuspend the formulation-stimulated dendritic cells with staining buffer and aseptically prepare a single-cell suspension.

6. Count the viable dendritic cells by flow cytometry.

7. Centrifuge the cell suspension (from step 3) for 8 min at $300 \times g$, 4°C, and discard the supernatant. Resuspend the cell pellet to 2×10^7 cells/mL in staining buffer, at 4°C.

8. Add 50-µl cell suspension (106 cells) to the wells of a 96-well, round-bottom microtiter plate.

9. Add 10 µl of appropriately diluted, labeled antibody to each tube or well containing cells and mix gently. Incubate 20 min in an ice bath.

10. Wash the cells two to three times with sample staining buffer.

11. Count the stained cells by flow cytometry (An example of dendritic cells uptake profile of different liposomal formulation by FACS analysis is shown in Fig. 4).

3.8.3. Spectral, Bioimaging Analysis for Internalization studies

1. Add 100 mL of each liposomal formulation (FITC/rhodamine-labeled antigen) to the cultured DCs (on the seven of culture) wells.

2. Remove the excess of formulations by washing with ice-cold, phosphate-buffered saline (PBS).

3. Harvest the cells.

4. After stimulation, keep the cells in the dark on the ice bath and wash three times in ice-cold, phosphate-buffered saline (PBS) or twice in ice-cold, ethanoic acid buffer followed by twice in cold PBS.

5. Resuspend the stimulated dendritic cells with ice-cold PBS containing 0.01% sodium azide and 5% FCS.

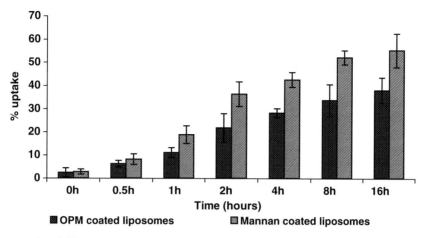

Fig. 4. Percent uptake of different liposomal formulations conjugated to Rhodamine by human DCs measured through flow cytometry. Cells were incubated with Rhodamine-MSP-1_{19}-loaded liposomal formulations and analyzed at different time intervals (0, 15, 30, 60, 120, 180, and 360 min). The kinetics of uptake has been presented with mean fluorescence intensity (MFI) vs. counts by FACS analysis and percentage uptake at various time intervals. Uptake of formulation on a per cell basis was quantified as fluorescence intensity per cell. Percentage of positive cells was determined as proportion of cells with fluorescence intensity higher than 99% of cells of the control sample (cells incubated with unconjugated rhodamine alone). Flow cytometric analysis revealed that the percentage of rhodamine-positive DCs increased rapidly and reached a plateau after 16 h of incubation (means of three independent experiments). A steady increase in the uptake percentage (%) was recorded and a maximum cell-associated fluorescence was observed at 16 h for OPM-coated cationic liposomes. Flow cytometric analysis of DCs revealed that plain liposomes did not significantly enhance the antigen uptake by DCs compared with the uptake recorded for mannan-coated liposomes

6. Fix the treated cells with 4% v/v *para*-formaldehyde for 30 min and finally wash with PBS.
7. Investigate the fixed cells by using an upright fluorescence microscope equipped with high-pressure mercury lamp (HBO 100) for excitation and triple bandpass filter set.
8. Capture the image using 100×-oil-immersion objective lens.
9. Perform images acquisition using Case Data Manager Software and Spectral Imaging 4.0 software (an example of spectral bio imaging of dendritic cells uptake is shown in Fig. 5).

3.9. Assessment of In Vivo Adjuvanticity of Mannosylated Liposomes

The adjuvanticity of the antigen-loaded liposomal formulations was assessed in BALB/c mice (6–8 weeks age) following the guidelines of Council for the Purpose of Control and Supervision of Experiments on Animals (CPCSEA), Ministry of Social Justice and Empowerment, Government of India.

3.9.1. Immunization of Mice

1. Anesthetize a mouse using pentobarbital (60 mg/kg body weight) by intraperitoneal injection.
2. Immunize the mouse subcutaneously with liposomal preparations equivalent to 25 μg of antigen (Malaria antigen, merozoite surface protein MSP-1_{19}) on zeroth and second week.

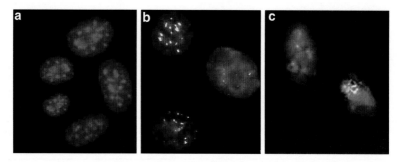

Fig. 5. Spectral bioimaging analysis for internalization studies of various formulations into the dendritic cells: (a) mock-treated DCs, (b) plain liposomes-treated DCs, (c) OPM-coated liposomes-treated DCs. Besides the quantitative determination by FACS analysis, the qualitative uptake of FITC–MSP-1$_{19}$-loaded liposomal systems by murine DCs after 180-min incubation was studied using spectral bio-imaging system, using an upright fluorescence microscope (Axioscope, Carl Zeiss, Germany) equipped with high-pressure mercury lamp (HBO 100) for excitation and triple, bandpass filter set. In the spectral range from 400 to 700 nm, the objective lens has minimal fluctuations ranging from 85 to 90%. The optical head attached to the microscope is composed of a Sagnac commonpath interferometer and imaging optics including a cooled CCD camera (Hamamatsu, Japan). Images acquisition was performed using Case Data Manager Software and the spectral analysis was done with Spectral Imaging 4.0 software (Applied Spectral Imaging, Israel). Spectral bioimage clearly shows the intracellular delivery efficacy of various liposomes in DCs. The intensity of fluorescence in the fluorescence photomicrographs of the DCs revealed intracellular fluorescence. The fluorescence observed in case of uncoated liposomes was relatively low than the fluorescence intensity of the OPM-coated formulation. It should be noted that fluorescence in solution does not penetrate inside APC nor inside intracellular vacuoles. This reveals that the fluorescence observed was due to the fluorescence present in the liposomes, which were phagocytized by dendritic cells. The increased fluorescence intensity produced by the OPM coated liposomes could be attributed to the specificity and affinity of the polysaccharide ligand (OPM) toward the mannose receptors of the macrophages/dendritic cells (APC) of the spleen. Fluorescence microscopic studies revealed that mannosylated liposomes complexes showed a better uptake by the specialized DCs via mannose receptor-mediated endocytosis

3.9.2. Collection of Serum

1. Anesthetize a mouse using pentobarbital (60 mg/kg body weight) by intraperitoneal injection.
2. Collect the blood from the retroorbital plexus of mouse under anesthesia on second, fourth, sixth, and eighth week.
3. Separate the serum by centrifugation and store at −20°C until assayed for quantification of antigen-specific antibodies.

3.9.3. Determination of Antibody

Antibody responses in immunized animals were assayed for anti-MSP-1$_{19}$ antibodies by microplate enzyme-linked immunosorbent assay (ELISA).

1. Prepare the antigen solution at 10 μg/mL in coating buffer.
2. Coat each well of 96-well microtiter plates with 100 μl of antigen (MSP-119) for overnight at 4°C.
3. Block the coated plate with PBS–BSA (3% (w/v)) for 1 h at 37°C (see Note 9).
4. Wash the plate three times with 200 μl of washing buffer.
5. Add 100 μl of serially diluted serum to each wells and incubate for 2 h at 37°C.

6. After further washing with washing buffer, add 100 μl of enzyme-labeled goat, antimouse antibodies (Horse-radish, peroxidase-labeled antimouse IgG) to each well.

7. Incubate the plate for 1 h at 37°C and wash with 200 μl of washing buffer.

8. Add 100 μl of substrate solution containing H_2O_2 and O-phenylenediamine dihydrochloride (1 mg/mL) as chromogen to each well.

9. Stop the reaction using 1N H_2SO_4.

10. Read the optical density at 490 nm of the reaction product by using a microplate reader.

4. Notes

1. The choice of the phospholipid is dependent on the actual needs of the antigen delivery system. Cationic lipid with a combination of pH-sensitive lipid can be used for DNA-based vaccines delivery or intracellular cytosolic delivery of protein molecules. They may also be used in combination vaccines or in combination with other adjuvants like MPL-A (monophosphoryl lipid A), and cytokines (IL-2, IL-4, IFN-gamma).

2. Purity of phospholipid is essential to develop stable liposomes. Moreover, nature of the dry lipids, its surface area, its porosity, and hydration of dry lipid film (temperature, pH and ionic strength of hydration medium) affect the lamelarity and morphology of formulated liposomes.

3. All solutions and equipment must be sterile, and aseptic technique should be used accordingly.

4. Use sterile Milli-Q water during the development of liposomal formulations. Sterilize by filtration. All buffers and reagents are stored at 4°C and are used within four weeks of preparation.

5. During a synthesis of OPM, washing the precipitate with absolute ethanol and dry diethyl ether is important for the synthesis of OPM.

6. Concentration of OPM and coating time onto the surface of liposomes can be optimized by measuring changes in size and zetapotenital.

7. All culture incubations should be performed in a humidified, 37°C, 5%-CO_2 incubator, unless otherwise specified. Some media (e.g. DMEM) may require altered levels of CO_2 to maintain pH 7.4.

8. When subculturing cells, add a sufficient number of cells to give a final concentration of ~2×10^6 cells/mL in each new culture. Cells plated at too low a density may be inhibited or delayed in entry in the growth stage.

9. During ELISA, BSA is used as a blocking reagent to help reduce background, owing to nonspecific binding. However, antibodies generated in sheep may crossreact with IgG found in normal BSA. When using sheep antibodies, use IgG-free BSA.

References

1. Foged C, Arigita C, Sundblad A, Jiskoot W, Storm G, Frokjaer S (2004) Interaction of dendritic cells with antigen-containing liposomes: effect of bilayer composition. Vaccine 22:1903–1913
2. Gregoriadis G, Bacon A, Caparros-Wanderley W, McCormack B (2002) A role for liposomes in genetic vaccination. Vaccine 20:B1–B9
3. Hattori Y, Kawakami S, Lu Y, Nakamura K, Yamashita F, Hashida M (2006) Enhanced DNA vaccine potency by mannosylated lipoplex after intraperitoneal administration. J Gene Med 8:824–834
4. Kawakami S, Sato A, Nishikawa M, Yamashita F, Hashida M (2000) Mannose receptor mediated gene transfer into macrophages using novel mannosylated cationic liposomes. Gene Ther 7:292–299
5. Kawakami S, Yamashita F, Nishida K, Nakamura J, Hashida M (2002) Glycosylated cationic liposomes for cell-selective gene delivery. Crit Rev Ther Drug Carrier Syst 19:171–190
6. Vyas SP, Sihorkar V (2000) Endogenous ligands and carriers in non-immunogenic site-specific drug delivery. Adv Drug Deliv Rev 43:101–164
7. Yoshikawa T, Imazu S, Gao J-Q, Hayashi K, Tsuda Y, Shimokawa M, Sugita T, Niwa T, Oda A, Akashi M, Tsutsumi Y, Mayumi T, Nakagaw S (2004) Augmentation of antigen-specific immune responses using DNA-fusogenic liposome vaccine. Biochem Biophys Res Commun 325:500–505
8. Vyas SP, Singh A, Sihorkar V (2001) Ligand-receptor mediated drug delivery: an emerging paradigm in cellular drug targeting. Crit Rev Ther Drug Carrier Syst 18(1):1–76
9. Opanasopit P, Higuchi Y, Kawakami S, Yamashita F, Nishikawa M, Hashida M (2001) Involvement of serum mannan binding proteins and mannose receptors in uptake of mannosylated liposomes by macrophages. Biochim Biophys Acta 1511:134–145
10. Lu Y, Kawakami S, Yamashita F, Hashida M (2007) Development of an antigen-presenting cell-targeted DNA vaccine against melanoma by mannosylated liposomes. Biomaterials 28:3255–3262
11. Venkatesan N, Vyas SP (2000) Polysaccharide coated liposomes for oral immunization: development and characterization. Int J Pharm 203:169–177
12. Sihorkar V, Vyas SP (2001) Potential of polysaccharide anchored liposomes in drug delivery, targeting and immunization. J Pharm Pharmaceut Sci 4(2):138–158
13. Sunamoto J, Iwamoto K (1986) Protein anchored and polysaccharide-anchored liposomes as drug carriers. Crit Rev Ther Drug Carrier Syst 2:117–136
14. Moreira JN, Almeida LM, Geraldes CF, Costa ML (1996) Evaluation of *in vitro* stability of large unilamellar lipsomes anchored with a modified polysaccharide (O-palmitoyl Pullulan). J Mat Sci Mat Med 7:301–303

Chapter 13

Liposomes for Specific Depletion of Macrophages from Organs and Tissues

Nico van Rooijen and Esther Hendrikx

Abstract

A liposome mediated macrophage "suicide" approach has been developed, based on the liposome mediated internalization of the small hydrophilic molecule clodronate in macrophages J Leukoc Biol 62:702, 1997. This molecule has a very short half life when released in the circulation, but does not easily cross phospholipid bilayers of liposomes or cell membranes. As a consequence, once ingested by a macrophage in a liposome encapsulated form, it will be accumulated within the cell as soon as the liposomes are digested with the help of its lysosomal phospholipases. At a certain intracellular clodronate concentration, the macrophage is eliminated by apoptosis. Given the fact that, neither the liposomal phospholipids chosen, nor clodronate are toxic to other (non-phagocytic) cells, this method has proven its efficacy and specificity for depletion of macrophage subsets in various organs. In several cases, organ specific depletion can be obtained by choosing the right administration route for the clodronate liposomes.

Key words: Liposomes, Macrophages, Clodronate, Depletion of macrophages, Kupffer cells, Spleen, Liver, Lung, Lymph nodes

1. Introduction

1.1. Macrophages, Liposomes and Clodronate

From an evolutionary point of view, macrophages are ancient cells. They form the core of the innate immune system and did appear long before the cells forming together the complex immune system of the higher vertebrates. As a consequence, during evolution, they did acquire functions, both in innate immune reactions and in the regulation of activities of many non-phagocytic cells. The latter functions are mediated by soluble molecules such as cytokines and chemokines. Macrophages are also involved in *homoiostasis* of the body by ingesting and digesting foreign particles, such as microorganisms, senescent erythrocytes and macromolecules. Their intracellular digestion is mediated by their lysosomal enzymes. As a consequence of their multifunctionality, one of the first questions to be solved in many newly started studies in animal models of human diseases, is that about a possible macrophage-dependency of findings or phenomena. Depletion of macrophages from organs or tissues may then help to solve this question. Given that, earlier methods for macrophage depletion were often not macrophage-specific, not generally applicable or could even activate those macrophages that were not successfully depleted (1), a new more sophisticated approach was required. The "liposome mediated macrophage *suicide* technique" based on the introduction and accumulation of small, strongly hydrophilic molecules in macrophages, with the help of liposomes, did meet the main needs for such an approach.

Clodronate is a member of the family of bisphosphonates developed for the treatment of osteolytic bone diseases. It shows high affinity for calcium and as a consequence adheres to bone when administered to vertebrates. Osteoclasts, play a role in the physiology of bone by breaking it down, opposed to osteoblasts who are involved in its reconstruction. It appeared that the activity of osteoclasts could be affected by bisphosphonate molecules that are struck to the bone. Given that both osteoclasts and macrophages belong to the mononuclear phagocyte system (MPS), we decided to try clodronate as one of the first effector molecules to be tested in our planned "liposome mediated macrophage *suicide* technique" (2). Although, subsequently, we found several other hydrophilic molecules that were also suitable as effector molecules in the approach (e.g. (3, 4)), clodronate may still be considered the best choice because it shows maximum efficacy and minimal toxicity.

Moreover, both liposomes and clodronate have already been introduced in the clinic and it may be anticipated that their combination in the transient suppression of macrophage activity for human application would be easier to achieve than for any of the other candidate effector molecules which are known until now (5, 6).

Though we did develop the method for our own studies on functional activities of macrophages in the spleen, it was clear that possible applications of the "liposome mediated macrophage *suicide* technique" in research would not be limited to the various subsets of macrophages in that organ (7, 8).

1.2. Liposome Mediated Depletion of Macrophages

Strong hydrophilic molecules such as the negatively charged bisphosphonate dichloromethylene-bisphosphonate (clodronate) and the positively charged diamidine propamidine can be solved in aqueous solutions in substantial concentrations. As a consequence such molecules can be encapsulated in multilamellar liposomes with a high efficacy (3). Once encapsulated, they cannot easily escape from the liposomes, since they are unable to cross their phospholipid bilayers. Leakage remains very low for that reason. After administration of such liposomes in vivo, their natural fate is phagocytosis by macrophages. Once ingested by a macrophage, a liposome will be digested with the help of the lysosomal panel of lytic enzymes, among which are phospholipases that are able to break down the phospholipid bilayers. In this way, the encapsulated molecules are released within the cell.

Since, they cannot easily escape from the cell either, because its cell membrane is, in its most basic form, also consisting of phospholipid bilayers, these molecules will be accumulating in the cell as more liposomes are ingested and digested by the macrophage. At a certain intracellular concentration, molecules such as clodronate and propamidine, will eliminate the macrophage by initiating its programmed cell death (apoptosis) (9). Reversely, clodronate molecules released from dead macrophages will be rapidly cleared from the circulation by the renal system, since their half life – when free in the circulation – is in the order of minutes (10). Macrophages can be found in nearly all tissues of the body. By choosing the right administration route of clodronate liposomes, particular organs or tissues can be depleted of macrophages. In this way, i.e. by creating a macrophage depleted organ or tissue, macrophage functions can be studied in vivo. Moreover, from a therapeutic perspective, promising results were obtained by application of clodronate liposomes for suppression of macrophage activity in various models of autoimmune diseases, transplantation, neurological disorders and gene therapy (5). For more information and specific references, see the "clodronate liposomes" website: http://www.ClodronateLiposomes.org

1.3. Comparative Accessibility of Macrophages in Different Tissues

1.3.1. Administration Routes for Liposomes and Physical Barriers

The extent to which resident macrophage populations in different organs are accessible to single molecules, molecular complexes or particulate carriers such as liposomes, depends on both the position of the macrophages in the tissues and on the properties of the molecules or particles. In general, all macrophages can be reached by small molecules, if the latter are able to pass the walls of blood vessels, e.g. capillaries, in order to penetrate into the parenchyme tissues. Large molecules, molecular complexes or particles can reach a macrophage only if there is no physical barrier between the site of injection and the macrophage. Such a barrier can be formed, e.g. by endothelial cells in the wall of blood-vessels, by alveolar epithelial cells in the lung, by reticular fibers or collagen fibers in the spleen or by the presence of densely packed cells such as lymphocytes in the white pulp of the spleen or in the paracortical fields of lymph nodes. By choosing the right administration route for the materials to be injected, this barrier can be kept at a minimum.

The in vivo accessibility of various macrophages to liposomes is the main factor that determines the efficacy of the approach. The dose of clodronate liposomes required for depletion of macrophages, and the time interval between injection of liposomes and their depletion, both depend on this accessibility.

1.3.2. Intravenous Administration

Intravenously injected materials can reach macrophages in the liver (Kupffer cells), spleen and bone marrow. Kupffer cells in the liver sinuses, as well as marginal zone macrophages and red pulp macrophages in the spleen have a strategic position with respect to large molecular aggregates and particulate materials in the circulation. Liposomes have a nearly unhindered access to these macrophages as concluded from their fast and complete depletion within 1 day after intravenous injection of clodronate liposomes in mice and rats (11). Obviously, it is a little more difficult for intravenously injected liposomes to reach the marginal metallophilic macrophages in the outer periphery of the white pulp. Depletion of the white pulp macrophages in the periarteriolar lymphocyte sheaths (PALS) is incomplete emphasizing the barrier formed by the reticulin fiber network and/or the densely packed lymphocytes in the white pulp (12). Also macrophages in the bone marrow were reached by intravenously injected clodronate liposomes. However two consecutive injections with a time interval of 2 days were required to get a nearly complete depletion of macrophages from the bone marrow (13).

Kupffer cells in the liver play a key role in the homeostatic function of the liver. They form the largest population of macrophages in the body, make up 30% of the hepatic nonparenchymal cell population, and have easy access to particulate materials in the circulation. Consequently, a large proportion of all intravenously administered particulate carriers used for drug targeting or gene transfer will be prematurely destroyed before they reach their targets. Therefore, transient blockade of phagocytosis by Kupffer cells might be an important factor to optimize in drug targeting, gene transfer, xenogeneic cell grafting (5) and in some autoantibody mediated disorders in which macrophages consume the body's own platelets (14) or red blood cells (15). Also, transient suppression of the cytokine mediated activity of Kupffer cells might have a beneficial effect on various disorders of the liver (16).

1.3.3. Subcutaneous Administration

Subcutaneously injected clodronate-liposomes are able to deplete macrophages in the draining lymph nodes of mice and rats. Such liposomes, when, e.g. injected in the footpad of mice, led to the depletion of subcapsular sinus lining macrophages and medulla macrophages in the draining popliteal lymph nodes (17). Macrophages in the paracortical fields and those in the follicles were not affected, emphasizing the existence of a barrier formed by reticular fibers and/or densely packed lymphocytes in these lymph node compartments, comparable to that formed in the white pulp of the spleen. After passing the popliteal lymph nodes, the lymph flow is still filtered by consecutively draining lymph node stations such as lumbar lymph nodes (in the mouse). Macrophages in these lymph nodes were partially depleted. It was apparent that only macrophages had been depleted in those compartments that directly drained the popliteal lymph nodes. The blood flow entering the spleen by the *arteria lienalis* is evenly distributed over the entire spleen, whereas different parts in the lymph nodes are correspond with their own draining area and have their own afferent lymph vessels. As a consequence, particles such as liposomes are not equally distributed over all macrophages in the lymph nodes.

1.3.4. Intraperitoneal Administration

Macrophages from the peritoneal cavity and the omentum of the rat were depleted by two consecutive intraperitoneal injections with clodronate liposomes, given at an interval of 3 days (18). The peritoneal cavity is drained by the parathymic lymph nodes (in rats and mice). After passing these lymph nodes, the lymph flow reaches the blood circulation via the larger lymph vessels such as *ductus thoracicus*. As a consequence, intraperitoneally injected clodronate liposomes are also able to deplete the macrophages of parathymic lymph nodes, and once they arrive in the blood circulation, they may deplete macrophages in liver and spleen. Given the relatively large volume that can be administered via the intraperitoneal route, the total number of macrophages that can be affected is even higher than that affected by intravenous injection.

1.3.5. Intratracheal and Intranasal Administration

Alveolar macrophages form a first line of defense against microorganisms entering the lung via the airways. In contrast to the interstitial macrophages that are separated from the alveolar space by an epithelial barrier, alveolar macrophages which are located in the alveolar space have direct access to liposomes administered via the airways, for instance by intratracheal instillation, intranasal administration or by the application of aerosolized liposomes. The direct access of clodronate liposomes to alveolar macrophages is demonstrated by their ability to eliminate these cells in mice and rats (19). Alveolar macrophages make up about 80% of the total macrophage population in the lung. Given their presence in high numbers and the total mass of lung tissue, they form an important population of macrophages in the body.

1.3.6. Intraventricular Administration in the CNS

Stereotaxical injection of clodronate liposomes into the fourth ventricle of the central nervous system (CNS) of rats resulted in a complete depletion of perivascular and meningeal macrophages in the cerebellum, cerebrum, and spinal cord of these rats (20). These results confirm that, also, macrophages in the brain are accessible to liposomes if the latter are administered along the right route.

In other recent studies, it was shown that microglia can be depleted from cultured slices of brain tissue using clodronate liposomes. This approach has been used to demonstrate that, in addition to their phagocytic activity, microglia in the CNS promote the death of developing neurons engaged in synaptogenesis (21).

1.3.7. Intra-articular Injection in the Synovial Cavity of Joints

Phagocytic synovial lining cells play a crucial role in the onset of experimental arthritis induced with immune complexes or collagen type II. A single intra-articular injection with clodronate liposomes caused the selective depletion of phagocytic synovial lining cells in mice and rats, demonstrating that this administration route allows easy access of liposomes to the macrophages lining the synovial cavity (22). Recent experiments have confirmed that liposomes are also able to reach synovium lining macrophages in men (6).

1.3.8. Local Injection in the Testes

Local injection of a suspension of liposomes can be performed in most organs. However, whether or not the liposomes will be able to diffuse from the injection site over the rest of the tissue will largely depend on the tissue structure. In the testis of rats, a loosely woven tissue structure allows the liposomes to reach most of the testicular macrophages, as demonstrated by the finding that at least 90% of the testicular macrophages can be depleted by clodronate liposomes (23).

1.4. Specificity with Respect to Macrophages

1.4.1. Selective Depletion of Phagocytic Cells

Liposomes of more than a few hundred nanometer will not be internalized by non-phagocytic cells. This explains why other cells such as lymphocytes and granulocytes are not depleted by multilamellar clodronate liposomes (24). According to a recent publication, blood monocytes (the precursors of mature resident macrophages) can be depleted by intravenous injection of clodronate liposomes (25). This may explain why in quite a number of studies, clodronate liposomes appeared to affect macrophages in tissues, in spite of the presence of a vascular barrier between liposomes and macrophages (see references in: http://www.ClodronateLiposomes.org). In such cases, mature macrophages in these tissues might be prevented from substitution by new ones, since their precursors are killed in the circulation. In this way, the normal turn-over of resident macrophages could be blocked.

Normal dendritic cells (DC), localized in the T-cell areas in the spleen, will not be depleted by application of clodronate liposomes. However, a particular group of so called myeloid dendritic cells, localized at the border between marginal zone and red pulp, will be depleted as efficacious as macrophages (26). This is not surprising, since these cells are able to internalize particles of more than one micron. Since macrophages and dendritic cells show a considerable overlap in their activities, it remains an open question whether these cells should be considered as macrophages or dendritic cells.

1.4.2. Uptake of Liposomes by Macrophage Subsets

Although macrophages in general seem to prefer liposomes with an overall negative charge, e.g. achieved by incorporation of the anionic phospholipid phosphatidylserine in their bilayers, also neutral and cationic liposomes are rapidly taken up by macrophages. Several modifications of the original liposome formulations, such as the incorporation of amphipathic polyethylene glycol (PEG) conjugates in the liposomal bilayers have been proposed in order to reduce the recognition and uptake of liposomes by macrophages. Nevertheless, a large percentage of these so-called long-circulating liposomes will still be ingested by macrophages, emphasizing that macrophages form the logical target for all liposomes, irrespective of their surface molecules (27).

Given the fact that macrophages will ingest all types of non-self macromolecules and particulate materials, it is difficult to achieve specific targeting to only one macrophage subset, e.g. in the spleen. In studies, intended to reveal the conditions for monoclonal antibody-mediated specific targeting of enzyme molecules to marginal metallophilic macrophages in the spleen, we found that highly specific targeting of the enzyme molecules could be achieved only by using monomeric conjugates of the antibody and the enzyme. Larger conjugates lead to their uptake by all macrophage subsets in the spleen (28). As yet, the choice of an administration route for liposomes remains the main approach to achieve some degree of selectivity with respect to macrophage subsets.

2. Materials

2.1. Preparation of Clodronate Liposomes and Control Liposomes

1. 100 mg/mL Phosphatidylcholine (PC, Lipoid) solution in ethanol 100%, filtered through 0.2-μm pore nylon filter (Millipore) (see Note 1).
2. 10 mg/mL Cholesterol (Sigma) solution in ethanol 100%, filtered through 0.2-μm pore nylon filter (Millipore).
3. Clodronate solution: dissolve 187.5 g clodronate (BioIndustria) (see Note 2) in 750 mL purified water (see Note 3). Dissolve clodronate on magnetic stirrer. Centrifuge the solution at $10,500 \times g$ for 4 min. Filter the upper solution through 0.2-μm pore filter and store at 4°C.
4. Chloroform, analytical grade.
5. Argon gas (or other inert gas, e.g. nitrogen gas).
6. Sterile phosphate-buffered saline (PBS) for injection, containing 8.2 g NaCl, 1.9 g $Na_2HPO_4 \cdot 2H_2O$, 0.3 g $NaH_2PO_4 \cdot 2H_2O$ at pH 7.4 per liter (Braun).
7. Rotary evaporator.

2.2. Preparation of Mannosylated Clodronate Liposomes for CNS Research

For some studies in the CNS, e.g. for research on the role of macrophages in experimental allergic encephalomyelitis (EAE), a rodent model for multiple sclerosis (MS), clodronate liposomes should be mannosylated (29, 30).

1. 1.85 mg/mL p-Aminophenyl α-D-mannopyranoside (Sigma) solution in methanol (p.a., Riedel-de Haen).
2. 100 mg/mL Phosphatidylcholine (PC, Lipoid) solution in ethanol 100%, filtered through 0.2-μm pore nylon filter (Millipore) (see Note 1).
3. 10 mg/mL Cholesterol solution in ethanol 100%, filtered through 0.2-μm pore nylon filter (Millipore).
4. Clodronate solution: dissolve 187.5 g clodronate (BioIndustria) (see Note 2) in 750 mL purified water (see Note 3). Dissolve clodronate on magnetic stirrer. Centrifuge the solution at $10,500 \times g$ for 4 min. Filter the upper solution through 0.2-μm pore filter and store at 4°C.
5. Chloroform, analytical grade.
6. Argon gas (or other inert gas, e.g. nitrogen gas).
7. Sterile phosphate-buffered saline (PBS) for injection, containing 8.2 g NaCl, 1.9 g $Na_2HPO_4 \cdot 2H_2O$, 0.3 g $NaH_2PO_4 \cdot 2H_2O$ at pH 7.4 per liter (Braun).
8. Rotary evaporator.

2.3. Preparation of Control Liposomes Labeled with a Fluorochrome Marker

In order to study whether or not liposomes are taken up by particular macrophage subsets in tissues, it may be helpful to study the distribution of control liposomes (see Note 4).

Materials:

1. 2.5 mg/mL DiI solution in 100% ethanol (see Note 5).
2. Sterile phosphate-buffered saline (PBS) for injection, containing 8.2 g NaCl, 1.9 g $Na_2HPO_4 \cdot 2H_2O$, 0.3 g $NaH_2PO_4 \cdot 2H_2O$ at pH 7.4 per liter (Braun).

2.4. Determination of Clodronate Content

1. Standard clodronate solution: dissolve 10.0 mg/mL clodronate (BioIndustria) (see Note 2) in purified water (see Notes 3 and 6).
2. 4 mM $CuSO_4$ solution in purified water (see Note 3).
3. Sterile phosphate-buffered saline (PBS) for injection, containing 8.2 g NaCl, 1.9 g $Na_2HPO_4 \cdot 2H_2O$, 0.3 g $NaH_2PO_4 \cdot 2H_2O$ at pH 7.4 per liter (Braun).
4. 0.65% HNO_3 solution in purified water.
5. Purified water (see Note 3).
6. Saline.
7. Phenol 90%.
8. Chloroform, analytical grade.

9. 16-mL Glass tubes, caps with Teflon inlay (Kimble).
10. 10-mL Polystyrene tubes (see Note 8).
11. Spectrophotometer.

3. Methods

3.1. Preparation of Clodronate Liposomes and Control Liposomes

1. Add 4.30 mL phosphatidylcholine solution to 4.00 mL cholesterol solution in a 0.5-L round bottom flask (see Notes 9 and 10).

2. Remove the ethanol by low vacuum (58 mbar) rotary (150 rpm) evaporation at 40°C. At the end, a thin phospholipid film will form against the inside of the flask. Remove the condensed ethanol by aerating the flask three times.

3. Disperse the phospholipid film in 20 mL clodronate solution (for clodronate liposomes) or 20 mL PBS (for empty liposomes) by gentle rotation (max. 100 rpm) at room temperature. Development of foam should be avoided by reducing the speed of rotation.

4. Keep the milky white suspension at room temperature for about 2 h (see Note 12).

5. Shake the solution gently (development of foam should be avoided). Put the suspension in a 50-mL plastic tube and sonicate in a water bath (55 kHz) for 3 min.

6. Keep the suspension at room temperature for 2 h (or overnight at 4°C). In order to limit the maximum diameter of the liposomes *for intravenous injection*, the suspension can be filtered using sterile membrane filters with 3.0-μm pores (Millipore).

7. Before using the clodronate liposomes:
   (a) Remove the non-encapsulated clodronate by centrifuging the liposomes at $24,000 \times g$ and 10°C for 60 min. The clodronate liposomes will form a white band at the top of the suspension, whereas the suspension itself will be nearly clear.
   (b) Carefully remove the clodronate solution under the white band of liposomes with a 10-mL pipet (about 1% will be encapsulated). Resuspend the liposomes in approximately 45 mL PBS (see Notes 7 and 13).

8. Wash the liposomes 4–5 times using centrifugation at $24,000 \times g$ and 10°C for 25 min. Remove each time the upper solution and resuspend (see Note 7) the pellet in approximately 45 mL PBS.

9. Resuspend (see Note 7) the final liposome pellet in PBS and adjust to a final volume of 20.0 mL. The suspension should be shaken (gently) before administration to animals or before dispensing, in order to achieve a homogeneous distribution of the liposomes in suspension.

3.2. Preparation of Mannosylated Clodronate Liposomes for CNS Research

1. Add 0.710 mL phosphatidylcholine solution, 2.00 mL mannopyranoside solution and 1.08 mL cholesterol solution in a 0.5-L round bottom flask (see Note 9).

2. Remove the ethanol, and methanol by low vacuum (58 mbar) rotary (150 rpm) evaporation at 40°C. At the end, a thin phospholipid film will form against the inside of the flask. Remove the condensed ethanol by aerating the flask three times.

3. Add 5 mL chloroform and dissolve the lipid film by gentle rotation.

4. Remove the chloroform by low vacuum (58 mbar) rotary (150 rpm) evaporation at 40°C. At the end, a thin phospholipid film will form against the inside of the flask. Remove the condensed chloroform by aerating the flask three times.

5. Disperse the phospholipid film in 4 mL clodronate solution (for clodronate liposomes) or 4 mL PBS (for empty liposomes) by gentle rotation (max.100 rpm) at room temperature. Development of foam should be avoided by reducing the speed of rotation.

6. Keep the milky white suspension at room temperature for about 2 h (or overnight at 4°C).

7. Shake the solution gently and sonicate it in a waterbath (55 kHz) for 3 min.

8. Keep the suspension at room temperature for 2 h (or overnight at 4°C) to allow swelling of the liposomes.

9. Before using the clodronate liposomes:

    (a) Remove the non-encapsulated clodronate by centrifuging the liposomes at $24,000 \times g$ and 10°C for 25 min. The clodronate liposomes will form a white band at the top of the suspension, whereas the suspension itself will be nearly clear.

    (b) Carefully remove the clodronate solution under the white band of liposomes with a 10-mL pipet (about 1% will be encapsulated). Resuspend (see Note 7) the liposomes in approximately 8 mL PBS.

10. Wash the liposomes 4–5 times using centrifugation at $24,000 \times g$ and 10°C for 15 min. Remove, each time, the upper solution and resuspend the pellet in approximately 8 mL PBS.

11. Resuspend (see Note 7) the final liposome pellet in PBS and adjust to a final volume of 4.00 mL. The suspension should

be shaken (gently) before administration to animals or before dispensing, in order to achieve a homogeneous distribution of the liposomes in suspension.

3.3. Preparation of Control Liposomes Labeled with a Fluorochrome Marker

1. Add 10 μL DiI solution per milliliter liposome suspension.
2. Shake liposome suspension thoroughly.
3. Incubate 10 min at room temperature (dark).
4. Centrifugate liposomes at $24,000 \times g$ for 10 min.
5. Remove supernatant.
6. Add sterile PBS and resuspend.
7. Centrifugate liposomes at $24,000 \times g$ for 10 min.
8. Add sterile PBS to original volume.
9. Store labeled liposomes dark at 4°C.

3.4. Determination of Clodronate

3.4.1. Extraction of Clodronate from Liposomes

1. Dispense in separate glass tubes: 1 mL of the clodronate-liposome suspension, 1 mL standard clodronate solution, and 1 mL PBS.
2. Add 8 mL of phenol/chloroform (1:2) to each tube (see Notes 14 and 15).
3. Vortex and shake the tubes extensively for about 15 s.
4. Hold the tubes at room temperature for at least 15 min.
5. Centrifuge ($1,100 \times g$) the tubes at 10°C for 10 min.
6. Hold the tubes at room temperature until clear separation of both phases (at least 10 min).
7. Transfer the aqueous (upper) phase to clean glass tubes using a Pasteur pipette. This phase contains the clodronate. Do not take the interphase.
8. Add 6 mL chloroform per tube: re-extract the solution by extensive vortexing (see Note 15).
9. Hold the tubes for at least 5 min at room temperature.
10. Centrifugate ($1,100 \times g$) the tubes at 10°C for 10 min.
11. Transfer the aqueous phase (without any chloroform) to 10-mL plastic tubes using a Pasteur pipette. These are the samples for determination of clodronate concentration.

3.4.2. Determination of Clodronate Concentration

1. Prepare a standard curve using 0, 10, 20, 40, 50, 70, and 80 μL of the extracted standard clodronate solution added with saline to a total volume of 1 mL per tube.
2. Dilute the samples with saline to a total volume of 1 mL per tube until they are within range of the standard curve. (A suspension of clodronate liposomes prepared according to protocol above contains about 4 or 5 mg clodronate per 1 mL suspension).

3. Add 2.25 mL 4 mM $CuSO_4$ solution, 2.20 mL purified water (see Note 3) and 0.05 mL HNO_3 solution to each tube, containing 1 mL sample or standard.
4. Vortex all tubes vigorously.
5. Read the standard curve and samples at 240 nm using a spectrophotometer. Determine the clodronate concentration.

4. Notes

1. Phosphatidylcholine can be stored at −20°C, dry and in aliquots under argon or nitrogen.
2. In the past 20 years we did use clodronate from several sources. Most of that period our clodronate was obtained from Roche Diagnostics, Mannheim, Germany. However quite recently we had to change the source of our clodronate from Roche to BioIndustria in Italy, due to the fact that the stock kept by Roche became gradually exhausted whereas their new production had been finished. In spite of the completely identical data sheets, clodronate from both sources did show slight differences. In spite of slight differences in the concentrations of clodronate, finally encapsulated in liposomes, the efficacy of the newly prepared clodronate liposomes appeared not to be markedly reduced. At present, clodronate can be purchased from Sigma, St. Louis, USA.
3. Purified water is used (Millipore), type II. Properties are: >18 MOhm, conductivity max 0.05 µS/0.05 µMho.
4. Control liposomes may be labeled, e.g. with the fluorochrome DiI, since they do not affect macrophages. As a result the label will show the distribution pattern of the liposomes within tissues and their uptake by macrophages. We recommend not to use DiI labeled clodronate liposomes for the following reasons: Clodronate liposomes will kill the macrophages. As a consequence, the DiI label will be redistributed as soon as the macrophages are dying and from that time on, the label does no longer represent the actual distribution of the liposomes. So, liposomes should either contain clodronate to eliminate macrophages or DiI to demonstrate the uptake of liposomes by macrophages. Combination may lead to misinterpretation.
5. It may take a while for DiI to dissolve. Vortexing may enhance the process. To prevent using crystals: centrifugate a moment at $10,000 \times g$ and use only the supernatant.

6. This can be divided into aliquots and stored at −20°C for longer time.

7. Resuspending the liposomes can be done by taking and removing the non-homogeneous suspension from 10-mL pipet. Repeat this many times, till the liposomes are distributed evenly through the whole suspension.

8. These tubes can be replaced by other tubes with a min. volume of 10 mL.

9. Before using glassware, it should be cleaned and rinsed. Rinsing must be done in following order: ethanol 70% (three times), ethanol 96% (three times) and at last ethanol 100% (three times).

10. Advised is here to use a round bottom flask of 500 mL. A smaller volume, for instance 100 mL, is also possible. The liquids, however, may boil and come into the bottle neck more easily. The vacuum must be reduced in time to stop the boiling and preventing a phospholipid film in the apparatus.

11. Try to work as clean as possible. Use the flow cabinet and use autoclaved or sterile materials if possible.

12. It is also possible to keep the suspension over night at 7°C. Clodronate liposomes can be kept up to a few days at this step. PBS liposomes can be stored for 1 day.

13. The clodronate solution can be used up to four times. It has to be centrifuged ($20,000 \times g$ for 40 min), and the upper white band (liposomes) can than be removed with a pipet. The clodronate solution should be filtered (0.45-μm filter) before use.

14. Chloroform and phenol are toxic and should be handled very carefully.

15. When PBS, water or ethanol and chloroform are mixed, pressure will build up. Be sure to open the tubes once in a while, especially when liquids come together.

References

1. Van Rooijen N, Sanders A (1997) Elimination, blocking and activation of macrophages: three of a kind? J Leukoc Biol 62:702
2. Van Rooijen N (1993) Extracellular and intracellular action of clodronate in osteolytic bone diseases: a hypothesis. Calcif Tissue Int 52:407
3. Van Rooijen N, Sanders A (1996) Kupffer cell depletion by liposome-delivered drugs: comparative activity of intracellular clodronate, propamidine and ethylenediaminetetraacetic acid (EDTA). Hepatology 23:1239
4. Van Rooijen N, Sanders A (1994) Liposome mediated depletion of macrophages: mechanism of action, preparation of liposomes and applications. J Immunol Methods 174:83
5. Van Rooijen N, Van Kesteren-Hendrikx E (2002) Clodronate liposomes: perspectives in research and therapeutics. J Liposome Res 12:81
6. Barrera P, Blom A, Van Lent PLEM, Van Bloois L, Storm G, Beijnen J, Van Rooijen N, Van De Putte LBA, Van Den Berg WB (2000) Synovial macrophage depletion with clodronate

containing liposomes in rheumatoid arthritis. Arthritis Rheum 43:1951

7. Van Rooijen N, Van Nieuwmegen R (1984) Elimination of phagocytic cells in the spleen after intravenous injection of liposome encapsulated dichloromethylene-diphosphonate. An enzyme-histochemical study. Cell Tissue Res 238:355

8. Van Rooijen N (1989) The liposome mediated macrophage 'suicide' technique. J Immunol Methods 124:1

9. Van Rooijen N, Sanders A, Van den Berg T (1996) Apoptosis of macrophages induced by liposome-mediated intracellular delivery of drugs. J Immunol Methods 193:93

10. Fleisch H (1993) Bisphosphonates in bone disease: from the laboratory to the patient

11. Van Rooijen N, Kors N, Van den Ende M, Dijkstra CD (1990) Depletion and repopulation of macrophages in spleen and liver of rat after intravenous treatment with liposome encapsulated dichloromethylene diphosphonate. Cell Tissue Res 260:215

12. Van Rooijen N, Kors N, Kraal G (1989) Macrophage subset repopulation in the spleen: differential kinetics after liposome-mediated elimination. J Leukoc Biol 45:97

13. Barbé E, Huitinga I, Dopp EA, Bauer J, Dijkstra CD (1996) A novel bone marrow frozen section assay for studying hematopoietic interactions in situ: the role of stromal bone marrow macrophages in erythroblast binding. J Cell Sci 109:2937

14. Alves-Rosa F, Stanganelli C, Cabrera J, Van Rooijen N, Palermo MS, Isturiz MA (2000) Treatment with liposome-encapsulated clodronate as a new strategic approach in the management of immune thrombocytopenic purpura (ITP) in a mouse model. Blood 96:2834

15. Jordan MB, Van Rooijen N, Izui S, Kappler J, Marrack P (2003) Liposomal clodronate as a novel agent for treating autoimmune hemolytic anemia in a mouse model. Blood 101:594

16. Schumann J, Wolf D, Pahl A, Brune K, Papadopoulos T, Van Rooijen N, Tiegs G (2000) Importance of Kupffer cells for T cell-dependent liver injury in mice. Am J Pathol 157:1671

17. Delemarre FGA, Kors N, Kraal G, Van Rooijen N (1990) Repopulation of macrophages in popliteal lymph nodes of mice after liposome mediated depletion. J Leukoc Biol 47:251

18. Biewenga J, Van der Ende B, Krist LFG, Borst A, Ghufron M, Van Rooijen N (1995) Macrophage depletion and repopulation in the rat after intraperitoneal administration of Cl_2MBP-liposomes: depletion kinetics and accelerated repopulation of peritoneal and omental macrophages by administration of Freund's adjuvant. Cell Tissue Res 280:189

19. Thepen T, Van Rooijen N, Kraal G (1989) Alveolar macrophage elimination in vivo is associated with an increase in pulmonary immune responses in mice. J Exp Med 170:499

20. Polfliet MMJ, Goede PH, Van Kesteren-Hendrikx EML, Van Rooijen N, Dijkstra CD, Van den Berg TK (2001) A method for the selective depletion of perivascular and meningeal macrophages in the central nervous system. J Neuroimmunol 116:188

21. Marin-Teva JL, Dusart I, Colin C, Gervais A, Van Rooijen N, Mallat M (2004) Microglia promote the death of developing Purkinje cells. Neuron 41:535

22. Van Lent PLEM, Van De Hoek A, Van Den Bersselaar L, Spanjaards MFR, Van Rooijen N, Dijkstra CD, Van De Putte LBA, Van Den Berg WB (1993) In vivo role of phagocytic synovial lining cells in onset of experimental arthritis. Am J Pathol 143:1226

23. Bergh A, Damber J-E, Van Rooijen N (1993) Liposome-mediated macrophage depletion: an experimental approach to study the role of testicular macrophages. J Endocrinol 136:407

24. Claassen I, Van Rooijen N, Claassen E (1990) A new method for removal of mononuclear phagocytes from heterogenous cell populations 'in vitro', using the liposome-mediated macrophage 'suicide' technique. J Immunol Methods 134:153

25. Sunderkotter C, Nikolic T, Dillon MJ, Van Rooijen N, Stehling M, Drevets DA, Leenen PJM (2004) Subpopulations of mouse blood monocytes differ in maturation stage and inflammatory response. J Immunol 172:4410

26. Leenen PJM, Radosevic K, Voerman JSA, Salomon B, Van Rooijen N, Klatzmann D, Van Ewijk W (1998) Heterogeneity of mouse spleen dendritic cells: *in vivo* phagocytic activity, expression of macrophage markers and subpopulation turnover. J Immunol 160:2166

27. Litzinger DC, Buiting AMJ, Van Rooijen N, Huang L (1994) Effect of liposome size on the circulation time and intraorgan distribution of amphipathic polyethylene-glycol containing liposomes. Biochim Biophys Acta 1190:99

28. Van Rooijen N, Ter Hart H, Kraal G, Kors N, Claassen E (1992) Monoclonal antibody mediated targeting of enzymes. A comparative

study using the mouse spleen as a model system. J Immunol Methods 151:149

29. Huitinga I, Van Rooijen N, De Groot CJA, Uitdehaag BMJ, Dijkstra CD (1990) Suppression of experimental allergic encephalomyelitis in Lewis rats after elimination of macrophages. J Exp Med 172:1025

30. Tran EH, Hoekstra K, Van Rooijen N, Dijkstra CD, Owens T (1998) Immune invasion of the central nervous system parenchyma and experimental allergic encephalomyelitis, but not leukocyte extravasation from blood, are prevented in macrophage-depleted mice. J Immunol 161:3767

Chapter 14

Vesicular Phospholipid Gels

Martin Brandl

Abstract

Highly concentrated phospholipid dispersions of vesicular morphology, Vesicular phospholipid Gels (VPGs) are of importance for sustained release of drugs upon implantation, or, upon transfer into SUV-dispersions, for drug delivery upon i.v. injection. Here the formation of homogeneous lipid and lipid/drug-blends is described as well as the preparation of VPGs by high-pressure homogenization, along with remote loading via "passive loading".

Key words: Phos pholipid gel, Sustained drug release, High-pressure homogenization, High-pressure filter extrusion, Passive loading

1. Introduction

Phosphatidylcholine hydrates and swells instantaneously upon contact with water, forming lamellar structures. For complete hydration of the lipid crystals, a minimum ratio of water : lipid of 45:55 is needed (1). In contrast, classical liposome dispersions contain excess water, i.e. lipid concentrations would not exceed 250 mg/g (325 mM). We could demonstrate that at high lipid concentrations as well, vesicles rather than multi-lamellar-type structures form if swelling is done under mechanical stress. Furthermore, intense mechanical stress, as for example experienced during high-pressure homogenisation, leads to small and unilamellar vesicles. This is not contradicted by the fact that at high phospholipid concentrations semisolid, gel-like masses are obtained. Since their morphology is vesicular, we talk about Vesicular Phospholipid Gels (VPGs) (2). VPGs entrap aqueous compartments not only within but also inbetween the vesicular structures. Typically, such Vesicular Phospholipid Gels contain 250–600 mg/g or 325–780 mM of lipid. VPGs can be prepared by high-pressure homogenization (2), high-pressure filter extrusion (3) or ball-milling (4). They may be loaded with water-soluble or lipid-soluble drugs. They retain their drug load within the core of the vesicles during autoclaving (5) and long-term storage, due to the lack of concentration gradient (6). VPGs may serve as depots to release drugs in a controlled manner, e.g. upon implantation or injection (7–9). VPGs may easily be transferred into "conventional" small-sized liposome (SUV) dispersions by mixing with excess aqueous medium and gentle mechanical agitation (10, 11). Since they retain entrapped drugs during dilution, an unusually high encapsulation efficiency within SUVs may be achieved by preparation of VPGs and subsequent dilution. Such high ratio of entrapped to unentrapped drug may render removal of free drug unnecessary.

2. Materials

2.1. Lipid blends made by freeze-drying from organic solutions

1. Phosphatidylcholine, natural or hydrogenated, soy or egg (e.g. E80, Lipoid GmbH).
2. Other lipids, e.g. cholesterol.
3. If applicable: lipophilic or amphiphilic drug.
4. Tert-Butanol, p.a. grade.
5. Chloroform Lichrosolv-grade.
6. Ethanol-bath cooled by dry-ice or liquid nitrogen.
7. Freeze dryer with cooling system that allows plate temperatures of −55°C, e.g. Beta 2–16 (Martin Christ GmbH), equipped with cooling-trap and hybrid-pump (see **Note 1**).
8. 40 mM phosphate buffer, pH 7.4 (or aqueous drug solution).

2.2. VPGs made by high-pressure homogenization using an APV MicronLab 40

1. Phosphatidylcholine (e.g. E80, Lipoid GmbH), or freeze-dried lipid blend, prepared as described in Subheading 2.1.
2. Glass bottle with wide neck and screw-cap, 100 mL.
3. Dough scraper.z
4. Phosphate buffer 40 mM, pH 7.4 (**see Note 1**); containing hydrophilic drug or marker (if applicable).
5. High-pressure homogenizer Micron Lab 40 (APV Homogenizer) (*see* **Note 2**).
6. Thermostated water bath.

2.3. Passive Loading with Cytostatic Drug

1. Drug solution, e.g. Gemcitabine (2′,2′-difluoro-2′-deoxycytidine; dFdC) – solution, 15 mg dFdC–HCl in 1 mL phosphate buffer, pH 7.3 (**see Note 3**).
2. 1.5–2mL vials with screw cap (e.g. Eppendorf tubes).
3. Thermostated shaking water bath.

2.4. Transfer of VPGs to SUV-dispersions

1. 1.5–2mL vials with screw cap.
2. Phosphate buffer 40 mM (pH 7.4).
3. Glass beads (Ø 1–2 mm).
4. Oscillating ball mill (MM200 Retsch) with adapter for 2-mL vials.
5. Bench top centrifuge (Biofuge Pico, Heraeus Instruments).

2.5. Release testing of VPGs

1. Custom made release cell with a rectangular donor compartment (5×5×50 mm), to take up about 1 g of the VPG and an acceptor compartment of semicircular cross section (radius 2.5 mm) and 50mm length (For a layout of cell see ref. 7).
2. Pulse-free pump, providing a constant flow rate of 10 mL/h, e.g. piston pump P-500 (Pharmacia Biotech).
3. Water bath set to 37°C.
4. Fraction collector and tubes, suitable to collect 10 ml fractions per hour (or multiples of that), e.g. RediFrac (Pharmacia Biotech).
5. Buffer reservoir, e.g. 2 L flask with acceptor medium, e.g. the buffer which has been used for VPG preparation.

3. Methods

3.1. Lipid blends made by freeze drying from organic solutions

For preparing VPGs consisting of two or more lipids, it is essential to prepare a homogeneous blend of the lipids first by transferring the lipids into a solid solution in order to avoid inhomogeneous distribution of the lipids over the bilayer upon swelling (12). The commonly used thin film hydration technique is not suited because of the big amounts of lipids needed for VPGs. A freeze-drying approach is described instead:

1. Dissolve the lipids (and, if applicable drug) in a blend of (organic) solvents (see **Note 4**) under gentle warming (max temperature well below the boiling point of the solvent) using injection vials. For an equimolar blend of phosphatidylcholine and cholesterol use 2.1 g of egg PC (Lipoid E80) and 1.1 g of cholesterol. To dissolve this mixture, approximately 10 mL of a solvent blend of t-butanol/chloroform (1:1) and gentle warming is suited.
2. Shock-freeze the solution by carefully dipping the vial into an ethanol/dry ice-bath, and store in a −80°C freezer until the freeze drying process is to commence.
3. Prepare the freeze-dryer with a cooling trap with liquid nitrogen installed between the chamber and the vacuum pump (see **Note 3**). The following initial settings are recommended:

shelf temperature −55°C and ice condenser temperature below −70°C.

4. Transfer the container(s) with the frozen mass to the shelf of the freeze dryer.

5. Start freeze drying with the above settings and a vacuum setting of 20 hPa. In order to avoid thawing and cooking of the sample, the shelf temperature is kept constant at −55°C until the evaporation of the solvent is slowing down, i.e. until the product temperature approximates the shelf temperature. Then reduce the pressure to its technical lower limit (approx. 2–3 Pa), and increase the shelf temperature gradually (over 8 h) to +40°C.

6. A spongy dry cake should be obtained. Flood the chamber with nitrogen gas, and close the container(s) with rubber stoppers via the remote closure device (if available). Crimp the vials with aluminium caps. Store the vial(s) in the fridge at 2–8°C until used.

7. Disperse the cake in 4.8 mL of aqueous medium, and continue as described in Subheading 3.2.

3.2. VPGs made by high-pressure homogenization

The following protocol, which is derived from the one-step liposome preparation technique, originally described in (12) is suited. A slurry is made from a single phospholipid or a freeze-dried cake of lipid-blend and buffer (drug/marker solution). The slurry is processed using an APV MicronLab 40 lab-scale homogenizers. A microfluidizer M110 may be used as well.

1. Prepare a slurry of the phospholipid in aqueous medium (e.g. 16 g of egg phosphatidylcholine, Lipoid E80 in 24 g of phosphate buffer or buffered drug solution, respectively) by manual agitation in a stoppered bottle, and let the lipid soak for 15 min under occasional shaking. For phospholipids (lipid blends) with a phase transition temperature above room temperature, incubate in a pre-heated water bath at temperatures well (≥5°C) above the phase transition temperature.

2. Pre-heat the homogenizer if desired. Process temperatures well (≥5°C) above the phase transition temperature of the phospholipid should be maintained. Transfer the highly viscous or cream-like slurry to the high-pressure homogenizer. Use a dough scraper for complete transfer. Process the slurry using the pre-selected pressure for the desired number of cycles. An intermediate pressure of, e.g. 70 MPa (700 bar) and five repetitive homogenization cycles are recommended.

3. Transfer the VPG to a tight container, such that the container is well–filled, and drying of the surface due to evaporation of water is minimized. Store the VPG at room temperature until used.

4. Loading of VPGs with drugs

Different approaches have been described to load VPGs with drugs. The choice of the proper technology depends on the physicochemical characteristics of the drug: Water soluble substances which, upon loading into VPGs are truly entrapped or encapsulated within the aqueous compartments of the VPG, may be directly loaded during VPG preparation (according to Subheading 3.2) or passively loaded into pre-formed VPGs (see protocol below). Amphiphilic or lipophilic substances expected to be incorporated within the bilayer are treated like lipids, i.e. a homogeneous drug/lipid-blend is formed by freeze drying (according to Subheading 3.1).

The process of passive loading has originally been described by Massing et al. (18). In principle, it comprises the incubation of "empty" VPG with drug until the drug is equally distributed throughout the preparation by diffusion. When added to pre-formed "empty" (i.e. drug-free) Vesicular Phospholipid Gels, the drug permeates according to the concentration gradient through the bilayers into the vesicles until equilibrium between the interior of the vesicles and the surrounding medium is achieved. Gentle warming can further facilitate the equilibration process. Passive loading has a number of advantages: No active ingredient is present during the preparation of the VPG. This reduces the extent of safety precautions. The composition of the medium of the VPG on one hand and of the drug solution on the other may be chosen independently. Drug and liposomes do not come in contact with each other until shortly before application, and possible interactions, which induce degradation, can thus be avoided (13). This approach may allow loading of VPGs in a hospital pharmacy setting.

1. Prepare a drug-free VPG (empty VPG) as described in Subheading 3.2.
2. Transfer approx. 350 mg of the VPG into vial. Add the appropriate volume of the drug solution (e.g. 50 µL of the Gemcitabine solution) and mix thoroughly using a sterile spatula. Close the container (*see* **Note 5**).
3. Incubate the mixture in a water bath at an appropriate temperature for an appropriate period of time (e.g. for maximum trapping efficiency, incubate the above mixture of Gemcitabine solution and empty VPG for 4 h at 60°C).
4. Store the drug-loaded VPG in a fridge at 2–8°C until further use.

4.1. Transfer of VPGs to SUV-dispersions

Upon addition of excess aqueous medium, VPGs can be transferred into dispersions of liposomes by gentle mechanical agitation.

Shaking by a ball mill is most appropriate for this. Both intensity and duration of mechanical agitation have an influence on the liposome size and the loss of entrapped drug, i.e. encapsulation efficiency measured for the overall process (VPG formation and dilution, see below) (26). A detailed description of the dilution process is given in the following protocol.

1. Transfer approximately 1 g of a VPG gained by one of the above protocols into a 2mL vial, add 3–5 glass beads (Ø 1–2 mm) and 100 μL of aqueous medium, e.g. 40 mM phosphate buffer (pH 7.4). Close the vial tightly and set it on an oscillating ball mill. Agitate at maximum speed for 2–3 min.

2. Add another 100μL aliquot of aqueous medium, and shake again for 2–3 min.

3. Repeat **step 2** 3–8 times, or, until the desired final dilution is achieved. A slightly turbid opalescent liposome dispersion should be obtained.

4. Spin the resultant SUV dispersion at $2,500 \times g$ in a bench centrifuge for 20 min at room temperature to remove eventual bigger particles or aggregates.

4.2. Release testing of VPGs

For assessment of the in vitro release of drug from VPG formulations a test method based on a custom-made flow-through cell has been established (8). An acceptor medium (buffer) is run through the cell at a rate of 10 mL/h, to mimic the flow of tissue fluid at the site of injection or implantation. As in this model, the donor and acceptor compartments are *not* separated by a (semi-permeable) membrane; not only drug-release via diffusion through the matrix, but also erosion of the lipid matrix can be followed. Fractions collected over distinct time intervals are analysed for both drug and phospholipid. The drug in the eluate is quantified by HPLC or another appropriate method. Differentiation between drug released in free and liposomal form can be done by (sub) fractionation of the eluate using size exclusion chromatography on Sephadex G-50 gel. The amount of lipid released can be quantified gravimetrically upon freeze-drying, as described in (8).

1. Place the buffer reservoir with the acceptor medium in the water bath for equilibration at 37°C.

2. Determine the tara of the empty, dry release cell and fill the donor compartment of the cell with the VPG sample bubble-free, using an ointment spatula. Clamp or screw the two halves of the cell tightly together.

3. Weigh the filled release cell to determine the accurate mass of the VPG loaded into the cell.

4. Mount the tubing and place the cell in the water bath (see Note 2); expel all air from the tubing by pumping acceptor medium through the system.

5. Start the pump and fraction collector; pump the acceptor medium through the system at a flow rate of 10 mL/h and collect fractions (1 fraction/h, or alternative settings).

6. Analyze an aliquot of each fraction for overall drug and for lipid content.

7. Fractionate an aliquot on Sephadex G-50 gel and analyse the sub-fractions for liposomal, and free drug content, respectively.

8. At the end of the experiment, collect the VPG remaining in the cell (if any) by rinsing it off with excess acceptor medium. Analyze it for its drug and lipid content.

5. Notes

1. For calcein (or any drug that is sensitive to traces of heavy metals), 10 mM EDTA (ethylenedinitrilotetraacetic acid disodium salt) should be included within the buffer in order to complexate traces of cobalt ions, which are released from the stainless steel material of the homogenizer during processing. Alternatively, a valve made of ceramic may be used.

2. For phospholipids (lipid blends) that have a phase transition temperature above room temperature, it is recommended to modify the homogenizer such that all parts that are in contact with the VPG (educt-reservoir, valve and product-reservoir) are thermostated at temperatures well above the phase transition temperature. This may be achieved by fitting a heating coil around the tower.

3. For handling of cytostatic drugs, a dedicated work space providing appropriate shielding such as a Class II biological safety bench, or an isolator should be used.

4. All parts of the freeze-dryer that come in contact with solvent vapour must be solvent-resistant. The solvent blend must dissolve all components of the formula and should have its freezing point well above the shelf temperature of the freeze dryer. Blends of tert-butanol and chloroform (from 2:1 to 1:4 mixing ratio) are suited for most lipid blends.

5. Appropriate safety precautions are to be followed when handling cytotoxic compounds.

References

1. Small DM (1967) Phase equilibria and structure of dry and hydrated egg lecithin. J Lipid Res 8(6):551–557
2. Brandl M, Drechsler M, Bachmann D, Bauer K-H (1997) Morphology of semisolid aqueous phosphatidylcholine dispersions, a freeze fracture electron microscopy study. Chem Phys Lipids 87(1):65–72
3. Schneider T, Sachse A, Roessling G, Brandl M (1994) Large-scale production of liposomes of defined size by a new continuous high pressure extrusion device. Drug Dev Ind Pharm 20(18):2787–2807
4. Massing U, Cicko S, Ziroli V (2008) Dual asymmetric centrifugation (DAC)—A new technique for liposome preparation. J Contr Rel 125:16–24
5. Tardi C, Drechsler M, Schubert R, Brandl M (1996) Three-dimensional networks of liposomes: change of properties upon autoclaving. Proc Int Symp Control Release Bioactive Mater 23:27–28
6. Brandl M, Massing U (2003) Vesicular phospholipid gels. In: Torchilin VP, Weisssig V (eds) Liposomes, 2nd edn. Oxford Press, Oxford, pp 353–372
7. Brandl M, Reszka R (1995) Preparation and characterization of phospholipid membrane gels as depot formulations for potential use as implants. Proc Int Symp Control Release Bioactive Mater 22:472–473
8. Tardi C, Brandl M, Schubert R (1998) Erosion and controlled release properties of semisolid vesicular phospholipid dispersions. J Control Release (2–3):261–270
9. Grohganz H (2004) Vesicular phospholipid gels as a potential implant system for cetrorelix. Dr. Scient.-thesis, University of Tromsoe, Tromsoe
10. Brandl M, Drechsler M, Bachmann D, Tardi C, Schmidtgen M, Bauer K-H (1998) Preparation and characterization of semi-solid phospholipid dispersions and dilutions thereof. Int J Pharm 170(2):187–199
11. Brandl M, Bachmann D, Reszka R, Drechsler M, inventors; (Max-Delbrueck-Centrum Fuer Molekulare Medizin, Germany). assignee. Unilamellar vesicular lipids for liposomal preparations with high active substance content. Application: US. US patent 97-803435 6399094. 2002 19970220
12. Brandl M, Bachmann D, Drechsler M, Bauer KH (1990) Liposome preparation by a new high pressure homogenizer Gaulin Micron LAB 40. Drug Dev Ind Pharm 16(14):2167–2191
13. Moog R, Brandl M, Schubert R, Unger C, Massing U (2000) Effect of nucleoside analogs and oligonucleotides on hydrolysis of liposomal phospholipids. Int J Pharm 206(1–2):43–53

Chapter 15

Environment-Responsive Multifunctional Liposomes

Amit A. Kale and Vladimir P. Torchilin

Abstract

Liposomal nanocarriers modified with cell-penetrating peptide and a pH-sensitive PEG shield demonstrate simultaneously a better systemic circulation and site-specific exposure of the cell-penetrating peptide. PEG chains were incorporated into the liposome membrane via the PEG-attached phosphatidylethanolamine (PE) residue with PEG and PE being conjugated with the lowered pH-degradable hydrazone bond (PEG-HZ-PE), while cell-penetrating peptide (TATp) was added as TATp-PEG-PE conjugate. Under normal conditions, liposome-grafted PEG "shielded" liposome-attached TATp moieties, since the PEG spacer for TATp attachment (PEG(1000)) was shorter than protective PEG(2000). PEGylated liposomes accumulate in targets via the EPR effect, but inside the "acidified" tumor or ischemic tissues lose their PEG coating because of the lowered pH-induced hydrolysis of HZ and penetrate inside cells via the now-exposed TATp moieties. pH-responsive behavior of these constructs is successfully tested in cell cultures in vitro as well as in tumors in experimental mice in vivo. These nanocarriers also showed enhanced pGFP transfection efficiency upon intratumoral administration in mice, compared to control pH nonsensitive counterpart. These results can be considered as an important step in the development of tumor-specific stimuli-sensitive drug and gene delivery systems.

Key words: pH-sensitive liposomes, Cell penetrating peptide, TATp, Hydrazone, PEG-PE, Enhanced permeability and retention

1. Introduction

Cancer chemotherapy is often complicated by serious systemic effects of anticancer actives. Therefore, despite new advances in the discovery of new potent anticancer agents, they still suffer the limitations in terms of dose regimen and usage in the patients. Site-specific release of drug from the long circulating carrier at the tumor site while maintaining minimal release during circulation, which leads to higher drug levels at tumor sites and less side effects, is of great interest in tumor chemotherapy.

There are several approaches to this problem, including the use of stimuli-sensitive pharmaceutical nanocarriers, which is based on the fact that many pathological sites, including tumors, demonstrate hyperthermia or acidification (1–3). In general, environmentally sensitive carriers exhibit dramatic changes in their swelling behavior, network structure, permeability, or stability in response to changes in the pH or ionic strength of the surrounding fluid or temperature (4).

Researchers working in the area of the development of environment-responsive drug delivery systems have architectured numerous carriers or conjugate systems to selectively deliver actives to pathological sites. Kataoka's group has prepared doxorubicin (DOX) loaded poly(beta-benzyl-L-aspartate) copolymer micelles and evaluated their pharmaceutical properties and biological significance (5). Accelerated DOX release was observed after lowering the surrounding pH from 7.4 to 5.0, suggesting a pH-sensitive release of DOX from the micelles. DOX loaded in the micelle showed a considerably higher antitumor activity compared to free DOX against mouse C26 tumor by i.v. injection, indicating a promising feature for PEG-PBLA pH-sensitive micelle as a long-circulating carrier system useful in modulated drug delivery.

Hydrophobically modified copolymers of N-isopropylacrylamide bearing a pH-sensitive moiety were investigated for the preparation of pH-responsive liposomes and polymeric micelles (6). The copolymers having the hydrophobic anchor randomly distributed within the polymeric chain were found to more efficiently destabilize egg phosphatidylcholine (EPC)/cholesterol liposomes than the alkyl-terminated polymers. Release of both a highly water-soluble fluorescent contents marker, pyranine, and an amphipathic cytotoxic anticancer drug, DOX, from copolymer-modified liposomes was shown to be dependent on pH. Also, polymeric micelles were studied as a delivery system for the photosensitizer aluminum chloride phthalocyanine, (AlClPc), currently evaluated in photodynamic therapy. pH-Responsive polymeric micelles loaded with AlClPc were found to exhibit increased cytotoxicity against EMT-6 mouse mammary cells in vitro than the control Cremophor EL formulation (7, 8). Drug carriers containing weak acids or bases can promote cytosolic delivery of macromolecules by exploiting the acidic pH of the endosome. Asokan et al. have prepared two pH-sensitive mono-stearoyl derivatives of morpholine, one with a (2-hydroxy)-propylene (ML1) linker and the other, an ethylene (ML2) linker. The pK(a) values of lipids ML1 and ML2, when incorporated into liposomes, are 6.12 and 5.91, respectively. Both lipids disrupt human erythrocytes at a pH equal to or below their pK(a) but show no such activity at pH 7.4. This group has also synthesized two Gemini surfactants or "bis-detergents" by cross-linking the headgroups of single-tailed, tertiary amine detergents through oxyethylene (BD1) or acid-labile acetal (BD2) moieties (9). As evidenced by thin-layer chromatography, BD2 was hydrolyzed under acidic conditions (pH 5.0) with an approximate half-life of 3 h at 37°C, while BD1 remained stable. Low pH-induced collapse of liposomes containing acid-labile BD2 into micelles was more facile than that of BD1. With BD1, the process appeared to be reversible in that aggregation of micelles was observed at basic pH. The irreversible lamellar-to-micellar transition observed with BD2-containing liposomes can possibly be attributed to acid-catalyzed hydrolysis of the acetal cross-linker, which generates two detergent monomers within the bilayer. Liposomes composed of 75 mol% bis-detergent and 25 mol% phosphatidylcholine were readily prepared and could entrap macromolecules such as polyanionic dextran of MW 40 kDa with moderate efficiency. The ability of BD2-containing liposomes to promote efficient cytosolic delivery of antisense oligonucleotides was confirmed by their diffuse intracellular distribution seen in fluorescence micrographs, and the up-regulation of luciferase in an antisense functional assay. Bae et al. formulated pH-sensitive polymeric mixed micelles composed of poly(L-histidine) (polyHis; M(w) 5000)/PEG (M(n) 2000) and poly(L-lactic acid) (PLLA) (M(n) 3000)/PEG (M(n) 2000) block copolymers with or without folate conjugation (10, 11). The polyHis/PEG micelles showed accelerated adriamycin release as the pH decreased from 8.0. In order to tailor the triggering pH of the polymeric micelles to the more acidic extracellular pH of tumors, while improving the micelle stability at pH 7.4, the PLLA/PEG block copolymer was blended with polyHis/PEG to form mixed micelles. Blending shifted the triggering pH to a lower value. Depending on the amount of PLLA/PEG, the mixed micelles were destabilized in the pH range of 7.2–6.6 (triggering pH for adriamycin release). When the mixed micelles were conjugated with folic acid, the in vitro results demonstrated that the micelles were more effective in tumor cell kill due to accelerated drug release and folate receptor-mediated tumor uptake. In addition, after internalization, polyHis was found to be effective for cytosolic ADR delivery by virtue of fusogenic activity. Certain pH-sensitive linkages have been popularly used to allow the drug release, protective "coat" removal, or new function appearance because of their fast degradation in acidified pathological sites (12–14). These include cis-aconityls (15, 16), electron-rich trityls (17), polyketals (18), acetals (19, 20), vinyl ethers (21, 22), hydrazones (23–25), poly(ortho-esters) (26), and thiopropionates (27). Such constructs may turn out to be useful for the site-specific delivery of drugs at the tumor sites(2), infarcts (28), inflammation zones (29) or cell cytoplasm or endosomes (30), since at these "acidic" sites, pH drops from the normal physiologic value of pH 7.4–6.0 and in the following section. A pH-sensitive cis-aconityl linkage has been used to make immunoconjugates of daunorubicin by Shen et al. (31) and Diener et al. (32) while DOX was conjugated to murine monoclonal antibodies (MoAb) raised against human breast tumor cells (33) or murine monoclonal antibody (MAb) developed against human pulmonary adenocarcinoma (34). The trityl group has been used in organic chemistry as an acid-cleavable protecting group for amino and hydroxyl groups. Patel group at Lilly Research Laboratories have established structure-stability relationship of different trityl-nucleoside derivatives by using NMR-spectroscopy (35). In general, the acid-sensitivity of these compounds increases with the electron-donating effects of the substituents (e.g., methoxy groups) that stabilize the intermediatory formed carbocation in the hydrolysis step. In vitro activity in a human colon carcinoma cell line showed that the antibody conjugates with the most pronounced acid lability exhibited the strongest inhibitory effects. However, the most stable conjugates were 20–30 times less active than the free nucleoside antimetabolite (36, 37). These structure-activity relationship also confirmed in animal experiments (35). Also, acetal linkages have the potential to be used as linkages for a range of alcohol functionalities, because their hydrolysis is generally first order relative to the hydronium ion, making the expected rate of hydrolysis 10 times faster with each unit of pH decrease (38) and, by altering their chemical structure, it is possible to tune their hydrolysis rate. In addition, acetals can be formed using a variety of types of hydroxyl groups including primary, secondary, tertiary and syn-1,2- and -1,3-diols, and the rate of hydrolysis can be tuned by varying the structure of the acetal. Gillies et al. synthesized a four different acetal-based conjugates using model drugs and PEO polymer (39). The hydrolysis kinetics of the conjugates had half-lives ranging from less than 1 min to several days at pH 5.0, with slower hydrolysis at pH 7.4 in all cases. Encrypted polymers containing pH-sensitive acetal linkage between either oligonucleotide or macromolecule and PEG showed direct vesicular escape and efficiently deliver oligonucleotides and macromolecules into the cytoplasm of hepatocytes (40). Acetal-based acid-degradable protein-loaded microgels also have showed promising results for the delivery of protein-based vaccines (41). Murthy group has introduced an acid-sensitive hydrophobic nanoparticle based on a new polymer, poly(1,4-phenyleneacetone dimethylene ketal) (PPADK), which complements existing biodegradable nanoparticle technologies (42). This polymer has ketal linkages in its backbone and degrades via acid-catalyzed hydrolysis into low molecular weight compounds that can be easily excreted. PPADK forms micro- and nanoparticles, via an emulsion procedure, and can be used for the delivery of hydrophobic drugs and potentially proteins (43). Acid-labile polyethylene glycol (PEG) conjugated vinyl ether lipids were synthesized and used at low molar ratios to stabilize the nonlamellar, highly fusogenic lipid, dioleoylphosphatidyl ethanolamine, as unilamellar liposomes (22). Acid-catalyzed hydrolysis of the vinyl ether bond destabilized these liposomes by removal of the sterically stabilizing PEG layer, thereby promoting contents release on the hours timescale at pH < 5. pH-Sensitive amphiphilic hydrogels were synthesized by radiation copolymerization of ethylene glycol vinyl ether (EGVE), butyl vinyl ether (BVE) and acrylic acid (AA) in the presence of crosslinking agent, diethylene glycol divinyl ether (DEGDVE) (44, 45). The results of the swelling experiments indicated that the hydrogel which has 60:40:5 comonomer ratio (mol% of EGVE:BVE:AA in monomeric mixture) is pH-sensitive. While the hydrogel is in a fully hydrated form at pH > 6, it extensively dehydrates below pH 6. A two-stage volume phase transition was observed in the range of pH 6.0–7.0 and 7.5–8.0. In 1980, Hurwitz and co-workers reported for the first time that hydrazone-based polymer-daunorubicin conjugates have substantial cytotoxicity than the analogues containing noncleavable linkers between those conjugates which appeared to be completely inactive (46). In 1989, the Lilly labs reported the use of hydrazone linkages to target MoAb to potent cytotoxic DAVLB hydrazide (47). In vivo studies of antitumor activity showed that the efficiency and safety of the conjugate was increased over that of the unconjugated. The Kratz group has prepared trasnferin and albumin as carriers for targeting of chlorambusil, an anticancer active (48, 49). In vitro studies with both conjugates demonstrated them to be as active or more active than the free drug, whereas they had reduced toxicities. Toncheva et al. have prepared amphiphilic AB and ABA block copolymers from poly (ortho esters) and poly (ethylene glycol). The micelles formed by these co-block polymers were stable in PBS at pH 7.4 and 37°C for 3 days and in a citrate buffer at pH 5.5 and 37°C for 2 h (26). The remarkably enhanced gene silencing in hepatoma cells was achieved by assembling lactosylated-PEG-siRNA conjugates bearing acid-labile beta-thiopropionate linkages into polyion complex (PIC) micelles through the mixing with poly(L-lysine) (50). The PIC micelles with clustered lactose moieties on the periphery were successfully transported into hepatoma cells in a receptor-mediated manner, releasing hundreds of active siRNA molecules into the cellular interior responding to the pH decrease in the endosomal compartment. Eventually, almost 100 times enhancement in gene silencing activity compared to that of the free conjugate was achieved for the micelle system, facilitating the practical utility of siRNA therapeutics. Kataoka group (51) also architectured three types of newly engineered block copolymers forming polyplex micelles useful for oligonucleotides and siRNA delivery: (1) PEG-polycation diblock copolymers possessing diamine side-chain with distinctive pKa for siRNA encapsulation into polyplex micelles with high endosomal escaping ability, (2) Lactosylated PEG-(oligonucleotide or siRNA) conjugate through acid-labile beta-thiopropionate linkage to construct pH-sensitive PIC micelles, and (3) PEG-poly(methacrylic acid) block copolymer for the construction of organic/inorganic hybrid nanoparticles encapsulating siRNA. Recently, N-ethoxybenzylimidazoles (NEBI) linkers were introduced as potential pH-sensitive linkages. Kinetic analysis of eight derivatives of NEBIs showed that their rates of hydrolysis are accelerated in mild aqueous acidic solutions compared to in solutions at normal, physiological pH. A derivative of NEBI carrying DOX, a widely used anticancer agent, also showed an increased rate of hydrolysis under mild acid compared to that at normal physiological pH. The DOX analogue resulting from hydrolysis from the NEBI exhibited good cytotoxic activity when exposed to human ovarian cancer cells (52).

We have demonstrated the utility of highly pH-sensitive hydrazone bond-based PEG-PE conjugates in preparing double-targeted stimuli-sensitive pharmaceutical nanocarriers (53, 54). Two important temporal characteristics of such carriers include their sufficiently long life-time under normal physiological conditions and their sufficiently fast destabilization within the acidic target. Since real practical tasks may require different times for such carriers to stay in the blood and to release their contents (or "develop" an additional function) inside the target, we have synthesized a series of PEG-HZ-PE conjugates with different substituents at the hydrazone bond and evaluated their hydrolytic stability at normal and slightly acidic pH values. These conjugates differed from each other with respect to the exact structure of groups forming the hydrazone linkage between phospholipid and PEG. The characterization of the in vitro behavior of these conjugates has provided important information useful for future design and development of pH-sensitive nanocarriers with controlled properties.

2. Materials

2.1. Chemicals

1. 1,2-dioleoyl-*sn*-glycero-3-phosphoethanolamine, DOPE; 1,2-dipalmitoyl-*sn*-glycero-3-phosphothioethanolamine (Sodium Salt), DPPE-SH; 1,2-dimyristoyl-*sn*-glycero-3-phosphoethanolamine-N-(lissamine rhodamine B sulfonyl) (ammonium salt), Rh-PE (all from Avanti Polar Lipids).

2. (N-e-maleimidocaproic acid) hydrazide, EMCH; 4-(4-N-maleimidophenyl) butyric acid hydrazide hydrochloride, MPBH; N-(k-maleimidoundecanoic acid)hydrazide, KMUH; succinimidyl 4-(N-maleimidomethyl) cyclohexane-1-carboxylate, SMCC (all from Pierce Biotechnology Inc., Rockford, IL).

3. 2-acetamido-4-mecrcapto butanoic acid hydrazide, AMBH (Molecular Probes).

4. Methoxy poly(ethylene) glycol butyraldehyde (MW 2,000), mPEG-SH (MW 2,000) (all from Nektar Therapeutics, Huntsville, AL).

5. Triethylamine.

6. 4-succinimidyl formylbenzoate (SFB) (Molbio, Boulder, Colorado).

7. Egg phosphatidylcholine (EPC), cholesterol (Ch), mPEG$_{2000}$-DSPE, 1,2-dioleoyl-3-trimethylammonium-propane (DOTAP), Rhodamine-PE (Rh-PE), and phosphatidylthio-ethanolamine (DPPE-SH) (all from Avanti Polar Lipids).

8. mPEG$_{2000}$-SH (Nektar Therapeutics, Huntsville, AL).

9. Maleimide-PEG$_{1000}$-NHS (Quanta Biodesign, Powell, OH).

10. TATp-cysteine (Research Genetics, Huntsville, AL).

11. Succinimidyl 4-(N-maleimidomethyl)cyclohexane-1-carboxylate hydrazide (SMCCHz) (Molecular Biosciences Boulder, CO).

12. 4-acetyl phenyl maleimide.

13. Sephadex G25m.

14. Sepharose CL4B.

15. Lewis Lung Carcinoma (LLC) cell line (ATCC, Rockville, MD).

16. Delbecco's minimal essential medium, complete serum free medium and fetal bovine serum (Cellgro, Kansas City, MO).

2.2. Syntheses

1. All reactions are monitored by TLC using 0.25 mm × 7.5 cm silica plates with UV-indicator (Merck 60F-254), and mobile phase chloroform:methanol (80:20% v/v).

2. Phospholipid and PEG alone or their conjugates are visualized by phosphomolybdic acid and Dragendorff spray reagents (see Note 1).

3. Silica gel (240–360 μm) and size exclusion media, Sepharose CL4B (40–165 μm) and Sephadex G25m (Sigma-Aldrich) are used for silica column chromatography and size exclusion chromatography respectively (see Note 2).

2.3. Preparation of the TATp-Bearing, Rhodamine-Labeled Liposomal Formulations

1. The pH-sensitive or pH-insensitive, Rh-labeled, TATp-bearing liposomes are prepared by the lipid film hydration method.

2. A mixture of PC:Chol (7:3), TATp-PEG$_{1000}$-PE, Rh-PE and either mPEG$_{2000}$-HZ-PE (pH-sensitive) or mPEG$_{2000}$-DSPE (pH-insensitive) at molar ratio 10:0.25:0.1:15 is evaporated under reduced pressure (see Note 3).

3. The dry lipid formed is hydrated with phosphate buffer saline, pH 7.4. The liposomal suspension is filtered through 0.2 μm polycarbonate filters and stored at 4°C until use.

4. The liposome particle mean size and size distribution are observed using a Coulter N4 Plus submicron particle analyzer.

2.4. Preparation of the TATp-Bearing, Rhodamine Labeled, pGFP Complexed Liposomal Formulations

1. The pH-sensitive or pH-insensitive, TATp-bearing pGFP-complexed liposomes are prepared by the spontaneous vesicle formation (SVF) method adopted from (55) with few modifications.

2. A plasmid solution is prepared by combining pGFP and 10 mM Tris-EDTA (TE) buffer, pH 7.4. A lipid solution in ethanol is prepared by dissolving EPC:Chol (7:3) in anhydrous ethanol, and then adding DOTAP, TATp-PEG$_{1000}$-PE and either mPEG$_{2000}$-HZ-PE (**22**, pH-sensitive) or mPEG$_{2000}$-DSPE (pH-insensitive) at 10:0.25:15 molar ratio. The charge (±) ratio is 10:1.

3. The lipid and plasmid solutions are preheated to 37°C before mixing together. After mixing these solutions for 10 min, ethanol is evaporated under the reduced pressure. The samples are filtered through 0.2 μm polycarbonate filters and stored at 4°C until use.

4. The liposomal formulations are subjected to the agarose gel electrophoresis to test for the quantitative presence and intactness of the plasmid within the liposomes (56). In a typical case, the pGFP concentration is 3.22 μg/mg of total lipid.

5. The liposome particle mean size and size distribution are observed using a Coulter N4 Plus submicron particle analyzer.

3. Methods

3.1. Synthesis of Hydrazone-Based mPEG-HZ-PE Conjugates (54, 57)

3.1.1. Synthesis of Aliphatic Aldehyde-Derived Hydrazone-Based mPEG-HZ-PE Conjugates

Step 1: Synthesis of hydrazide-activated phospholipids

1. 22 μmol of phosphatidylthioethanolamine, **2**, are mixed with 1.5 molar excess of each acyl hydrazide linker (Table 1) in 3 mL anhydrous methanol containing 5 molar excess of triethylamine over lipid (Scheme 1). The reaction is performed at 25°C under argon for 8 h (*see* **Note 4**).

2. Solvent is removed under reduced pressure, and the residue is dissolved in chloroform and applied to a 5-mL silica gel column which had been activated (150°C overnight) and prewashed with 20 mL of chloroform. The column is equilibrated

Table 1
List of acyl hydrazide cross-linkers

Linker used	Mol. wt.	Length of spacer arm
AMBH 2-acetamido-4-mercapto butanoic acid hydrazide	191.25	–
EMCH (N-e-maleimidocaproic acid) hydrazide	225.24	11.8Å
MPBH 4-(4-N-maleimidophenyl)butyric acid hydrazide	309.5	17.9Å
KMUH N-(k-maleimido undecanoic acid) hydrazide	295.8	19.0Å
SMCCH Succinimidyl 4-(N-maleimidomethyl) cyclohexane-1-carboxylate hydrazide	365.31	–

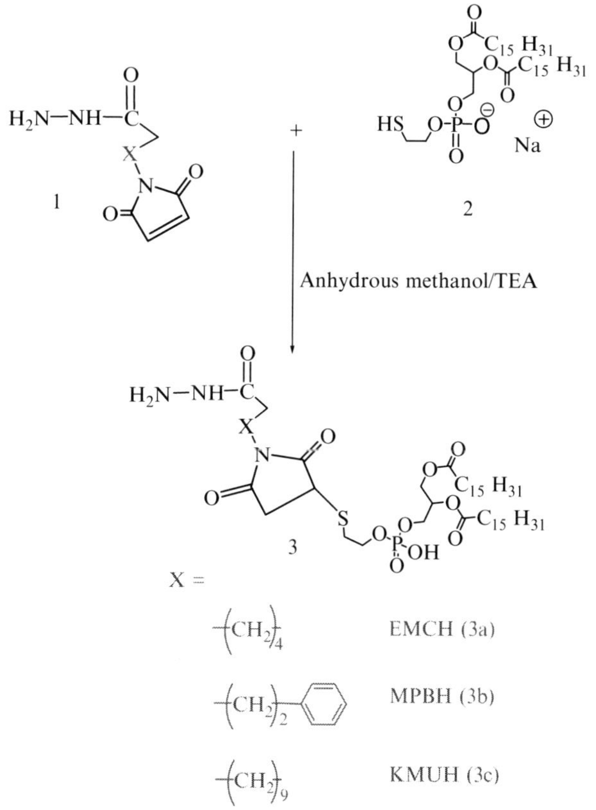

Scheme 1. Synthesis of acyl hydrazide-activated phospholipids

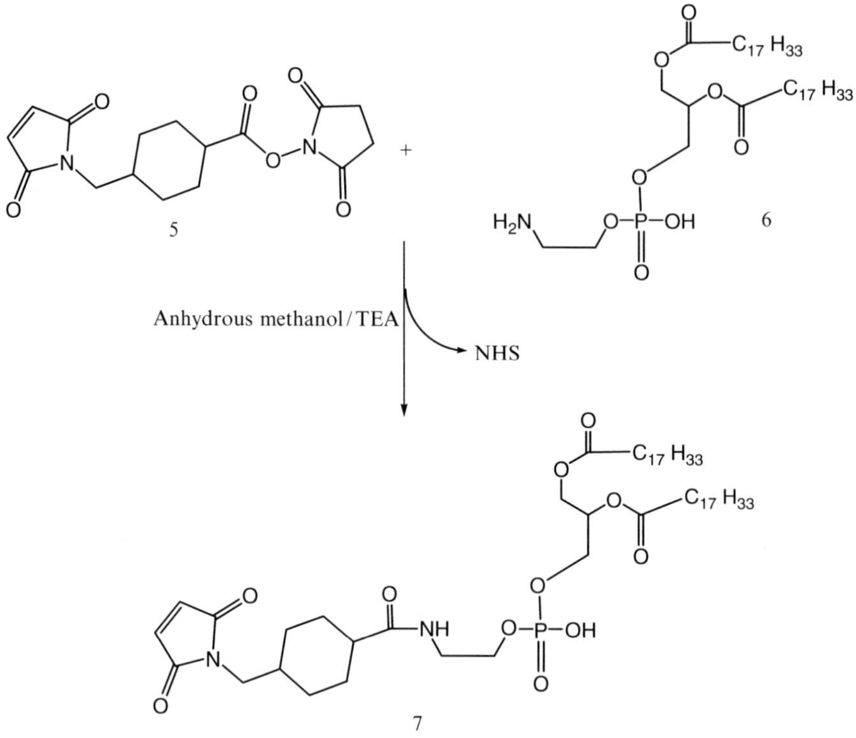

Scheme 2. Maleimide activation of phosphatidylethanolamine

with an additional 15 mL of chloroform followed by 5 mL of each of the following chloroform:methanol mixtures 4:0.25, 4:0.5, 4:0.75, 4:1, 4:2 and, finally, with 6 mL of 4:3 v/v. The phosphate-containing fractions eluting in 4:1, 4:2 and 4:3 chloroform:methanol (v/v) are pooled and concentrated under reduced pressure. The product is stored in glass ampoules as chloroform solution under argon at –80°C.

3. For the activation of phospholipid with AMBH, a maleimide derivative of phosphatidylethanolamine, **7**, is prepared using SMCC (Scheme 2). In brief, phosphatidylethanolamine, **6**, in chloroform was reacted with 1.5 molar excess of SMCC, **5**, over lipid in presence of 5 molar excess of TEA under argon for 5 h. The maleimide-derivative is separated from excess SMCC on silica gel column using chloroform:methanol (4:0.2 v/v) mobile phase. The elution fractions containing Ninhydrin-negative and phosphorus-positive fractions are pooled and concentrated under reduced pressure. DOPE-

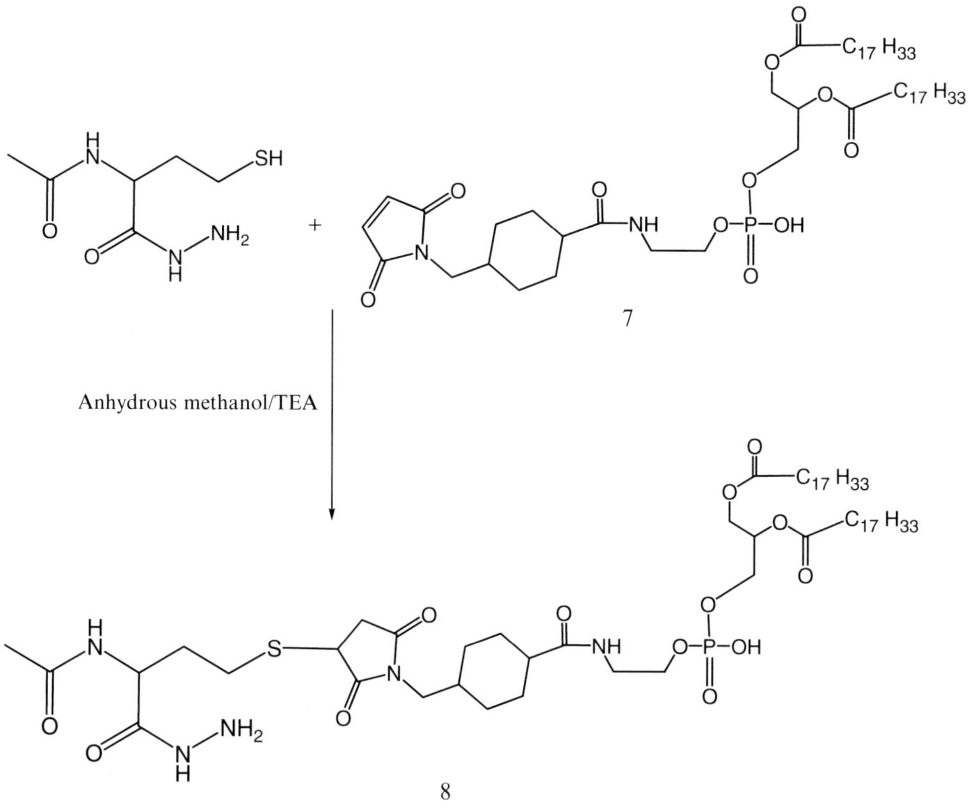

Scheme 3. AMBH-derivatized phospholipid via sulfhydryl-maleimide addition reaction

maleimide is further used to synthesize AMBH-activated derivative of phospholipid, **8**, by reacting with 1.5 molar excess of AMBH using TEA as catalyst (Scheme 3).

Step 2: Synthesis of mPEG-HZ-PE conjugates

1. 21 μmol of mPEG$_{2000}$-butyraldehyde are reacted with 14 μmol of linker-activated phospholipid in 2 mL chloroform at 25°C in a tightly closed reaction vessel (Schemes 4 and 5).

2. After an overnight stirring, chloroform is evaporated under vacuum in rotary evaporator. The excess mPEG$_{2000}$-butyraldehyde is separated from PEG-HZ-PE conjugates using gel filtration chromatography. The gel filtration chromatography is performed using sepharose-CL4B equilibrated overnight in pH 9–10 degassed ultra pure water (elution medium) in 1.5 × 30 cm glass column.

3. The thin film formed in round bottom flask after evaporating chloroform is hydrated with the elution medium and applied to the column. The micelles formed by PEG-HZ-PE conjugate are the first to elute from the column (*see* **Note 5**).

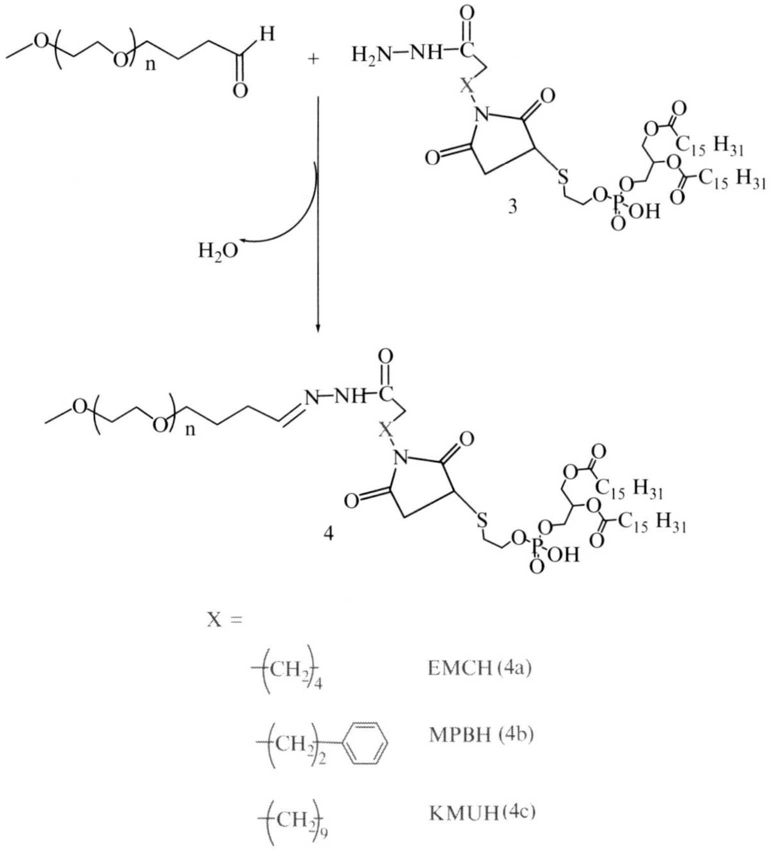

Scheme 4. Synthesis of aliphatic aldehyde-based hydrazone-derived mPEG-HZ-PE

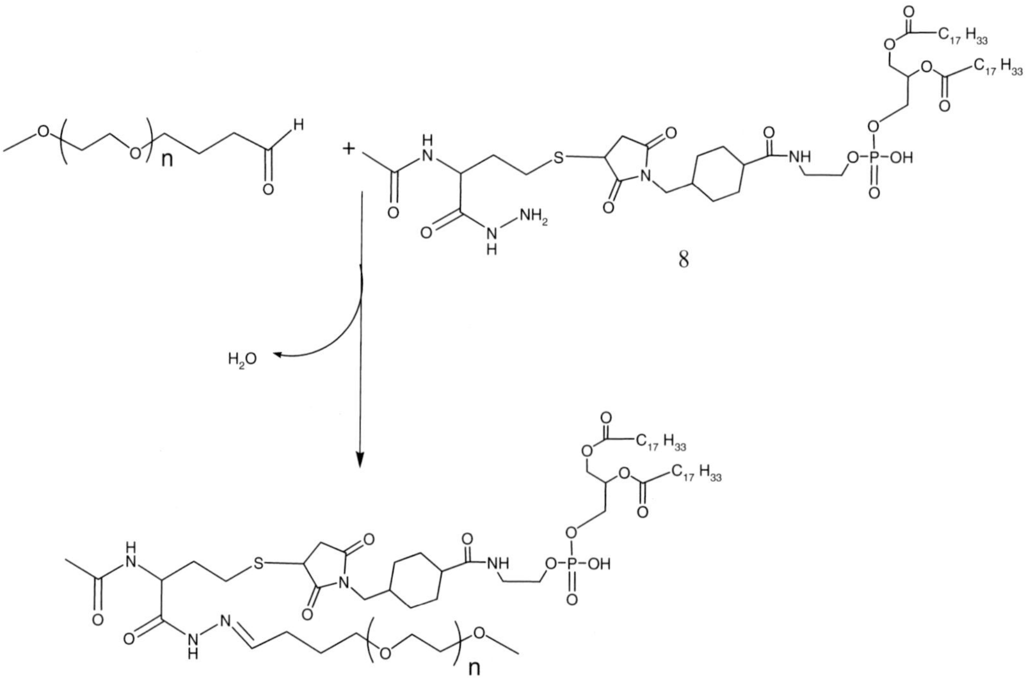

Scheme 5. Synthesis of PEG-HZ-PE conjugate using AMBH-activated phospholipid

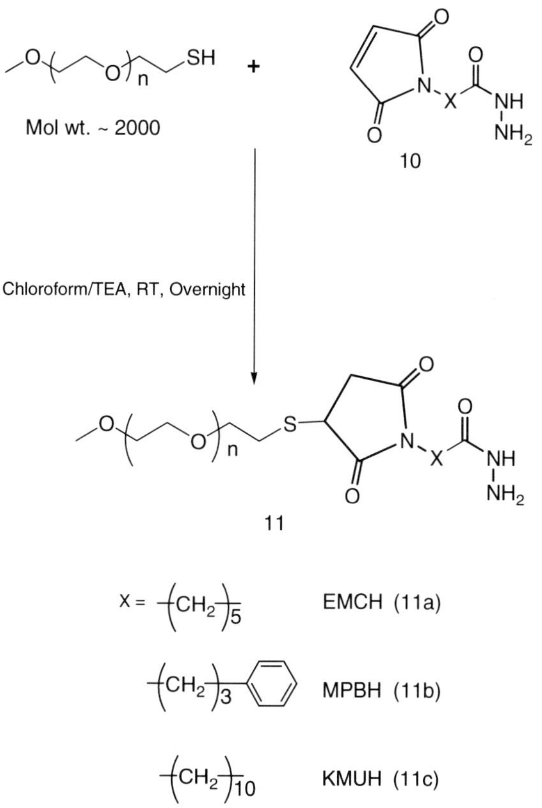

Scheme 6. Synthesis of acyl hydrazide activated PEG

X = $-(CH_2)_5-$ EMCH (11a)

$-(CH_2)_3-\text{C}_6\text{H}_5$ MPBH (11b)

$-(CH_2)_{10}-$ KMUH (11c)

4. Micelle containing fractions are identified by Dragendorff spray reagent and pooled together, kept in freezer at −80°C overnight before subjecting to freeze drying.

5. The freeze dried PEG-HZ-PE conjugates are weighed and stored at −80°C as chloroform solution.

3.1.2. Synthesis of Aromatic Aldehyde-Derived Hydrazone-Based mPEG-HZ-PE Conjugates

Step 1: Synthesis of hydrazide-activated PEG derivatives

1. 40 μmol of mPEG-SH in chloroform are mixed with two molar excess of acyl hydrazide cross-linkers: EMCH (**10a**), MPBH (**10b**), KMUH (**10c**) presence of 5 molar excess of triethylamine over lipid (*see* **Note 5**), (Scheme 6).

2. The excess EMCH is separated from the product by size exclusion chromatography using Sephadex G25m media.

3. The acyl hydrazide derivatives of PEG, (**11a**), (**11b**), (**11c**) are freeze dried and stored as chloroform solution at −80°C.

Step 2: Synthesis of aromatic aldehyde-activated phospholipid

1. 35 μmol of phosphatidylethanolamine, DOPE-NH$_2$, **12**, in chloroform are mixed with 2 molar excess of 4-succinimidyl-

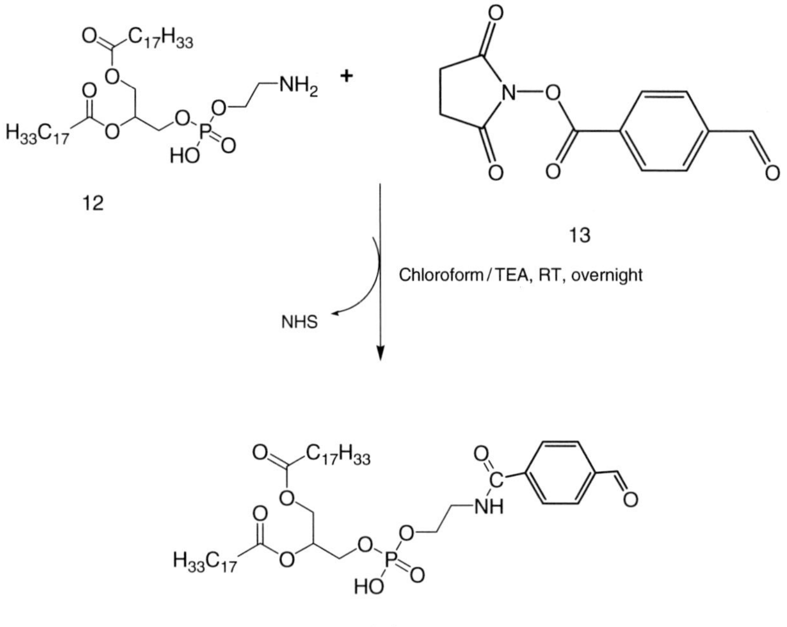

Scheme 7. SFB activation of phosphatidylethanolamine

formyl benzoate, SFB, **13**, in presence of 3 molar excess triethylamine over lipid (Scheme 7).

2. After stirring for 3 h, solvent is evaporated, residue is redissolved in chloroform and product is separated on silica gel column using acetonitrile:methanol mobile phases: 4:0, 4:0.25, 4:0.5, 4:0.75 and 4: 1 v/v.

3. The fractions containing product are identified by TLC analysis, pooled and concentrated. The product is stored as chloroform solution at −80°C.

Step 3: Synthesis of mPEG-HZ-PE conjugates

1. 1.5 molar excess of SFB activated phospholipid, **14**, are reacted with acyl hydrazide derivatized PEGs, **11a**, **11b**, and **11c** respectively, in chloroform at room temperature (Scheme 8).

2. After overnight stirring, chloroform is evaporated under reduced pressure.

3. The PEG-HZ-PE conjugate is purified using size exclusion chromatography using Sepharose CL4B as described before.

3.1.3. Synthesis of Aromatic Ketone-Derived Hydrazone-Based mPEG-HZ-PE Conjugates

Step 1: Synthesis of hydrazide derivative of PEG

1. mPEG-SH (MW 2000), **16**, is reacted with 2 molar excess of SMCCHz, **17**, in presence of triethylamine for 8 h in dry chloroform (Scheme 9).

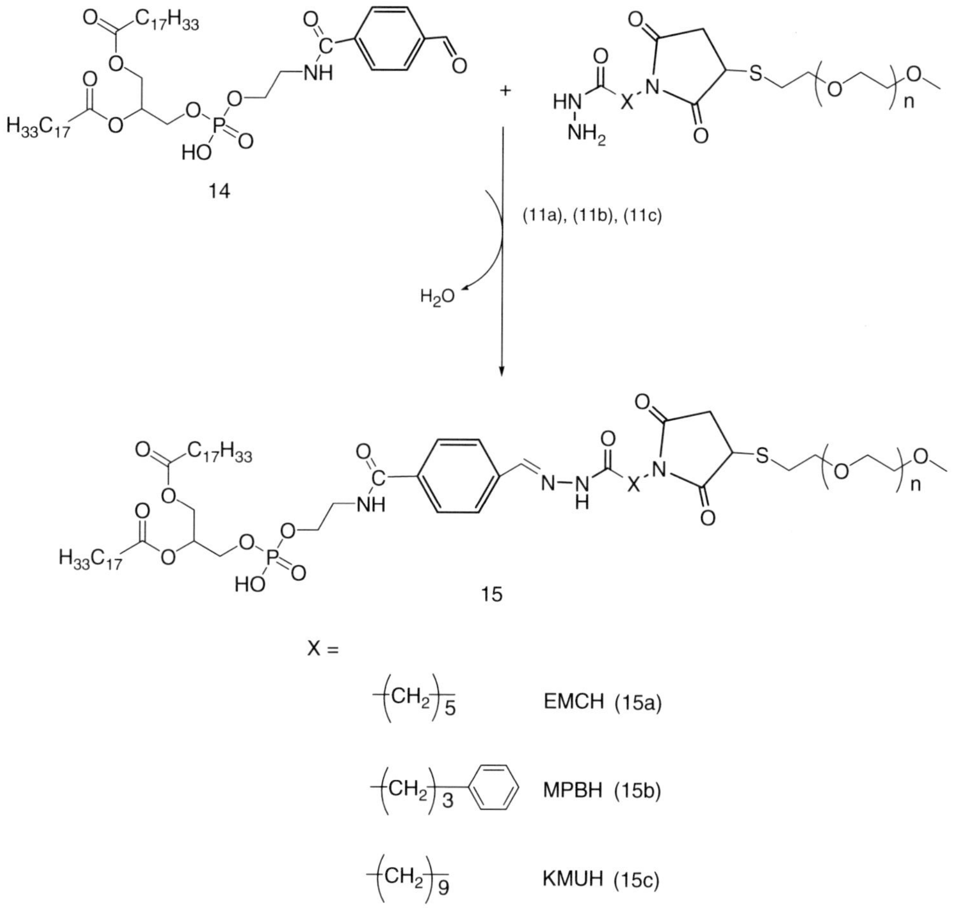

Scheme 8. Synthesis of PEG-HZ-PE conjugate

2. Chloroform is evaporated, and the residue is dissolved in water.

3. The PEG-hydrazide derivative, **18**, is separated and purified by the size exclusion gel chromatography using Sephadex G25m media.

4. The product is freeze dried and stored as chloroform solution at −80°C.

Step 2: Activation of phospholipid with 4-acetyl phenyl maleimide

1. 40 µmol of 4-acetyl phenyl maleimide, **19**, are reacted with 27 µmol of phosphatidylthioethanol (DPPE-SH), **20**, in presence of triethylamine overnight with constant stirring under inert atmosphere of argon (Scheme 9).

2. The product, **21**, is separated on a silica gel column using chloroform:methanol mobile phase (4:1 v/v).

3. The fractions containing product are identified by TLC analysis, pooled and concentrated.

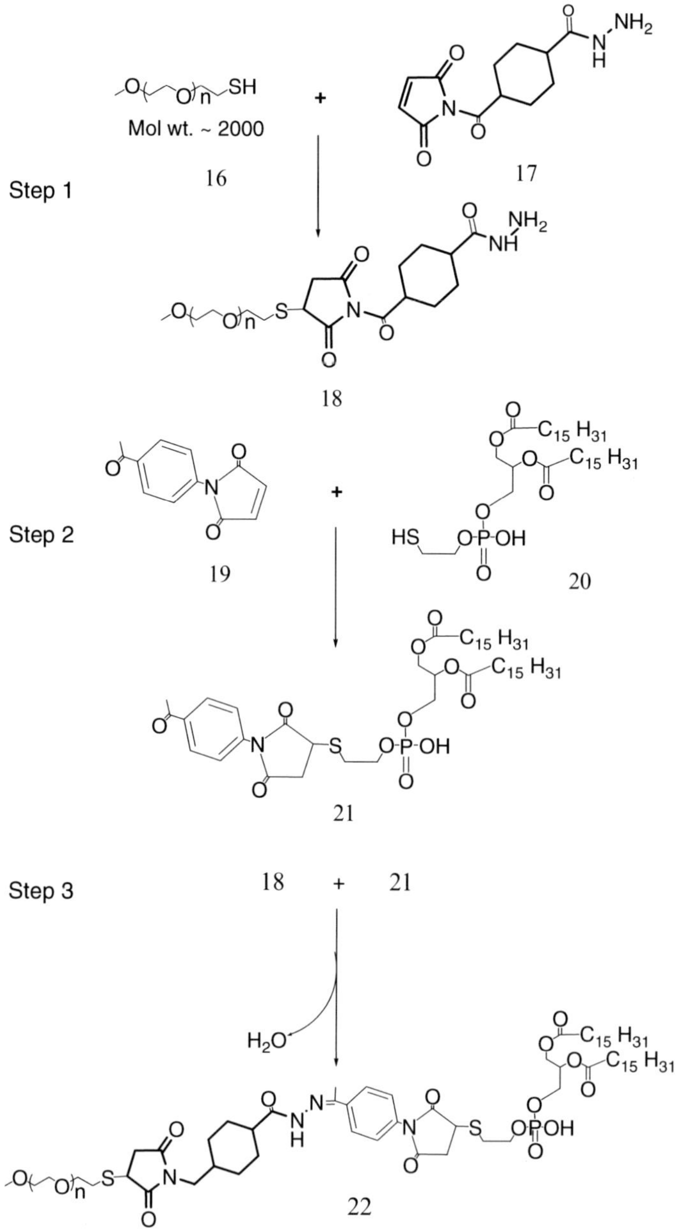

Scheme 9. Synthesis of aromatic ketone-derived hydrazone based mPEG-HZ-PE

4. Aromatic ketone-activated phospholipid is stored as chloroform solution at −80°C.

Step 3: Synthesis of mPEG-HZ-PE conjugate

1. Hydrazide-activated PEG derivative, 18, is reacted with 1.5 molar excess of the aromatic ketone-derivatized phospholipid, 21, overnight under the constant stirring at room temperature (Scheme 9).

2. The PEG-HZ-PE conjugate, 22, is separated and purified by size exclusion gel chromatography using Sepharose-CL4B media.

3.2. Synthesis of PE-PEG$_{1000}$-TATp Conjugate (57)

Step 1: Synthesis of PE-PEG$_{1000}$-maleimide

1. 1.5 molar excess of DOPE-NH$_2$, 23, is reacted with NHS-PEG$_{1000}$-maleimide, 24, in chloroform under argon at room temperature in the presence of 3 molar excess triethylamine overnight with stirring.
2. The product PE-PEG$_{1000}$-maleimide, 25, is separated on the Sephadex G25m column equilibrated overnight with the degassed double deionized water.
3. The product is freeze dried and stored under chloroform at −80°C.

Step 2: Synthesis of PE-PEG$_{1000}$-TATp

1. Twofold molar excess of TATp-SH is mixed with PE-PEG$_{1000}$-maleimide, 25, in chloroform under inert atmosphere with gentle shaking for 8 h.
2. The excess TATp-SH is separated from the product, 26, by gel filtration chromatography using Sephadex G25m media.
3. The freeze-dried product is stored under chloroform at −80°C until further use.

3.3. In vitro pH-Dependant Degradation of PEG-HZ-PE Conjugates

1. The time-dependant degradation of PEG-HZ-PE micelles incubated in buffer solutions (Phosphate buffer saline, pH 7.4 and 5.0) maintained at 37°C is followed by HPLC using Shodex KW-804 size exclusion column.
2. The elution buffer used is pH 7.0, Phosphate buffer (100 mM phosphate, 150 mM sodium sulfate), run at 1.0 mL/min. For fluorescent detection (Ex 550 nm/Em 590 nm) of micelle peak, Rh-PE (1 mol% of PEG-PE) is added to the PEG-PE conjugate in chloroform.
3. A film is prepared by evaporating the chloroform under argon stream and hydrated with the phosphate buffer saline, pH 7.4 or 5.0 (adjusted by pre-calculated quantity of 1 N HCl).
4. A peak that represents micelle population appears at the retention time between 9–10 min.
5. The degradation kinetics of micelles is assessed by following the area under micelle curve.

3.4. Avidin–Biotin Affinity Chromatography

1. To check the pH-sensitivity, biotin-containing micelles are formulated by mixing mPEG$_{2000}$-HZ-PE (60% mol), PEG$_{750}$-PE (37% mol), Rhodamine-PE (0.5% mol, fluorescent marker), and biotin-PE (2.5% mol, biotin component) in chloroform.

2. Chloroform is evaporated and a thin film is formed using rotary evaporator.
3. To test the binding of biotin-bearing Rh-PE-labeled, TATp-bearing liposomes before and after incubation at lowered pH values, the corresponding samples are kept for 3 h at pH 7.4 or 5.0 and then applied onto the Immobilized NeutrAvidin protein column (*see* **Note 6**).
4. The degree of the retention of the corresponding preparation on the column is estimated following the decrease in the sample rhodamine fluorescence at 550/590 nm after passing through the NeutrAvidin column (58).

3.5. In Vitro Cell Culture Studies

1. H9C2 rat embryonic cardiomyocytes in 10% fetal bovine serum DMEM are grown on coverslips in 6-well plates, then treated with various Rh-PE-labeled liposome samples (with and without preincubation for 3 h at pH 5.0) in serum-free medium (2 mL/well, 30 mg total lipid/mL).
2. After a 1 h incubation period, the media are removed and the plates washed with serum-free medium three times.
3. Individual coverslips are mounted cell-side down onto fresh glass slides with PBS (*see* **Note 7**). Cells are viewed with a Nikon Eclipse E400 microscope under bright light or under epifluorescence with rhodamine/TRITC filter (58) (*see* **Note 8**).
4. The images are analyzed using ImageJ 1.34I software (NIH) for integrated density comparison of red fluorescence between two groups (*see* **Note 9**).

3.6. In Vivo Studies

1. LLC tumors are grown in nu/nu mice (Charles River Breeding Laboratories, MA) by the s.c. injection of 8×10^4 LLC cells per mouse into the left flank (protocol #05-1233R, approved by the Institutional Animal Care and Use Committee at Northeastern University, Boston).
2. When tumors reach 5–10 mm in diameter, they are injected at four to five different spots with 150 μL of Rh-labeled, TATp-bearing pH-sensitive or pH-insensitive liposomes in phosphate-buffered saline, pH 7.4 (*see* **Note 10**).
3. Mice are killed 6 h later by cervical dislocation, and excised tumors are cryo-fixed as described earlier.
4. Microtome cut sections are washed thoroughly with phosphate buffer saline (pH 7.4), dried and fixed on slides using Fluor Mounting medium.
5. These sections are observed under fluorescence microscopy using TRITC filter (59).
6. Further, the images are analyzed using ImageJ 1.34I software (NIH) for integrated density comparison of red fluorescence between pH-sensitive and pH-insensitive groups.

3.7. In Vivo Transfection with pGFP

1. LLC tumors are grown in nu/nu mice (Charles River Breeding Laboratories, MA) by the s.c. injection of 8×10^4 LLC cells per mouse into the left flank (protocol # 05-1233R, approved by the Institutional Animal Care and Use Committee at Northeastern University, Boston).

2. When tumors reach 5–10 mm in diameter, they are injected at four to five different spots with 150 µL of pGFP-loaded, TATp-bearing pH-sensitive or pH-insensitive liposomes in phosphate-buffered saline, pH 7.4.

3. Mice are killed 72 h later by cervical dislocation, and excised tumors are fixed in a 4% buffered paraformaldehyde overnight at 4°C, blotted dry of excess paraformaldehyde and kept in 20% sucrose in PBS overnight at 4°C.

4. Cryofixation is done by the immersion of tissues in ice-cold isopentane for 3 min followed by freezing at –80°C. Fixed, frozen tumors are mounted in Tissue-Tek OCT 4583 compound (Sakura Finetek, Torrance, CA) and sectioned on a Microtome Plus (TBS).

5. Sections are mounted on slides and analyzed by the fluorescence microscopy using FITC filter and with hematoxylin-eosin staining (*see* **Note 11**).

6. The images are analyzed using ImageJ 1.34I software (NIH) for integrated density comparison of green fluorescence between pH-sensitive and non-pH-sensitive groups.

4. Results and Discussion

4.1. Synthesis of Hydrazone-Based mPEG-HZ-PE Conjugates

Hydrazone-linkages have been very instrumental for the use as "pH-sensitive connections" because of their wide range of hydrolytic degradation kinetics strictly controlled by the nature of hydrazone bond formed. Hydrazones are much more stable than imines as a result of the delocalization of the π-electrons in the former. In fact, parent hydrazones are too stable for the application in drug delivery systems, and an electron withdrawing group has to be introduced to moderate the stability by somewhat disfavoring electron delocalization throughout the molecule as compared to the parent hydrazone. Hydrazones can be prepared from aldehydes or ketones and hydrazides under very mild conditions including aqueous solutions. Hydrazone bond formation can take place even in vivo from separate fragments which self-assemble under physiological conditions (60).

A set of different synthetic methods were designed based on the use of various aldehydes that can produce the hydrazone linkage between PEG and PE (54). Synthesis of aliphatic aldehyde-derived hydrazone containing PEG-PE conjugate was pursued in two steps. First, phospholipid was activated with four different acyl hydrazides. The sulfhydryl reactive group of phosphatidylthioethanolamine was reacted with maleimide end of maleimido acyl hydrazides (refer Table 1) through Michael addition, thus providing acyl hydrazide activated PE. mPEG-butyraldehyde, an aliphatic aldehyde, was then reacted with acyl hydrazide activated PE to get hydrazone based PEG-PE conjugate. To synthesize aromatic aldehyde-derived hydrazone, an aromatic aldehyde moiety was introduced into the phospholipid by reacting succinimidyl 4-formylbenzoate (SFB) with phosphatidylethanolamine under mild alkaline conditions. The acyl hydrazide-PEG derivatives were synthesized using mPEG-SH and maleimido acyl hydrazides (EMCH, MPBH, and KMUH). The SFB-activated phospholipid was then reacted with acyl hydrazide derivatized PEG. Aromatic ketone-derived hydrazone-based PEG-PE conjugates were synthesized by reacting aromatic ketone-activated phospholipids with acyl hydrazide-activated PEG (57).

4.2. Synthesis of PE-PEG$_{1000}$-TATp Conjugate

TATp-SH was attached to the heterobifunctional PEG via the two step synthesis as shown in the Scheme 10. First, Mal-PEG-PE conjugate was synthesized by reacting DOPE-NH$_2$ with the NHS end of heterobifunctional PEG derivative, 23, NHS-PEG$_{1000}$-maleimide. PE-PEG$_{1000}$-maleimide was then reacted with TATp-SH to form PE-PEG$_{1000}$-TATp conjugate. The conjugate was separated by gel chromatography using the Sephadex G25m media.

4.3. In Vitro pH-Dependent Degradation of PEG-HZ-PE Conjugates

All PEG-HZ-PE derivatives spontaneously form micelle in aqueous surroundings (61). The stability of hydrazone-based PEG-PE conjugates incubated at physiological pH 7.4 and acidic pH 5.0 in buffer solutions maintained at 37°C was investigated by HPLC. For this purpose, the area under the micelle peak of PEG-HZ-PE (R_t 9–10 min) was observed over a period of time. PEG-HZ-PE conjugates derived from an aliphatic aldehyde and different acyl hydrazides were found to be highly unstable under acidic conditions, with the micelle peak was completely disappearing within 2 min incubation at pH 5.0. At the same time, these conjugates were relatively stable at physiological pH: the PEG-HZ-PE conjugate, **9,** with AMBH as cross-linker showed the half-life of 150 min followed by EMCH, **4a,** (120 min), MPBH, **4b,** (90 min), and KMUH, **4c,** (20 min) (Table 2). The rate of hydrolysis among the aliphatic aldehyde-derived hydrazone-based PEG-PE conjugates (4a, 4b, 4c, and 9) at pH 7.4 seems to be dependent on carbon chain length of acyl hydrazide. The increase in number of carbon atoms in acyl hydrazide led to increase in rate of hydrolysis (PEG-PE conjugate 4c, acyl hydrazide with 10-C atoms >4a, acyl hydrazide with 5-C atoms >9, acyl hydrazide with 3-C atoms). Introducing an aromatic character within carbon chain of acyl hydrazide led to increase in hydrolysis as observed in case of 4b and 4a (rate of hydrolysis of 4b > 4a).

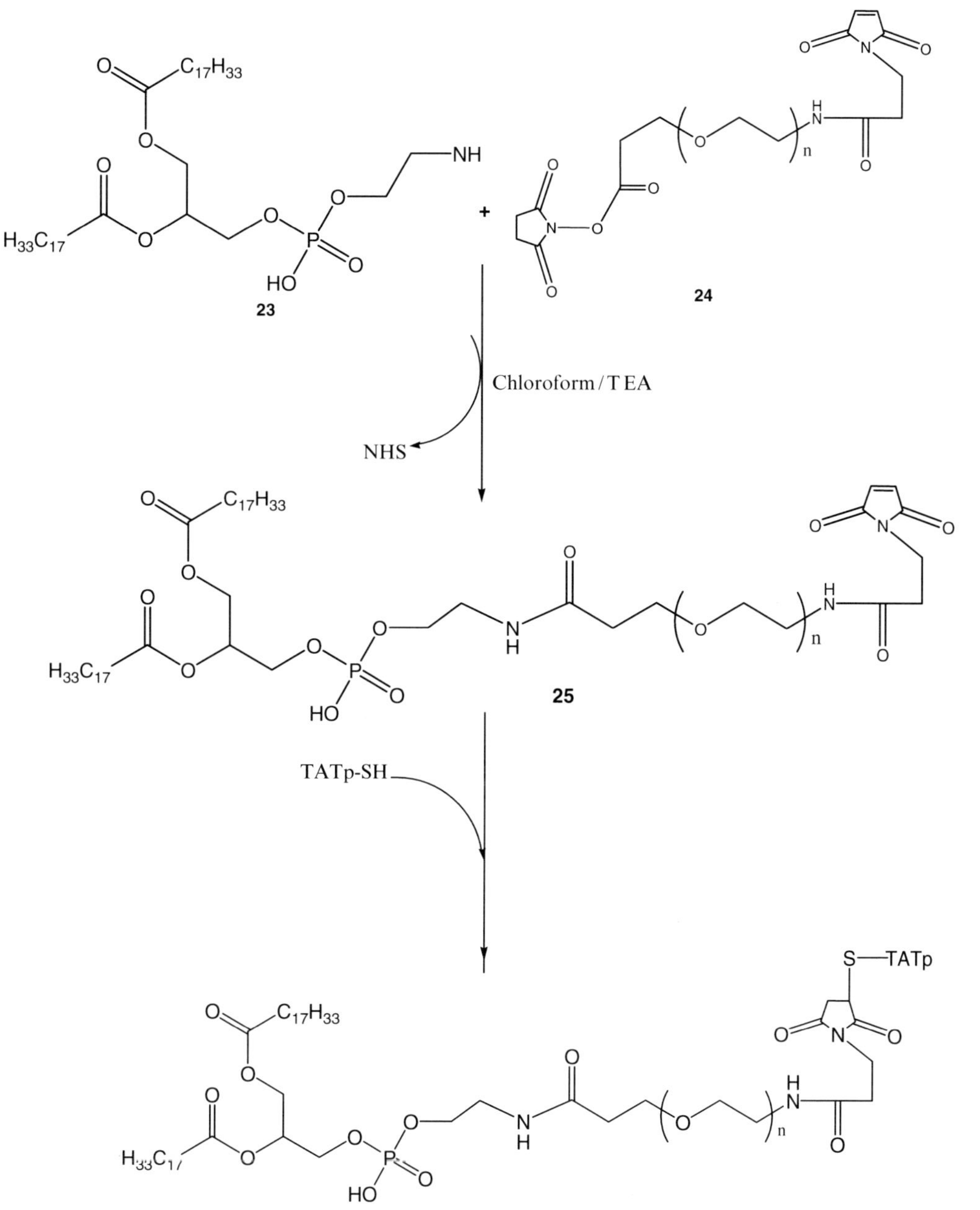

Scheme 10. Synthesis of PE-PEG-TATp conjugate

Alternatively, the PEG-HZ-PE conjugates derived from an aromatic aldehyde and acyl hydrazides were found to be highly stable at pH 7.4 and 5.0 (Table 2). The half-life values were not attained at either of those pH values even at the end of incubation period of 72 h in pH 7.4 and 48 h in pH 5.0 buffer solutions maintained at 37°C. The resistance to hydrolysis exhibited by

Table 2
Half-lives of different hydrazone-based mPEG-HZ-PE conjugates incubated in phosphate buffered saline, pH 7.4 and pH 5.0 at 37°C over a period of time, h

mPEG-HZ-PE Conjugate	Half-life (h)	
	pH 7.4	pH 5.0
4a	2	<0.03
4b	1.5	<0.03
4c	0.33	<0.03
9	2.5	<0.03
15a	>72	>48
15b	>72	>48
15c	>72	>48
22	40	2.0

hydrazones derived from aromatic aldehydes can be attributed to the conjugation of the π bonds of $-C=N-$ bond of the hydrazone with the π bonding benzene ring. Thus, it supports the finding that hydrazones formed from aromatic aldehydes are more stable to acidic hydrolysis than those formed from aliphatic ones (62, 63). The hydrazone hydrolysis involves the protonation of the $-C=N$ nitrogen followed by the nucleophilic attack of water and cleavage of C–N bond of tetrahedran intermediate (64). Any of these steps is determining and dependant on the pH. The substituents on the carbonyl reaction partner influence the rate of hydrolysis through altering the pKa of the hydrazone with electron donating substituents facilitating protonation of the $-C=N$ nitrogen (65).

This would support the fact that PEG-HZ-PE conjugates containing hydrazone bond derived from the aliphatic aldehyde are more prone to hydrolytic degradation. Aromatic aldehyde-derived hydrazone bond is too stable for the purpose of pH-triggered drug release. Careful selection of an aldehyde and an acyl hydrazide would be necessary for the application of the hydrazone-based chemistry for the development of pH-sensitive pharmaceutical nanocarriers.

As Scheme 9 shows, an aromatic ketone-derived hydrazone bond was introduced between PEG and PE. The presence of a methyl group (electron donating) on the carbonyl functional group would provide a sufficient lability of the hydrazone bond under mildly acidic conditions while an immediate aromatic ring (electron withdrawing) next to the hydrazone bond would offer

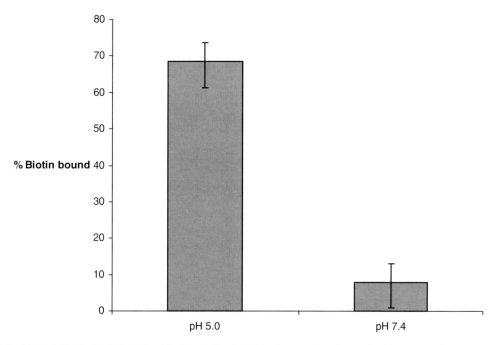

Fig. 1. Binding of pH-sensitive biotin-micelles to NeutrAvidin columns after the incubation at room temperature at pH 5.0 and 7.4

4.4. Avidin–Biotin Affinity Chromatography

the stability under acidic and neutral conditions. mPEG-HZ-PE conjugate, wherein the hydrazone bond is derived from an aromatic ketone, exhibited the half-lives of 2–3 h at slightly acidic pH values, and much higher stability (up to 40 h) at the physiological pH (Table 2).

To determine the pH-sensitivity of mPEG-HZ-PE conjugates, biotin-embedded micelles shielded by cleavable $mPEG_{2000}$-HZ-PE, were eluted through avidin immobilized gel media columns. The control micelle formulation (incubated at pH 7.4 at 37°C for 3 h) showed only a minimal biotin binding against 69% biotin binding of test micelle formulation (incubated at pH 5.0 at 37°C for 3 h), Fig. 1. This proves shielding effect of $mPEG_{2000}$-HZ-PE conjugate under physiological pH condition and de-shielding after exposure to acidic environment.

4.5. In Vitro Cell Culture Studies

To study shielding/de-shielding effect of mPEG-HZ-PE under the influence of acidic pH, internalization of Rh-labeled, TATp-bearing, mPEG-HZ-PE-shielded liposomes pre-incubated at pH 7.4 and 5.0 was followed using H9C2 cells. As seen in Fig. 2a, b, Rh-labeled TATp-bearing, pH-sensitive liposomes incubated at pH 5.0 showed 2.5 times (ImageJ 1.34I data) more internalization than when incubated at pH 7.4 because of better accessibility of TATp for its action after detachment of pH-sensitive PEG corona from liposomal surface under the influence of "acidic" pH.

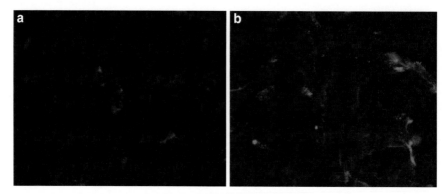

Fig. 2. Fluorescence microscopy showing the internalization of Rh-PE-labeled TATp-modified pH-sensitive liposomes by H9C2 cells after the incubation at pH 7.4 2(**a**), and pH 5.0 2(**b**)

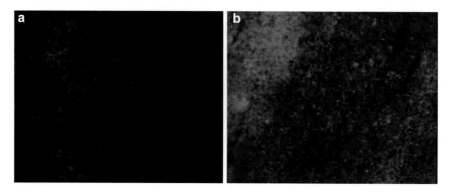

Fig. 3. TRITC image of frozen tissue section treated with intratumoral injection of Rh-labeled/TATp/pH-insensitive liposome 3(**a**) or Rh-labeled/TATp/pH-sensitive liposome 3(**b**) into LLC tumor bearing mice

4.6. In Vivo Studies

We attempted intratumoral injections of Rh-labeled, TATp-bearing pH-sensitive or pH-insensitive liposomes into LLC tumor-bearing mice to cover different physiological conditions. An "acidic" pH at the tumor site is a well-known fact which is of interest while developing physiology-based targeted delivery systems. Under the fluorescence microscope with TRITC filter, samples prepared 6 h post-injection from tumors injected with TATp-bearing, Rh-labeled, pH-sensitive liposomes demonstrated intensive and bright red fluorescence which was 4 times (as per ImageJ 1.34I data) more than that observed in the samples obtained from the tumors injected with TATp-bearing, Rh-labeled, pH-insensitive liposomes (Fig. 3a, b).

4.7. In Vivo pGFP Transfection Experiment

Also, we attempted a localized transfection of tumor cells by the direct intratumoral administration of sterically shielded with pH-sensitive (containing mPEG-HZ-PE, 25) or pH-insensitive (containing mPEG-DSPE) conjugates TATp-liposome-pGFP complexes into the tumor tissue by the intratumoral injections.

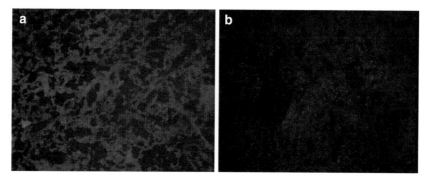

Fig. 4. Fluorescence microscopy images of the LLC tumor sections from the tumors injected with pGFP-loaded TATp-bearing liposomes with the pH-cleavable PEG coat 4(**a**), and with the pH-non-cleavable PEG coat 4(**b**). Notice enhanced GFP expression in 4(a) case

Histologically, hematoxylin/eosin-stained tumor slices in animals injected with both preparations showed the identical typical pattern of poorly differentiated carcinoma (polymorphic cells with basophilic nuclei forming nests and sheets and containing multiple sites of neoangiogenesis). However, under the fluorescence microscope with FITC filter, samples prepared 72 h post-injection from tumors injected with pH-sensitive PEG-TATp-liposome-pGFP complexes demonstrated intensive and bright green fluorescence compared to only minimal GFP fluorescence observed in the samples obtained from the tumors injected with pH-insensitive PEG-TATp-liposome-pGFP complexes (Fig. 4a, b).

The enhanced pGFP transfection by using pH-sensitive PEG-TATp-liposome-pGFP complexes is an ultimate result of the removal of mPEG-HZ-PE coat under the decreased pH of the tumor tissue, and better accessibility of de-shielded TATp moieties in TATp-liposome-pGFP complexes for internalization by the cancer cells allowing for the increased interactions of pGFP with cancer cell nuclei.

Owing to their physico-chemical properties, the long-circulating (PEGylated) liposomal carriers have the ability to accumulate inside the tumor tissue via the EPR effect, without further escape into undesired non-target sites. The pH at tumor sites is "acidic" (2, 3). Therefore, when TATp-pGFP-liposomes with an additional pH-sensitive PEG coating accumulate in the tumor tissue, the lowered pH-mediated removal of the protective PEG coat takes place, and TATp moieties become exposed and accessible for the interaction with cells. This leads to rapid pGFP pay-load delivery into the cancer cells as a result of the extensive TATp-mediated internalization of liposomes, and thereby enhanced transfection. The ImageJ analysis indicated a three times less transfection in the case of PEG-TATp-pGFP-pH-insensitive liposomes as non-detachable PEG coat interferes and sterically hinders the interactions between TATp and target cancer cells.

5. Conclusions

pH-sensitive mPEG-HZ-PE conjugates based on hydrazone bond chemistry were synthesized. The pH-dependant hydrolytic kinetics could be tuned using appropriate aldehyde or ketone and acyl hydrazide. These conjugates have immense applications in targeted drug delivery systems e.g., the development of the targeted drug carriers carrying the temporarily hidden function (e.g., cell penetrating peptide, TATp), and a detachable PEG-HZ-PE which, in addition to prolonging circulation half-life of carriers, can expose TATp function only under the action of certain local stimuli (such as lowered pH), represent a significant step on the way toward "smart" multifunctional pharmaceutical nanocarriers for target accumulation by EPR effect and intracellular penetration in a controlled fashion.

6. Notes

1. Phosphomolybdic acid and Dragendorff are usually used for visualization of PEG and phospholipids respectively, but iodine fumes could be used as the universal developing agent for visualization of both and their conjugates as well on reversible basis without damaging the developed TLC plate.
2. If other grades of silica gel are intended for the separation purpose, then conduct a series of suitability experiments before use with respect to volume of media required, sample volume and typical elution profile.
3. While handling all materials stored in chloroform under nitrogen at –80°C, allow them stabilize at room temperature before opening and using them. This is good lab safety practice.
4. Keep the reaction vessel continuously flushed with inert gases like argon or nitrogen to remove headspace oxygen which might affect reaction rate or degrade the reaction vessel contents.
5. Use standardized suitable Sepharose CL4B gel filtration column for proper separation of micelles from excess PEG. Use degassed water (pH >8, adjusted with 0.1 N NaOH) as elution medium to protect "acid labile" PEG-PE conjugates.
6. Prepare Immobilized NeutrAvidin protein column well ahead so that it would stabilize as per manufacturer's (Pierce) manual for at least 30 min of application of samples.
7. Formation of air bubbles is highly discouraged. Careful mounting of slide on stage is necessary.

8. This protocol can be used for other cell lines as well.

9. Quantification may be desirable for the interpretation of the results and can be performed in several different ways, we have used ImageJ software (NIH) for this purpose, but other techniques could be used.

10. If tumors do not reach the desired size, wait for more time as it takes longer to grow in some animals but 2–3 weeks is average time for tumor cell used in the current protocol.

11. Hemotoxylin-eosin staining protocol is used to observe histology of cancer tissue of treated and non-treated animals. Other parallel methods could be followed to reach the similar outcome.

Acknowledgements

This work was supported by the NIH grants RO1 HL55519 and RO1 CA121838 to VPT.

References

1. Jayasundar R, Singh VP (2002) In vivo temperature measurements in brain tumors using proton MR spectroscopy. Neurol India 50:436–439
2. Engin K, Leeper DB, Cater JR, Thistlethwaite AJ, Tupchong L, McFarlane JD (1995) Extracellular pH distribution in human tumours. Int J Hyperthermia 11:211–216
3. Ojugo AS, McSheehy PM, McIntyre DJ, McCoy C, Stubbs M, Leach MO, Judson IR, Griffiths JR (1999) Measurement of the extracellular pH of solid tumours in mice by magnetic resonance spectroscopy: a comparison of exogenous (19)F and (31)P probes. NMR Biomed 12:495–504
4. Khare AR, Peppas NA (1993) Release behavior of bioactive agents from pH-sensitive hydrogels. J Biomater Sci Polym Ed 4:275–289
5. Kataoka K, Matsumoto T, Yokoyama M, Okano T, Sakurai Y, Fukushima S, Okamoto K, Kwon GS (2000) Doxorubicin-loaded poly(ethylene glycol)-poly(beta-benzyl-l-aspartate) copolymer micelles: their pharmaceutical characteristics and biological significance. J Control Release 64:143–153
6. Leroux J, Roux E, Le Garrec D, Hong K, Drummond DC (2001) N-isopropylacrylamide copolymers for the preparation of pH-sensitive liposomes and polymeric micelles. J Control Release 72:71–84
7. Le Garrec D, Taillefer J, Van Lier JE, Lenaerts V, Leroux JC (2002) Optimizing pH-responsive polymeric micelles for drug delivery in a cancer photodynamic therapy model. J Drug Target 10:429–437
8. Taillefer J, Brasseur N, van Lier JE, Lenaerts V, Le Garrec D, Leroux JC (2001) In-vitro and in-vivo evaluation of pH-responsive polymeric micelles in a photodynamic cancer therapy model. J Pharm Pharmacol 53:155–166
9. Asokan A, Cho MJ (2004) Cytosolic delivery of macromolecules. 3. Synthesis and characterization of acid-sensitive bis-detergents. Bioconjug Chem 15:1166–1173
10. Lee ES, Na K, Bae YH (2003) Polymeric micelle for tumor pH and folate-mediated targeting. J Control Release 91:103–113
11. Lee ES, Shin HJ, Na K, Bae YH (2003) Poly(l-histidine)-PEG block copolymer micelles and pH-induced destabilization. J Control Release 90:363–374
12. Braslawsky GR, Kadow K, Knipe J, McGoff K, Edson M, Kaneko T, Greenfield RS (1991) Adriamycin(hydrazone)-antibody conjugates require internalization and intracellular acid hydrolysis for antitumor activity. Cancer Immunol Immunother 33:367–374

13. Yoo HS, Lee EA, Park TG (2002) Doxorubicin-conjugated biodegradable polymeric micelles having acid-cleavable linkages. J Control Release 82:17–27
14. Lee ES, Na K, Bae YH (2005) Super pH-sensitive multifunctional polymeric micelle. Nano Lett 5:325–329
15. Shen WC, Ryser HJ (1981) cis-Aconityl spacer between daunomycin and macromolecular carriers: a model of pH-sensitive linkage releasing drug from a lysosomotropic conjugate. Biochem Biophys Res Commun 102:1048–1054
16. Ogden JR, Leung K, Kunda SA, Telander MW, Avner BP, Liao SK, Thurman GB, Oldham RK (1989) Immunoconjugates of doxorubicin and murine antihuman breast carcinoma monoclonal antibodies prepared via an N-hydroxysuccinimide active ester intermediate of cis-aconityl-doxorubicin: preparation and in vitro cytotoxicity. Mol Biother 1:170–174
17. Patel VF, Hardin JN, Mastro JM, Law KL, Zimmermann JL, Ehlhardt WJ, Woodland JM, Starling JJ (1996) Novel acid labile COL1 trityl-linked difluoronucleoside immunoconjugates: synthesis, characterization, and biological activity. Bioconjugate Chem 7:497–510
18. Heffernan MJ, Murthy N (2005) Polyketal nanoparticles: a new pH-sensitive biodegradable drug delivery vehicle. Bioconjugate Chem 16:1340–1342
19. Gillies ER, Frechet JM (2005) pH-Responsive copolymer assemblies for controlled release of doxorubicin. Bioconjugate Chem 16:361–368
20. Gillies ER, Jonsson TB, Frechet JM (2004) Stimuli-responsive supramolecular assemblies of linear-dendritic copolymers. J Am Chem Soc 126:11936–11943
21. Gumusderelioglu M, Kesgin D (2005) Release kinetics of bovine serum albumin from pH-sensitive poly(vinyl ether) based hydrogels. Int J Pharm 288:273–279
22. Shin J, Shum P, Thompson DH (2003) Acid-triggered release via dePEGylation of DOPE liposomes containing acid-labile vinyl ether PEG-lipids. J Control Release 91:187–200
23. Kratz F, Beyer U, Roth T, Schutte MT, Unold A, Fiebig HH, Unger C (1998) Albumin conjugates of the anticancer drug chlorambucil: synthesis, characterization, and in vitro efficacy. Arch Pharm (Weinheim) 331:47–53
24. Beyer U, Roth T, Schumacher P, Maier G, Unold A, Frahm AW, Fiebig HH, Unger C, Kratz F (1998) Synthesis and in vitro efficacy of transferrin conjugates of the anticancer drug chlorambucil. J Med Chem 41:2701–2708
25. Di Stefano G, Lanza M, Kratz F, Merina L, Fiume L (2004) A novel method for coupling doxorubicin to lactosaminated human albumin by an acid sensitive hydrazone bond: synthesis, characterization and preliminary biological properties of the conjugate. Eur J Pharm Sci 23:393–397
26. Toncheva V, Schacht E, Ng SY, Barr J, Heller J (2003) Use of block copolymers of poly(ortho esters) and poly (ethylene glycol) micellar carriers as potential tumour targeting systems. J Drug Target 11:345–353
27. Oishi M, Nagasaki Y, Itaka K, Nishiyama N, Kataoka K (2005) Lactosylated poly(ethylene glycol)-siRNA conjugate through acid-labile beta-thiopropionate linkage to construct pH-sensitive polyion complex micelles achieving enhanced gene silencing in hepatoma cells. J Am Chem Soc 127:1624–1625
28. Steenbergen C, Deleeuw G, Rich T, Williamson JR (1977) Effects of acidosis and ischemia on contractility and intracellular pH of rat heart. Circ Res 41:849–858
29. Frunder H (1949) The pH changes of living tissue during activity and inflammation. Pharmazie 4:345–355
30. Mellman I, Fuchs R, Helenius A (1986) Acidification of the endocytic and exocytic pathways. Annu Rev Biochem 55:663–700
31. Shen WC, Ryser HJ (1981) cis-Aconityl spacer between daunomycin and macromolecular carriers: a model of pH-sensitive linkage releasing drug from a lysosomotropic conjugate. Biochem Biophys Res Commun 102:1048–1054
32. Diener E, Diner UE, Sinha A, Xie S, Vergidis R (1986) Specific immunosuppression by immunotoxins containing daunomycin. Science 231:148–150
33. Ogden JR, Leung K, Kunda SA, Telander MW, Avner BP, Liao SK, Thurman GB, Oldham RK (1989) Immunoconjugates of doxorubicin and murine antihuman breast carcinoma monoclonal antibodies prepared via an N-hydroxysuccinimide active ester intermediate of cis-aconityl-doxorubicin: preparation and in vitro cytotoxicity. Mol Biother 1:170–174
34. Sinkule JA, Rosen ST, Radosevich JA (1991) Monoclonal antibody 44–3A6 doxorubicin immunoconjugates: comparative in vitro antitumor efficacy of different conjugation methods. Tumour Biol 12:198–206
35. Patel VF, Hardin JN, Mastro JM, Law KL, Zimmermann JL, Ehlhardt WJ, Woodland JM, Starling JJ (1996) Novel acid labile

COL1 trityl-linked difluoronucleoside immunoconjugates: synthesis, characterization, and biological activity. Bioconjug Chem 7:497–510

36. Patel VF, Hardin JN, Starling JJ, Mastro JM (1995) Novel trityl linked drug immunoconjugates for cancer therapy. Bioorg Med Chem Lett 5:507–512

37. Patel VF, Hardin JN, Grindey GB, Schultz RM (1995) Tritylated oncolytics as prodrugs. Bioorg Med Chem Lett 5:513–518

38. Fife T, Jao L (1965) Substituent effects in acetal hydrolysis. J Org Chem 30:1492–1495

39. Gillies ER, Goodwin AP, Frechet JM (2004) Acetals as pH-sensitive linkages for drug delivery. Bioconjug Chem 15:1254–1263

40. Murthy N, Campbell J, Fausto N, Hoffman AS, Stayton PS (2003) Design and synthesis of pH-responsive polymeric carriers that target uptake and enhance the intracellular delivery of oligonucleotides. J Control Release 89:365–374

41. Murthy N, Xu M, Schuck S, Kunisawa J, Shastri N, Frechet JM (2003) A macromolecular delivery vehicle for protein-based vaccines: acid-degradable protein-loaded microgels. Proc Natl Acad Sci U S A 100:4995–5000

42. Heffernan MJ, Murthy N (2005) Polyketal nanoparticles: a new pH-sensitive biodegradable drug delivery vehicle. Bioconjug Chem 16:1340–1342

43. Lee S, Yang SC, Heffernan MJ, Taylor WR, Murthy N (2007) Polyketal microparticles: a new delivery vehicle for superoxide dismutase. Bioconjug Chem 18:4–7

44. Gümüsderelioglu M, Kesgin D (2005) Release kinetics of bovine serum albumin from pH-sensitive poly(vinyl ether) based hydrogels. Int J Pharm 288:273–279

45. Gümüsderelioglu M, Topal IU (2005) Vinyl ether/acrylic acid terpolymer hydrogels synthesized by [gamma]-radiation: characterization, thermosensitivity and pH sensitivity. Radiat Phys Chem 73:272–279

46. Hurwitz E, Wilchek M, Pitha J (1980) Soluble macromolecules as carriers for daunorubicin. J Appl Biochem 2:25

47. Laguzza BC, Nichols CL, Briggs SL, Cullinan GJ, Johnson DA, Starling JJ, Baker AL, Bumol TF, Corvalan JR (1989) New antitumor monoclonal antibody-vinca conjugates LY203725 and related compounds: design, preparation, and representative in vivo activity. J Med Chem 32:548–555

48. Beyer U, Roth T, Schumacher P, Maier G, Unold A, Frahm AW, Fiebig HH, Unger C, Kratz F (1998) Synthesis and in vitro efficacy of transferrin conjugates of the anticancer drug chlorambucil. J Med Chem 41:2701–2708

49. Kratz F, Beyer U, Roth T, Schutte MT, Unold A, Fiebig HH, Unger C (1998) Albumin conjugates of the anticancer drug chlorambucil: synthesis, characterization, and in vitro efficacy. Arch Pharm (Weinheim) 331:47–53

50. Oishi M, Nagasaki Y, Itaka K, Nishiyama N, Kataoka K (2005) Lactosylated poly(ethylene glycol)-siRNA conjugate through acid-labile beta-thiopropionate linkage to construct pH-sensitive polyion complex micelles achieving enhanced gene silencing in hepatoma cells. J Am Chem Soc 127:1624–1625

51. Kataoka K, Itaka K, Nishiyama N, Yamasaki Y, Oishi M, Nagasaki Y (2005) Smart polymeric micelles as nanocarriers for oligonucleotides and siRNA delivery. Nucleic Acids Symp Ser (Oxf) 49:17–18

52. Kong SD, Luong A, Manorek G, Howell SB, Yang J (2007) Acidic hydrolysis of N-Ethoxybenzylimidazoles (NEBIs): potential applications as pH-sensitive linkers for drug delivery. Bioconjug Chem 18:293–296

53. Sawant RM, Hurley JP, Salmaso S, Kale AA, Tolcheva E, Levchenko T, Torchilin VP (2006) "Smart" drug delivery systems: double-targeted pH-responsive pharmaceutical nanocarriers. Bioconjugate Chem 17:943–949

54. Kale AA, Torchilin VP (2007) Design, synthesis, and characterization of pH-sensitive PEG-PE conjugates for stimuli-sensitive pharmaceutical nanocarriers: the effect of substitutes at the hydrazone linkage on the ph stability of PEG-PE conjugates. Bioconjug Chem 18:363–370

55. Jeffs LB, Palmer LR, Ambegia EG, Giesbrecht C, Ewanick S, MacLachlan I (2005) A scalable, extrusion-free method for efficient liposomal encapsulation of plasmid DNA. Pharm Res 22:362–372

56. Torchilin VP, Levchenko TS, Rammohan R, Volodina N, Papahadjopoulos-Sternberg B, D'Souza Gerard GM (2003) Cell transfection in vitro and in vivo with nontoxic TAT peptide-liposome-DNA complexes. Proc Natl Acad Sci U S A 100:1972–1977

57. Kale AA, Torchilin VP (2007) Enhanced transfection of tumor cells in vivo using "Smart" pH-sensitive TAT-modified pegylated liposomes. J Drug Target 15:538–545

58. Sawant RM, Hurley JP, Salmaso S, Kale AA, Tolcheva E, Levchenko T, Torchilin VP (2006) "Smart" drug delivery systems: double-targeted pH-responsive pharmaceutical nanocarriers. Bioconjug Chem 17:943–949

59. Torchilin VP, Levchenko TS, Rammohan R, Volodina N, Papahadjopoulos-Sternberg B,

D'Souza GGM (2003) Cell transfection in vitro and in vivo with nontoxic TAT peptide-liposome-DNA complexes. Proc Natl Acad Sci U S A 100:1972–1977

60. Rideout D (1994) Self-assembling drugs: a new approach to biochemical modulation in cancer chemotherapy. Cancer Invest 12:189–202, discussion 268–269

61. Lukyanov AN, Gao Z, Torchilin VP (2003) Micelles from polyethylene glycol/phosphatidylethanolamine conjugates for tumor drug delivery. J Control Release 91:97–102

62. Apelgren LD, Bailey DL, Briggs SL, Barton RL, Guttman-Carlisle D, Koppel GA, Nichols CL, Scott WL, Lindstrom TD, Baker AL et al (1993) Chemoimmunoconjugate development for ovarian carcinoma therapy: preclinical studies with vinca alkaloid-monoclonal antibody constructs. Bioconjugate Chem 4:121–126

63. Baker MA, Gray BD, Ohlsson-Wilhelm BM, Carpenter DC, Muirhead KA (1996) Zyn-Linked colchicines: controlled-release lipophilic prodrugs with enhanced antitumor efficacy. J Control Release 40:89–100

64. Cordes EH, Jencks WP (1963) The Mechanism of hydrolysis of schiff's bases derived from aliphatic amines. J Am Chem Soc 85:2843–2848

65. Harnsberger HF, Cochran EL, Szmant HH (1955) The basicity of hydrazones. J Am Chem Soc 77:5048–5050

Chapter 16

Functional Liposomal Membranes for Triggered Release

Armağan Koçer

Abstract

Shortly after the discovery of liposomes (J Mol Biol 13:238–252, 1965), Gregoriadis et al. (Lancet 1:1313–1316, 1974) suggested their use as drug delivery vesicles. Since then there have been many developments in liposomal composition, efficient drug encapsulation and retention, stability, and targeting (Biochim Biophys Acta 1113:171–199, 1992). However, even though some of the very potent drug formulations in liposomes were clinically approved, in most cases the amount of drug passively released from such ideal, long-circulating, sterically stable liposomes was not enough to show a therapeutic effect (Cancer Chemother Pharmacol 49:201–210, 2002; Cancer Chemother Pharmacol 48:266–268, 2001; Eur J Cancer 37:2015–2022, 2001; Breast Cancer Res Treat 77:185–188, 2003; Lung Cancer 34:427–432, 2001; Cancer Chemother Pharmacol 50:131–136, 2002). It has been hypothesized that the enhanced release at the target site will significantly improve the specificity and efficacy of a liposomal drug (J Liposomes Res 8:299–335, 1998; Pharmaco Rev 51:691–744, 1999; Curr Opin Mol Ther 3:153–158, 2001). To solve this challenge, more research efforts were directed toward a triggered release, in response to a specific stimulus at a target site. Here, we present an engineered, bacterial channel protein as a remote-controlled nanovalve in sterically stable liposomes for a triggered release of the liposomal content on command.

Key words: Triggered liposomal release, Membrane channel protein, Mechanosensitive channel of large conductance (MscL), Membrane protein reconstitution into liposomes, Calcein efflux assay, Remote-controlled nanovalve

1. Introduction

An ideal liposomal drug delivery system should have a long circulation time in the blood, accumulation at the target site, and controlled drug release that matches the efficacy profile of the drug.

Long-circulating and stable liposomes have been obtained at the cost of a very low release profile. The dilemma of having a stable liposome structure during circulation but having a leaky structure at the target site, is only now beginning to be solved by the use of specific liposome compositions. However, these efforts, for the most part, have been meeting with limited success [13]. In our approach, we have kept the stable liposome structure as it is and reconstituted a pore-forming bacterial membrane protein into the liposome and used it as the release mechanism. The channel protein has been engineered by using custom-designed chemical modulators that respond to specific signals present in the target site [14–16]. This strategy provides a much higher degree of flexibility for fine-tuning the liposome's response to its environment. For example, the pH in the environment of a solid tumor is about 6.5. Technically, it is difficult to design liposomes that can be stable at physiological pH 7.4, but leaky at pH 6.5 [17]. However, we have shown that with our system it is possible to fine-tune the pH release profile of liposomes by fine-tuning the pH response of the reconstituted channel protein via chemical modulators.

The channel protein used in this study is the Mechanosensitive channel of large conductance (MscL) from *Escherichia coli*. It is a very attractive candidate as a release valve in liposomal delivery systems, because it keeps its functionality when it is reconstituted into artificial lipid bilayers. It has a large and nonselective pore that allows the passage not only for ions but also for small molecules, and last but not least, it is one of the best studied channel proteins [18–23]. Although, normally, the channel opens in response to tension, its opening is influenced also by the polarity of its hydrophobic constriction zone [24]. An increase in the polarity or hydrophilicity of the 22nd amino acid, located in this part of the protein, results in channel openings even in the absence of tension [25]. Therefore, in order to operate it more easily in liposomal delivery systems, we used this charge-induced opening principle by rationally designing sulfhydryl-reactive chemical modulators and covalently and specifically attaching them to engineered cysteines in the pore. When placed in this hydrophobic region of the channel, these modulators would alter the hydration and open the channel, by creating charge but only in response to an external stimulus.

The chemical modification of MscL requires a free cysteine residue at a critical part of the channel, namely the 22nd amino acid. A 6-histidine(His) tag at the C-terminal of the protein allows a one-step isolation of MscL, which is >98% pure as analyzed by SDS-PAGE and N-terminal sequencing. The protein can be labeled in its pure isolated form, or alternatively, in its nickel–nitriloacetic acid (Ni–NTA) column-attached form. ESI–MS is a method to follow the labeling conditions and efficiency. After the protein is chemically labeled and reconstituted into desired liposomes, a simple fluorescent, dye efflux experiment is used to test the triggered liposomal release.

2. Materials

2.1. MscL Production

1. *E. coli* strain PB104 having G22C-MscL with C-terminal 6His tag (Biomade Technology Foundation, The Netherlands).
2. TY liquid medium: For a 1 l medium, mix 10 g bactotryptone (BD), 5 g yeast extract (BD), and 5 g NaCl. After sterilization, add 100 mg ampicillin (Sigma) and 10 mg chloramphenicol (Fluka).
3. TY-agar: As TY medium with additional 15 g/l agar (Difco) before sterilization.
4. Isopropyl-b-D-thiogalactopyranoside (IPTG) (Sigma).
5. 25 mM Tris-HCl (pH 8.0).
6. 1 mM Dithiothreitol (DTT) (Sigma).
7. 5 mM $MgSO_4$, 6.25 mg DNase I (Sigma), and 6.25 mg RNase (Roche Diagnostics).
8. French Press (Kindler).
9. Beckman L7-55 ultracentrifuge.

2.2. MscL Isolation

1. Membrane vesicles from *E. coli* strain PB104 having G22C-MscL with C-terminal 6-His tag (see Subheading 2.1, item 1).
2. Solubilization buffer: 10 mM Na_2HPO_4/NaH_2PO_4, pH 8.0, 300 mM NaCl, 35 mM imidazole, and 2% (vol/vol) Triton X-100 (see Notes 1 and 2).
3. Wash buffer: the solubilization buffer with only 1% (vol/vol) TritonX-100.
4. Histidine buffer: 10 mM Na_2HPO_4/NaH_2PO_4, 300 mM NaCl, 50 mM histidine, and 0.2% (vol/vol) TritonX-100.
5. Elution buffer: histidine buffer containing 235 mM histidine.
6. Ni–NTA metal-affinity matrix (Qiagen).
7. 1.5×10 cm column for chromatography.
8. Rotary mixer.
9. Beckman L7-55 ultracentrifuge.

2.3. Bradford Assay

1. Bradford reagent: 50 mg Coomassie Brilliant Blue G-250 (Serva Blue G, Serva, cat. no. 35050) is dissolved in 50 ml ethanol (95%), mixed well, and added to 100 ml phosphoric acid (85%). After the dye has completely dissolved, the volume is completed to 1 l and is filtered through Whatman #1 paper. It is stored in the dark at room temperature.
2. BSA stock solution: 1 mg BSA (Sigma) is dissolved in 1 ml water and stored frozen in 50 ml aliquots.

	3. Multiwell, microtiter plate reader equipped with a 595 nm filter (Power Wave X, Bio-Tek Instruments Inc.).
	4. Microtiter plates (Omnilabo International B.V.).
2.4. MscL Labeling	1. Isolated protein (see Subheading 3.2).
2.4.1. Labeling of Isolated Protein with Methylthiosulfonate (MTS) Compounds	2. MTS compounds (commercial or custom designed compounds that will transform into a charged form in response to an external stimulus).
	3. Efflux buffer: 10 mM sodium phosphate, pH 8, 150 mM NaCl, and 1 mM EDTA.
	4. Lipid buffer: 10 mM sodium phosphate buffer, pH 8.0, and 150 mM NaCl.
	5. pD-10 column.
2.4.2. Labeling During Isolation	1. Cysteine-reactive iodides or bromides (commercial or custom-designed compounds that will transform into a charged form in response to an external stimulus).
	2. Membrane vesicles from *E. coli* strain PB104, having G22C–MscL with C-terminal 6-His tag (see Subheading 3.1, step 16).
	3. Solubilization buffer (see Subheading 2.2, item 2).
	4. Wash buffer (see Subheading 2.2, item 3).
	5. Pre-labeling buffer: It is the wash buffer without imidazole in it.
	6. Histidine buffer (see Subheading 2.2, item 4).
	7. Elution buffer (see Subheading 2.2, item 5).
	8. Ni–NTA metal-affinity matrix (Qiagen).
	9. 1.5 × 10 cm column for chromatography.
	10. Rotary mixer.
	11. Beckman L7-55 ultracentrifuge.
2.5. Mass Spectrum Sample Preparation	1. Biobeads SM-2 adsorbents (Bio-Rad): 30 mg of biobeads are weighed and placed into a 500-ml flask. It is washed three times with distilled water. Biobeads are left to settle down and the small floating particles are removed. The biobead slurry is placed into a vacuum flask and degased under vacuum while stirring gently for about 30 min. Water is decanted until 1 cm above the biobead bed and is stored at 4°C. In our experience, the biobeads can be kept safely under these conditions for at least 2 months.
	2. 2 ml microtube.
	3. Formic acid (Merck, proanalse, 98–100%).
	4. Acetonitrile (Biosolve, HPLC supragradient, 99.97%).

2.6. Liposome Preparation for MscL Reconstitution

1. 1,2-Dioleoyl-sn-glycero-3-phosphocholine (DOPC) (Avanti Polar Lipids Inc.).
2. Cholesterol (Avanti Polar Lipids Inc.).
3. 3.1,2-Distearoyl-sn-glycero-3-phosphoethanolamine-N-[methoxy(polyethylene glycol)2000] (DSPE-PEG-2000; Avanti Polar Lipids Inc.).
4. 400 nm pore size polycarbonate filter (Avestin).
5. Lipid buffer (see Subheading 2.4.1., item 4).
6. Rotary evaporator.

2.7. MscL Reconstitution into Liposomes

1. Biobeads SM-2 adsorbents (Bio-Rad) (see Subheading 2.5., item 1).
2. Calcein (Na salt) (Sigma) solution: 200 mM calcein in 10 mM sodium phosphate buffer (pH 8.0).
3. Efflux buffer (see Subheading 2.4.1., item 3).
4. Isolated MscL protein.
5. Triton X-100.
6. 400 nm pore size polycarbonate filter (Avestin).
7. Mini-Extruder (Aventi Polar Lipids).

2.8. Calcein Efflux Assay

1. Efflux buffer (see Subheading 2.4.1., item 3).
2. [2-(Trimethylammonium)ethyl] methanethiosulfonate bromide (MTSET; Anatrace, cat. no. T110MT): 160 mM in efflux buffer (see Note 3).
3. Sephadex G50.
4. Column for chromatography (1.5 cm i.d. × 30 cm length).

3. Methods

3.1. MscL Production

1. A loopful of cells from the stock culture of *E. coli* strain that expresses G22C–MscL with C-terminal 6-His tag is stroked onto the surface of TY-agar and the agar plate is incubated overnight at 37°C.
2. 10 ml of TY liquid medium in a sterile tube is inoculated with a single colony from the streaked plate and incubated overnight at 37°C with shaking at 200 r.p.m.
3. 1 l of TY liquid medium in a 5-l flask is inoculated with 10 ml of overnight culture (totally four flasks).
4. The optical density of the growing culture is measured every 30 min by taking 1 ml sample from the culture and measuring the absorbance at 600 nm.

5. When the absorbance is around 0.550, IPTG is added into each flask to a final concentration of 1 mM in order to start the expression of *mscl* gene. The culture is grown for two more hours at 37°C with shaking at 200 r.p.m.

6. The *E. coli* cells are centrifuged and pelleted down at 7,000 g for 15 min at 4°C.

7. The supernatant is decanted. The pellet is washed by adding 25 mM Tris-HCl (pH 8.0) and 1 mM DTT.

8. The *E. coli* cells are centrifuged and pelleted down at 7,000 g for 15 min at 4°C.

9. The cells are resuspended in 48 ml of 25 mM Tris-HCl (pH 8.0), 5 mM $MgSO_4$, 6.25 mg DNase, and 6.25-mg RNase, mixed well, and incubated at 4°C for 30 min with continuous stirring.

10. The suspension is French pressed two times at 1,000 bar.

11. The cell debris is centrifuged at 4,000 g for 1 h at 4°C.

12. The supernatant is pipetted into a clean beaker and EDTA is added to a final concentration of 10 mM. It is centrifuged at 16,600 g for 10 min at 4°C.

13. The supernatant is pipetted into a clean beaker. $MgSO_4$ (2 mM final concentration) is added.

14. It is centrifuged in preweighed centrifuge tubes at 118,000 g for 2 h at 4°C.

15. The supernatant is removed. The centrifuge tubes are reweighed to calculate the wet weight of the pellet.

16. The pellet is dissolved in 25 mM Tris-HCl (pH 8.0) to 0.7 g/ml final concentration of membrane vesicles. They are stored at −80°C and used within six months.

3.2. MscL Isolation

1. 3 g wet weight of frozen stock membrane vesicles are thawed at 4°C and are suspended in 30 ml solubilization buffer in a 50 ml sterile tube.

2. It is incubated at 4°C for 30 min while continuously mixing it in a rotary mixer.

3. The suspension is centrifuged at 118,000 g at 4°C for 45 min and the solubilized fraction is transferred into a precooled (4°C) 50-ml tube.

4. 4-ml Ni–NTA matrix slurry is put into a 1.5 × 10 cm column. Once ethanol flows out, the column is washed with 10 ml of water.

5. The column is equilibrated with 30 ml of solubilization buffer.

6. The matrix is transferred from the column into the solubilized fraction in a 50 ml tube (Step 3) and gently mixed by rotating at 4°C for 30 min.

7. The matrix is transferred back into the 1.5 × 10-cm column.
8. After the solubilization buffer flows through, the matrix is washed with 30 ml of wash buffer.
9. The column is washed with 15 ml of histidine buffer. The elution speed is adjusted to 0.5 ml/min.
10. The MscL protein is eluted with 10 ml elution buffer with an elution speed of 0.5 ml/min. 1 ml fractions are collected.
11. The protein concentration in each fraction is determined by using Bradford assay.
12. Aliquot the protein-containing fractions into 300 ml fractions and save at −80°C (see Note 4).

3.3. Bradford Assay

1. 0, 1, 2, 3, and 4 μl of BSA (1 mg/ml) is pipetted into the assigned wells of a 96-well plate (Omnilabo International B.V.). 10, 9, 8, 7, and 6 μl of elution buffer is added into each well, respectively, to make up the volume to 10 μl.
2. 10 μl from each protein elution fraction is pipetted into individual wells of a 96-well plate.
3. 200 μl of Bradford reagent is added into all wells containing standard or sample.
4. The absorbance is read at 595 nm.
5. The concentration of MscL is determined from a standard curve of absorbance versus concentration of BSA.

3.4. MscL Labeling

3.4.1. Labeling of Isolated Protein with Methylthiosulfonate (MTS) Compounds

1. A desired MTS label is dissolved in an efflux buffer at 80 mM final concentration (see Note 3).
2. Dissolved MTS label is added into 500 μl of ~0.2 mg/ml pure protein at a final concentration of 40 mM.
3. The protein–label mixture is gently mixed by rotating at room temperature for 15 min.
4. A pD10 column is equilibrated with 25 ml of lipid buffer supplemented with 0.1% (v/v) Triton X-100.
5. The protein–label mixture is applied onto the column and 0.5 ml fractions are collected.
6. The protein concentration in each fraction is determined by using Bradford assay (see Subheading 3.3).
7. 0.5 ml of sample is stored at −80°C for mass spectroscopy.
8. The labeled protein is used directly for reconstitution.

3.4.2. Labeling During Isolation

1. Follow the procedure for MscL isolation (see Subheading 3.2) until Step 8. Then continue as explained here.
2. After the solubilization buffer flows through, the matrix is washed first with 15 ml of wash buffer, and then with 15 ml of prelabeling buffer. At the end, enough buffer is left in the column to have ~2.5 ml matrix slurry.

3. The slurry is then transferred into a smaller plastic container that has lead.

4. A 10 mM stock solution of iodides or bromides is prepared in prelabeling buffer or in DMSO, depending on the solubility of the compound (see Note 5).

5. The compound is added to the matrix at a final concentration of 1 mg/ml and incubated at room temperature for 15 min while gently rotating (see Note 6).

6. The matrix is poured into a clean 1.5 × 10 cm column. The rest of the experiment is proceeded at 4°C.

7. The entire buffer that contains unbound label is eluted. The column is washed with 15 ml of histidine buffer with the elution speed of 0.5 ml/min.

8. For elution, protein concentration determination, and storage steps, follow the same procedure as explained in Subheading 3.2 (Steps 10–12).

3.5. Mass Spectrum Sample Preparation

1. 500 µl of labeled MscL protein is put into a sterile 2 ml microtube and 100 mg wet weight of biobeads SM-2 adsorbents is added to it.

2. It is incubated at 40°C for 45 min to adsorb the detergent.

3. The solution is transferred into a new microtube and incubated at 60°C for 30 min.

4. The sample is cooled on ice and centrifuged at 20,800 g for 15 min at 4°C.

5. The white pellet is washed with ice-cold sterile water by adding 2 ml of water and centrifuging as in Step 4.

6. The water is removed by pipetting and Step 5 is repeated once more.

7. The water is removed again as much as possible.

8. The pellet is dissolved in 300 µl of 50% formic acid and 50% acetonitrile shortly before ESI–MS analysis.

3.6. Liposome Preparation for MscL Reconstitution

1. 1.2 ml of 50 mg/ml DOPC, 0.425 ml of 20 mg/ml cholesterol, and 1.56 ml of 20 mg/ml DSPE-PEG-2000 are put in a round-bottomed flask. A thin lipid film is obtained on the glass walls by evaporating the chloroform in a rotary evaporator under vacuum, while rotating the flask with the maximum speed at room temperature for 45 min.

2. The lipid film is rehydrated in 5 ml of lipid buffer to obtain 20 mg/ml lipid stock.

3. 1 ml aliquots of the liposome suspension is put into 2 ml Eppendorf tubes with screw caps.

4. The liposomes are frozen by immersing the Eppendorf tubes into liquid nitrogen and then thawed in a 60°C water bath. This step is repeated at least five times (see Note 7).

3.7. MscL Reconstitution into Liposomes

1. 1 ml of lipid stock is thawed in a 60°C water bath. 500 ml of it is extruded by 11 passes through a 400 nm pore size polycarbonate filter by using mini-extruder.

2. 30 μl of 10% (v/v) Triton X-100 is added into 300 μl of the extruded 20 mg/ml liposomes and mixed well (see Notes 8–10).

3. The necessary amount of protein is added to the detergent-saturated liposomes in order to reach a protein:lipid ratio of 1:120 (wt:wt) (see Note 11).

4. It is mixed by pipetting and incubated at 60°C in a water bath for 30 min.

5. 1 vol of calcein buffer is added to the liposome–protein mixture and mixed by pipetting.

6. 200 mg (wet weight) biobeads are added into the lipid, protein, and buffer mixture. It is covered with aluminum foil and incubated in a rotary mixer at room temperature for 4 h, or alternatively is incubated at 4°C overnight with a gentle rotation.

7. The resulting proteoliposomes are transferred with a pipette into a clean 2 ml Eppendorf tube.

3.8. Calcein Efflux Assay

3.8.1. Separation of Free Calcein from Proteoliposome/Calcein Mixture

1. A size-exclusion column is prepared by pouring 50 ml of Sephadex G50 matrix slurry in an efflux buffer into a 1.5 cm i.d. × 30 cm length column (the height of the matrix bed is about ~25 cm).

2. The column is equilibrated with the efflux buffer.

3. 400-μl proteoliposome and calcein mixture (see Subheading 3.7, step 7) is applied onto the column.

4. After the sample has soaked into the column, the proteoliposomes are eluted from the column with the efflux buffer under gravity. The proteoliposomes proceed in the column as a discrete dark orange band in the elution front, which is collected into an Eppendorf tube.

3.8.2. Efflux Assay

1. 5-μl proteoliposomes are added into 2 ml efflux buffer in a 4 ml cuvette. It is mixed continuously with a magnetic bar.

2. The fluorescence of calcein is monitored continuously at 520 nm (excitation at 490 nm) in a spectrofluorometer.

3. (a) Proteoliposomes containing unlabeled G22C–MscL: After about a min of recording, the channels are activated by

adding 25 μl of 160 mM MTSET. If the reconstitution is successful and the protein is active, the fluorescence increases (see Notes 3 and 12). (b) Proteoliposomes containing labeled G22C-MscL: After about 1 min of recording, the channels are activated by a desired trigger, for instance, a change in pH, illumination, addition of lysolipids, etc. (see Note 13).

4. After the signal becomes stable (~15–20 min), all the liposomes in the cuvette are burst by adding 100 μl Triton-X-100 (final concentration of 8 mM).

5. The % release through the channels is calculated from the following formula:

$$\% \text{ Release} = (I_t - I_0) \times 100 / (I_{100} - I_0),$$

where I_t is the measured fluorescence intensity at a given time; I_0 is the initial fluorescence intensity, which is caused by the initial free calcein in the sample before stimulation of the channel and the residual fluorescence resulting from the quenched liposomal calcein; I_{100} is the fluorescence intensity from total liposomal calcein, which is obtained by bursting all liposomes by the addition of Triton-X-100.

4. Notes

1. Prepare 10% of Triton-X-100 in water from a new bottle of Triton-X-100 and store in 1 ml aliquots at −20°C.

2. Prepare fresh in order to prevent detergent aging.

3. MTSET is not very stable for long time. Prepare the stock solution just before starting the experiments. Keep it on ice up to 30 min. Then, if you need more, prepare a fresh one.

4. The fractions can be stored safely in this fashion for at least 6 months.

5. If the chemical compound is light sensitive, then avoid exposure to light during the whole procedure as much as possible. Wrap the columns and Eppendorf tubes in an aluminum foil.

6. Incubation time can vary from minutes up to overnight. It is necessary to optimize this for each new label by using wilde-type MscL as a negative control for nonspecific labeling. WT–MscL does not have any cysteine residues so that the time of incubation should be long enough to allow all the subunits of G22C–MscL to be labeled but short enough to prevent any of the subunits of the WT–MscL from being labeled.

7. Frozen liposomes can be stored safely at −80°C up to 6 months.

8. The amount of detergent added depends on the lipid composition and the type of detergent used. Generally, the amount of detergent required should be sufficient enough to achieve detergent-saturated liposomes. This can be tested by titrating the desired liposomes with a desired detergent and following the optical density at 540 nm. Briefly, after freeze–thaw cycles, the lipid mixture is diluted to 4 mg/ml in lipid buffer and extruded eleven times through 400 nm filters. A 1 ml volume of these liposomes is put into a cuvette and placed in a spectrophotometer. The absorbance at 540 nm is followed upon stepwise addition of small portions of a desired detergent. Liposomes solubilize in three stages: first, the non-micellar detergent partitions between the aqueous buffer and the liposomal bilayer and induce turbidity changes. In the case of Triton-X-100, the first stage appears as an increase in the optical density. Further addition of detergent causes saturation of liposomes with detergent, and the liposomes will start to lyse gradually. This second stage is observed as a decrease of optical density. In the third or last stage, all liposomes solubilize and form mixed micelles with detergent and the optical density reaches its lowest value. Therefore, for a reconstitution as described here, it is necessary to titrate liposomes until the start of the second stage.

9. Use fresh detergent to prevent the possible interference of unknown chemicals produced in the aging process of the detergent. Beware that there can be batch-to-batch variation in even, theoretically, the best detergents. It is important to check every new batch of detergent before using for experiments.

10. After the addition of indicated amount of detergent into lipids, the solution should be clearer. If it still stays opaque, either the detergent or the lipid stock is not good anymore. This affects reconstitution dramatically and one can lose all the activity. If this is the case, the first test is another batch of detergent and if it is still not working make a fresh lipid stock.

11. In this lipid composition, the protein: lipid ratio can be varied between 1:20 and 1:300 (wt:wt) in order to get a signal in the calcein efflux experiment.

12. This control experiment gives information on the maximum channel activity one may expect from the particular preparation of proteoliposomes. MTSET is a positively charged reagent that reacts very rapidly and specifically with cysteine groups via a disulfide bond while having no effect on the liposomal membrane. In this way, it covalently attaches five positive charges to the G22C–MscL channel, one for each subunit, and forces its opening.

13. For every new trigger, it is important to check the effect of the trigger on liposomes alone. For that, the reconstitution procedure is followed as explained except that at Step 3.7.3 instead of protein, the same amount of protein elution buffer (see Subheading 2.2, item 5) is added into the detergent saturated liposomes. Additionally, it is also important to check the specificity of the trigger toward labeled MscL. For that, instead of the labeled G22C–MscL, WT and/or unlabeled G22C-MscL is reconstituted into liposomes (see Subheading 3.7).

Acknowledgments

The author would like to thank Professor George Robillard for his critical reading of the chapter. This work was supported by Biomade Technology Foundation, NanoNed, and The Netherlands Organization for Scientific Research (NWO-VIDI).

References

1. Bangham AD, Standish MM, Watkins JC (1965) Diffusion of univalent ions across the lamellae of swollen phospholipids. J Mol Biol 13:238–252
2. Gregoriadis G, Wills EJ, Swain CP, Tavill AS (1974) Drug-carrier potential of liposomes in cancer chemotherapy. Lancet 1:1313–1316
3. Woodle MC, Lasic DD (1992) Sterically stabilized liposomes. Biochim Biophys Acta 1113:171–199
4. Terwogt JM, Groenewegen G, Pluim D, Maliepaard M, Tibben MM, Huisman A et al (2002) Phase I and pharmacokinetic study of SPI-77, a liposomal encapsulated dosage form of cisplatin. Cancer Chemother Pharmacol 49:201–210
5. Thomas AL, O'Byrne K, Furber L, Jeffery K, Steward WP (2001) phase II study of Caelyx, liposomal doxorubicin: lack of activity in patients with advanced gastric cancer. Cancer Chemother Pharmacol 48:266–268
6. Harrington KJ, Lewanski C, Northcote AD, Whittaker J, Peters AM, Vile RG, Stewart JSW (2001) Phase II study of pegylated liposomal doxorubicin (Caelyx™) as induction chemotherapy for patients with squamous cell cancer of the head and neck. Eur J Cancer 37:2015–2022
7. Rimassa L, Carnaghi C, Garassino I, Salvini P, Ginanni V, Gullo G et al (2003) Unexpected low efficacy of stealth liposomal doxorubicin (Caelyx) and vinorelbine in metastatic breast cancer. Breast Cancer Res Treat 77:185–188
8. Kim ES, Lu C, Khuri FR, Tonda M, Glisson BS, Liu D et al (2001) A phase II study of STEALTH cisplatin (SPI-77) in patients with advanced non-small cell lung cancer. Lung Cancer 34:427–432
9. Vail DM, Kurzman ID, Glawe PC, O'Brien MG, Chun R, Garrett LD et al (2002) STEALTH liposome-encapsulated cisplatin (SPI-77) versus carboplatin as adjuvant therapy for spontaneously arising osteosarcoma (OSA) in the dog: a randomized multicenter clinical trial. Cancer Chemother Pharmacol 50:131–136
10. Bally MB, Lim H, Cullis PR, Mayer LD (1998) Controlling the drug delivery attributes of lipidbased drug formulations. J Liposomes Res 8:299–335
11. Drummond DC, Meyer O, Hong K, Kirpotin DB, Papahadjopoulos D (1999) Optimizing liposomes for delivery of chemotherapeutic agents to solid tumors. Pharmaco Rev 51:691–744

12. Fenske DB, MacLachlan L, Cullis PR (2001) Long-circulating vectors for the systemic delivery of genes. Curr Opin Mol Ther 3:153–158
13. Guo X, Szoka FC (2003) Chemical approaches to triggerable lipid vesicles for drug and gene delivery. Acc Chem Res 36:335–341
14. Kocer A, Walko M, Meijberg W, Feringa BL (2005) A light-actuated nanovalve derived from a channel protein. Science 309:755–758
15. Kocer A, Walko M, Bulten E, Halza E, Feringa BL, Meijberg W (2006) Rationally designed chemical modulators convert a bacterial channel protein into a pH-sensory valve. Angew Chem Int Ed Engl 45:3126–3130
16. Kocer A, Walko M, Feringa BL (2007) Synthesis and Utilization of reversible and irreversible light activated nanovalves derived from the channel protein MscL. Nat Protoc 2:1426–1437
17. Andresen TL, Jensen SS, Jorgensen K (2005) Advanced strategies in liposomal cancer therapy: problems and prospects of active and tumor specific drug release. Prog Lipid Res 44:68–97
18. Delcour AH, Martinac B, Adler J, Kung C (1989) Modified reconstitution method used in patch-clamp studies of Escherichia coli ion channels. Biophys J 56:631–636
19. Sukharev SI, Blount P, Martinac B, Blattner FR, Kung C (1994) A large-conductance mechanosensitive channel in *E. coli* encoded by *mscL* alone. Nature 368:265–268
20. Perozo E, Cortes DM, Sompornpisut P, Kloda A, Martinac B (2002) Open channel structure of MscL and the gating mechanism of mechanosensitive channels. Nature 418:942–948
21. Perozo E, Kloda A, Cortes DM, Martinac B (2002) Physical principles underlying the transduction of bilayer deformation forces during mechanosensitive channel gating. Nature Struct Biol 9:696–703
22. Van den Bogaart G, Krasnikov V, Poolman B (2007) Dual-colo fluorescence-burst analysis to probe protein efflux through the mechanosensitive channel. MscL Biophys J 92:1233–1240
23. Sukharev S, Anishkin A (2004) Mechanosensitive channels: what can we learn from 'simple' model systems? Trends Neurosci 27:345–351
24. Anishkin A, Chiang CS, Sukharev S (2005) Gain-of-function mutations reveal expanded intermediate states and a sequential action of two gates in MscL. J Gen Physiol 125:155–170
25. Yoshimura K, Batiza A, Kung C (2001) Chemically charging the pore constriction opens the mechanosensitive channel MscL. Biophys J 80:2198–2206

Chapter 17

A "Dock and Lock" Approach to Preparation of Targeted Liposomes

Marina V. Backer and Joseph M. Backer

Abstract

We developed a strategy for covalent coupling of targeting proteins to liposomes decorated with a standard adapter protein. This strategy is based on the "dock and lock" interactions between two mutated fragments of human RNase I, a 1–15-aa fragment with the R4C amino acid substitution, (Cys-tag), and a 21–127-aa fragment with the V118C substitution, (Ad-C). Upon binding to each other, Cys-tag and Ad-C spontaneously form a disulfide bond between the complimentary 4C and 118C residues. Therefore, any targeting protein expressed with Cys-tag can be easily coupled to liposomes decorated with Ad-C. Here, we describe the preparation of Ad-liposomes followed by coupling them to two Cys-tagged targeted proteins, human vascular endothelial growth factor expressed with N-terminal Cys-tag, and a 254-aa long N-terminal fragment of anthrax lethal factor carrying C-terminal Cys-tag. Both proteins retain functional activity after coupling to Ad-C-decorated drug-loaded liposomes. We expect that our "dock and lock" strategy will open new opportunities for development of targeted therapeutic liposomes for research and clinical use.

Key words: Recombinant targeting proteins, Self-assembled protein complex, Targeted drug delivery, Liposomes, Dock and lock

1. Introduction

Coupling proteins to drug-loaded liposomes requires conjugation of a lipid moiety to the protein, which might be detrimental to protein/target interactions. This is particularly true for relatively small proteins, like growth factors, cytokines, and hormones. One approach of "safe" protein conjugation is based on site-specific modification of cysteines that are not directly involved in target recognition (1, 2). This method, described for monoclonal antibodies, is hardly applicable to proteins of small size.

Another approach is introducing an additional cysteine residue in the N- or C-terminus of a targeting protein (3–6). However, direct coupling of a small targeting protein to a bulky liposome might interfere with the protein/target interactions. To solve this problem, we developed a "dock and lock" strategy for decorating liposomes with targeting proteins that does not require direct lipidation of targeting proteins and provides for additional space between a targeting protein and liposome. This strategy is based on the ability of two mutant complimentary fragments of human RNase I to form a disulfide bond upon mixing (Fig. 1a and ref. 7). One fragment, named Cys-tag, is a 15-aa long N-terminal peptide of human RNase I with the R4C amino acid substitution, and can be genetically fused to any targeting protein (8). The second peptide, named Ad-C, is a 21–127-aa fragment of human RNase I with the V118C substitution that serves as an adapter between a liposome and a Cys-tagged targeting protein. To perform this function, Ad-C is modified with a pegylated phospholipid and inserted into the lipid membrane of drug-loaded liposomes (Fig. 1b). Ad-C decorated liposomes are produced independently and can be used as plug-and-play components for modular assembly of targeted liposomes. Here, we describe the construction of Ad-liposomes and coupling to two vastly different recombinant proteins, human vascular endothelial growth factor (VEGF) expressed with N-terminal Cys-tag and a 254-aa long N-terminal fragment of anthrax lethal factor (LFn) carrying C-terminal Cys-tag. After "docking and locking," both proteins retain their full functional activity. We expect that the "dock and lock" strategy will provide new opportunities for decorating therapeutic and imaging liposomes with targeting proteins.

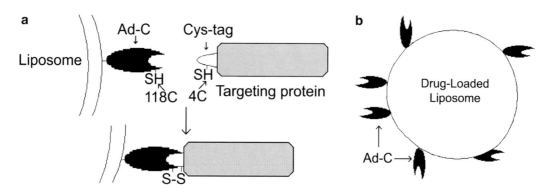

Fig. 1. The "dock and lock" system. (a) Two mutated fragments of human RNase I spontaneously form a complex where complimentary cysteines form a disulphate bond. (b), purified Ad-liposome can be coupled to a Cys-tagged targeting protein of choice.

2. Materials

2.1. Immobilization of Hu-peptide on Thiol-Sepharose 4B

1. Macro Spin, empty 2-mL columns with filter from Nest Group (Southborough, MA).
2. Hu-peptide, CA-extended: CA-KESRAKKFQRQHMDS, synthesized by Genemed Synthesis (South San Francisco, CA).
3. Tris(2-carboxyethyl)phosphine (TCEP) is from Pierce (Rockford, IL).
4. Activated Thiol-Sepharose 4B.
5. PD-10 gel-filtration columns from GE Healthcare (Barrington, IL).
6. 0.5 M Ethylendiamine tetraacetic acid (EDTA).
7. Trifluoroacetic acid (TFA).
8. Sodium monobasic phosphate.
9. Di-basic phosphate.
10. Sodium acetate.
11. 5 M NaCl solution (Invitrogen, Carlsbad, CA).
12. Stock buffer 1 M NaOAc, pH 6.5.
13. Stock buffer 1 M sodiumphosphate, pH 8.0.
14. Conjugation buffer: 50 mM sodiumphosphate, pH 8.0, 150 mM NaCl, 1 mM EDTA.

2.2. Limited Digestion of BH-RNase V188C with Subtilisin

1. BH-RNase V118C mutant, 10 mg/mL (SibTech, Inc. Brookfield, CT).
2. Subtilisin Carlsberg serine protease and protease inhibitor phenylmethylsulphonyl fluoride (PMSF) are from Sigma (St. Louis, MO).
3. Lauda water bath, or any waterbath operating at 4°C.
4. Fast Flow SP-Sephasose, 1-mL prepacked columns from GE Healthcare (Barrington, IL).
5. Buffers for ion-exchange chromatography on SP-column: Buffer A, 20 mM NaOAc pH 6.5, and Buffer B, 20 mM NaOAc pH 6.5, 1 M NaCl.

2.3. Modification of Ad-C with PEGylated Lipid

1. 1 M sodiumphosphate, pH 7.2.
2. Traut's reagent and dimethylsulfoxide (DMSO) are from Sigma (St. Louis, MO).
3. Poly(ethylenglycol)-α-Distearoyl Phosphatidylethanolamine,-ω-maleimide FW 3,400 (mPEG-DSPE-maleimide) from Shearwater Polymers (Huntsville, AL). Dissolve mPEG-DSPE-maleimide in DMSO at 10 mg/mL immediately before reaction.

2.4. Affinity Purification of Ad-C/ Lipid Conjugate and Insertion It into Doxil

1. Tween-20 and 1 M citric acid solution are from Fisher Scientific.
2. Binding buffer: 20 mM NaOAc pH 6.5, 150 mM NaCl.
3. Washing buffer: 20 mM NaOAc pH 6.5, 150 mM NaCl 0.1% Tween-20.
4. Elution solution: 0.1 M citric acid.
5. Doxil® (doxorubicin HCl liposome injection, 2 mg/mL) from Ortho Biotech.
6. 1 M HEPES buffer solution, pH 7.2 (Invitrogen).
7. Liposome running buffer: 10 mM HEPES pH 7.2, 0.15 M NaCl, 0.1 mM EDTA. Sterilize running buffer by filtration through 45 µm disposable filter.

2.5. Coupling Doxil-Ad-C to Cys-tag-Targeting Proteins

Cys-VEGF (2 mg/mL) and LFn-Cys (2 mg/mL) are from SibTech, Inc. (Brookfield, CT). Plasmids for bacterial expression of recombinant proteins with N- or C-terminal Cys-tag (KESCAKKFQRQHMDS) are commercially available from SibTech, Inc.

3. Methods

Making targeted liposomes via the "dock and lock" procedure includes two steps. The first step is the preparation of liposomes decorated with Ad-C (Fig. 1b). Purified Ad-liposomes can be stored in a refrigerator for several weeks without loss of functional activity, and be coupled, as needed, to a Cys-tagged targeting protein of choice. The second step, actual "docking and locking," is accomplished by simple mixing of Ad-liposomes and Cys-tagged protein. Covalent "locking" of the complex takes from several minutes to several hours, depending on the nature of targeting protein, and is effective for both N- and C-terminal Cys-tags.

Making Ad-liposomes includes limited digestion of BH-RNase V118C mutant with subtilisin, random lipidation of the resulting peptide fragments, selection of Ad-C–lipid conjugate capable of binding to Cys-tag and finally, insertion of the selected conjugate into preformed liposomes. To simplify the protocol and to maximize the yield of the final product, we do not purify the Ad-C fragment after subtilisin reaction. All BH-RNase fragments generated during subtilisin digestion are modified with Trout's reagent followed by SH-directed lipidation in a convenient "single-pot" format, without purification of intermediate products. Upon completion of lipidation, the entire reaction mixture is passed through an affinity column with immobilized Hu-peptide, which is a part of native human RNase I (N-terminal 1–15 amino acids). Native Hu-peptide binds the mutated RNase fragment

Ad-C noncovalently because it does not have a complimentary cysteine in position 4. By default, this column binds only those Ad-C molecules that retain the ability to bind to Cys-tag.

3.1. Immobilization of Hu-peptide on Activated Thiol-Sepharose 4B

1. Remove lyophilized CA-extended Hu-peptide from −20°C freezer and let it adjust to room temperature for 5–10 min. Centrifuge the tube with peptide for 1 min at $10,000 \times g$ to collect all powder at the bottom of the tube. Dissolve peptide in conjugation buffer to make a final concentration of 4 mg/mL.

2. To ensure accessibility of the thiol group in CA-extended Hu-peptide, add 1/10th volume of 1 M TCEP to a final TCEP concentration of 0.1 M, and incubate the mixture at room temperature for 30 min (see Note 1).

3. Equilibrate PD-10 column by passing through ten column volumes of conjugation buffer by gravity flow.

4. To prepare Activated Thiol-Sepharose 4B for coupling, gently resuspend the Sepharose solution, take out 4 mL and put it into a 50-mL conical tube. Add 20 mL of conjugation buffer, mix by inverting the tube several times upside-down and centrifuge at $4,000 \times g$ for 15 min. Carefully remove the supernatant, add 1 mL of conjugation buffer to the pellet, and resuspend it gently by pipeting.

5. Load peptide/TCEP reaction mixture on equilibrated PD-10 column, collect the peptide-containing peak that appears immediately after the column void volume, and mix the purified peptide with equilibrated Activated Thiol-Sepharose 4B at 1:1 (v/v) ratio. Incubate the peptide/Sepharose mixture at room temperature for 1 h with occasional gentle mixing.

6. Pour the peptide/Sepharose mixture into an empty 2-mL plastic column with filter, let it settle down and pass through ten column volumes of conjugation buffer by gravity flow to wash off unbound Hu-peptide. Place the column in a refrigerator and let it cool down for 2–4 h. The column is ready for affinity chromatography. For storage, pass through the column five column volumes of conjugation buffer supplemented with 10% EtOH.

3.2. Preparation of Ad-C for by Limited Digestion of BH-RNase V118C Mutant

1. Before starting subtilisin digestion, make all preparations for digestion and SP-column chromatography: set the water bath at 4°C, place on ice a 15-mL conical tube with sterile DI water for dissolving subtilisin, 1.5-mL microcentrifuge tubes for weighing and reconstitution of subtilisin, 1 mL of BH-RNase V118C, 0.13 mL 10× PMSF concentrate, and 0.13 mL 10% TFA. Equilibrate the SP-column with 5 mL of Buffer A, followed by 5 mL of buffer B, and finally, 10 mL of 15% (v/v)

Buffer B supplemented with 0.5% (v/v) TFA. When 4°C in the water bath is reached, prepare 10 mg/mL subtilisin in ice-cold DI water (see Note 2).

2. Add 0.1 mL of ice-cold subtilisin (10 mg/mL) to 1 mL of ice-cold BH-Rase V118C (10 mg/mL) to make a final w/w ratio of 1:10. Mix both components by pipeting up-and-down several times and place immediately in the 4°C water bath for 15 min.

3. Add 0.13 mL of ice-cold 10× PMSF and 0.13 mL of ice-cold 10% TFA to the reaction mixture, mix well by inverting and place the reaction tube on ice for 5 min (see Note 3). Save a 20-μL aliquot for SDS-PAGE.

4. Load the reaction mixture on the equilibrated SP-column. Sample loading and the following chromatography are done at room temperature. Collect the flow-through fraction and save a 20-μL sample for SDS-PAGE analysis.

5. Wash the column with 10 mL (ten column volumes) of 20% buffer B containing 0.5% (v/v) TFA to ensure complete removal of subtilisin. To remove TFA, wash the column with 10 mL 20% (v/v) Buffer B. Optional: save a 20-μL aliquot of wash for SDS-PAGE analysis.

6. Elute Ad-C/BH-RNase mixture by isocratic flow with 60% Buffer B. Typically, the volume of eluted fraction is 2.1 mL with a total protein concentration of 2 mg/mL and an average of 70% (3 mg) of Ad-C in the mixture. Analyze all collected fractions by reducing SDS-PAGE as shown in Fig. 2.

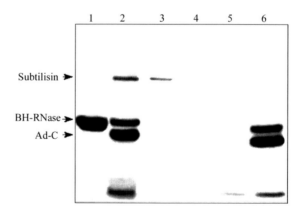

Fig. 2. Preparation of Ad-C by limited digestion of BH-RNase V118C mutant. Samples (20 μL volumes) were mixed with 2× Laemmli buffer, boiled for 2 min, centrifuged at 10,000 × g for 5 min, and loaded on 17.5% polyacrylamide gel, 20 μL each. After 1-h run at 200 V, gel was stained with Bio-Safe Coomassie (Bio-Rad Laboratories, Hercules, CA). Lane 1, initial BH-RNase V118C; lane 2, after 15 min of subtilisin digestion; lanes 3–6, SP-column chromatography; lane 3, flow-through fraction; lane 4, TFA-containing wash; lane 5, wash with 20% Buffer B, lane 6, elution with 60% Buffer B. Note complete removal of subtilisin from reaction (compare lane 2 and lane 3).

7. Wash the SP-column with 5 mL Buffer B, followed by 5 mL Buffer A, and finally, with 10 mL 20% EtOH for storage.

3.3. Modification with PEGylated Lipid

1. Add 1/10th volume of 1 M NaPi pH 7.2 to Ad-C/BH-RNase mixture eluted from SP-column to adjust the pH for Trout's reaction.

2. Weigh 2–4 mg of Trout's reagent and dissolve it in conjugation buffer to a final concentration of 7 mg/mL (50 mM, a 50× solution). Dilute the 50× solution with conjugation buffer to a final of 1 mM and add it to Ad-C/BH-RNase mixture to obtain a twofold molar excess of the reagent over protein (see Note 4). Incubate for 1 h at room temperature.

3. Weigh 2–4 mg of mPEG-DSPE-maleimide, and dissolve it in DMSO to make a final concentration of 10 mg/mL (2.9 mM). Add mPEG-DSPE-maleimide directly to Trout's reaction for a final protein-to-lipid molar ratio of 1:2. Incubate the reaction at room temperature for 1 h.

3.4. Affinity Purification of Lipid–Ad-C Conjugate and Insertion It into Doxil

1. During 1-h lipidation reaction, equilibrate the Hu-column by passing through ten volumes of cold binding buffer. Dilute lipidation reaction mixture tenfold with cold binding buffer and load it on Hu-column at a slow rate, not exceeding 0.1 mL/min in refrigerator (see Note 5).

2. Wash nonspecifically bound BH-RNase and its fragments by passing through five column volumes of washing buffer at 1 mL/min, and elute bound Ad-C with 0.1 M citric acid. Neutralize eluted protein immediately by mixing with 1/10th volume of 1 M NaPi pH 7.2.

3. Insertion into Doxil and purification of protein-decorated liposomes are done as recently described in great detail (8). This procedure usually results in concentrations of liposome-associated Ad-C in a micromolar range (2–10 µM).

3.5. Coupling Cys-tagged Protein to Ad-Liposome

1. Mix Ad-C-Doxil with 1 mg/mL Cys-tag protein at a 1:1 (v/v) ratio and incubate at RT. Depending on the 3D structure of targeting protein, the time required for complex formation may vary from 30 min to 12 h and should be optimized for each Cys-tagged protein.

2. To purify liposomes, pass them through Sepharose 4B column, as recently described in great detail (8). Purified liposomes can be analyzed by SDS-PAGE under reducing and non-reducing conditions (Fig. 3a) to ensure the coupling of Cys-tagged protein to Ad-C via a disulfide bond (see Note 6). Run an appropriate functional activity test to ensure that the targeting protein retains its functional activity after coupling (Fig. 3b, c).

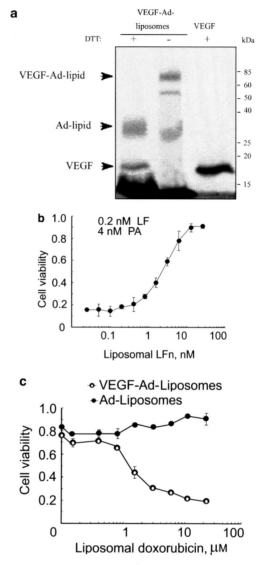

Fig. 3. Cys-tagged proteins conjugated to Ad-liposomes retain their functional activity. (a) SDS-PAGE analysis of coupling of Cys-tagged VEGF to Ad-liposomes. Note that the band corresponding to free VEGF is present only under reducing conditions (+DTT) and is absent in the absence of DTT. (b) Competition of LFn-Ad-liposome with the full-length LF in the presence of PA on RAW 264.7 mouse monocytes (ATCC, Rockville, MD) was done as described (7). LFn-Ad-liposomes successfully compete with the full-length LF for binding to ATR2/CMG2 receptor in RAW cells. In this 4-h long competition assay, LFn-Ad-liposomes rescued RAW cells from LF-induced toxicity with an IC_{50} value of 2 nM, which is equal to the activity of free LFn (7) indicating that liposomal LFn retains full functional activity. (c) Targeting Doxil with VEGF leads to efficient delivery into the cells expressing VEGF receptors (293/KDR cells). Cells were plated on 96-well plates, 1,000 cell/well. Twenty hours later, varying amounts of VEGF-Ad-Doxil or equivalent amounts of nontargeted Ad-Doxil were added to cells in triplicate wells. After a 2-h incubation at 37°C, liposome-containing media was removed, cells were shifted to complete culture medium, and allowed to grow for 96 h under normal culture conditions. Cells were quantitated by CellTiter 96® Cell Proliferation kit (Promega). (Reprinted with permission from ref. (7). Copyright 2006 American Chemical Society).

4. Notes

1. Due to the presence of N-terminal cysteine, even freshly prepared CA-Hu-peptide usually contains 25–50% dimers, therefore reducing with TCEP should be always done prior to mixing peptide with Activated Thiol-Sepharose. To maximize coupling to Sepharose and minimize peptide dimerization, the time between incubation with TCEP and coupling to Activated Thiol-Sepharose should be as short as possible. Therefore, we recommend preparing Sepharose during the 30 min of TCEP treatment. Hu-columns can be safely stored in a refrigerator for at least 1 year without loss in binding capacity.

2. Protein digestion by subtilisin is very fast, and to maintain the site-specific pattern of BH-RNase cleavage, it should be kept under strict temperature and time control. We recommend performing subtilisin cleavage in a 4°C water bath, rather than in a refrigerator, where the temperature can vary from 4 to 6°C. The difference of two degrees would lead to overdigestion of BH-RNase V118C, and as a result, to a substantial loss of Ad-C. For the same reason, keep all reagents, water, and plastic tubes for subtilisin reaction on ice. Lyophilized subtilisin is stored at −20° C, and should be adjusted to room temperature for 10–15 min before opening.

3. We found that the high subtilisin to BH-RNase ratio of 1:10 provides for the best yield of Ad-C. Typically, the level of selective BH-RNase cleavage is 70–75%, as judged by SDS-PAGE analysis (Fig. 2). Reaction conditions leading to 100% cleavage of BH-RNase V188C usually yield much less Ad-C. The addition of ice-cold PMSF and TFA followed by 5-min incubation on ice provides a partial inhibition of subtilisin, which allows running SP-column chromatography without substantial protein loss. SP-column chromatography provides for complete separation of subtilisin, because it does not bind to SP-column equilibrated with 15% (v/v) Buffer B and 0.5% (v/v) TFA (Fig. 2). As an additional advantage, eluted BH-RNase fragments are at a high enough concentration to proceed directly to Trout's reaction followed by lipidation.

4. Trout's reagent and mPEG-DSPE-maleimide are not stable in solution, and should be prepared immediately before the reaction. Since it is hardly reliable to weigh an amount of less than 2 mg, making solutions with high initial concentration of each compound is an inevitable intermediate step. The unused reagents should be disposed as soon as the reaction has been started. To calculate the amounts of the reagents necessary for the reaction, use the molecular weights of BH-RNase (14.3 kDa) and Ad-C (9.4 kDa), total protein concentration

in eluted fraction, and percentage of Ad-C estimated by SDS-PAGE gel. In our experience, such a rough estimation is enough, because both reagents are taken in excess and usually there is no problem with the described modifications.

5. Since Ad-C harbors the V118C mutation, it binds weakly to Hu-peptide, which is a fragment of native human RNase I. Therefore, to collect all reactive Ad-C from the lipidation reaction mixture, we load it on a Hu-column in a refrigerator, at the lowest possible flow rate, preferably reloading the flow-through several times. Make sure that the duration of loading is not less than 15–17 h (overnight). Note that both Ad-C–lipid and unmodified Ad-C will bind to Hu-column. Unmodified Ad-C will be separated later, after purification of Ad-liposomes by gel-filtration on Sepharose 4B.

6. Since Ad-C binds Cys-tag via disulfide bond, it is easy to follow by reducing and non-reducing SDS-PAGE. Under non-reducing conditions (in the absence of DTT in loading buffer), Ad-C/Cys-tagged protein conjugate migrates with an apparent molecular weight of roughly the sum of two components, while in the presence of DTT in loading buffer, Ad-C and Cys-tagged protein migrate as two separate bands according to their molecular weights, due to the reduction of disulfide bonds, in the conjugate (Fig. 3a).

References

1. Sun MM, Beam KS, Cerveny CG, Hamblett KJ, Blackmore RS, Torgov MY, Handley FG, Ihle NC, Senter PD, Alley SC (2005) Reduction-alkylation strategies for the modification of specific monoclonal antibody disulfides. Bioconjugate Chem 16:1282–1290

2. Hamblett KJ, Senter PD, Chace DF, Sun MM, Lenox J, Cerveny CG, Kissler KM, Bernhardt SX, Kopcha AK, Zabinski RF, Meyer DL, Francisco JA (2004) Effects of drug loading on the antitumor activity of a monoclonal antibody drug conjugate. Clin Cancer Res 10:7063–7070

3. Durek T, Becker CF (2005) Protein semi-synthesis: new proteins for functional and structural studies. Biomolecular Eng 22:153–172

4. Wood RJ, Pascoe DD, Brown ZK, Medlicott EM, Kriek M, Neylon C, Peter L, Roach PL (2004) Optimized conjugation of a fluorescent label to proteins via intein-mediated activation and ligation. Bioconjugate Chem 15:366–372

5. Li L, Olafsen T, Anderson AL, Wu A, Raubitschek AA, Shively JE (2002) Reduction of kidney uptake in radiometal labeled peptide linkers conjugated to recombinant antibody fragments. Site-specific conjugation of DOTA-peptides to a Cys-diabody. Bioconjugate Chem 13:985–995

6. Albrecht H, Burke PA, Natarajan A, Xiong CY, Kalicinsky M, DeNardo GL, DeNardo SJ (2004) Production of soluble ScFvs with C-terminal-free thiol for site-specific conjugation or stable dimeric ScFvs on demand. Bioconjugate Chem 15:16–26

7. Backer MV, Patel V, Jehning B, Backer JM (2006) Self-assembled "Dock and Lock" system for linking payloads to targeting proteins. Bioconjugate Chem 17:912–919

8. Backer MV, Levashova Z, Levenson R, Blankenberg FG, Backer JM (2008) Cysteine-containing fusion tag for site-specific conjugation of therapeutic and imaging agents to targeting proteins. In: Otvos L (ed) Peptide-based Drug Design. Methods in molecular medicine. Humana Press, New York, NY. doi:10.1007/978-1-59745-419-3_16.

Conjugation of Ligands to the Surface of Preformed Liposomes by Click Chemistry

Benoît Frisch, Fatouma Saïd Hassane, and Francis Schuber

Abstract

Click chemistry represents a new bioconjugation strategy that can be used to conveniently attach various ligands to the surface of preformed liposomes. This efficient and chemoselective reaction involves a Cu(I)-catalyzed azide-alkyne cycloaddition, which can be performed under mild experimental conditions in aqueous media. Here, we describe the application of a model click reaction to the conjugation, in a single step of unprotected α-1-thiomannosyl ligands, functionalized with an azide group to liposomes containing a terminal alkyne-functionalized lipid anchor. Excellent coupling yields were obtained in the presence of bathophenanthrolinedisulphonate, a water soluble copper-ion chelator, acting as a catalyst. No vesicle leakage was triggered by this conjugation reaction and the coupled mannose ligands were exposed at the surface of the liposomes. The major limitation of Cu(I)-catalyzed click reactions is that this conjugation is restricted to liposomes made of saturated (phospho)lipids. Efficient copper-free azide-alkyne click reactions are, however, being developed, which should alleviate this constraint in the future.

Key words: Liposome, Azide-alkyne cycloaddition, Bioconjugation chemistry, Click chemistry, Mannose

1. Introduction

Liposomes can be surface-modified with a variety of molecules that carry out a number of functions such as promoting the targeting of the vesicles to specific tissues, cell types and/or modulate, e.g., by PEGylation, their biodistribution and pharmacokinetic properties (1). Targeting, which represents a major issue to increase the specificity and efficiency of bioactive molecules (e.g., drugs, genes,...) delivery involves, in most cases, the use of ligands that are recognized by receptors expressed at the surface of target cells (2–4). These ligands are either relatively small molecules, such as folic acid, peptides or carbohydrate clusters, which trigger receptor-mediated endocytosis, or proteins such as monoclonal antibodies, and their fragments that are directed against specific antigens. The design of targeted liposomes is much dependent on the development of well–controlled bioconjugation reactions, and numerous methods have been developed for attaching ligands to the surface of liposomes; for reviews see (2, 5–8). They fall into two major categories: (1) conjugation of ligands to hydrophobic anchors and incorporation of the lipidated ligands into liposomes either during the preparation of the vesicles or by post-insertion into preformed vesicles (reviewed in (8)), or (2) covalent coupling of ligands to the surface of preformed vesicles that carry functionalized (phospho)lipid anchors. The most popular conjugations involve the reaction of thiolcontaining ligands with anchors carrying thiol-reactive functions such as maleimide, bromoacetyl, or 2-pyridyldithio linkages, generating thioether or disulfide bonds (9–11). Amide and carbamate bonds were also used, and more recently peptide ligands were coupled to the surface of liposomes via hydrazone and α-oxo hydrazone linkages (12). With the inception of sterically stabilized liposomes (13), ligands were also coupled at the distal end of a PEG spacer-arm linked to a hydrophobic anchor (1, 14, 15). Chemically controlled conjugation between preformed liposomes and ligands should ideally combine several features, such as mild reaction conditions in aqueous media, high yields and chemoselectivity. In this respect, the application of the "click chemistry" concept that involves a copper(I)-catalyzed Huisgen 1,3-dipolar cycloaddition reaction of azides and alkynes yielding 1,4-disubstituted 1,2,3-triazole linked conjugates (16), are very attractive for the development of new bioconjugation strategies. This reaction, which has attracted considerable interest during these recent years (for recent reviews see (17–19)), is particularly appealing because of its high regiospecificity, chemoselectivity and tolerance to a wide variety of other functional groups. In this Chapter, we describe the application of "click chemistry" to the conjugation in a single step of an unprotected α-D-mannosyl derivative carrying a spacer-arm functionalized with an azide group to the surface of liposomes that incorporate a synthetic lipid carrying a terminal triple bond (Scheme 1). Reaction conditions were optimized for this model reaction, and mannosylated vesicles were obtained in excellent yield. As assessed by agglutination experiments with Concanavalin A, the mannose residues were perfectly accessible on the surface of the vesicles and could engage into multivalent interactions. Thus, "click chemistry" can be added to the tool box of reactions available to the conjugation of ligands at the surface of carriers such as liposomes (20, 21) or nanoparticles (22). The limitations of "click chemistry" reactions in the field of liposomes that involve copper ions as catalysts are discussed and some perspectives are given.

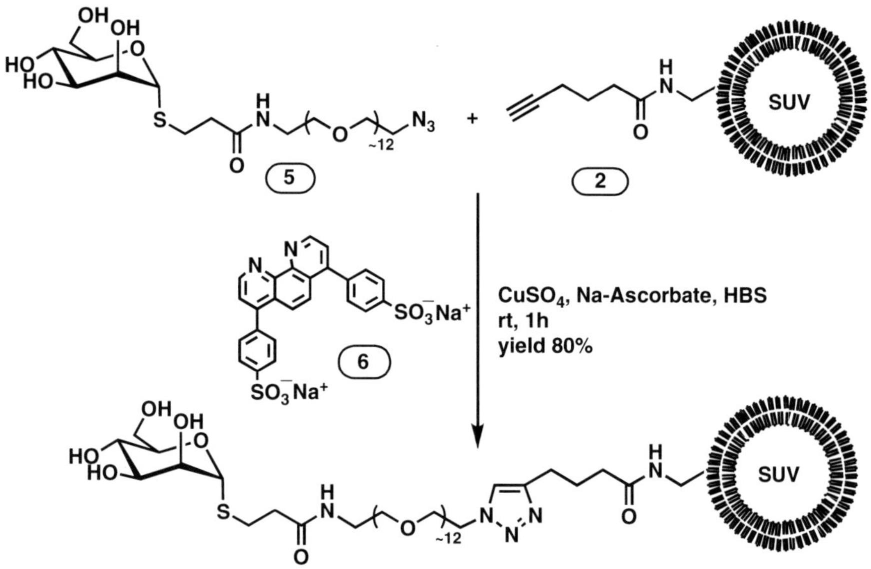

Scheme 1. Coupling by click reaction of an azido-functionalized mannosyl ligand to preformed liposomes that incorporate a terminal alkyne-functionalized lipid anchor

2. Materials

2.1. Synthesis

1. Dicyclohexylcarbodiimide (DCC), N-hydroxysuccinimide (NHS), 5-hexynoic acid and di-isopropylethylamine (DIEA) were purchased from Sigma-Aldrich (Saint-Quentin-Fallavier, France).
2. Dipalmitoylglycero-ethoxy-ethoxy-ethoxy-ethylamine (1) was synthesized as described previously (23).
3. Compound (3) was synthesized as outlined in (24).
4. Compound (4) was synthesized according to (25).

2.2. Liposome Preparation

1. 1,2-Dipalmitoyl-sn-glycero-3-phosphocholine (DPPC) and 1,2-dipalmitoyl-rac-glycero-3-phospho-rac-(1-glycerol) sodium salt (DPPG) (from Sigma-Aldrich) were stored at −20°C as solutions in chloroform/methanol (9/1 v/v). The purity of the phospholipids (over 99%) was assessed by TLC (see Note 1).
2. Cholesterol (Sigma-Aldrich) was recrystallized in methanol.
3. HBS: 10 mM HEPES, 145 mM NaCl, pH 6.5.
4. HBS-CF: 10 mM HEPES, 40 mM 5(6)-carboxyfluorescein, 100 mM NaCl, pH 6.5. Store at 4°C.
5. Sonicator equipped with a 3 mm diameter titanium probe (Vibra Cell, Sonics and Materials Inc., Danubury, CT).

6. The size of the liposomes was determined by dynamic light scattering using a Sub-micron Particle Analyzer (Coulter, Hialeah, Fl).

2.3. Azide-alkyne Coupling Reaction by "Click Chemistry"

1. l-Ascorbic acid sodium salt (Acros-Organics Noisy-le-Grand, France).
2. Bathophenanthrolinedisulphonic acid disodium salt (6) (Alpha Aesar, Strasbourg-Bischheim, France).
3. Solution A: 8 mM $CuSO_4·5H_2O$ in HBS (prepare fresh, store at 4°C).
4. Solution B: 145 mM sodium ascorbate in HBS (prepare fresh, store at 4°C).
5. Solution C: 28 mM compound 6 in HBS (prepare fresh, store at 4°C).

3. Methods

3.1. Synthesis of the Terminal Alkyne-functionalized Lipid Anchor (2) (Scheme 2)

1. DCC (52 mg, 0.25 mmol) and NHS (12 mg, 0.105 mmol) were added to a solution of 5-hexynoic acid (23 mg, 0.21 mmol) in CH_2Cl_2 (4 mL).
2. **1** (150 mg, 0.21 mmol) and DIEA (43 µL, 0.25 mmol) were then added to the mixture.
3. After 22 h of reaction at room temperature under stirring and under argon, the formed precipitate was removed by filtration and the organic phase was washed with 2 × 10 mL of a citric

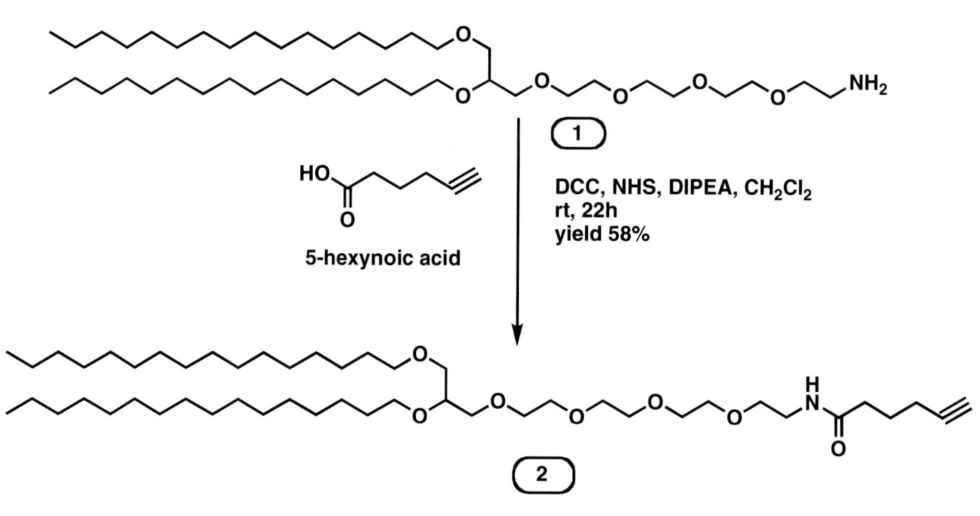

Scheme 2. Synthesis of the terminal alkyne-functionalized lipid anchor (**2**)

acid solution (5%, w/v) followed by 2 × 10 mL of brine. After passage over $MgSO_4$, the solvent was evaporated to dryness. The reaction product was obtained (98 mg; yield 58%) after purification by chromatography on silica gel eluted with CH_2Cl_2:AcOEt (9:1 to 7:3).

3.2. Synthesis of the Azido-functionalized Mannosyl Ligand (5) (Scheme 3)

1. To a solution of 3 (149 mg, 0.342 mmol) in 3 mL of CH_2Cl_2 were added DCC (85 mg, 0.41 mmol) and NHS (47 mg, 0.41 mmol).

2. After 45 min of stirring at room temperature under argon, the amine 4 (210 mg, 0.342 mol) in 1 mL of CH_2Cl_2 containing DIEA (47.5 µL, 0.273 mol) was added to the reaction mixture. After 18 h, 20 µL of DIEA and 0.2eq. of amine 4 were again added. The stirring was continued for 48 h.

3. The formed precipitate was then removed by filtration and the organic phase was washed with 2 × 10 mL of a citric acid solution (5%, w/v) followed by water. After passage over $MgSO_4$, the solvent was evaporated to dryness.

4. The reaction product 5 (129 mg; yield 30%, yellow oil) was obtained after purification by chromatography on silica gel eluted with CH_2Cl_2:MeOH (30:1). (See Note 2).

3.3. Liposome Preparation

1. Small unilamellar vesicles (SUV) were prepared by sonication. Briefly, phospholipids (DPPC, DPPG) and cholesterol (70/20/50 molar ratio) dissolved in chloroform/methanol (9:1, v/v) were mixed in a round-bottom flask.

2. For functionalized vesicles, 2 dissolved in chloroform/methanol (9:1, v/v) was added at given concentrations (between 5 and 10 mol%).

3. After solvent evaporation under high vacuum, the dried lipid film was hydrated by the addition of 1 mL HBS to obtain a final concentration of 10 µmol lipid/mL.

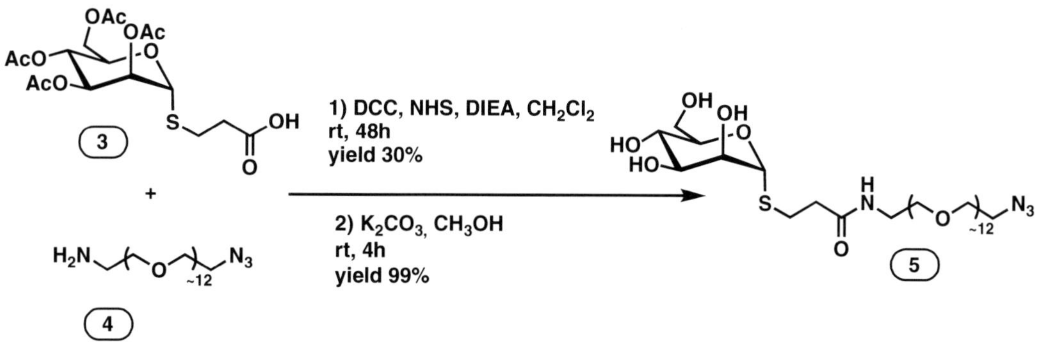

Scheme 3. Synthesis of the azido-functionalized 1-thiomannose derivative (5)

4. The mixture was vortexed and the resulting suspension was sonicated for 1 h at 60°C, i.e., above the T_m of the lipids, using a 3 mm diameter probe sonicator.

5. The liposome preparations were then centrifuged at $10,000 \times g$ for 10 min to remove the titanium particles originating from the probe.

3.4. Liposome Characterization

1. The size of the liposomes was determined by dynamic light scattering. The different vesicle preparations were homogenous in size and exhibited an average diameter between 90 and 130 nm.

2. The phospholipid content of liposomes was determined according to Rouser (26) with sodium phosphate as standard.

3.5. Conjugation of the Azido-functionalized Ligand to Liposomes by Click Reaction (Scheme 3)

1. To a 200 µL suspension in HBS of liposomes containing the alkyne-functionalized lipid anchor **2** that was adjusted to 1 mM of alkyne group (i.e., ~0.5 mM surface available alkyne group), were added in that precise order : sol. A (286 µL), sol. B (355 µL) followed by 164 µL of sol. C. Finally, 15 µL of 13.9 mM solution of **5** in water were added to the reaction mixture. (See Notes 3–5).

2. The reaction mixture was gently stirred under argon at room temperature for 1 h.

3. After the conjugation step, the liposomes were purified by exclusion chromatography on a 1×18 cm Sephadex G-75 (Amersham Biosciences) column equilibrated in HBS.

3.6. Quantification of the Conjugation Reaction

1. The mannose residues coupled to the liposomes were quantified by the resorcinol-sulfuric acid method (27). The standard curve was established as follows: to a 12.2×100 mm glass tube containing 2–12 µg of D-mannose in 200 µL water were added successively 20 µL of 0.1% (w/v) Triton X-100, 200 µL of a 6 mg/mL aqueous solution of resorcinol and 1 mL of 75% (v/v) H_2SO_4.

2. The tubes were vortexed and covered with aluminum foil before heating in a boiling water-bath for 12 min and then cooled to room temperature. The optical density was then recorded at 430 nm.

3. Aliquots of conjugated liposomes (about 0.4 µmol lipid) were first dried under vacuum using a Speed Vac.

4. HBS (200 µL) was then added to the tubes and the mixtures, after vortex mixing, were treated as indicated above for the standards. Before optical density reading the samples were filtered through a 0.45 µm PTFE filter. (See Note 7).

3.7. Liposome Stability under Coupling Conditions

1. Liposomes were prepared as above in HEPES-CF.
2. Non-encapsulated dye was eliminated by gel filtration (see above 3.5.2).
3. The dye-loaded liposomal suspensions were treated as above for the click chemistry coupling step.
4. The leakage of 5(6)-carboxyfluorescein was assessed by determining the increase of fluorescence (λ_{ex} 490 nm; λ_{em} 520 nm) observed in the presence of excess detergent. To measure the total fluorescence intensity corresponding to 100% dye release, Triton X-100 (0.1% w/v final) was added to the vesicles.
5. The percentage of dye release caused by the coupling conditions was calculated using the equation: $(F - F_0) \times 100 / (F_t - F_0)$, where F is the fluorescence intensity measured after exposing the vesicles to the coupling conditions and F_0, F_t are the intensities obtained before the coupling conditions and after Triton X-100 treatment respectively (9). F_t values were corrected for dilutions caused by the Triton X-100 addition. (See Notes 6, 8 and 9).

4. Notes

1. The purity of DPPC and DPPG was determined by TLC on silica gel plates (Merck 0.25 mm, Kieselgel 60F_{254}, 40–60 μm) eluted with respectively $CHCl_3$:MeOH:H_2O (65:25:4, v/v) and $CHCl_3$:MeOH:CH_3COOH:H_2O (25:15:4:2, v/v). Both phospholipids showed a single spot revealed either by a fluorescamine spray reagent, UV (254 nm) or by I_2 vapors.
2. Compound 5 was built with a relatively long PEG spacer arm of defined length (12 ethylene glycol units) (25). The purpose was to provide an optimal accessibility of the liposomal mannose ligands to their receptors (28). However, the PEG spacer arm length can be changed if needed, e.g., for stealth vesicles.
3. In contrast to the nearly quantitative click reactions observed between small molecular weight molecules carrying alkyne and azide functions in the presence of catalytic amounts of the ascorbate/$CuSO_4$ couple, in the present case, where the terminal alkyne was exposed at the surface of vesicles of ~100 nm diameter, much higher ascorbate/$CuSO_4$ concentrations were needed to achieve conjugation reactions. Nevertheless, even under these conditions, the coupling yields remained relatively modest, i.e., about 25% conjugation was observed in the presence of 2.3 mM $CuSO_4$ and 51.5 mM sodium ascorbate. This observation is in agreement with other works in literature

which showed that for bioconjugation reactions involving for example large molecular complexes, catalytic quantities of the ascorbate/$CuSO_4$ system are not enough to drive click reactions to completion (29).

4. To increase the yield of the click reaction, we have added a ligand of copper ions. Indeed, stabilization of Cu(I) oxidation state by specific heterocyclic ligands was shown to largely accelerate the 1,3-dipolar cycloaddition reaction between azides and alkynes(29). In our case, bathophenanthrolinedisulfonate (6) provided a highly water-soluble and potent chelate catalyst for the click reaction (29). When used in a twofold molar excess over copper, 6 allowed to reduce the agglutination of the vesicles, to largely increase the yield of mannosylation and to decrease the reaction time. For example, under standard conditions (24 h) the coupling yield increased from 23%, in the absence of 6, to nearly completion in its presence. Moreover, reaction times could also be decreased because within 1 h in the presence of 6, the observed conjugation of 5 was already about 80%. Altogether, we have routinely conjugated – in high yield – ligands such as 5 (using a twofold molar excess over surface exposed alkyne groups) to the surface of preformed liposomes carrying alkyne functions in the presence of $CuSO_4$/Na-ascorbate/6 (2.3, 50 and 4.6 mM) in an aqueous buffer (pH 6.5) at room temperature for 6 h (standard conditions).

5. In the liposome field, it is of importance to verify the integrity of the vesicles after the coupling steps. To that end, using a dynamic light scattering technique, we have first verified whether the reaction conditions described earlier altered the size of the vesicles. Some changes were noticeable using our standard conditions, i.e., about 50% diameter increases were noted even under control conditions in the absence of 6. However, we found that this effect could be efficiently limited just by changing the order of reactants added to the liposome suspensions. Thus, when ligand 6 was added before the $CuSO_4$/ascorbate mixture followed by the mannosyl ligand 5, no significant increase in size of the vesicles could be observed and, importantly, an identical yield of conjugation was measured.

6. The experimental conditions used for the click reaction could, in principle, provoke some leakage of the vesicles. To test this point, we have exposed to our standard coupling conditions the same type of liposomes having encapsulated self-quenching concentrations of 5(6)-carboxyfluorescein. Using well established methods based on fluorescence quenching determinations (9), we could demonstrate that essentially no leakage was triggered by the conjugation reaction.

7. To determine whether the conjugated mannose residues were easily accessible at the surface of the liposomes, we have

incubated suspensions of targeted vesicles, that contained 10 mol% of mannosyl residues exposed at their surface, in the presence of concanavalin A. Addition of the lectin resulted in a gradual increase of turbidity, as assessed by an increase of the OD at 360 nm which reached a plateau after about 15 min. In contrast, no absorbance change was observed with control liposomes. The subsequent addition of an excess of free D-mannose (5.0 mM) triggered a prompt decrease in turbidity confirming that the aggregation was due to a specific recognition of the mannose residues on the surface of the vesicles by the lectin that resulted in an aggregation due to multivalent interactions.

8. The use of copper catalysts in click chemistry could represent a limitation; indeed, vesicles made of unsaturated phospholipids are known to be readily (per)oxidized by copper ions in the presence of oxygen (i.e., via formation of reactive oxygen species) and to become leaky (30, 31). This restricts the application of the conjugation technique discussed in this Chapter to liposomes made of saturated phospholipids. Alternative means of promoting catalyst-free ligation reactions between azides and alkynes have been described recently. They involve an activation of the alkyne partner either in strain-promoted activation of cycloalkynes in [3+2] cycloaddition reactions with azides (32, 33) or 1,3-dipolar cycloadditions of azides with electron-deficient alkynes (34). Recently, a "traceless Staudinger ligation" based on the reduction of azides by phosphines via iminophosphorane intermediates has also emerged as a new ligation strategy (35). This bioconjugation reaction which has been used for example for the chemical synthesis of proteins could constitute an attractive extension for the click reactions to biological systems that are sensitive to the action of copper ions.

9. Another potential problem is "Cu(I) saturation" (18). In order for the click reaction to take place, the Cu(I)-acetylide complex intermediate must have physical contact with the azide. If this complex is however closely surrounded by other terminal alkynes the possibility exists that these alkynes will also chelate with the complex, thereby "saturating" it. This effectively prevents any azide group from reaching the complex and performing the displacement reaction. One example came from the work of Ryu and Zhao (36) which described a substrate containing four terminal alkynes in close proximity and that was unable to undergo a click reaction. However, when the alkynes were replaced by azide functional groups, the substrate readily reacted. In our case, Cu(I) saturation could also be invoked to explain vesicle aggregation under certain reaction conditions.

References

1. Torchilin VP (2006) Multifunctional nanocarriers. Adv Drug Deliv Rev 58:1532–1555
2. Forssen E, Willis M (1998) Ligand-targeted liposomes. Adv Drug Deliv Rev 29:249–271
3. Allen TM (2002) Ligand-targeted therapeutics in anticancer therapy. Nat Rev Cancer 2:750–763
4. Sapra P, Allen TM (2003) Ligand-targeted liposomal anticancer drugs. Prog Lipid Res 42:439–462
5. Schuber F (1995) Chemistry of ligand-coupling to liposomes. In: Philippot JR, Schuber F (eds) Liposomes as tools in basic research and industry. CRC Press, Boca Raton, pp 21–39
6. Nobs L, Buchegger F, Gurny R, Allemann E (2004) Current methods for attaching targeting ligands to liposomes and nanoparticles. J Pharm Sci 93:1980–1992
7. Torchilin VP (2005) Recent advances with liposomes as pharmaceutical carriers. Nat Rev Drug Discov 4:145–160
8. Schuber F, Said Hassane F, Frisch B (2007) Coupling of peptides to the surface of liposomes – Application to liposome-based synthetic vaccines. In: Gregoriadis G (ed) Liposome technology, 3rd edn. Informa Healthcare, New York, USA, pp 111–130
9. Barbet J, Machy P, Leserman LD (1981) Monoclonal antibody covalently coupled to liposomes: specific targeting to cells. J Supramol Struct Cell Biochem 16:243–258
10. Martin FJ, Papahadjopoulos D (1982) Irreversible coupling of immunoglobulin fragments to preformed vesicles. J Biol Chem 257:286–288
11. Schelté P, Boeckler C, Frisch B, Schuber F (2000) Differential reactivity of maleimide and bromoacetyl functions with thiols: application to the preparation of liposomal diepitope constructs. Bioconj Chem 11:118–128
12. Bourel-Bonnet L, Pecheur EI, Grandjean C, Blanpain A, Baust T, Melnyk O, Hoflack B, Gras-Masse H (2005) Anchorage of synthetic peptides onto liposomes via hydrazone and alpha-oxo hydrazone bonds. Preliminary functional investigations. Bioconj Chem 16:450–457
13. Immordino ML, Dosio F, Cattel L (2006) Stealth liposomes: review of the basic science, rationale, and clinical applications, existing and potential. Int J Nanomedicine 1:297–315
14. Zalipsky S, Mullah N, Harding JA, Gittelman J, Guo L, DeFrees SA (1997) Poly(ethylene glycol)-grafted liposomes with oligopeptide or oligosaccharide ligands appended to the termini of the polymer chains. Bioconj Chem 8:111–118
15. Sudimack J, Lee RJ (2000) Targeted drug delivery via the folate receptor. Adv Drug Deliv Rev 41:147–162
16. Kolb HC, Sharpless KB (2003) The growing impact of click chemistry on drug discovery. Drug Discov Today 8:1128–1137
17. Lutz JF, Zarafshani Z (2008) Efficient construction of therapeutics, bioconjugates, biomaterials and bioactive surfaces using azide-alkyne "click" chemistry. Adv Drug Deliv Rev 60:958–970
18. Hein CD, Liu XM, Wang D (2008) Click chemistry, a powerful tool for pharmaceutical sciences. Pharm Res 25:2216–2230
19. Moorhouse AD, Moses JE (2008) Click chemistry and medicinal chemistry: a case of "cyclo-addiction". Chem Med Chem 3:715–723
20. Said Hassane F, Frisch B, Schuber F (2006) Targeted liposomes: Convenient coupling of ligands to preformed vesicles using "click chemistry". Bioconj Chem 17:849–854
21. Cavalli S, Tipton AR, Overhand M, Kros A (2006) The chemical modification of liposome surfaces via a copper-mediated [3+2] azide-alkyne cycloaddition monitored by a colorimetric assay. Chem Commun (Camb):30 3193–3195
22. Sun EY, Josephson L, Weissleder R (2006) "Clickable" nanoparticles for targeted imaging. Mol Imaging 5:122–128
23. Espuelas S, Haller P, Schuber F, Frisch B (2003) Synthesis of an amphiphilic tetraantennary mannosyl conjugate and incorporation into liposome carriers. Bioorg Med Chem Lett 13:2557–2560
24. Ponpipom MM, Bugianesi RL, Robbins JC, Doebber TW, Shen TY (1981) Cell-specific ligands for selective drug delivery to tissues and organs. J Med Chem 24:1388–1395
25. Iyer SS, Anderson AS, Reed S, Swanson B, Schmidt JG (2004) Synthesis of orthogonal end functionalized oligoethylene glycols of defined lengths. Tetrahedron Lett 45:4285–4288
26. Rouser G, Fleisher J, Yamamoto A (1970) Two dimensional thin layer chromatographic separation of plant lipids and determination of phospholipids by phosphorus analysis of spots. Lipids 5:494–496

27. Monsigny M, Petit C, Roche AC (1988) Colorimetric determination of neutral sugars by a resorcinol sulfuric acid micromethod. Anal Biochem 175:525–530
28. Engel A, Chatterjee SK, Alarifi A, Riemann D, Langner J, Nuhn P (2003) Influence of spacer length on interaction of mannosylated liposomes with human phagocytic cells. Pharm Res 20:51–57
29. Lewis WG, Magallon FG, Fokin VV, Finn MG (2004) Discovery and characterization of catalysts for azide-alkyne cycloaddition by fluorescence quenching. J Am Chem Soc 126:9152–9153
30. Gal S, Pinchuk I, Lichtenberg D (2003) Peroxidation of liposomal palmitoyllinoleoylphosphatidylcholine (PLPC), effects of surface charge on the oxidizability and on the potency of antioxidants. Chem Phys Lipids 126:95–110
31. Lee LV, Mitchell ML, Huang S-J, Fokin VV, Sharpless KB, Wong C-H (2003) A potent and highly selective inhibitor of human a-1, 3-fucosyltransferase via click chemistry. J Am Chem Soc 125:9588–9589
32. Baskin JM, Prescher JA, Laughlin ST, Agard NJ, Chang PV, Miller IA, Lo A, Codelli JA, Bertozzi CR (2007) Copper-free click chemistry for dynamic in vivo imaging. Proc Natl Acad Sci USA 104:16793–16797
33. Ning X, Guo J, Wolfert MA, Boons GJ (2008) Visualizing metabolically labeled glycoconjugates of living cells by copper-free and fast Huisgen cycloadditions. Angew Chem Int Ed Engl 47:2253–2255
34. Li Z, Seo TS, Ju J (2004) 1,3-Dipolar cycloaddition of azides with electron-deficient alkynes under mild condition in water. Tetrahedron Lett 45:3143–3146
35. Tam A, Soellner MB, Raines RT (2007) Water-soluble phosphinothiols for traceless staudinger ligation and integration with expressed protein ligation. J Am Chem Soc 129:11421–11430
36. Ryu EH, Zhao Y (2005) Efficient synthesis of water-soluble calixarenes using click chemistry. Org Lett 7:1035–1037

Chapter 19

Targeted Magnetic Liposomes Loaded with Doxorubicin

Pallab Pradhan, Rinti Banerjee, Dhirendra Bahadur, Christian Koch, Olga Mykhaylyk, and Christian Plank

Abstract

Targeted delivery systems for anticancer drugs are urgently needed to achieve maximum therapeutic efficacy by site-specific accumulation and thereby minimizing adverse effects resulting from systemic distribution of many potent anticancer drugs. We have prepared folate receptor targeted magnetic liposomes loaded with doxorubicin, which are designed for tumor targeting through a combination of magnetic and biological targeting. Furthermore, these liposomes are designed for hyperthermia-induced drug release to be mediated by an alternating magnetic field and to be traceable by magnetic resonance imaging (MRI). Here, detailed preparation and relevant characterization techniques of targeted magnetic liposomes encapsulating doxorubicin are described.

Key words: Targeted, Magnetic nanoparticles, Magnetic liposomes, Folate, Doxorubicin

1. Introduction

Chemotherapy is the standard treatment for many cancers. However, the full potential of many anticancer drugs cannot be exploited due to their severe side effects. A delivery system that is selective for tumors might potentiate the therapeutic efficacy of anticancer drugs and minimize side effects.

Liposomes have emerged as efficient drug delivery systems for anticancer agents. A liposomal formulation of doxorubicin, Caelyx®, is used in routine clinical use for cancer treatment (1, 2). Compared with free doxorubicin, Caelyx® provides preferred accumulation in tumors and consequently reduced side effects. Caelyx® is a non-targeted stealth liposome formulation of encapsulated doxorubicin. Due to a long circulation half-life, Caelyx® accumulates in tumors by the EPR (enhanced permeability and retention) effect in higher amounts than free doxorubicin given at similar doses.

However, Caelyx® has still some side effects like stomatitis, hand-foot syndrome, mild myelosuppression and vomiting, which is due to uncontrolled biodistribution of Caelyx® to healthy organs (3, 4). Hence, there is a continued need of targeted delivery systems for doxorubicin and other drugs.

Folate receptor-mediated targeting of liposomal anticancer drugs e.g., doxorubicin or paclitaxel has been found to be promising (5–8). Magnetic liposomes (magnetic nanoparticles encapsulated within liposomes) appear to be a versatile delivery system due to biocompatibility, chemical functionality and their potential for combination of drug delivery and hyperthermia treatment of cancers (9, 10). Magnetic liposomes encapsulating anti cancer drugs can be physically targeted to tumors by magnetic fields (using the principle of MDT, magnetic drug targeting) (11, 12). Also, along with drug delivery, magnetic liposomes can be further used for magnetic hyperthermia (with exposure of the target tissue to alternating current magnetic fields), which itself has been found to be very effective for cancer treatment (13, 14). The bio-distribution of the magnetic liposomes can also be monitored by magnetic resonance imaging (MRI) due to the T2 shortening effect of magnetic nanoparticles encapsulated within liposomes (15). Our research interest is to develop multifunctional magnetic liposomal formulations of the anticancer drugs like doxorubicin, which can be targeted to tumors both physically by magnetic field and biologically via the folate receptor and thereby to get a better therapeutic efficacy due to selective drug distribution to the tumor. Furthermore, such liposomes would be suitable for combined therapy by magnetic hyperthermia and chemotherapy and provide the known synergistic effects of the two treatment modalities (16, 17).

Here, we describe a detailed procedure for the preparation of the folic acid tagged doxorubicin magnetic liposomes and their relevant characterizations and few representative results.

2. Materials

2.1. Synthesis and Characterization of DSPE-PEG$_{2000}$-Folate

1. Folic acid
2. Anhydrous dimethyl sulfoxide (DMSO)
3. 1,2-Distearoyl-*sn*-glycero-3-phosphoethanolamine-*N*-(amino(polyethylene glycol)2000) (DSPE-PEG$_{2000}$-amine, Avanti Polar Lipids, USA)
4. Dicyclohexylcarbodiimide (DCC)
5. Pyridine

6. Ninhydrin
7. Silica Gel TLC plate (Kieselgel 60 F_{254}, Merck, Darmstadt, Germany)
8. Ethanol
9. 0.22 μ PTFE syringe filter for organic solvents (Sartorius Minisart SRP 25 hydrophobic, Sartorius, Germany)
10. Spectra/Por Dialysis Membrane (MWCO 1000, Spectrum Laboratories, Inc., USA)
11. Sodium chloride
12. Reverse phase HPLC column: Vydac 214TP510 protein C4 column 25 × 1 cm (Grace, USA)
13. Methanol
14. Sodium phosphate

2.2. Preparation of Doxorubicin and Magnetic Nanoparticles Loaded Liposomes

1. Doxorubicin (Sigma-Aldrich, USA) – wear protective gloves and glasses before handling doxorubicin as it is very toxic
2. Ammonium sulphate
3. DPPC (1, 2-dipalmitoyl-*sn*-glycero-3-phosphocholine, Avanti Polar Lipids, USA)
4. Cholesterol (Avanti Polar Lipids, USA)
5. DSPE-PEG$_{2000}$ (1, 2-distearoyl-*sn*-glycero-3-phosphoethanolamine-*N*-(methoxy(polyethylene glycol)-2000), Avanti Polar Lipids, USA)
6. Chloroform
7. FluidMag-HS (Chemicell, Germany)
8. 400 nm, 200 nm and 100 nm polycarbonate membranes (Avanti Polar Lipids, USA)
9. Extruder (Avanti Polar Lipids, USA)
10. Phosphate buffered saline (PBS) –20× buffer is prepared by dissolving 16 g sodium chloride (NaCl), 0.4 g potassium chloride (KCl), 2.88 g di-sodium hydrogen phosphate (Na_2HPO_4), 0.48 g potassium di-hydrogen phosphate (KH_2PO_4) 100 ml autoclaved deionized water. Twenty times PBS is diluted to 1× using autoclaved deionized water before use.
11. Sephadex G50 (Sigma-Aldrich, USA)

2.3. Determination of Doxorubicin and Phospholipid Concentration in Liposomes

1. Triton-X 100 (AppliChem, Germany)
2. Microcentrifuge tubes (Eppendorf, Germany)
3. Ninety-six well cell culture plate (Techno Plastic Products, Switzerland)
4. Phospholipid assay kit (Wako chemicals, Germany)

2.4. Determination of the Iron Concentration in FluidMag-HS and Magnetic Liposomes

1. 10% hydroxylamine hydrochloride solution was prepared in deionised water.
2. Ammonium acetate buffer: Dissolve 25 g ammonium acetate in 10 ml deionised water (see Note 1), then add 70 ml glacial acetic acid and fill up 100 ml with deionised water.
3. 0.1% Phenanthroline solution: Dissolve 100 mg 1,10-phenanthroline monohydrate in 100 ml deionised water and then add 2 drops of concentrated hydrochloric acid (Fluka, USA) and warm solution (if necessary) to get a clear solution.
4. 0.05 N $KMnO_4$ solution: Dissolve 0.790 g $KMnO_4$ (MW-158.03) in 100 ml deionised water.
5. Stock iron solution: Dissolve 392.8 mg ammonium iron (II) sulfate hexahydrate in a mixture of 2 ml concentrated sulfuric acid and 10 ml distilled water and add 0.05 N $KMnO_4$ dropwise till a faint pink colour persists and adjust the volume to 100 ml with deionised water.
6. Standard iron solution: Dilute stock iron solution 1:50 with deionised water just before calibration measurements.

2.5. In vitro Evaluation of Targeting Efficiency by Fluorescence Activated Cell Sorting

1. KB (human epidermoid carcinoma) and HeLa (human cervical carcinoma) cell lines are purchased from DSMZ, Germany.
2. Roswell Park Memorial Institute 1640 medium (RPMI 1640) w/o folic acid (Sigma-Aldrich, USA)
3. FCS (fetal calf serum, Biotech)
4. Penicillin/streptomycin (Biochrom, Germany)
5. Twenty-four well cell culture plates (Techno Plastic Products, Switzerland)
6. Twenty-four well plate format magnetic plate (OZ Biosciences, France or chemicell, Germany)
7. PBS-Dulbecco's without Ca^{2+} and Mg^{2+} (PBS, Biochrom, Germany)
8. Trypsin-EDTA solution (0.25%, Biochrom, Germany)
9. Fluorescence Activated Cell Sorting (FACS) buffer – PBS-Dulbecco's solution supplemented with 1% FCS

2.6. In vitro Evaluation of Targeting Efficiency of Folate Receptor Targeted Liposomes by Spectrofluorometry

1. Commercially available doxorubicin liposome (Caelyx®, Schering-Plough, Australia)
2. Bicinchoninic acid (BCA) protein assay kit (Pierce, USA)

2.7. Cytotoxicity Study

1. MTS assay kit (The Promega Cell Titer 96® Aqueous Non-Radioactive Cell Proliferation Assay (MTS), Promega, Germany).
2. Ninety-six well format magnetic plate (OZ Biosciences, France or chemicell, Germany)

3. Methods

3.1. Synthesis and Characterization of DSPE-PEG$_{2000}$-Folate

1. 1,2-Distearoyl-*sn*-glycero-3-phosphoethanolamine-*N*-((polyethylene glycol) 2000)-folate (DSPE-PEG$_{2000}$-folate) is synthesized by a previously described method (18).

2. First, 12.5 mg of folic acid is completely dissolved in 1 ml of anhydrous DMSO (see Note 2).

3. 50 mg DSPE-PEG$_{2000}$-amine and 16.25 mg of DCC and 250 µl of pyridine are added to the DMSO solution of folic acid.

4. The reaction is allowed to proceed for 4 h at room temperature.

5. Progression of the reaction is monitored by the disappearance of a ninhydrin positive DSPE-PEG$_{2000}$-amine spot on a silica gel TLC plate. A 2 µl drop of the reaction mixture is applied to a TLC plate. The plate is immersed in a 5% solution of ninhydrin in ethanol and heated on a heating plate set to 100°C or with a hair dryer.

6. The pyridine is removed by rotary evaporation and 6.25 ml of deionized water is then added to the solution. Then the solution is filtered through a syringe filter having 0.22 µm PTFE membrane to remove insoluble by-products.

7. The filtered solution is dialyzed with a Spectra/Por dialysis membrane (MWCO 1000) twice against 2,000 ml of 50 mM sodium chloride and thrice against 2,000 ml of deionised water for a total of 72 h time.

8. The solution is then lyophilized and stored at −20°C.

9. The final product is analyzed by reverse phase HPLC (Biocad Sprint, USA). The HPLC machine was equipped with a Vydac C4 column. Analysis of DSPE-PEG$_{2000}$-folate is performed with isocratic elution using a solvent mixture of 92:8 v/v methanol and 10 mM sodium phosphate (pH 7) at a flow rate of 1 ml/min (18).

3.2. Preparation of Doxorubicin and Magnetic Nanoparticles Loaded Liposomes

1. Co-encapsulation of magnetic nanoparticles and doxorubicin within the liposome is done by the ammonium sulphate gradient method with some modifications (6).

2. 20 mg of total lipid of DPPC/Cholesterol/DSPE-PEG$_{2000}$/DSPE-PEG$_{2000}$-Folate at 80:20:4.5:0.5 molar ratio (this particular composition has been optimized for a thermo-sensitive liposome formulation using a calcein release assay) is dissolved in chloroform/methanol (2:1 v/v) solution in a round bottom flask and the organic solvent is evaporated under vacuum in a rotary evaporator.

3. A thin lipid film can be observed following the evaporation of the solvents. To remove the traces of the solvents completely,

the round bottom flask containing thin lipid film is put under high vacuum for one more hour.

4. An aqueous magnetic fluid (FluidMag-HS, see Note 3) containing 5 mg of total iron in 1 ml of 250 mM ammonium sulphate (pH 4–4.5, see Note 4) is added to the round bottom flask with the thin lipid film and rotated at 120 rpm in the rotary evaporator at 60°C for 15–20 min for hydration. The magnetic liposomes thus formed are large in size (>1 µM).

5. Downsizing of the liposomes is done through sequential extrusion through 400 nm, 200 nm and 100 nm polycarbonate membranes (11 times for each of the membranes) using an extruder ((Avanti Polar Lipids, USA, see Note 5).

6. Then, the ammonium sulphate outside the liposomes is exchanged with PBS (see Note 6) by gel filtration through a sephadex G50 column. For this purpose, Sephadex G50 bulk material is suspended in water and filled in a GE Healthcare, previously Pharmacia, HR 10/30 column. After equilibrating the column with PBS, the sample is separated on the column with PBS as eluting buffer at a flow rate of 1 ml/min.

7. The product fraction from the gel filtration is transferred to a round bottom flask and preheated to 60°C using the water bath of a rotary evaporator. Subsequently, doxorubicin hydrochloride is added to the preheated suspension at a weight ratio of 1:10 with respect to the amount of original total lipid used for the liposome preparation. The round bottom flask is rotated using the rotary evaporator at 90 rpm for 1 h at 60°C.

8. Thereafter, the liposomes suspension is centrifuged at 1,000×g for 15 min to remove unencapsulated magnetic nanoparticles (12).

9. Unencapsulated doxorubicin is removed from the liposomes through sephadex G50 gel filtration (same conditions as mentioned earlier) using PBS as running buffer at a flow rate of 1 ml/min (see Note 7).

10. The magnetic liposome containing doxorubicin thus prepared is stored at 4°C and is used for different in vitro studies within the 2–3 weeks.

11. The magnetic responsiveness of the folate targeted doxorubicin containing magnetic liposomes (sample code – MagFolDox) can be observed using fluorescence microscopy (Zeiss Axiovert 135; Carl Zeiss, Germany). For this purpose, the magnetic liposomes are diluted 100 times with PBS and then a drop of the diluted magnetic liposomes is put on a clean glass slide and observed under 40× magnification. A small magnet (e.g., a neodymium-iron-boron permanent magnet) is placed on the glass slide next to the drop. Then magnetic liposomes can be seen rushing towards the magnet and accumulate at the air-liquid interface of the drop

next to the magnet. In phase contrast mode, the liposomes appear brown because of the presence of magnetic particles. In fluorescence mode, the liposomes can be seen by their red doxorubicin fluorescence. Typical photographs of the magnetic liposome accumulation under magnetic field are shown in Fig. 1.

3.3. Determination of Doxorubicin and Phospholipid Concentration in Liposomes

1. 10–20 µl of liposome sample is added to 1 ml of 1% Triton-X 100 in a microcentrifuge tube and vortexed for 30 s to 1 min.
2. The microcentrifuge tube is incubated for 1 h at 60°C.
3. The tube is again vortexed shortly and the microcentrifuge tube is allowed to cool down to room temperature.
4. Thereafter, 200 µl of the doxorubicin containing triton-X 100 solution is added to at least three wells for each sample of a flat bottom 96 well cell culture plate. Only 1% triton-X 100 is taken as a blank. The fluorescence intensities are measured using a spectrofluorimeter (Wallac Victor 2 Multi-label Counter, PerkinElmer) using 485 nm excitation and 590 nm emission filters.
5. The concentration of doxorubicin in the liposome samples is determined from a calibration curve which is obtained from a dilution series of doxorubicin in 200 µl 1% triton-X 100. Then, fluorescence intensity is plotted against the doxorubicin

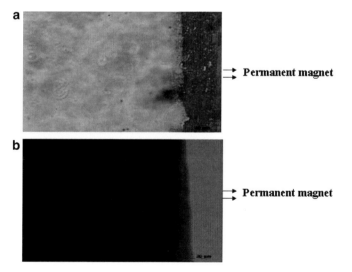

Fig. 1. Accumulation of folate targeted doxorubicin containing magnetic liposomes (MagFolDox) under the influence of a magnetic field. The magnetic liposomes accumulate at the air liquid interface next to a permanent magnet as can be observed from the photographs taken in (**a**) phase contrast as well as (**b**) fluorescence mode. This confirms that the magnetic liposomes respond well to the magnetic field, which is a prerequisite for magnetic drug targeting. The material accumulating next to the magnet is red fluorescent, confirming that the material contains doxorubicin

concentration. The doxorubicin concentration in the liposome samples is calculated from the linear fit of the calibration curve.

6. The phospholipid concentration of the liposomes is determined using a phospholipid assay kit (Wako Chemicals, Germany) according to the instructions of the manufacturer.

7. The doxorubicin encapsulation efficiency is calculated using the following formula:

$$\text{dox encapsul}(\%) = \frac{w/w \text{ dox / phospholipid after encapsulation}}{\text{initial } w/w \text{ ratio dox / phospholipid}} \times 100$$

3.4. Determination of the Iron Concentration in FluidMag-HS and Magnetic Liposomes

1. Iron concentration of the MagFluid-HS and magnetic liposomes is determined spectrophotometrically by complexation with 1,10-phenanthroline using a reported method with slight modifications (19).

2. Preparation of a calibration curve for iron determination.

3. Increasing amounts of standard iron solution are added to microcentrifuge tubes (e.g., 50, 70, 90 up to 150 μl) and the volume is adjusted to 150 μl with deionised water.

4. For blank, 150 μl deionised water are used instead of standard iron solution.

5. To each tube, 20 μl concentrated hydrochloric acid, 20 μl 10% hydroxylamine hydrochloride solution, 200 μl ammonium acetate buffer, 80 μl 0.1% 1,10-phenanthroline solution and 730 μl deionised water are added.

6. After vortexing, the samples are incubated for 20 min at room temperature.

7. The absorbance is measured at 510 nm against the blank.

8. The absorbance at 510 nm is plotted as a function of the iron concentration in the standard samples. A linear fit is done to get the formula to calculate the iron concentration in the magnetic fluid and magnetic liposome samples.

9. To measure the iron concentration in fluidMag-HS, 10–20 μl of magnetic fluid are mixed with 200 μl concentrated hydrochloric acid and 50 μl deionized water. When the magnetic nanoparticles are dissolved (see Note 8), the volume is adjusted to 5 ml with deionized water. For magnetic liposome samples, 50–100 μl of liposome samples are mixed with 200 μl concentrated hydrochloric acid and incubated until the magnetic liposomes are completely dissolved (see Note 9).

10. Twenty micro litre of the above solutions are added to microcentrifuge tubes followed by the addition of 20 μl concentrated hydrochloric acid, 20 μl hydroxylamine hydrochloride solution, 200 μl ammonium acetate buffer, 80 μl 1,10-phenanthroline solution and 860 μl deionized water. The samples are mixed by vortexing and incubated for 20 min at room temperature.

11. A blank sample is prepared by mixing 20 μl concentrated hydrochloric acid, 20 μl hydroxylamine hydrochloride solution, 200 μl ammonium acetate buffer, 80 μl 1,10-phenanthroline solution and 880 μl deionized water.

12. The absorbance of the samples at 510 nm is measured against the blank using a spectrophotometer (here: Beckman DU 640 spectrophotometer).

13. Using the calibration curve, the iron concentrations of magnetic fluid and magnetic liposome samples are calculated.

14. The magnetic nanoparticle encapsulation efficiency in the magnetic liposomes is calculated using the following formula (*MNP* magnetic nanoparticles).

$$\text{MNP encapsul}(\%) = \frac{w/w \text{ MNP / phospholipid after encapsulation}}{\text{initial } w/w \text{ ratio MNP / phospholipid}} \times 100$$

3.5. In vitro Evaluation of Targeting Efficiency by FACS

1. The in vitro targeting efficiency of folate targeted magnetic liposomes is evaluated on KB (human epidermoid carcinoma) and HeLa (human cervical carcinoma) cell lines. Both cell lines are known to highly express the α-folate receptor, which has also been confirmed in our laboratory. The two cell lines are cultured in RPMI 1460 media without folic acid (see Note 10) but supplemented with 10% FCS, 100 U/ml penicillin and 100 μg/ml streptomycin at 37°C in 5% CO_2 atmosphere. The cells are split every 3–4 days with a split ratio of 1:3 to 1:4. All the experiments are done with cells of passage numbers below 25. Different samples (at the concentration of 25 μM liposomal doxorubicin) are added.

2. KB/HeLa cells are seeded in 24 well culture plates at a density of 3×10^5 cells per well and cultured overnight.

3. The samples to be examined (doxorubicin containing liposomes) are added to result in a doxorubicin concentration in the cell culture supernatant of 25 μM.

4. The plates are incubated for 2 h at 37°C in the CO_2 incubator.

5. To evaluate the effect of magnetic field targeting of folate targeted magnetic liposomes, the culture plate with the respective samples is positioned on a 24 well format magnetic plate.

6. After 2 h incubation, the cells are washed twice with PBS-Dulbecco's, detached with trypsin-EDTA solution and pelleted at 1,200 g for 3 min.

7. The cell pellets are resuspended in FACS buffer and re-centrifuged twice.

8. Finally, the cell pellet is suspended in 500 µl FACS buffer and analysed for doxorubicin fluorescence using a FACS instrument (here: FACS Vantage microflow cytometer, Beckton Dickinson) using the 488 nm line of an air cooled argon laser as the excitation source. Fluorescence from cell associated doxorubicin is detected using a 550 nm long pass emission filter.

3.6. In vitro Evaluation of Targeting Efficiency of Folate Receptor Targeted Liposomes by Spectrofluorometry

1. KB/HeLa cells are seeded in 24 well culture plates at a density of 4×10^5 cells per well and are grown overnight at 37°C in a CO2 incubator.

2. Different doxorubicin liposome samples including a commercially available doxorubicin liposome (Caelyx®) and free doxorubicin are added to result in a final doxorubicin concentration of 50 µM in the cell culture supernatant.

3. In free folate competition studies (see Note 11), 1 mM folic acid (see Note 12) is added to the incubation medium.

4. The plates are incubated for 2 h at 37°C in the CO_2 incubator.

5. As described before, cells incubated with magnetic liposome formulations are positioned on a 24 well format magnetic plate during the first hour of incubation to see the effects of magnetic field on targeting.

6. After a total of 2 h incubation, cells are washed twice with PBS-Dulbecco's and then detached with trypsin-EDTA (0.25%) solution. The cells are pelleted at 1,200 g for 3 min and are resuspended in PBS. This process is repeated twice.

7. After discarding supernatant, the cells are lysed with 1 ml 1% triton X-100.

8. Then, 200 µl of cell lysate each is transferred to at least three different wells of a flat bottom 96 well cell culture plate and the fluorescence intensity is measured for doxorubicin fluorescence in a spectrofluorimeter using 485 nm excitation and 590 nm emission filters.

9. The amount of doxorubicin is calculated using a pre-determined standard curve derived from a serial dilution of free doxorubicin.

The amount of doxorubicin per cell can be calculated from a pre-determined calibration curve where the cellular protein content from cell lysates is plotted against the cell number. The cellular protein content is determined using the bicinchoninic acid (BCA) protein assay (Pierce) according to the instructions of the manufacturer.

3.7. In vitro Evaluation of Targeting Efficiency of Folate Receptor Targeted Liposomes by Fluorescence Microscopy

1. For this study, a similar protocol is followed as for the spectrofluorimetric study (steps 1–5 mentioned earlier).

2. However, after 2 h of incubation, the cells are washed twice with PBS-Dulbecco's and are visualized under a fluorescence microscope (here: Zeiss Axiovert 135; Carl Zeiss, Germany) using the rhodamine filter.

3. For the cells treated with magnetic liposomes under 1 h magnetic field exposure, the cells are trypsinized and centrifuged twice additionally with PBS-Dulbecco's to remove loosely adhered magnetic liposomes on the cells.

4. Photographs of the cells are taken in both phase contrast and fluorescence mode using identical exposure times to compare visually the fluorescence intensities of the cells (see Fig. 2).

3.8. Cytotoxicity Study

1. KB/HeLa cells are seeded in flat bottom 96 well cell culture plates at a density of 2×10^4 cells per well and are incubated over night before the addition of different samples.

2. The medium is replaced with serially diluted liposome samples and free doxorubicin (from 0.7 to 50 µM doxorubicin) in 200 µl fresh media.

3. Then the plates are incubated for 2 h at 37°C in the CO_2 incubator.

4. To evaluate the cytotoxicity of folate targeted magnetic liposomes under permanent magnetic field, the plate with magnetic liposomes is put on a 96 well format magnetic plate for the first hour of incubation by another hour incubation without magnetic field (total 2 h incubation).

5. Thereafter, the cells are washed two times with PBS-Dulbecco's, and finally 200 µl fresh media is added. The cells are then cultivated for further 46 h.

6. After a total 48 h, the cell viability is measured by MTS assay using a commercial kit. For this purpose, the cell culture supernatants are replaced with 100 µl each of fresh medium containing 10 µl of "96® Aqueous One Solution Reagent" each.

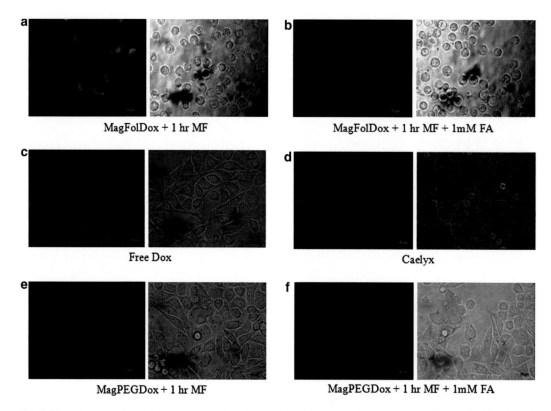

Fig. 2. Fluorescence (*left panel*) and corresponding phase contrast (*right panel*) micrographs of HeLa cells following 2 h incubation of different liposome formulations and free doxorubicin at 50 μM doxorubicin concentration in the supernatants at 37°C. It can be seen that cells treated with folate targeted magnetic liposomes under the influence of a permanent magnetic field (MagFolDox + MF) display more fluorescence than cells treated with Caelyx and nontargeted magnetic liposomes under magnetic field influence (MagPEGDox + MF) and slightly more fluorescence than the free doxorubicin treated cells. However, in presence of free FA, the fluorescence intensity of MagFolDox treated cells under magnetic field (i.e., MagFolDox + MF + 1 mM FA) is significantly lower, which is due to competitive inhibition of receptor mediated endocytosis of MagFolDox liposomes. This also suggests that the MagFolDox liposomes are taken up by the folate receptor mediated endocytosis (20). On the other hand, the uptake of nontargeted magnetic liposomes (MagPEGDox) is not changed by free folic acid competition. A quantitative fluorometric study shows that in HeLa cells, MagFolDox + MF treated cells have fivefold higher uptake of doxorubicin than free doxorubicin treated cells, 9.5-fold higher than MagFolDox − MF (i.e., without exposure of magnetic field) treated cells and 118-fold higher than the Caelyx® treated cells. In KB cells, doxorubicin uptake with MagFolDox liposomes under 1 h magnetic field exposure is 1.5-fold higher than with free doxorubicin, 3.5-fold higher than with MagFolDox − MF (i.e., without exposure of magnetic field) and 50-fold higher than with Caelyx®. KB cells show a comparatively higher uptake of the liposomes and free doxorubicin than HeLa cells. Higher uptake of MagFolDox under magnetic field exposure is likely due to a combined targeting effect of magnetic field and folate receptor, where the magnetic field enables avid binding of MagFolDox liposomes with the folate receptors on the cells and thereby possibly assists in clustering of folate receptors to facilitate the process of receptor mediated endocytosis (21).

7. The plates are then incubated for 30 min at 37°C. Thereafter, the absorbance is measured at 490 nm wavelength in a plate reader (here: Wallac Victor², Perkin Elmer, USA). The percentage of cell viability is calculated using the following formula

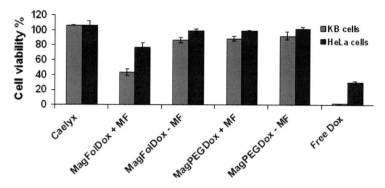

Fig. 3. Cytotoxicity profile of different magnetic liposome samples and free doxorubicin following 2 h exposure at a doxorubicin concentration of 25 μM (the results are expressed as mean ± S.D.; $n = 3$). The MagFolDox + MF sample is more toxic than other liposomal formulations including the Caelyx in both KB and HeLa cells. The higher cytotoxicity of MagFolDox + MF (under magnetic field exposure) is due to higher uptake of liposomal doxorubicin by magnetic field assisted folate receptor mediated endocytosis, which has been already observed in the FACS and spectrofluorimetric studies. In contrast, although the amount of free doxorubicin uptake is lower than with MagFolDox + MF samples, the cytoxicity is much higher. This could be due to rapid nuclear localization of free doxorubicin which is not observed with MagFolDox liposomal doxorubicin, which is mostly confined to the cytoplasm (see Fig. 2). However, when the MagFolDox will be used in vivo, targeted delivery may increase the accumulation of liposomal doxorubicin in the tumor area due to both magnetic and folate receptor mediated targeting and thereby increase therapeutic efficacy. Furthermore AC magnetic field induced hyperthermia ought to induce intracellular release of doxorubicin which should increase the potency of MagFolDox liposomes

$$\text{cell viability}(\%) = \frac{(\text{absorbance supernatant treated cells}) - (\text{blank})}{(\text{absorbance supernatant control cells}) - (\text{blank})} \times 100$$

where blank is the absorbance of 10 μl "96® Aqueous One Solution Reagent" in 100 μl culture medium and supernatant control cells is the supernatant of untreated cells cultivated in the same manner as the sample cells (see Fig. 3).

4. Notes

1. Deionized water used for all the experiments is water having resistance of 18.2 MΩ purified by a Milli Q water purification system.
2. 1 ml of anhydrous DMSO is to be added to 12.5 mg folic acid in a 1.5 ml microcentrifuge tube and the microcentrifuge tube should be stirred on a vortexer for 4–6 h to dissolve the folic acid completely in anhydrous DMSO.
3. FluidMag-HS is an aqueous based magnetic fluid containing superparamagnetic nanoparticles having a core size of 10 nm and hydrodynamic diameter of about 60 nm.

4. The pH of the ammonium sulphate solution is adjusted to 4–4.5 by using 1N HCl. Normally, the pH of 250 mM ammonium sulphate solution in water is around 5.5.

5. During extrusion of magnetic liposomes mainly through 200 and 100 nm pore size membranes, care should be taken to extrude very slowly to avoid rupture of the membrane due to high back pressure. Also, extrusion should be done at 55–60°C, at which temperature the liposomal membrane is in liquid crystalline state.

6. Exchange of ammonium sulphate outside the liposomes with PBS is required to create a gradient of ammonia between the inner aqueous space of liposomes and surrounding media to facilitate the loading of doxorubicin inside the liposomes. The detail principle of the ammonium sulphate gradient method of loading is described elsewhere (22).

7. In this step, it can be seen that most of the doxorubicin (which is red in colour) will be eluted with the liposomal fraction, the retention time of which is normally 6–10 min. Unencapsulated doxorubicin can be seen on the column as faint red colour be retarded on the column.

8. When magnetic fluid (dark brown) is completely dissolved, the colour of the solution will be clear light yellow. Sometimes the solution may be heated 80–90°C to facilitate the process of acid decomposition of magnetic nanoparticles.

9. A similar procedure as acid decomposition of magnetic nanoparticles can be applied to dissolve the magnetic liposomes.

10. Folic acid deficient media helps up-regulating of folate receptor expression and presentation at cell surfaces (5).

11. Free folic acid is added for competitive inhibition of folate targeted liposome binding to the folate receptor. Consequently, receptor-mediated uptake of these liposomes will also be inhibited (5).

12. To dissolve folic acid in water/PBS, the pH of the solution is adjusted to 7–7.2 by 1N NaOH.

Acknowledgments

Financial support of the German Excellence Initiative via the "Nanosystems Initiative Munich (NIM)" is gratefully acknowledged. Pallab Pradhan gratefully acknowledges DAAD (Deutscher Akademischer Austausch Dienst) for awarding the "DAAD Sandwich fellowship" for pursuing the research in Germany. Support from DST, Govt. of India is also gratefully acknowledged.

References

1. Northfelt DW, Martin FJ, Working P, Volberding PA, Russell J, Newman M, Amantea MA, Kaplan LD (1996) Doxorubicin encapsulated in liposomes containing surface-bound polyethylene glycol: pharmacokinetics, tumour localization, and safety in patients with AIDS-related Kaposi's sarcoma. J Clin Pharmacol 36(1):55–63
2. Muggia F, Hamilton A (2001) Phase III data on Caelyx in ovarian cancer. Eur J Cancer 37:S15–S18
3. Cady FM, Kneuper-Hall R, Metcalf JS (2006) Histologic patterns of polyethylene glycol-liposomal doxorubicin-related cutaneous eruptions. Am J Dermatopathol 28:168–172
4. Hubert A, Lyass O, Pode D, Gabizon A (2000) Doxil (Caelyx): an exploratory study with pharmacokinetics in patients with hormone-refractory prostate cancer. Anti-Cancer Drugs 11:123–127
5. Lee RJ, Low PS (1995) Folate-mediated tumor cell targeting of liposome-entrapped doxorubicin in vitro. Biochim Biophys Acta 1233:134–144
6. Gabizon A, Horowitz AT, Goren D, Tzemach D, Shmeeda H, Zalipsky S (2003) In vivo fate of folate-targeted polyethylene-glycol liposomes in tumor-bearing mice. Clin Cancer Res 9:6551–6559
7. Turk MJ, Waters DJ, Low PS (2004) Folate-conjugated liposomes preferentially target macrophages associated with ovarian carcinoma. Cancer Lett 213:165–172
8. Wu J, Liu Q, Lee RJ (2006) A folate receptor-targeted liposomal formulation for paclitaxel. Int J Pharmaceut 316:148–153
9. Gonzales M, Krishnan MK (2005) Synthesis of magnetoliposomes with monodisperse iron oxide nanocrystal cores for hyperthermia. J Magn Magn Mater 293:265–270
10. Pradhan P, Giri J, Banerjee R, Bellare J, Bahadur D (2007) Preparation and characterization of manganese ferrite-based magnetic liposomes for hyperthermia treatment of cancer. J Magn Magn Mater 311:208–215
11. Alexiou C, Arnold W, Klein RJ, Parak FG, Hulin P, Bergemann C, Erhardt W, Wagenpfeil S, Lübbe AS (2000) Locoregional cancer treatment with magnetic drug targeting. Cancer Res 60:6641–6648
12. Nobuto H, Sugita T, Kubo T, Shimose S, Yasunaga Y, Murakami T, Ochi M (2004) Evaluation of systemic chemotherapy with magnetic liposomal doxorubicin and a dipole external electromagnet. Int J Cancer 109:627–635
13. Jordan A, Scholz R, Maier-Hauff K, Johannsen M, Wust P, Nadobny J, Schirra H, Schmidt H, Deger S, Loening S, Lanksch W, Felix R (2001) Presentation of a new magnetic field therapy system for the treatment of human solid tumors with magnetic fluid hyperthermia. J Magn Magn Mater 225:118–126
14. Hilger I, Andrä W, Hergt R, Hiergeist R, Schubert H, Kaiser WA (2001) Electromagnetic heating of breast tumors in interventional radiology: in vitro and in vivo studies in human cadavers and mice. Radiology 18:570–575
15. Martina MS, Fortin JP, Ménager C, Clément O, Barratt G, Grabielle-Madelmont C, Gazeau F, Cabuil V, Lesieur S (2005) Generation of superparamagnetic liposomes revealed as highly efficient MRI contrast agents for in vivo imaging. J Am Chem Soc 127:10676–10685
16. Hahn GM, Braun J, Her-Kedcar I (1975) Thermochemotherapy synergism between hyperthermia (42°–43°C) and adriamycin (or bleomycin) in mammalian cell inactivation. Proc Natl Acad Sci USA 72:937–940
17. Hahn GM, Strande DP (1976) Cytotoxic effects of hyperthermia and adriamycin on Chinese hamster cells. J Natl Cancer Inst 57:1063–1067
18. Gabizon A, Horowitz AT, Goren D, Tzemach D, Mandelbaum-Shavit F, Qazen MM, Zalipsky S (1999) Targeting folate receptor with folate linked to extremities of poly (ethylene glycol)-grafted liposomes: in vitro studies. Bioconjug Chem 10:289
19. Mykhaylyk O, Antequera YS, Vlaskou D, Plank C (2007) Generation of magnetic nonviral gene transfer agents and magnetofection in vitro. Nat Prot 2:2391–2411
20. Yoo HS, Park TG (2004) Folate receptor targeted biodegradable polymeric doxorubicin micelles. J Contr Release 96:273–283
21. Mannix RJ, Kumar S, Cassiola F, Montoya-Zavala M, Feinstein E, Prentiss M, Ingber DE (2008) Nanomagnetic actuation of receptor-mediated signal transduction. Nat Nanotech 3:36–40
22. Haran G, Cohen R, Bar LK, Barenholz Y (1993) Transmembrane ammonium sulfate gradients in liposomes produce efficient and stable entrapment of amphipathic weak bases. Biochim Biophys Acta 1151:201–215

Chapter 20

Liposomes for Drug Delivery to Mitochondria

Sarathi V. Boddapati, Gerard G.M. D'Souza, and Volkmar Weissig

Abstract

Efficacy of therapeutically active drugs known to act on intracellular targets can be enhanced by specific delivery to the site of action. Triphenylphosphonium cations can be used to create subcellular targeted liposomes that efficiently deliver drugs to mitochondria, thus enhancing their therapeutic action.

Key words: Mitochondriotropics, Mitochondrial medicine, Mitochondrial nanocariers, Mitochondrial drug delivery

1. Introduction

With recent developments in mitochondrial biology and the identification of mitochondrial drug targets, Mitochondrial medicine is emerging as a rapidly growing field of biomedical research (1, 2). The generation of mitochondriotropic triphenylphosphonium cation that is conjugated with stearyl residues (STPP) to facilitate incorporation into liposomal bilayers has led to the creation of mitochondriotropic liposomes, which can be used to deliver therapeutically active drugs to the mitochondria (3). Exogenous ceramide is known to induce the formation of ceramide channels in the mitochondria, releasing cytochrome-C and hence leading to apoptosis (4–6), making ceramide a model drug for specific delivery to mitochondria. Methods to prepare and study the effect of mitochondria targeted delivery of drugs using STPP liposomes, with in vitro and in vivo models are outlined in this chapter.

2. Materials

2.1. Preparation and Characterization of STPP Liposomes

1. Di-oleoyl phasphatidyl choline; DOPC (Avanti Polar Lipids, Alabaster, Alabama).
2. Cholesterol (Avanti Polar Lipids, Alabaster, Alabama).
3. Stearyl triphenylphosphonium; STPP (synthesized in-house; synthesis and characterization of STPP published previously (3)).
4. Chloroform.
5. 5 mM HEPES, pH 7.4, prepared from HEPES (Fisher Scientific) and pH adjusted with 1 N HCl/1 N NaOH.
6. Sonic Dismembrator (model 100, Fischer Scientific).
7. Rotary evaporator (Labconco, Fisher Scientific).
8. Sephadex G-15.
9. Round bottom flasks 3–5 ml.

2.2. Cell Association and Uptake of STPP Liposomes

1. Rhodamine labeled PE (Avanti Polar Lipids, Alabaster, Alabama).
2. Mitofluor Green (Molecular probes, Eugene, Oregon).
3. Hoechst 33342 (Molecular probes, Eugene, Oregon).
4. Six-well culture plates.
5. 22-mm cover glass.
6. 25-cc cell-culture flasks.
7. Trypsin-EDTA containing 0.25% Trypsin and 0.1% EDTA (Mediatech cellgro, Fisher Scientific).
8. Dulbecco's Modification of Eagle's (Mod.) 1× DMEM (Mediatech cellgro, Fisher Scientific).
9. Fetal Bovine Serum (Fisher Scientific).
10. 4T1 mouse mammary tumor cells (ATCC).
11. 10× PBS.
12. 4% paraformaldehyde.
13. Glass slides.
14. Fluoromount G, fluorescent cell-mounting medium (Trevigen).
15. DOTAP (Avanti Polar Lipids, Alabaster, Alabama).

2.3. Preparation and Evaluation of Drug Incorporated STPP Liposomes

1. Ceramide C6; (6-Hexanoyl-d-erythro-sphingosine; Fisher Scientific).
2. MTS reagent; CellTiter 96® AQ$_{ucous}$ One Solution Cell Proliferation Assay (Promega).
3. DNeasy Tissue Kit (Qiagen).

4. Agarose.

5. Ethidium Bromide.

6. DNA electrophoresis system (Owl electrophoresis systems, Fisher Scientific).

2.4. Preparation and Characterization of STPP Liposomes for Animal Studies

1. PEG5000-PE (Avanti Polar Lipids, Alabaster, Alabama).
2. 5 mM HEPES-buffered saline, pH 7.4, prepared from HEPES (Fisher Scientific) and pH adjusted with 1 N HCl/1 N NaOH.
3. BalB/C mice (Charles river labs).
4. C57BL mice (Charles river labs).
5. LLC (Lewis Lung Carcinoma) cells (ATCC).

3. Methods

3.1. Preparation and Characterization of STPP Liposomes (see Note 1)

1. STPP liposomes are prepared by probe sonication (7). A mixture of DOPC:Cholesterol:STPP in the molar ratio 83.5/15/1.5 (see Note 2) (total lipid 25 mg/ml) is prepared in chloroform in an appropriate sized flask; 3–5-ml flask to prepare 1 ml of liposomes.

2. Taking care to avoid foaming, the lipids are dried in the presence of a vacuum, at 37 °C to yield a thin film of lipids at the bottom of the flask. Dry the lipids completely.

3. Hydrate the film completely by adding 5 mM HEPES (pH 7.4) to the dry lipid film.

4. Immerse the tip of the probe sonicator into the sample adjusting it so that the tip does not touch the bottom of the flask.

5. Sonicate at a power output of approximately 10 Watts for 30 min. To remove any titanium particles which have leaked from the probe during sonication, centrifuge the sample for 10 min at about $3,000 \times g$.

6. Carefully collect the supernatant to obtain formed liposomes.

7. Prepare an approximately 10-cm column of Sephadex G-15. Separate the non-incorporated STPP from the formed liposomes by passing through the Sephadex G-15 column.

8. Size distribution and zeta potential are measured using Brookhaven Instruments Zeta Plus. Particle size measurements confirm the formation of SUV's (54 ± 22 nm). Measure the zeta potential to establish the STPP incorporation into the liposomes. Zeta potential values of $+30 \pm 12$ mV were

obtained from measurements performed on STPP liposomes. In addition, ^{31}P NMR can be used to confirm the presence of STPP on the liposomes (3).

3.2. Cell Association and Uptake of STPP Liposomes

1. To study cell association and uptake, liposomes with a fluorescently labeled phospholipid should be prepared. Rhodamine-PE (Rh-PE) a commonly used liposomal membrane marker is used to prepare STPP liposomes for cell uptake studies.

2. STPP liposomes are prepared in the presence of 0.5 mole% Rh-PE. The method of preparation is the same as described in Subheading 3.1 earlier.

3. Wrap the round bottom flask with aluminium foil during liposome preparation, to protect the fluorescent marker from exposure to light.

4. 4T1 cells are grown in 25-cc culture flasks using DMEM containing 10% fetal bovine serum (FBS).

5. Cells are harvested when they are at 75–80% confluence by incubating with 5 ml Trypsin-EDTA for 2–5 min till the cells start to separate from the flask upon gentle tapping. Harvested cells are then passaged into 25-cc flasks (for cell association experiments) and into six-well plates (for cell uptake experiments).

6. Cell association experiments are performed in 25-cc flasks. Cells are incubated with Rh-PE-labeled STPP liposomes in the presence of serum-free DMEM for time periods of 5–30 min (see Note 3).

7. Incubated cells are washed three times with 1× PBS and harvested.

8. The cells are fixed with 4% paraformaldehyde in PBS. Flow cytometry is performed on BD FACS Calibur Flow Cytometer. An example of results obtained from this analysis has recently been published (8).

9. For uptake experiments, cells are grown on 22-mm cover slips in six-well plates to 75–80% confluence.

10. The cells are incubated with Rh-PE-labeled STPP liposomes in serum-free DMEM for periods of time up to 2 h.

11. Incubated cells are washed three times with 1× PBS.

12. Organelle specific dyes, Hoechst 33342 and Mitofluor Green are used to label the nuclei and mitochondria in the cells, respectively.

13. Following staining with dyes, cells are washed three times with 1× PBS, cover slips are mounted on slides using Fluoromount G mounting medium and examined on a Zeiss Meta 510 LSM with Zeiss software (see Note 4).

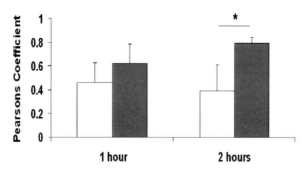

Fig. 1. Analysis of fluorescence colocalization: Pearsons coefficient ± standard deviation ($n=6$) for colocalization of rhodamine fluorescence (liposomal marker) with Mitofluor green fluorescence (mitochondrial marker) obtained with ImageJ. *Open bars* indicate nontargeted liposomes, *shaded bars* indicate STPP liposomes. (*Asterisk* indicates a *P* value of <0.005). (Reproduced with permission from ref. 8)

14. Confocal fluorescence micrographs obtained upon incubation of cells with STPP liposomes (see Note 5) are subjected to fluorescence colocalization analysis using ImageJ. Pearsons coefficient values obtained for STPP liposomes are compared to non-targeted liposomes prepared by replacing 1.5% STPP with 1.5% DOTAP (Fig. 1).

3.3. Preparation and Evaluation of Drug-Incorporated STPP Liposomes

1. Drug-incorporated STPP liposomes are prepared by adding 10 mole% ceramide C6 to the lipid composition mentioned in step 1 in Subheading 3.1 earlier.

2. Ceramide-incorporated STPP liposomes are evaluated for efficacy of targeted delivery by measuring cell cytotoxicity and DNA fragmentation.

3. For cell cytotoxicity, 4T1 cells are grown in 96-well plates to 50–60% confluence for the incubation with drug containing liposomes.

4. The cells are washed with serum free DMEM and liposomes diluted to the appropriate ceramide concentration, using serum-free DMEM.

5. The cells are incubated with liposomes in serum-free DMEM for a time period of 48 h.

6. Following incubation the cells are washed with 1× PBS and 20 µl/well MTS reagent is added. After incubation for 4 h, the resultant absorbance is measured at 490 nm.

7. Mitochondria-targeted STPP liposomes significantly increased cell cytotoxicity by almost 20% when compared to non-targeted liposomes (8).

8. For DNA fragmentation analysis, 4T1 cells are grown in 25-cc culture flasks upto 60% confluence.

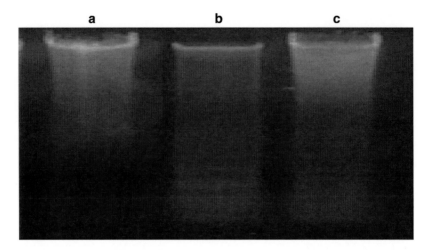

Fig. 2. Agarose gel electrophoresis of apoptotic DNA ladder formed by the incubation of cells with ceramide STPP liposomes. The cells were incubated with the various liposomal preparations for 18 h at a ceramide concentration equivalent to 25 µM, followed by DNA extraction and gel electrophoresis. *Lane* **a**: ceramide liposomes; *lane* **b**: ceramide STPP liposomes; *lane* **c**: untreated

9. Cells are incubated with liposomes at a ceramide concentration of 25 µM for a time period of 18 h.

10. Cells are harvested and total DNA is extracted using DNeasy Tissue Kit.

11. 2 µg of total DNA is run electrophoretically on a 1.4% agarose gel containing 0.5 µg/ml of ethidium bromide. The gel is visualized under UV and photographed.

12. The formation of a DNA ladder, characteristic of DNA fragmentation is observed only in the case of mitochondria-targeted STPP liposomes. A typical image is seen in Fig. 2.

3.4. Preparation and Characterization of STPP Liposomes for Animal Studies

1. STPP liposomes used for animal studies are prepared in the presence of 3 mole% PEG-5000-PE. The method of preparation is the same as described in Subheading 3.1 earlier (see Note 6).

2. Maximal tolerated dose for STPP liposomes is established by injecting Balb/C mice with 0.45, 1.5, 4.5, and 15 mg/kg of STPP in injection volume.

3. Mice are injected via tail vein every alternate day with liposomes for a total of 10 doses over a period of 21 days.

4. The animals are monitored for changes in body weight, hydration, ataxia and abnormal behavior (8).

5. For studies involving biodistribution in animals, 0.5 mol% DTPA-PE is added during liposome preparation.

6. Liposomal-DTPA-PE is supplemented with 1 M citrate buffer and incubated for 1 h with ^{111}In-citrate complex at RT, and then dialyzed overnight against HBS at 4°C to remove free label (9).

7. C57BL mice with subcutaneous solid tumors generated by injecting LLC (Lewis Lung Carcinoma) cells are used for the biodistribution study (see Note 7).

8. Tumor-bearing mice are injected i.v. (via tail vein) with 5 µCi of the radiolabeled (^{111}In) liposome dispersions.

9. The injected mice are sacrificed after 24 h and organs lung, heart, liver, spleen, kidney, tumor, muscle, skin, tail, and blood are removed.

10. The isolated organs are weighed and the amount of radioactivity was quantified as CPM using a Beckman 5500B gamma-counter.

11. The amount of radioactivity per gram of tissue is calculated and the percent uptake per gram organ is calculated.

12. STPP liposomes do not show a significantly different biodistribution than non-targeted liposomes.

3.5. STPP Liposomes for Tumor Growth Inhibition Studies

1. STPP liposomes for tumor growth inhibition studies containing ceramide and 3 mole% PEG5000PE are prepared as described in Subheading 3.4 earlier.

2. Tumors are generated in Balb/C mice by subcutaneously injecting 4T1 cells into the flank region of mice. Tumor growth is monitored on a regular basis till a measurable tumor is formed (see Note 8).

3. Mice are randomized and liposomes injected IV via tail vein injection. The injection schedule followed is as per a previous study conducted using ceramide liposomes (10), with mice injected at a frequency of one injection per 2 days.

4. STPP liposomes are injected at a dose of 6 mg/kg of ceramide.

5. At the end of 21 days from the first injection, mice are sacrificed and tumors extracted.

6. Tumor growth is measured and tumor volumes of the groups are compared over the period of the study.

7. Figure 3 shows results obtained from a tumor growth inhibition study with STPP liposomes. STPP liposomes at a dose of 6 mg/kg had significantly inhibited tumor growth when compared to either untreated control or empty STPP liposomes.

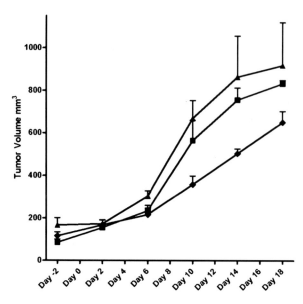

Fig. 3. Tumor Growth Inhibition: Tumor volume (mm^3) measured over a time period of treatment in Balb/c mice bearing murine 4T1 mammary carcinoma tumors ($n=6$); after treatment with buffer (*closed squares*), empty STPP nanocarrier (*closed triangles*), and ceramide in STPP nanocarrier (*closed diamonds*). (Reproduced with permission from ref. 8)

4. Notes

1. This method involves the use of a bath sonicator and a rotary evaporator equipped with the proper adaptor to hold a 3–5-ml round-bottom flask and connected to a vacuum pump. In the absence of a rotary evaporator, a desicator equipped with vacuum can be used to evaporate the chloroform used to dissolve the lipids.

2. For small quantities of STPP, prepare a fresh stock solution of STPP in chloroform as per the requirements of final liposome preparation.

3. Cell viability assay (MTS assay) was performed by incubating STPP liposomes with 4T1 cells grown in 96-well plates. Data obtained from this assay was used to establish the amount of STPP liposomes that can be safely used with cells (3).

4. Slides are stored at 4°C, protected from light until further processing.

5. The images are obtained with a 63× oil immersion objective. Appropriate control slides are used to set the levels of detection of dyes.

6. All preparations used for animal injections are prepared in HEPES-buffered saline in order to maintain isotonicity.

7. Solid murine tumor LLC (Lewis Lung Carcinoma) cell line is initiated in mice by a local subcutaneous injection. 2×10^5 cells in 0.2 cc of Hanks buffer salt solution are injected in each animal. Tumors are allowed to develop for 7–10 days after inoculation on reaching a surface diameter of 4–6 mm and a thickness of 2–3 mm.

8. The tumor growth is measured by vernier caliper and the tumor volume is calculated as per the formula 0.5 (length × width^2).

References

1. Larson N-G, Luft R (1999) Revolution in mitochondrial medicine. FEBS Lett 455: 199–202
2. Schon EA, DiMauro S (2003) Medicinal and genetic approaches to the treatment of mitochondrial disease. Curr Med Chem 10:2523–2533
3. Boddapati SV, Tongcharoensirikul P, Hanson RN, D'Souza GG, Torchilin VP, Weissig V (2005) Mitochondriotropic liposomes. J Liposome Res 15:49–58
4. Shabbits JA, Mayer LD (2003) Intracellular delivery of ceramide lipids via liposomes enhances apoptosis in vitro. Biochim Biophys Acta 1612:98–106
5. Siskind LJ (2005) Mitochondrial ceramide and the induction of apoptosis. J Bioenerg Biomembr 37:143–153
6. Siskind LJ, Davoody A, Lewin N, Marshall S, Colombini M (2003) Enlargement and contracture of C2-ceramide channels. Biophys J 85:1560–1575
7. Lasch J, Weissig V, Brandl M (2003) In: Torchilin VP, Weissig V (eds) Liposomes – a practical approach. Oxford University Press, Oxford, pp 3–30.
8. Boddapati SV, D'Souza GG, Erdogan S, Torchilin VP, Weissig V (2008) Organelle-targeted nanocarriers: specific delivery of liposomal ceramide to mitochondria enhances its cytotoxicity in vitro and in vivo. Nano Lett 8:2559–2563
9. Levchenko TS, Rammohan R, Lukyanov AN, Whiteman KR, Torchilin VP (2002) Liposome clearance in mice: the effect of a separate and combined presence of surface charge and polymer coating. Int J Pharm 240:95–102
10. Stover TC, Sharma A, Robertson GP, Kester M (2005) Systemic delivery of liposomal short-chain ceramide limits solid tumor growth in murine models of breast adenocarcinoma. Clin Cancer Res 11: 3465–3474

Chapter 21

Cytoskeletal-Antigen Specific Immunoliposomes: Preservation of Myocardial Viability

Vishwesh Patil, Tala Khudairi, and Ban-An Khaw

Abstract

Pathological conditions such as hypoxia and inflammation can lead to the development of cell membrane-lesions. The presence of these membrane-lesions leads to egress of intracellular macromolecules as well as exposure of intracellular microenvironment to the extracellular milieu resulting in necrotic cell death. An intracellular structure that becomes exposed to the extracellular environment is myosin, a cytoskeletal antigen. We had hypothesized that cell viability can be preserved in nascent necrotic cells if the cell membrane lesions were sealed and the injurious conditions removed. Cell membrane lesion sealing and preservation of cell viability were achieved by the application of Cytoskeletal-antigen Specific ImmunoLiposomes (CSIL) as molecular "Band-Aid" that initially plugs the holes with subsequent sealing of the lesions. Anti-myosin antibody was chosen as the cytoskeleton-antigen specific antibody to develop CSILs, because antimyosin antibody is highly specific for targeting myosin exposed through myocardial cell membrane lesions in various cardiomyopathies. Liposomes are biocompatible lipid bilayer vesicles that have been used in many biological applications for several decades.

This chapter will be limited to the description of CSIL therapy to ex vivo studies in adult mammalian hearts. Due to page limitations, cell culture, gene delivery and in vivo studies will not be included. Therapeutic efficacy of CSIL in preservation of myocardial viability as well as function (by left ventricular developed pressure measurements) as assessed in globally ischemic Langendorff instrumented hearts is both dose and time dependent. This approach of cell membrane lesion repair and sealing may have broader applications in other cell systems.

Key words: Immunoliposomes, Myocardial infarction, Preservation of myocardial viability, Lanvgen-dorff perfused isolated hearts

1. Introduction

The process of cell death is currently believed to be an interplay between oncosis (formerly referred to as necrosis) and apoptosis (1, 2). Apoptosis may result from low intensity ischemic stress, whereas, oncotic cell death is likely due to severe ischemia-reperfusion insults (3). Cell death in acute myocardial infarction is mainly due to oncosis (4). The hallmark of oncotic cell death which follows depletion of ATP is development of membrane lesions that leads to intracellular micro- and macro-molecular leakage (5). Normally, rapid self-sealing of ruptured cell membrane lesions is an innate property of certain cells including mammalian cells (6–10). The process of self-sealing of membrane lesions mobilizes vesicles such as lysosomes and endosomes in response to elevated physiological concentration of Ca++ in the extracytoplasmic environment. However, cell membrane lesion sealing and repair need not depend solely on and limited to innate mechanisms. Liposomes have been reported to seal cell membrane lesions non-specifically (11) as well as fuse with the membrane of various cell types (12). Liposomes that fuse with cell membranes have been developed for the delivery of genetic constructs or nanoparticles coated with DNA (13, 14). The existence of these non-specific methods for fusion of liposomal membrane with the cell membrane attests to the potential for further augmentation of membrane repair with targeted "Plug and Seal" (15) approach for cell membrane lesion sealing and prevention of oncotic cell death. Using this approach, cell membrane lesions in hypoxic embryonic cardiocytes in cultures were sealed with liposomes targeted to the intracellular cytoskeletal antigen, myosin with antimyosin, resulting in preservation of almost 100% cell viability (15). Fusion of CSIL with myocardial cell membrane was confirmed by the delivery of intra-liposomally entrapped genetic constructs to the cytosolic compartment of hypoxic cardiocytes that subsequently expressed the reporter gene products (16). High efficiency of cell membrane lesion sealing and fusion was verified by the expression of the reporter gene product in almost every treated cardiocyte (16, 17).

Since gene expression is the hallmark of cell viability, all cells expressing the reporter gene product must necessarily be considered viable (16, 17). Although no direct ex vivo evidence of cell membrane lesion sealing is demonstrated in the current report, the mechanism may be inferred from previous reports (15–17). Cardiocytes treated with fluorescent lipid incorporated CSIL showed integration of the fluorescent lipid into the cell membranes. Furthermore, fusion of CSIL with the cell membrane may be inferred from gene transfection studies. Reporter genes entrapped in the intraliposomal cavities of CSIL successfully transfected hypoxic cardiocytes. Only hypoxic cardiocytes treated with CSIL, and not with IgG-L, plain liposomes, nor placebo, resulted in highly efficient gene transfection (18, 19). If preservation of myocardial viability were by non-specific mechanisms, IgG-L treatment should result in the same extent of myocardial salvage.

This Chapter will be limited to our research in the development of cytoskeletal-antigen specific immunoliposome technique for the preservation of cardiocyte viability in adult mammalian hearts in ex vivo experiments (18, 19).

2. Materials

1. Bioreactors (Cell-max QUAD, Cellco Inc Laguna Hills, California).
2. Dulbecco's Modified Eagle's Medium (DMEM) (Sigma-Aldrich, MO, USA).
3. FETALCLONE (Hyclone, Logan, Utah).
4. Penicillin, streptomycin (100 units/mL each) and Amphotericin B (2.5 μg/mL) (Sigma-Aldrich, MO, USA).
5. Protein A Sepharose 4 Fast Flow (Amersham Pharmacia Biotech, Uppsala, Sweden).
6. Gel electrophoresis (XcellSurelock Mini-Cell with NuPage gel, 10% Bis-Tris, Invitrogen, Carlsbad, California).
7. N-glutarylphosphatidyl ethanolamine (NGPE; Avanti Polar Lipids, Hercules, California).
8. 0.2 μm polycarbonate-membrane syringe filters (Millex, Bedford, Massachusetts).
9. Coulter N4 + MD Submicron Particle Size Analyzer (Coulter Electronics, Miami, Florida).
10. Nitro blue tetrazolium (NBT; Fisher Scientific, Fair Lawn, New Jersey).

3. Methods

3.1. Production and Modification of Antibody

The protocols were approved by the Institutional Animal Care and Use Committee, Northeastern University (Boston, Massachusetts) and conform to the guidelines specified in the National Institutes of Health Guide for the Care and Use of Laboratory Animals.

1. Myosin-specific monoclonal 2G42D7, an immunoglobulin G (IgG)-1 murine antibody has an apparent affinity of approximately 1×10^9 L/mol [30, 31]. Antimyosin antibody 2G42D7 is produced in bioreactors (Cell-max QUAD, Cellco Inc. Laguna Hills, California) with Dulbecco's Modified Eagle's Medium (DMEM) (Sigma-Aldrich, MO,USA) supplemented with 10%

FETALCLONE (Hyclone, Logan, Utah), penicillin, streptomycin (100 units/mL each) and Amphotericin B (2.5 µg/mL) (Sigma) at 37°C in 5% CO_2.

2. The monoclonal antibodies (MAb) are purified by protein A-affinity chromatography from culture media or bioreactor fluid (20). Purity of the monoclonal antibody is assessed by sodium dodecyl sulfate polyacrylamide gel electrophoresis (XcellSurelock Mini-Cell with NuPage gel, 10% Bis-Tris, Invitrogen, Carlsbad, California).

3. Antibody activity is assessed by enzyme-linked immunosorbent assay (16–19, 21).

4. Antimyosin antibody is modified with N-glutarylphosphatidyl ethanolamine (NGPE; Avanti Polar Lipids, Hercules, California) (22). Briefly:

    (a) 2 mg of 2G4-2D7 in 2 mL of 137 mMNaCl, 2.7 mMKCl, 8.5 Na_2HPO_4, 1.4 mM KH_2PO_4, at pH 7.4 (phosphate buffered saline [PBS]), are mixed with solid HEPES (24 mg).

    (b) NGPE (0.3 mg) in 0.5 mL of 16 mM octyl-glucoside in 50 mM of 2-N-morpholino ethane sulfonic acid (pH 4.5) and 15 mg N-hydroxysulfosuccinimide are activated with 12 mg of 1-ethyl-3-3-dimethylaminopropyl carbodiimide and added slowly to the antibody solution. The pH is adjusted to ~8.0 with 1 M KOH and incubated overnight at 4°C with gentle stirring (16).

5. Murine IgG is similarly modified.

3.2. Immunoliposome Preparation

Unilamellar liposomes are prepared by the detergent dialysis method (23).

1. Phosphatidylcholine (30 mg) in chloroform and cholesterol (18 mg) (1:1 molar ratio) are dried for 2 h in a rotary evaporator.

2. Two milliliters of 50 mg octyl-glucoside/mL PBS are added to the dry lipid film, stirred, and sonicated for 5 min.

3. An aliquot containing the NGPE-antibody to lipid ratio of 1 mg antibody (antimyosin or murine IgG) to 24 mg lipid is sonicated for 2 min, then dialyzed against 4 L PBS (pH 7.4) overnight at 4°C.

4. The resulting liposomes are extruded serially through 0.8, 0.45 and 0.2 µm polycarbonate-membrane syringe filters (at least 11 times each) (Millex, Bedford, Massachusetts) (15, 19).

5. The mean (± SD) diameter of CSIL is 200 ± 35 nm, that of IgG-L is 210 ± 13.1 (Coulter N4 ± MD Submicron Particle Size Analyzer, Coulter Electronics, Miami, Florida).

3.3. Targeted Sealing of Cell Membrane Lesions: Model of Preservation of Cell Viability by Immunoliposome Therapy

1. The phenomenon of plug and seal to prevent necrotic cell death is demonstrated by using myosin as the cytoskeletal target antigen and the corresponding antimyosin antibody as the anchoring device incorporated in liposomes in a hypoxic model of injury in H9C2 rat embryonic cardiocytes (15). Similarly, delivery of intra-liposomally entrapped model drugs (15) or reporter genes (16) were achieved with high efficiency. The application of cell membrane lesion sealing for the preservation of myocardial viability in adult mammalian hearts is demonstrated as follows:

3.4. Experimental Acute Myocardial Infarction

Langendorff perfused isolated hearts. The Langendorff isolated perfused heart model is used to demonstrate adult ischemic myocardial preservation (24).

1. Hearts from CD-1 male rats (250–300 g; n = 4 each group, 34 total) are excised and perfused within 25 s to 2 min with non-recirculating oxygenated Krebs-Henseleit bicarbonate buffer (120 mM NaCl, 25 mM $NaHCO_3$, 1.2 mM $MgSO_4 \cdot 7H_2O$, 5 mM KCl, 1.7 mM $CaCl \cdot 2H_2O$, 10 mM glucose, pH 7.4, 37°C) at a constant coronary perfusion pressure (CPP) of 80 mm Hg (18, 19).

2. Each heart immersed in 0.9% NaCl at 37°C in a water-jacketed chamber is paced at 300 beats/min (5 Hz).

3. The left ventricular (LV) end-diastolic pressure is set at 10 mm Hg, utilizing a water-filled balloon-tipped catheter attached to a pressure transducer. Baseline hemodynamic measurements are recorded on a strip-chart recorder (Hewlett-Packard 7754A) for a 10-min stabilization period.

4. Global ischemia (25 min) is initiated by decreasing the coronary perfusion pressure (CPP) to zero within 60 s (see Note 1).

5. A 2-mL aliquot of freshly prepared 1 mg antimyosin-NGPE-CSIL, 1 mg non-specific IgG-NGPE-liposomes (IgG-L), or placebo (PBS) is infused at various times during ischemia.

6. A 3-mL syringe is used to deliver the test reagents via a three-way stopcock placed 8 cm above the aorta, enabling injection without turning on the perfusion pump, while maintaining the global ischemic condition.

7. During ischemia, the heart is paced for 5 min at zero CPP to ensure adequate myocardial injury, and then left un-paced for an additional 20 min.

8. Perfusion is restored to 80 mm Hg after 25 min of global ischemia, and pacing is re-initiated at 3 min of reperfusion.

9. Hemodynamic measurements are recorded for 30 min of reperfusion.

10. Hearts are weighed, sectioned transversely (5 or 6 slices), and stained for 20 min in 0.05% nitro blue tetrazolium (NBT; Fisher Scientific, Fair Lawn, New Jersey) at 60°C (18, 19, 25).

11. Sham-instrumented hearts underwent an identical procedure without ischemia.

3.5. Assessment of Myocardial Preservation

1. Myocardial function is assessed as left ventricular developed pressure (LVDP) defined as the difference between the LV end-systolic and diastolic pressures, measured at pre-ischemic, ischemic, and post-ischemic times (5, 10, 15, 20, 25, and 30 min of reperfusion) (18)(see Note 2).

2. The LVDP recovery during reperfusion is calculated as the mean percent LVDP of the pre-ischemic baseline values.

3. A single-blinded histochemical infarct size analysis is determined by computer planimetry (Adobe Photoshop version 4.0) of the digital photographs of NBT-stained heart slices.

4. The digital images are assigned numbers; total ventricular and infarcted areas of the right and left ventricular slices are quantitated by computer planimetry after adjusting the brightness and contrast for optimal differentiation of the infarcted regions (Fig. 1a).

5. After quantitation, the single-blinded code is broken, and the infarct sizes are correlated to various treatments.

6. For ultra-structural assessment, a sample from the mid-portion of the LV of each heart is cut into smaller pieces (≈ 0.5 mm^3); prefixed in 3% glutaraldehyde, 0.1 M sodium acidulate, and 2% formaldehyde; then postfixed in 1% osmium tetroxide and 0.1 M sodium cacodylate; dehydrated in a graded series of ethanol; and embedded in Spurr's resin and polymerized overnight at 60°C.

7. Ultra-thin sections (gold-silver) are cut using a diamond knife (Dupont Company) and ultramicrotome (Reichert Ultracut E, Austria), and the sections are contrasted with uranyl acetate and lead citrate (26) and examined with a JEOL (JEM1010) transmission electron microscope.

8. Two micrographs each ($\times$6,000 magnification) from two hearts of each group are randomly obtained, and all mitochondria in the micrographs are quantitated by computer planimetry. From each group, 225 ± 12 (mean$\pm$SEM) mitochondria are analyzed.

3.6. Statistical Analyses

1. Results are reported as mean values$\pm$SEM. Wilcoxon rank-sum distribution (27) is used to determine significant differences of the mean values for all statistical comparisons in this report.

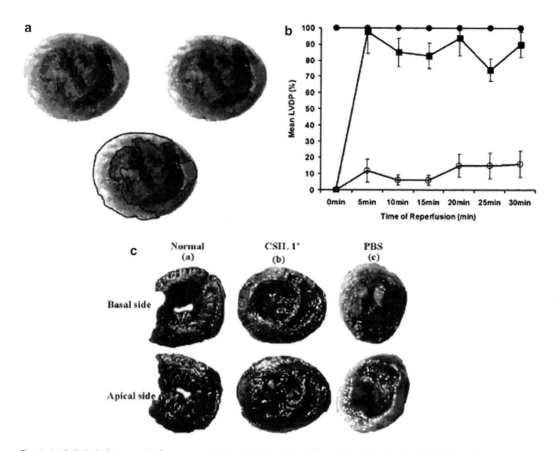

Fig. 1. (a) A digital photograph of a representative nitro blue tetrazolium-stained heart slice (*top left*), a color-contrasted version to highlight the infarcted regions (*top right*), and an outline of the planimetered infarct area for infarct sizing (*bottom*). (b) Mean left ventricular developed pressures (LVDPs) (% pre-ischemic values) at 5–30 min of reperfusion after 25 min of global ischemia of hearts treated with CSIL at 1 min of ischemia (CSIL 1′) (*squares*), PBS (*open circles*), or sham-instrumented hearts (*solid circles*). (c) Representative digital photographs of nitro blue tetrazolium-stained mid-slices of normal hearts (*a*) or hearts treated with CSIL 1′ (*b*) or PBS (*c*)

2. Alpha is set at ≤0.05. The LVDP of treatment groups (at 5 min of reperfusion, or the overall mean value of individual mean LVDPs at 5, 10, 15, 20, 25, and 30 min of each group for total time-function curves) is compared.

3. The same statistical analysis is employed in the time response study, comparison of single-blinded histochemical infarct sizes, mean mitochondrial sizes, and LVDP of hearts at the plateau phase (between 20 and 30 min of reperfusion).

3.7. Results

3.7.1. Treatment at 1 min of Ischemia

Treatment with CSIL is initiated at 1 min of global ischemia resulted in functional recovery in isolated rat hearts by 5 min of reperfusion (18). The LVDP (98 ± 14%) is similar to that of sham controls, but is greater than that of hearts treated with placebo (12 ± 7%, p = 0.01) (Fig. 1b). The total time-function LVDP curves show that recovery after CSIL treatment at 1 min of ischemia (87 ± 6%) is not significantly different from that of the sham group, but is greater than that of placebo-treated hearts (12 ± 2%, p = 0.01). The sufficiency of 25 min of global ischemia to induce extensive myocardial injury is demonstrated by the extensive lack of NBT staining in heart slices of placebo controls (Fig.1c, panel c). Hearts treated with CSIL at 1 min of ischemia are almost normally stained on both the basal and apical sides of the slices (Fig. 1c, panel b) and are similar to normal heart slices (Fig. 1c, panel a) (see Note 3).

3.8. Time Response

The mean LVDP of the total time-function curve of CSIL-treated hearts at 5, 10, and 20 min of ischemia (77 ± 3%, 70 ± 12%, and 48 ± 8%, respectively) is less than that of the sham-operated hearts but greater than that of IgG-L (44 ± 7%, 58 ± 4%, and 30 ± 4%, respectively) or placebo-treated hearts (12 ± 2%, p = 0.01) (Fig. 2a–c) (18). A sequential delay of 5, 10, and 15 min in the recovery of function to near normal LVDP is observed in hearts

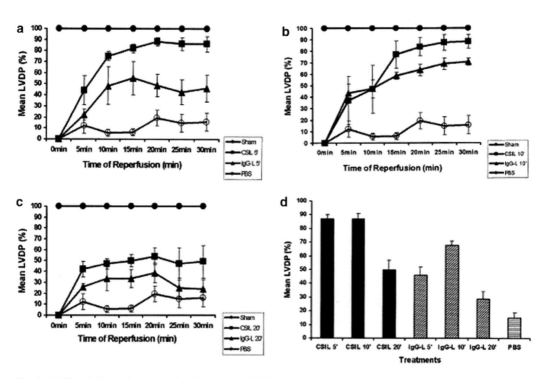

Fig. 2. (a) Mean left ventricular developed pressure (LVDP) during reperfusion of hearts treated with CSIL at 5 min of ischemia (CSIL 5′), IgG-L at 5 min of ischemia (IgG-L 5′), phosphate buffered saline (PBS), or sham instrumentation. (b) Mean LVDP during reperfusion of hearts treated with CSIL or IgG-L at 10 min of ischemia (CSIL 10′ and IgG-L 10′), PBS, or sham. (c) Mean LVDP during reperfusion of hearts treated with CSIL or IgG-L at 20 min of ischemia (CSIL 20′ and IgG-L 20′), PBS, or sham. (Athrough c) Sham = solid circles; CSIL = squares; IgG-L = triangles; PBS = open circles. (d) Mean LVDP assessed at 20–30 min of reperfusion (plateau phase) in hearts treated with CSIL, IgG-L, or PBS

treated with CSIL at 1, 5, and 10 min of ischemia, respectively (p = NS), and are greater than the VLDP of hearts treated with CSIL at 20 min (p = 0.05). The time-function curves for hearts treated with CSIL at 1 and 5 min of ischemia are greater than those at 20 min (p = 0.01) (18) (see Note 4).

3.8.1. Functional Assessment at 20–30 min of Reperfusion

In all treated hearts, mean LVDP recovery reach a plateau by 20 min of reperfusion (Figs. 1b, 2a–c). The mean plateau LVDP in hearts treated with CSIL at 5 and 10 min of ischemia ($87 \pm 3\%$ and $87 \pm 4\%$, respectively) is greater than that of the corresponding IgG-L controls ($46 \pm 6\%$ and $68 \pm 3\%$, p = 0.01 and 0.02, respectively). The plateau LVDP of hearts treated with CSIL at 20 min of ischemia ($50 \pm 7\%$) is greater than that of IgG-L controls($29 \pm 5\%$), but was not statistically different (p = 0.1); however, it is significantly greater than that of placebo controls ($15 \pm 4\%$, p = 0.01) (Fig. 2d). No difference in LVDP is observed between hearts treated with CSIL at 5 and 10 min of ischemia (18) (see Note 4).

3.9. Single-Blinded Histochemical Infarct Size Assessment

Minimal nitro-blue tetrazolium (NTB) unstained areas of the myocardium, indicative of minimal injury, are observed in hearts treated with CSIL at 5 min of ischemia (Fig. 3a, panel a). Myocardial injury increased with longer delays of CSIL treatment (at 10 and 20 min) (Fig. 3a, panels b and c). Hearts treated with IgG-L at 5, 10, and 20 min of ischemia (Fig. 3a, panels d–f) show more extensive injury than the corresponding CSIL-treated hearts. Infarct sizes of hearts treated with IgG-L at 5, 10, or 20 min of ischemia ($39 \pm 4\%$, $35 \pm 7\%$, and $45 \pm 6\%$, respectively) are the same (Fig. 3b).

Infarct sizes of hearts treated with CSIL at 1, 5, and 10 min ($4 \pm 1\%$, $8 \pm 3\%$, and $6 \pm 2\%$, respectively) are similar to sham hearts ($3 \pm 2\%$, p = NS), but are smaller than hearts treated with CSIL at 20 min ($19 \pm 3\%$, $p \leq 0.05$). Infarct sizes of hearts treated with CSIL at 5, 10, and 20 min of ischemia are smaller than those treated with IgG-L at the corresponding times (Fig. 3b) (18, 19) (see Note 5).

3.10. Mitochondrial Size

Average size of normal mitochondria ($1,441 \pm 146$ [mean number of pixels $\pm$ SEM]) is similar to that of hearts treated with CSIL at 1, 5, 10, and 20 min of ischemia ($1,496 \pm 103$, $1,496 \pm 66$, $1,845 \pm 147$, and $1,504 \pm 101$, respectively; $p \pm NS$). The mean mitochondria size of hearts treated with IgG-L at 5, 10, and 20 min of ischemia ($2,294 \pm 95$, $2,387 \pm 119$, and $2,667 \pm 37$, respectively) or placebo ($2,234 \pm 270$) is greater than that of CSIL-treated hearts ($p \leq 0.05$) (18) (see Notes 6 and 8).

3.10.1. Dose Response

The LVDP of rat hearts treated with CSIL at antibody concentrations of 1, 0.5 and 0.2 mg in 24, 12 and 6 mg liposomal lipids is

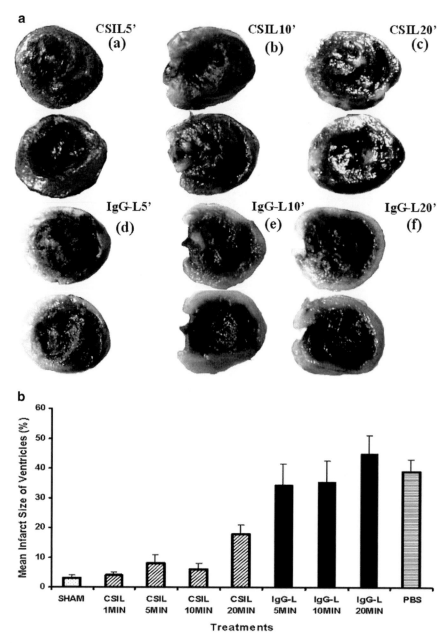

Fig. 3. (a) Digital photographs of mid-slices of nitro blue tetrazolium stained hearts treated with CSIL at 5 (a), 10 (b), and 20 (c) min of ischemia orIgG-L at 5 (d), 10 (e), and 20 (f) min of ischemia. (b) Mean infarct sizes are expressed as the percentage of right and left ventricles of hearts treated with CSIL, IgG-L, phosphate buffered saline (PBS), or sham instrumentation

shown in Fig. 4a (19). At 1 mg antibody dose with 24 mg lipids, mean LVDP during reperfusion is 87.4 ± 14% of sham control mean LVDP. AT 0.5 and 0.2 mg antibody doses (12 and 6 mg lipids respectively), the LVDP is 34.4 ± 17.5 and 25.7 ± 14.3% respectively. The mean infarct size (4.25 ± 3.2% of total ventricular

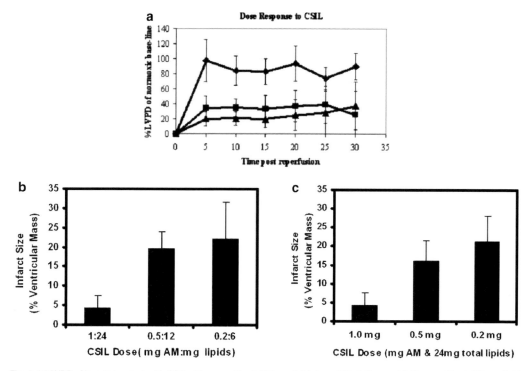

Fig. 4. (a) LVDP of hearts treated with CSIL at 1 mg antibody/24 mg lipid dose (*filled diamond*) 0.5 mg antibody/12 mg lipids (*filled rectangle*) and 0.2 mg antibody/6 mg lipids (*filled triangle*). (b) Mean infarct sizes of rat hearts treated with CSIL consisting of 1 mg antibody/24 mg lipids, 0.5 mg antibody/12 mg lipids or 0.2 mg antibody/6 mg lipids. (c) Mean infarct sizes of rat hearts treated with CSIL consisting of 1 mg, 0.5 or 0.2 mg antibody each concentration in 24 mg of lipid

mass) of rat hearts treated with 1 mg antibody/24 mg lipids is significantly smaller than the mean infarct sizes of rats treated with 0.5 mg antibody/12 mg lipids (19.8 ± 4.3 %, $P = 0.005$) or 0.2 mg antibody/6 mg lipids (22 ± 9.6%, $P = 0.04$) (Fig. 4b). Rat hearts treated with 0.5 or 0.2 mg antibody doses in 24 mg of lipids also show similar infarct size increase relative to the 1 mg antibody/24 mg lipid CSIL dose. At 0.5 mg antibody/24 mg lipid CSIL dose, the mean infarct size is 16.0 ± 5.5 ($P = 0.02$). At 0.5 mg antibody/24 mg lipid CSIL dose, the mean infarct size is 21.3 ± 7.0% ($P = 0.02$) (Fig. 4c) (see Notes 5 and 7).

3.11. Study Limitations

Our model of Langendorff perfused hearts used protein-free oxygenated perfusion buffer. It is a non-working heart model; therefore, the ischemic myocytes are not subjected to additional stresses. Furthermore, prolonged periods of reperfusion >30 min cannot be investigated in this model, because of the inability to maintain long-term steady-state myocardial function ex vivo. Future in vivo studies should permit investigation of prolonged ischemia and reperfusion with CSIL therapy. The results presented may not necessarily imply that CSIL therapy may be beneficial in a clinical scenario of AMI. Furthermore, no conclusions can be drawn regarding the impact of CSIL therapy on apoptotic myocardial cell death.

4. Conclusions

Our studies support the hypothesis that cardiac cell membrane lesion sealing with CSIL result in preservation of myocardial viability, as determined by function, histochemistry, and ultra-structural morphology. There is a time response to myocardial preservation with CSIL therapy. Early CSIL intervention after the onset of ischemia resulted in almost complete myocardial recovery (18). Even when the intervention was initiated at 20 min of global ischemia, myocardial preservation was still greater than that seen in hearts with IgG-L or placebo treatment. There is also a dose response to CSIL therapy. Sufficient concentration of CSIL is essential to achieve optimal cell membrane lesion sealing (19). Therefore, CSIL therapy may find therapeutic applications in preservation of myocardial viability and efficient non-viral gene therapy.

5. Notes

Current studies confirm preservation of myocardial viability in ex vivo adult ischemic hearts treated with CSIL. Treatment with CSIL initiated at 1 min of ischemia resulted in apparent complete preservation of function, viability, and mitochondrial integrity. A time-dependent relationship between CSIL administration and myocardial preservation is also evident. This beneficial effect is maintained even when treatment was initiated at longer durations of global ischemia (18). We also observed that myocardial preservation was dose dependent (19). Thus, CSIL treatment need not be initiated immediately, but may be rendered subsequent to the onset of acute myocardial infarction to reduce the infarct size.

1. In the coronary artery-occluded Langendorff rat heart model, irreversible injury occurred within 20 min of ischemia (28). Disruption of the sarcolemma was observed in canine hearts subjected to 15(29, 30) to 20 min of coronary occlusion and reperfusion (31).
2. The model of 25 min of global ischemia and 30 min of reperfusion was chosen to provide a more extensive irreversible myocardial injury than the 20-min localized ischemia model. The extent of myocardial injury is augmented by pacing the

hearts for an additional 5 min during global ischemia, which resulted in ~30% infarction of the total ventricles. The extent of recovery indicated by hemodynamic, histochemical, and mitochondrial ultra-structure indicates that myocardial preservation after CSIL treatment was achieved. Treatment with CSIL resulted in a rapid recovery of function; whether this is due to the reversal of myocardial stunning must await further studies. Greater functional recovery observed in IgG-L-treated hearts than in placebo hearts is not consistent with the infarct size data. The presence of IgG-L in the ischemic myocardium may provide a non-specific mechanism of increasing myocardial function (18, 19).

3. Preservation of myocardial viability with early CSIL treatment is consistent with clinical observations that maximal therapeutic benefits are associated with early intervention (32).

4. This benefit persisted even with late administration of CSIL. However, there is a time-dependent delay in the recovery to near normal LVDP with a delay in the initiation of CSIL therapy (Figs. 1b, 2a, b), which may be due to the need for more extensive myocardial cell membrane lesion sealing. The IgG-liposomes may temporarily plug membrane lesions without fusion with the cell membrane, ultimately leading to myocardial cell death, as determined by NBT. Nonetheless, the LVDP of IgG-L-treated hearts was still lower than that of CSIL treated hearts (17, 18).

5. In isolated perfused hearts, ventricular dysfunction may be due to myocardial stunning or lethal cell injury. In the Langendorff perfused ischemic rat hearts, ATP concentrations decrease rapidly to 60% in the first minute, with a rapid secondary decrease by 13 min due to contracture (33). Recovery from stunned to normal myocardium requires 24–48 h in the in vivo reperfused heart with coronary artery occlusion of 2–20 min (34). Dobrinina et al. (11) showed that neutral liposomes preserved liver integrity in rats subjected to hepatotropic poisons by non-specific mechanisms. However, histochemical infarct size data do not support this hypothesis in the myocardium (Fig. 3b).

6. Stunning is related to decreased availability of adenosine triphosphate (ATP) to the myofibrils; injury of the contractile apparatus; alterations in calcium homeostasis with calcium overload; and burst of free radicals. The earliest consequence of severe ischemia is contractile dysfunction, occurring within 6–10 s of ischemia. Loss of tetrazolium staining after ischemic injury occurs due to loss of cofactor NADH and reflects lack of dehydrogenase enzymatic activity. The total NAD and NADH contents are relatively stable in the early phases of ischemia but decrease after irreversible myocardial injury.

Pronounced mitochondria swelling with loss of cristae, development of amorphous matrix densities, and breaks in the sarcolemma are seen in irreversibly injured myocytes. Increased mitochondrial size is an early indication of ischemia. In hearts with 3–15 min of ischemia, morphologic injury is reversible, yet some mitochondria remain swollen even at 20 min of reperfusion. Therefore, a majority of the mitochondria in the ischemic myocardium is expected to be edematous at 30 min of reperfusion. However, early CSIL treatment of ischemic hearts (1 and 5 min) results in normal mitochondrial size, smaller than that of CSIL-treated hearts at later times (35). Cessation of mitochondrial electron transport has been observed 2 s after the onset of global ischemia in isolated rat hearts (36). The results of mitochondrial size determination are compatible with the delay in the recovery of function with later administration of CSIL. Thus, mitochondrial ultra-structural data agree with the functional and histochemical data. Preservation of cell membrane integrity after CSIL intervention results in a faster recovery of function and a reduction in infarct size. Whether this is associated with prevention of the influx of extracellular Ca_2^{++} that is associated with ischemia and reperfusion is not assessed in our studies. However, uncontrolled influx of Ca_2^{++} into the cytosol occurs after reperfusion and results in rigor (37, 38).

7. Hearts treated with CSIL at 1 min of ischemia are similar to sham hearts that have minimal injury, a result of the insertion of the Thebesian drainage port into the apex. Hearts treated with CSIL at earlier times (1, 5, and 10 min) show extensive NBT staining, consistent with maximal myocardial preservation. Hearts treated with CSIL at 20 min result in increased regions of negative NBT staining, indicating that myocardial recovery is not complete (Fig. 3a, panel c). However, the extent of histochemical myocardial injury is consistent with the extent of recovery of function in these hearts. Similar to the recovery of LVDP of hearts treated with CSIL, NBT staining also indicate that there is greater myocardial salvage in all hearts treated with CSIL than in IgG-L-treated controls (Fig. 3b) (18, 19).

8. Treatment with CSIL enables globally ischemic hearts to return to near normal function within 15 min of reperfusion, which is consistent with the prevention of the occurrence of uncontrolled myocardial Ca_2^{++} overload. The absence of mitochondrial swelling and the return of function to near normal in CSIL-treated hearts are also consistent with the maintenance of Ca_2^{++} homeostasis. Cell membrane lesion sealing with neutral immunoliposomes may also reduce injury mediated by acid and oxidative stress. However, plain liposomes in serum-free

perfusate augment this injury. Plain liposomes may mimic fatty acids that inhibit enzymes, which may result in the uncoupling of oxidative phosphorylation. They may also act as detergents, resulting in additional disruption of cell membranes (39).

Acknowledgments

We thank Drs. V. P. Torchilin, S. Gutmann, W. C. Hartner, and W. Fowle for the use of their equipment, advice on statistical analysis and electron microscopy, and helpful suggestions.

References

1. Leist M, Single B, Castoldi AF, Kühnle S, Nicotera P (1997) Intracellular adenosine triphosphate (ATP) concentration: a switch in the decision between apoptosis and necrosis. J Exp Med 185:1481–1486
2. Eguchi Y, Shimizu S, Tsujimoto Y (1997) Intracellular ATP levels determine cell death fate by apoptosis or necrosis. Cancer Res 57:1835–1840
3. Fischer S, MacLean AA, Liu A, Cardella JA, Slutsky AS, Suga M, Moreia JF, Keshavjee S (2000) Dynamic changes in apoptotic and necrotic cell death correlate with severity of ischemia–reperfusion in lung transplantation. Am J Respir Crit Care Med 162:1932–1939
4. Ohno M, Takemura G, Ohno A, Misao J, Hayakawa Y, Minatoguchi S, Fujiwara T, Fujiwara H (1998) "Apoptotic" myocytes in infarct area in rabbit hearts may be oncotic-myocytes with DNA fragmentation. Circulation 98:1422–1430
5. Majno G, Joris I (1995) Apoptosis, oncosis, and necrosis: an overview of cell death. Am J Pathol 146:3–15
6. Shi R, Qiao X, Emerson N, Malcom A (2001) Dimethylsulfoxide enhances CNS neuronal plasma membrane resealing after injury in low temperature or low calcium. J Neurocytol 30(9–10):829–839
7. McNeil PL (2002) Repairing a torn cell surface: makeway, lysosomes to the rescue. J Cell Sci 115(Pt 5):873–879
8. Togo T, Alderton JM, Steinhardt RA (2003) Long-term potentiation of exocytosis and cell membrane repair in fibroblasts. Mol Biol Cell 14:93–106
9. McNeil PL, Ito S (1989) Gastrointestinal cell plasma membrane wounding and resealing in vivo. Gastroenterology 96(5 Pt1):1238–1248
10. Walev I, Hombach M, Bobkiewicz W, Fenske D, Bhakdi S, Husmann M (2002) Resealing of large transmembrane pores produced by streptolysin O in nucleated cells is accompanied by NF-kappa B activation and downstream events. FASEB J 16(2):237–239
11. Dobrinina OV, Shatinina Z, Archakov AI (1990) Phosphatidylcholineliposomes in the repair of hepatocyte plasma membrane. Bull Exp Biol Med 110:94–96
12. Wang CY, Huang L (1987) pH-sensitive immunoliposomes mediate target-cell-specific delivery and controlled expression of a foreign gene in mouse. Proc Natl Acad Sci U S A 84:7851–7855
13. Kono K, Iwamoto M, Nishikawa R, Yanagie H, Takagishi T (2000) Design of fusogenic liposomes using a poly(ethylene glycol) derivative having amino groups. J Control Release 68:225–235
14. Kunisawa J, Masuda T, Katayama K, Yoshikawa T, Tsutsumi Y, Akashi M, Mayumi T, Nakagawa S (2005) Fusogenic liposome delivers encapsulated nanoparticles for cytosolic controlled gene release. J Control Release 105:344–353
15. Khaw BA, Torchilin VP, Vural I, Narula J (1995) Plug and seal: prevention of hypoxic cell death by sealing membrane lesions with cytoskeleton-specific immunoliposomes. Nat Med 1:1195–1198
16. Khaw BA, da Silva J, Vural I, Narula J, Torchilin VP (2001) Intracytoplasmic gene delivery for in vitro transfection with cytoskeleton-specific immunoliposomes. J Control Release 75:199–210
17. Khaw BA, Narula J, Vural I, Torchilin VP (1998) Cytoskeleton-specific immunoliposomes: sealing of hypoxic cells and intracellular delivery of DNA. Int J Pharm 162:71–76

18. Khudairi T, Khaw BA (2004) Preservation of ischemic myocardial function and integrity with targeted cytoskeleton-specific immunoliposomes. J Am Col Cardiol 43:1683–1689
19. Khaw BA, Khudairi T (2007) Dose-response to cytoskeletal-antigen specific immunoliposome therapy for preservation of myocardial viability and function in Langendorff instrumented rat hearts. J Liposome Res 17:1–15
20. Ey PL, Prowse SJ, Jenkin CR (1978) Isolation of pure IgG1, IgG2a and IgG2b immunoglobulins from mouse serum using protein A-sepharose. Immunochemistry 15:429–436
21. Engvall E, Perlmann P (1972) Enzyme-linked immunosorbent assay, ELISA. 3. Quantitation of specific antibodies by enzyme-labeled anti-immunoglobulin in antigen-coated tubes. J Immunol 109:129–135
22. Weissig V, Lasch J, Klibanov AL, Torchilin VP (1986) A new hydrophobic anchor for the attachment of proteins to liposomal membranes. FEBS Lett 202:86–90
23. Mori A, Huang L (1993) Immunoliposome targeting in a mouse model: optimization and therapeutic application. In: Gregoratos G (ed) Liposome technology, vol 3. CRC Press, Boca Raton, FL, pp 153–163
24. Neely JR, Rovetto MJ (1975) Techniques for perfusing isolated rat hearts. Methods Enzymol 6:43–60
25. Vivaldi MT, Kloner RA, Schoen FJ (1985) Triphenyltetrazolium staining of irreversible ischemic injury following coronary artery occlusion in rats. Am J Pathol 121:522–530
26. Bozzola JJ, Russell LD (1992) Electron microscopy: principles and techniques for biologists. Jones and Bartlett, Boston, MA
27. Zar JH (1999) Biostatistical analysis, 4th edn. Prentice-Hall, Upper Saddle River, NJ
28. Takashi E, Ashraf M (2000) Pathologic assessment of myocardial cell necrosis and apoptosis after ischemia and reperfusion with molecular and morphological markers. J Mol Cell Cardiol 32:209–224
29. Jennings RB, Schaper J, Hill ML, Steenbergen C, Reimer KA (1985) Effect of reperfusion late in the phase of reversible ischemic injury: changes involume, electrolytes, metabolites, and ultrastructure. Circ Res 56:262–278
30. Jennings RB, Sommers HM, Smyth GA, Flack HA, Linn H (1960) Myocardial necrosis induced by temporary occlusion of a coronary artery in the dog. Arch Pathol 70:68–78
31. Koba S, Konno N, Suzuki H, Katagiri T (1995) Disruption of sarcolemmal integrity during ischemia and reperfusion of canine hearts as monitored by use of lanthanum ions and a specific probe. Basic Res Cardiol 90:203–210
32. Shuster M, Dickinson D (1996) Recommendations for ensuring early thrombolytictherapy for acute myocardial infarction. Can Med Assoc J 154:483–487
33. Herse DJ, Garlick PB, Humphrey SM (1977) Ischemic contracture of the myocardium: mechanisms and prevention. Am J Cardiol 39:986–993
34. Roberts R, Morris D, Pratt CM (1994) Pathophysiology, recognition, and treatment of acute myocardial infarction and its complications. In: Schlant RC, Alexander RW (eds) Hurst's the heart: arteries and veins, 8th edn. McGraw Hill, New York, NY, pp 1107–1184
35. Reimer KA, Jennings RB (1992) Myocardial ischemia, hypoxia, and infarction. In: Fozzard HA, Haber E, Jennings RB (eds) The heart and cardiovascular system, 2nd edn. Raven Press, New York, NY, pp 1875–1953
36. Whitman G, Kieval R, Wetstein L, Seeholzer S, McDonald G, Harken A (1983) The relationship between global myocardial redox state and high energy phosphate profile: a phosphorous-31 nuclear magnetic resonance study. J Surg Res 33:332–339
37. Ganote CE, Nayler WG (1985) Contracture and the calcium paradox. J Mol Cell Cardiol 17:733–745
38. Murphy JG, Marsh JD, Smith TW (1987) The role of calcium in ischemic myocardial injury. Circulation 75(Suppl V):V15–V24
39. Wenzel DC, Hale TW (1978) Toxicity of free fatty acids for cultured rat heart muscle and endothelial cells. II. Unsaturated long-chain fatty acids. Toxicology 11(2):119–125

Chapter 22

Gadolinium-Loaded Polychelating Polymer-Containing Tumor-Targeted Liposomes

Suna Erdogan and Vladimir P. Torchilin

Abstract

Magnetic resonance (MR) is one of the most widely used imaging modalities in contemporary medicine to obtain images of pathological areas. Still, there is a big effort to facilitate the accumulation of contrast in the required zone and further increase a local spatial concentration of a contrast agent for better imaging. Certain particulate carriers able to carry multiple contrast moieties can be used for an efficient delivery of contrast agents to areas of interest and enhancing a signal from these areas. Among those carriers, liposomes draw special attention because of their easily controlled properties and good pharmacological characteristics. To enhance the signal intensity from a given reporter metal in liposomes, one may attempt to increase the net quantity of carrier-associated reporter metal by using polylysine (PLL)-based polychelating amphiphilic polymers (PAP). In addition to heavy load of reporter metal onto the pharmaceutical nanocarrier (liposome), the accumulation of the contrast nanoparticles in organs and tissues of interest (such as tumors) can be significantly enhanced by targeting such particles both "passively," via the so-called enhanced permeability and retention (EPR) effect, or "actively," using various target-specific ligands, such as monoclonal antibodies. Combining three different properties – heavy load with Gd via the liposome membrane-incorporated PAP and tumor specificity mediated by the liposome-attached mAb 2C5 – in a single nanoparticle of long-circulating (PEGylated) liposomes could provide a new contrast agent for highly specific and efficient tumor MRI.

Key words: Tumor targeting, MRI, Gd, Liposome, Polychelating amphihilic polymer, Cancer-specific monoclonal antibody, 2C5 antibody

1. Introduction

Magnetic resonance (MR) is one of the most widely used imaging modalities in contemporary medicine to obtain images of pathological areas. As well as the other medical diagnostic imaging techniques such as Gamma Scintigraphy (GS), Computed Tomography (CT), and Ultra-Sonography (US), Magnetic Resonance Imaging

(MRI) requires that the sufficient intensity of a corresponding signal from an area of interest is to be achieved in order to differentiate this area from surrounding tissues. Usually, the tissue concentration of a contrast agent (such as Mn or Gd) that must be achieved for successful MR imaging (MRI) of different organs and tissues for early detection and localization of numerous pathologies is relatively high (10^{-4} M) (1). To facilitate the accumulation of contrast in the required zone and further increase a local spatial concentration of a contrast agent for better imaging, it was a natural progression in the development of the effective contrast agents to use certain particulate carriers able to carry multiple contrast moieties for an efficient delivery of contrast agents to areas of interest and enhancing a signal from these areas. Various particulate carriers have been suggested as carriers for contrast agents. Among those carriers, liposomes draw special attention because of their easily controlled properties and good pharmacological characteristics (2–4).

MRI using contrast liposomes is quite well advanced. Normally, liposomal contrast agents for MRI act by shortening relaxation times (T1 and T2) of surrounding water protons resulting in the increase (T1 agents) or decrease (T2 agents) of the intensity of a tissue signal (5–8). Mn or Gd which provide sufficient changes in the relaxation times, are usually used to prepare liposomal contrasts for MR imaging. Still, because of toxicity and poor solubility of free paramagnetic heavy metal cations at physiologic pH, chelated complexes are used in most MRI-T1 contrast agent designs (9). Membranotropic chelating agents such as DTPA–stearylamine (DTPA–SA) (10), DTPA–phosphatidylamine (DTPA–PE) (11–13) or DTPA–bis(methylamide) (DTPA–BMA) (14–16) consist of the polar head containing chelated paramagnetic atom and the lipid moiety that anchors the metal–chelate complex in the liposome membrane. Anchoring the metal–chelate complex on the liposome surface provides the better relaxivity when compared with liposome-encapsulated paramagnetic ions. Liposomes with membrane-bound paramagnetic ion also have a reduced risk of the leakage of potentially toxic metals in the body (17, 18).

To enhance the signal intensity from a given reporter metal in liposomes, one may attempt to increase the net quantity of carrier-associated reporter metal. For this purpose, polylysine (PLL)-based polychelating amphiphilic polymers (PAP) have been suggested (19). In these polymers, polychelating polymers additionally modified with a hydrophobic residue to assure their firm incorporation into the liposome membranes. This polychelator easily incorporates into the membrane in the process of liposome preparation and sharply increases the number of bound reporter metal atoms per vesicle and to decrease the dosage without compromising the image signal intensity. In the case of MR, metal atoms chelated into these groups are directly exposed to the water environment that enhances the relaxivity of the paramagnetic ions and leads to the enhancement of the vesicle contrast properties (19–21).

In addition to heavy load of reporter metal onto the pharmaceutical nanocarrier (liposome), the accumulation of the contrast nanoparticles in organs and tissues of interest (such as tumors) can be significantly enhanced by targeting such particles both "passively," via the so-called enhanced permeability and retention (EPR) effect (22), or "actively," using various target-specific ligands, such as monoclonal antibodies (23, 24). Prolonged circulation of contrast liposomes in the blood, which could be achieved by coating liposomes with some soluble polymers, such as polyethylene glycol (PEG) (25), may improve the "passive" accumulation of contrast liposomes in tumors via the EPR effect (26).

Natural antinuclear autoantibodies (including the representative monoclonal 2C5 antibody, mAb 2C5) demonstrate nucleosome-restricted specificity and recognize various live cancer cells via the cancer cell surface-bound nucleosomes released from apoptotically dying cancer cells (27). Previous studies have clearly demonstrated that, in addition to their own anticancer activity (28), such antibodies can serve as tumor-targeting ligands. Combining these properties – heavy load with Gd via the liposome membrane-incorporated PAP and tumor specificity mediated by the liposome-attached mAb 2C5 – in a single nanoparticle of long-circulating (PEGylated) liposomes could provide a new contrast agent for highly specific and efficient tumor MRI.

2. Material

2.1. Synthesis of PAP, DTPA–Polylysyl-N-Glutaryl-Phosphatidyl-ethanolamine (DTPA–PLL–NGPE)

1. N-glutaryl-phosphatidyl ethanolamine (NGPE) (Avanti Polar Lipids, Inc.). Stored at −80°C.

2. N,N′-carbonyldiimidazole (Fluka Chemie GmbH).

3. N-hydroxysuccinimide (Sigma).

4. CBZ-protected PLL (MW 5400 Da) (Sigma).

5. Triethylamine (Sigma), (dry, bottle protected with septum).

6. Chloroform.

7. Dry ether.

8. Methanol:Chloroform (1:1 v/v).

9. DTPA anhydride (Sigma).

10. Succinic anhydride (Sigma).

2.2. Loading of Chelating Polymers with Gd Ions

1. $GdCl_3 \cdot 6H_2O$ (Sigma).
2. 0.1 M citrate buffer (pH 5.3).
3. Pyridine.
4. Dialysis Membrane (MWCO 3500) (Standard RC with glycerol for general dialysis) (Spectrum Medical Industries).

2.3. Synthesis of pNP-PEG-PE

1. pNP–PEG_{3400}–pNP (bis-paranitrophenyl carbonate polyethylene glycol), (SunBio).
2. DOPE (dioleoyl (18:1) phosphatidyl ethanolamine), (Avanti Polar Lipids, Inc.).
3. Triethylamine (TEA), (Sigma), (dry, bottle protected with septum).
4. Chloroform (dry, HPLC grade, no methanol).
5. CL-4B Sepharose (Sigma).
6. HCl (1.0 N).

2.4. Preparation of Gd-Loaded PAP-Containing PEGylated Liposomes

1. Egg Phosphatidylcholine (Avanti Polar Lipids, Inc.). Stored at –80°C.
2. Polyethylene glycol-phosphatidyl ethanolamine conjugate (PEG_{2000}-PE) (Avanti Polar Lipids, Inc.). Stored at –80°C.
3. Cholesterol (Sigma).
4. Octyl glucoside (Sigma).
5. HEPES-buffered saline (HBS): 10 mM Hepes, 150 mM NaCl, pH 7.4.
6. 5 mM Citrate buffer (pH 5.0) with 10 $mg.mL^{-1}$ octyl glucoside.
7. Tris buffered saline (TBS): 50 mM Tris–HCl, 150 mM NaCl, pH 7.4.
8. Cancer-specific monoclonal anti-nucleosome 2C5 antibody (mAb 2C5) (Harlan Bioproducts for Science).
9. Dialysis Membrane (MWCO 300 kDa) (Spectrum Medical Industries).

2.5. Antibody Activity Determination

1. 40 $\mu g.mL^{-1}$ poly-l-lysine (30,000–70,000).
2. Tris buffer saline (TBS): 50 mM Tris–HCl, pH 7.4-7.5, 150 mM NaCl.
3. TBST: TBS containing 0.05 % (w/w) Tween-20.
4. TBST–Casein: TBS containing 0.05 % (w/w) Tween-20 and 2 $mg.mL^{-1}$ casein.
5. 40 $\mu g.mL^{-1}$ nucleohistone (NH).
6. 1:5000 diluted anti-mouse IgG peroxidase conjugate.

7. Peroxidase substrate KBlue.
8. 0.5 M sulfuric acid.
9. 96 well polyvinylchloride microplates.

2.6. Interaction of Gd-Loaded PAP-Containing PEGylated Liposomes with Cancer Cells by Fluorescent Microscopy

1. LLC (Lewis Lung Carcinoma) (American Type Culture Collection).
2. MCF-7 (Human Breast Carcinoma) (American Type Culture Collection).
3. BT20 (Human Mammary Adenocarcinoma) (American Type Culture Collection).
4. Eagle's Minimum Essential Medium (MEM) (Mediatech, Inc.).
5. Dulbecco's Modified Eagle medium (DMEM) (Mediatech, Inc.).
6. Trypsin (Mediatech, Inc.).
7. Hank's Balanced Salt Solution (HBSS) (Mediatech, Inc.).
8. 1% Bovine serum albumin (Sigma) containing serum free medium (Mediatech, Inc.).
9. The fluorescence-free glycerol-based Trevigen® mounting medium (Trevigen Inc.).
10. Six well tissue plates.

2.7. In Vitro Cell Binding of 2C5 Modified-Gd-Loaded PAP-Containing PEGylated Liposomes

1. LLC (Lewis Lung Carcinoma) (American Type Culture Collection).
2. MCF-7 (Human Breast Carcinoma) (American Type Culture Collection).
3. BT20 (Human Mammary Adenocarcinoma) (American Type Culture Collection).
4. Eagle's Minimum Essential Medium (MEM) (Mediatech, Inc.).
5. Dulbecco's Modified Eagle medium (DMEM) (Mediatech, Inc.)
6. Trypsin (Mediatech, Inc.)
7. 0.25% trypsin–2 mM EDTA in Hank's Balanced Salt Solution (HBSS).
8. 1% Bovine serum albumin (Sigma) containing serum free medium (Mediatech, Inc.).
9. UPC10 (Antimyeloma antibody used as an isotype-matching nonspecific control) (ICN Pharmaceuticals).
10. 2 N NaOH.
11. Six well tissue plates.
12. 5 mHz RADX NMR Proton Spin Analyzer for reading of relaxivity value.

3. Methods

3.1. Synthesis of PAP, DTPA–Polylysyl-N-Glutaryl-Phosphatidylethanol-amine (DTPA–PLL–NGPE)

1. 25 mg N-glutaryl-phosphatidylethanolamine (NGPE) are activated with 20 mg N,N'-carbonyldiimidazole in the presence of 13 mg N-hydroxysuccinimide (NHS) for 16 h at room temperature.

2. At this point of time, 186 mg N-carbobenzoxy poly-L-lysine (CBZ-PLL MW 5400) and 5 µL triethylamine are added to the initial mixture and the reaction is allowed to proceed for another 5 h at room temperature with stirring. After about 1 h, additional 200 mL chloroform are added to the reaction mixture (see Note 1).

3. The whole reaction mixture is dried a rotary evaporator and suspended in about 25 mL double distilled water, filtrated and freeze-dried.

4. 174 mg CBZ–PLL–NGPE is dissolved in 8 mL of 30% hydrogen bromide in glacial acetic acid and the reaction is allowed to proceed for 2 h at room temperature.

5. Deprotected PLL–NGPE is precipitated with about 20 mL dry ethyl ether, washed with the same solvent and freeze-dried.

6. 150 mg PLL–NGPE is suspended in about 2 mL chloroform: methanol (1:1) mixture and reacted with 559 mg DTPA anhydride in 2 mL DMSO in the presence of 200 µL of triethylamine for 16 h at room temperature with stirring.

7. At this point of time, 485 mg succinic anhydride in 1 mL DMSO is added to block remaining polymer amino groups and the reaction is allowed to proceed for 1 h at room temperature.

8. The reaction mixture is purified from water soluble compounds by dialysis against deionized water (MWCO 3500) and then freeze-dried (see Note 2).

3.2. Loading of Chelating Polymers with Gd Ions

1. $GdCl_3$ $6H_2O$ (150 mg in 0.25 mL of 0.1 M citrate buffer, pH 5.3) is added to 25 mg of DTPA–PLL–NGPE suspended in about 2 mL of dry pyridine.

2. After 2 h of incubation at room temperature with stirring, the reaction mixture is dialyzed against deionized water (MWCO 3500) and freeze-dried (see Notes 2 and 3).

3.3. Synthesis of pNP–PEG–PE

1. Firstly, pNP–PEG_{3400}–pNP (800 mg, 213 µmole, 4.8× excess over PE) is dissolved in 10 mL dry $CHCl_3$ (80 mg.mL^{-1}) in a 25 mL pear-shaped flask.

2. DOPE (33 mg, 44 µmole, 1,310 µL of $CHCl_3$ solution (at 25 mg.mL^{-1})) is added to solution with magnetic stirring at room temperature.

3. Fresh TEA (20 μL, 143 μmole, 3× excess over PE) is added to solution with magnetic stirring.

4. The mixture is incubated overnight at RT under argon, with stirring (see Note 4).

5. The solvent is removed by rotary evaporator and then placed on freeze-drier overnight to eliminate any further residual solvent (see Note 5).

6. pNP-PEG-PE product is purified by Gel Filtration. For this purpose, firstly, a column that contains about 220 mL of clean CL-4B Sepharose media is set up and equilibrated in 0.001 M HCl (pH 3.0). Also, set-up automatic fraction collector using 13×100 mm test tubes and collecting 75 drops (~4 mL) per fraction is set up.

7. 4 mL of 1 mM HCl (pH 3.0) is added to dry reaction flask (should contain ~180 mg of product, and ~630 mg excess PEG) to hydrate the film, and form micelles (at a total PEG concentration ~200 mg.mL^{-1}) (see Note 6).

8. The 4 mL product solution is applied to column, and eluted with 1 mM HCl (pH 3), 300 ml maximum (see Note 7).

9. TLC is used to confirm the product in the fractions (using ½ of a 20×20 cm aluminum-backed sheet, and using a larger tank with 10 min development, and 65/25/4 or 80/20/2 compositions). When it is confirmed that the entire product has been eluted, no more fractions need to be collected.

10. Finally, pNP–PEG–PE product is isolated by freeze-drying. For this purpose, all fractions that contain product are pooled and placed in flasks (keep volume less than half full) and then flasks are placed on freeze-drier (Maintain on freeze-drier for at least 3 days).

3.4. Preparation of Gd-Loaded PAP-Containing PEGylated Liposomes

1. 13.38 mg of phosphatidyl choline, 2.92 mg cholesterol, 3.72 mg PEG2000-PE and 6.764 mg of Gd–DTPA–PLL–NGPE (1.75% mol of lipid), are introduced in a 50 mL round-bottomed flask and are solubilized in chloroform.

2. All lipids are mixed well and dried down on a rotary evaporator. The temperature of drying should be above the highest transition temperature of the individual lipid components. At this stage, lipids should dry onto wall of the vessel as a completely clear glassy film.

3. After removing the residual chloroform, dry lipid film is hydrated carefully with B-octylglucoside (50 mg.mL^{-1}) in HBS (pH 7.4) by gently swirling around the wall of the vessel. This must be carried out above the lipid transition temperature.

4. The mixture is homogenized by sonication at about 20 W for 30 min using a probe sonicator (see Note 8).

5. The mixture is introduced in dialysis bag (MWCO 3500) and dialyzed overnight against HBS at 4°C (see Note 2).

6. To prepare the 2C5 modified DTPA–PLL–NGPE containing PEGylated liposomes, firstly, 2.24 mg.mL^{-1} (10–40 molar excess) of pNP–PEG–PE dispersed in 10 mg.mL^{-1} micellar solution of octyl glucoside in 5 mM Na–citrate, 150 mM NaCl (pH 5.0) is added to an equal volume of a 2 mg.mL^{-1} of antibody solution in TBS (pH 8.5). The mixtures were incubated at pH 8.5 for 48 h at 4°C.

7. After 48 h incubation, the reaction mixtures are mixed in equal volumes with DTPA–PLL–NGPE containing PEGylated liposomes and incubated overnight at 4°C.

8. Following the incubation, the remaining octyl glucoside and free, non-incorporated into liposomes mAb 2C5 were removed by dialysis using cellulose ester dialysis tubes with a cutoff size of 300 kDa against 4 L HBS (pH 7.4) for 2 days at 4°C.

3.5. Antibody Activity Determination

1. Initially, the 96 well polyvinylchloride microplates are coated with 50 μl of 40 μg.mL^{-1} poly-L-lysine (30,000–70,000) dissolved in TBS and kept overnight at 4°C.

2. The poly-L-Lysine solution present in the wells is discarded and the wells are blocked with 200 μL of TBST–Casein for 1 h at room temperature.

3. Then the wells are coated with 50 μL of 40 μg.mL^{-1} nucleohistone (NH) in TBST–Casein for 1 h at room temperature.

4. After incubation with NH for 1 h, the wells are washed three times with 200 μL of TBST.

5. 10 mg.mL^{-1} of 2C5 in TBST–Casein is prepared and 73 μL is put in the first well.

6. 50 μL of TBST–Casein is put in all other wells in the column.

7. Sequential transfer of 23 μL from one well to the next well is made to get serial dilution of the antibody. The last well in the column is left free of antibody as a background control. The plate is incubated at room temperature for 1 h.

8. After incubation, the wells are washed three times with 200 μL of TBST.

9. The wells are incubated for 1 h at room temperature with 50 μL of 1:5000 diluted anti-mouse IgG peroxidase conjugate in TBST Casein.

10. The wells are washed three times with 200 μL of TBST.

11. Read the plate after 15 min of incubation with 100 μL of peroxidase substrate KBlue into each well a 630 nm with the

reference filter at 490 nm. If needed, stop the reaction by adding 50 μL of 0.5 M sulfuric acid and read the plate at 450 nm.

3.6. Interaction of 2C5 Modified-Gd-Loaded PAP-Containing PEGylated Liposomes with Cancer Cells by Fluorescent Microscopy

1. Adherent LLC, MCF-7 and BT20 cell lines were grown on glass cover slips placed into six well tissue plates.
2. Cells were incubated at 37°C, 5% CO_2 in a humidified atmosphere until reached a confluence of 70–80%.
3. After the cells reached a confluence of 70–80%, the plates were washed twice with Hank's buffer.
4. Then, treated with 1% BSA in a serum free media and incubated for 1 h at 37°C, 5% CO_2 in a humidified atmosphere.
5. After discarding the BSA-containing serum-free media, the cells were incubated with 0.5 mol% of the amphiphilic fluorescent label, Rh–PE containing 2C5-modified and plain Gd–DTPA–PLL–NGPE-containing PEG–liposomes in Hank's buffer (see Subheading 3.3 for preparation of liposome dispersion).
6. After 2 h incubation at 37°C, 5% CO_2, the cells were washed twice with Hank's buffer.
7. Then, the cover slips were mounted individually cell-side down on fresh glass slides by using the fluorescence-free glycerol-based Trevigen® mounting medium.
8. Mounted slides were observed with the Nikon Eclipse microscope under the bright light or under the epi-fluorescence using Rhodamine/TRITC filter (Fig. 1).

3.7. In Vitro Cell Binding of 2C5 Modified-Gd-Loaded PAP-Containing PEGylated Liposomes

1. Adherent LLC, MCF-7 and BT20 cell lines are eeded at (1×10^6 cells/well) into six-well tissue culture plates.
2. Cells are incubated at 37°C, 5% CO_2 in a humidified atmosphere until they reach a confluence of 70–80%.
3. After 24 h incubation, the cells are treated with 1% BSA in a serum free media for 1 h to prevent a nonspecific binding.
4. Then, the medium is replaced with various concentrations of Gd-loaded PAP-containing antibody-free liposomes or mAb 2C5 and UPC10-modified immunoliposomes for 2 h at 37°C.
5. The cells are washed three times with serum free media to remove the unbounded liposomes.
6. Then, cells are detached using 0.25% trypsin–2 mM EDTA in HBSS (Hank's Balanced Salt Solution) and incubated at 55°C for 30 min with a solution of 2 N NaOH to lyses.
7. After lyses, the cells are resuspended in 1 mL serum-free media and their relaxation parameters (i.e., the quantity of cell-associated Gd) were determined (Figs. 2a–c).

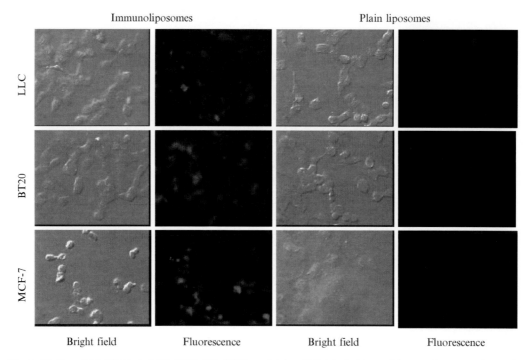

Fig. 1. Binding of Rh-PE-labeled Gd–DTPA–PLL–NGPE-containing PEGylated plain and immunoliposomes to murine Lewis lung carcinoma (LLC) and human mammary adenocarcinoma (BT-20) and breast adenocarcinoma (MCF-7) cells (With permission from (21))

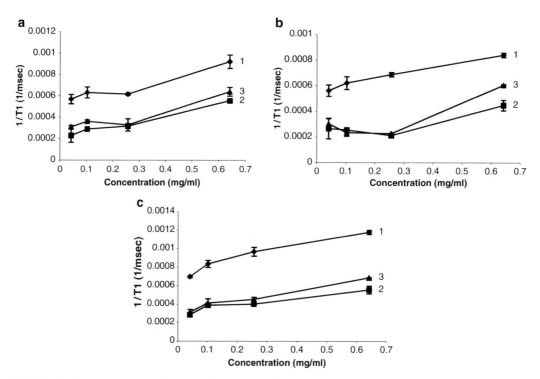

Fig. 2. (**a–c**) Relaxation parameters (reflecting the quantity of the cell-associated Gd and MR signal intensity) of various cancer cells incubated with 2C5-immunoliposomes loaded with Gd via DTPA–PLL–NGPE (1) and with control Gd–liposomes of the same composition [(UPC-10-Gd-liposomes (2); plain Gd–liposomes (3)]: LLC (a), BT20 (b) and MCF-7 (c) cells ($n = 6$) (With permission from (21))

3.8. Tumor Visualization by MRI

1. MR imaging was performed on a 9.4 T Bruker horizontal bore scanner (Billerica, MA) equipped with a home-built RF transmit and receive 3×4 cm elliptical surface coil and using ParaVision 3.0 software.

2. Tumor-bearing animals were imaged prior and 4, 24, and 48 h after intravenous injection of 2C5-modified and unmodified Gd–PAP-containing PEGylated liposomes.

3. T1 maps were acquired using a RARE inversion recovery sequence with the following parameters: TE = 7.253 ms, TR = 10,000 ms, TI = 0.001, 200, 400, 800, 1,600, 3,200, and 6,400 ms. FOV = 25.6 × 25.6 mm², in-plane spatial resolution = 0.2 × 0.2 mm².pixel^{-1}, matrix size = 128 × 128, slice thickness = 1 mm, and a total imaging time of 16 min 5 s (Fig. 3).

4. For quantitative analysis of T1 relaxation, T1 color-coded maps were constructed using Marevisi 3.5 software (Institute for Biodiagnostics, National Research Council, Canada).

5. Tumors were manually segmented and subjected to region-of-interest (ROI) analysis for the determination of tumor-associated T1 relaxation times (Fig. 4).

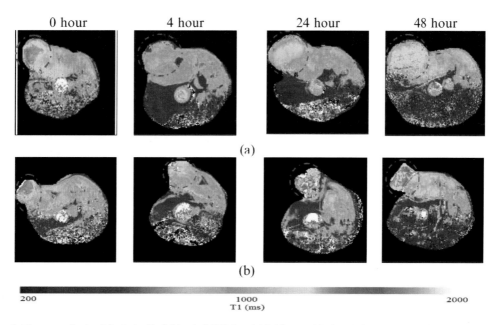

Fig. 3. T1 maps of mice injected with Gd-loaded PEGylated (**a**) 2C5-modified, and (**b**) unmodified Gd-PAP-containing liposomes (With permission from (29))

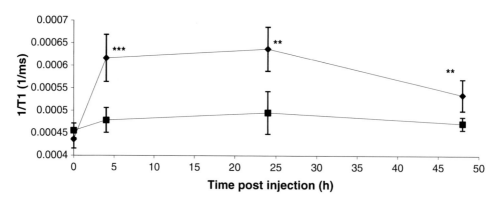

Fig. 4. Tumor-associated R1 values of mice injected with 2C5-modified, (*filled diamond*), and unmodified, (*filled square*), Gd-PAP-containing PEGylated liposomes at different post-injection times (***$P<0.01$; **$P<0.05$) (With permission from (29))

4. Notes

1. Conversion of initial NGPE is checked by Thin Layer Chromatography.
   Stationary Phase: Silica Gel 60 F_{254} TLC plate
   Mobil phase: Chloroform/Methanol/Water = 65: 25:4
   UV light is used for general visualization. Rf value of initial NGPE = 0.37
   NGPE into product N,α-(ε-CBZ-PLL) NGPE = 0.59

2. Dialysis overnight, extensive change of buffer.

3. Suspension might be very viscous in this case, it should be got warm.

4. The reaction is monitored by TLC (Silica Gel 60 F_{254}).
   Stationary Phase: Silica Gel 60 F_{254} TLC plate
   Mobil phase; 65/25/4 (or 80/20/2) Chloroform/Methanol/Water
   UV light is used for general visualization. For specific visualization, use dragendorff reagent, molybdenum reagent and ninhydrin reagent are used for PEG spots, PE spots (phosphorus detection) and primary amine spots, respectively.

5. To evaporation of solvent, magnetic stir-bar is removed, the solution is transferred to a round-bottom flask and the flask is placed on a rotary evaporator. The flask is tilted slightly so solution does not pool at bottom of flask during evaporation. And, evaporation is maintained until no more solvent will evaporate. If a viscous liquid still remains, Argon stream is used to dry the remaining viscous liquid as much as possible.

6. The flask is shaken and swirled by hand, then vortexed until dissolution. PEG and PEG–PE must be fully dispersed within the loading solution for Gel separation to occur. If full dispersion is questioned, the solution is carefully sonicated (e.g., in a water bath for not more than 5 min at 15 W).

7. The fraction collector is not activated until about 25% of the column volume has passed (~55 mL). The first 55 mL (holding volume) is collected separately in a beaker.

8. It should be avoided from overheating of the lipid suspension causing degradation. Sonication tips also tend to release titanium particles into the lipid suspension which must be removed by centrifugation prior to use.

Acknowledgment

This study was supported by NIH Grant R01-EB002995 to Vladimir P. Torchilin.

References

1. Wolf GL (1999) Delivery of diagnostic agents: achievements and challenges. Adv Drug Deliv Rev 37(1–3):1–12
2. Gregoriadis G (1993) Liposome technology, vol 1–3. CRC Press, Boca Raton, FL
3. Lasic DD (1993) Liposomes from physics to applications. Elsevier Science Publishers, Amsterdam
4. Torchilin VP (1997) Pharmacokinetic considerations in the development of labeled liposomes and micelles for diagnostic imaging. Q J Nucl Med 41:141–153
5. Tóth É, Helm L (2002) Relaxivity of MRI contrast agents. In: Krause W (ed) Topics in current chemistry, contrast agent 1, vol 221. Springer-Verlag, Berlin Heidelberg, pp 61–63
6. Unger E, Shen DK, Wu GL, Fritz T (1991) Liposomes as MR contrast agents: pros and cons. Magn Reson Med 22(2):304–308
7. Barsky D, Pütz B, Schulten K, Magin RL (1992) Theory of paramagnetic contrast agents in liposome systems. Magn Reson Med 24(1):1–13
8. Unger E, Tilcock C, Ahkong QF, Fritz T (1990) Paramagnetic liposomes as magnetic resonance contrast agents. Invest Radiol 25(Suppl 1):S65–S66
9. Gries H (2002) Extracellular MRI contrast agents based on gadolinium. In: Krause W (ed) Topics in current chemistry, contrast agent 1, vol 221. Springer-Verlag, Berlin Heidelberg, pp 3–29
10. Strijkers GJ, Mulder WJ, van Heeswijk RB, Frederik PM, Bomans P, Magusin PC, Nicolay K (2005) Relaxivity of liposomal paramagnetic MRI contrast agents. MAGMA 18(4):186–192
11. Kabalka G, Buonocore E, Hubner K, Moss T, Norley N, Huang L (1987) Gadolinium-labeled liposomes: targeted MR contrast agents for the liver and spleen. Radiology 163(1):255–258
12. Kabalka GW, Davis MA, Moss TH, Buonocore E, Hubner K, Holmberg E, Maruyama K, Huang L (1991) Gadolinium-labeled liposomes containing various amphiphilic Gd-DTPA derivatives: targeted MRI contrast enhancement agents for the liver. Magn Reson Med 19(2):406–415
13. Trubetskoy VS, Cannillo JA, Milshtein A, Wolf GL, Torchilin VP (1995) Controlled delivery of Gd-containing liposomes to lymph nodes: surface modification may enhance MRI contrast properties. Magn Reson Imaging 13(1):31–37
14. McDannold N, Fossheim SL, Rasmussen H, Martin H, Vykhodtseva N, Hynynen K (2004)

Heat-activated liposomal MR contrast agent: initial in vivo results in rabbit liver and kidney. Radiology 230(3):743–752

15. Løkling KE, Fossheim SL, Skurtveit R, Bjørnerud A, Klaveness J (2001) pH-sensitive paramagnetic liposomes as MRI contrast agents: in vitro feasibility studies. Magn Reson Imaging 19(5):731–738

16. Løkling KE, Skurtveit R, Fossheim SL, Smistad G, Henriksen I, Klaveness J (2003) pH-sensitive paramagnetic liposomes for MRI: assessment of stability in blood. Magn Reson Imaging 21(5):531–540

17. Fritz T, Unger E, Wilson-Sanders S, Ahkong QF, Tilcock C (1991) Detailed toxicity studies of liposomal gadolinium–DTPA. Invest Radiol 26(11):960–968

18. Unger EC, Fritz TA, Tilcock C, New TE (1991) Clearance of liposomal gadolinium: in vivo decomplexation. J Magn Reson Imaging 1(6):689–693

19. Trubetskoy VS, Torchilin VP (1994) New approaches in the chemical design of Gd-containing liposomes for use in magnetic resonance imaging of lymph nodes. J Lipid Res 4:961–980

20. Weissig V, Babich J, Torchilin VP (2000) Long-circulating gadolinium-loaded liposomes: potential use for magnetic resonance imaging of the blood pool. Colloids Surf B Biointerfaces 18:293–299

21. Erdogan S, Aruna R, Sawant R, Hurley J, Torchilin VP (2006) Gadolinium-loaded polychelating polymer-containing cancer cell-specific immunoliposomes. J Lipid Res 16:45–55

22. Maeda H, Wu J, Sawa T, Matsumura Y, Hori K (2000) Tumor vascular permeability and the EPR effect in macromolecular therapeutics: a review. J Control Release 65:271–284

23. Gupta H, Weissleder R (1996) Targeted contrast agents in MR imaging. Magn Reson Imaging Clin N Am 4:171–184

24. Morawski AM, Lanza GA, Wickline SA (2005) Targeted contrast agents for magnetic resonance imaging and ultrasound. Curr Opin Biotechnol 16:89–92

25. Klibanov AL, Maruyama K, Torchilin VP, Huang L (1990) Amphipathic polyethyleneglycoles effectively prolong the circulation time of liposomes. FEBS Lett 268:235–237

26. Bertini I, Bianchini F, Calorini L et al (2004) Persistent contrast enhancement by sterically stabilized paramagnetic liposomes in murine melanoma. Magn Reson Med 52:669–672

27. Iakoubov LZ, Torchilin VP (1998) Nucleosome-releasing treatment makes surviving tumor cells better targets for nucleosomespecific anticancer antibodies. Cancer Detect Prev 22:470–475

28. Chakilam AR, Pabba S, Mongayt D, Iakoubov LZ, Torchilin VP (2004) A single monoclonal antinuclear autoantibody with nucleosome-restricted specificity inhibits growth of diverse human tumors in nude mice. Cancer Ther 2:353–364

29. Erdogan S, Medarova ZO, Aruna R, Moore A, Torchilin VP (2008) Enhanced tumor MR imaging with gadolinium-loaded polychelating polymer-containing tumor-targeted liposomes. J Magn Reson Imaging 27:574–580

Chapter 23

Angiogenic Vessel-Targeting DDS by Liposomalized Oligopeptides

Tomohiro Asai and Naoto Oku

Abstract

Liposomal oligopeptides are one of the promising nanocarriers to deliver a drug, DNA or siRNA to target tissues. In this chapter, we describe our methodology to develop liposomal oligopeptides targeting to tumor angiogenic vessels. At first, we introduce our strategies to identify objective peptides. We performed in vivo biopanning using a phage-displayed peptide library and identified Ala-Pro-Arg-Pro-Gly (APRPG) peptide as a ligand for angiogenic vessels. To modify APRPG peptide on the surface of PEGylated liposomes, we synthesized a novel lipid derivative of the peptide, distearoylphosphatidylethanolamine–polyethyleneglycol–APRPG (DSPE–PEG–APRPG). The lipid derivative of APRPG peptide is expected to be readily incorporated into liposomal membrane and enables to present the peptides on the surface of PEGylated liposomes.

We next describe how to evaluate the advantages of liposomal oligopeptides using specific examples; (1) Intratumoral distribution of APRPG–PEG-modified liposomes, (2) Therapeutic efficacy of adriamycin encapsulated in APRPG–PEG-modified liposomes, (3) Preparation of 5′-*O*-dipalmitoylphosphatidyl 2′-*C*-cyano-2′-deoxy-1-β-D-*arabino*-pentofuranosylcytosine (DPP–CNDAC) liposomes modified with APRPG–PEG, and (4) Therapeutic experiment with APRPG–PEG-modified liposomal DPP–CNDAC.

Key words: Oligopeptides, Liposomes, Polyethyleneglycol, Angiogenic vessels, APRPG

1. Introduction

Active targeting to specific tissues such as tumors is achieved by modification of drug carriers with certain ligands (oligopeptides, proteins, antibodies, glycoconjugates, and so on). In this chapter, we describe the usefulness of liposomal oligopeptides for a drug delivery system. Liposomalization of oligopeptides requires only very simple technique by using lipid derivative of oligopeptides that is easily synthesized and readily incorporated into the liposomal membrane (1–4). In addition, oligopeptides can be bound to a polyethyleneglycol (PEG)–lipid conjugate such as, PEG–distearoylphosphatidylethanolamine (PEG–DSPE) and can be presented on the surface of liposomes (5–7). PEG-coating of liposomes has been used in a liposomal DDS, since PEGylated liposomes characteristically remain in the blood circulation longer than non-modified ones through avoidance of reticuloendothelial system (RES)-trapping of drug carriers (8). PEG-coating of the liposomal surface is known to form a fixed aqueous layer around the liposome due to the interaction between the PEG and water molecules. Thus, PEGylated liposomes prevent the binding of certain serum proteins and opsonins that are responsible for the RES-trapping (9). In case of PEGylated liposomes modified with oligopeptides, this long circulating characteristic in the bloodstream increases the opportunity for specific binding of ligand-modified liposomes to target tissues. Here, we present methodologies and results from our recent studies on liposomal oligopeptides. We have developed liposomal oligopeptides to construct angiogenic vessel-targeting DDS for cancer chemotherapy. Angiogenic vessel-targeting DDS has become a focus of interest, since angiogenic vessels have properties different from those of the preexisting systemic vessels (10) and certain drugs or drug carriers first meet angiogenic vessels before extravasation into tissues such as tumors. Angiogenic endothelial cells express specific address molecules that are not or little expressed on preexisting ones (10). Therefore, specific oligopeptides against these address molecules are applicable for active targeting to angiogenic vessels.

For constructing angiogenic vessel-targeting DDS, we firstly isolated a peptide that specifically bound to tumor angiogenic vessels from a phage-displayed peptide library. The epitope sequence of the peptide is determined to be Ala-Pro-Arg-Pro-Gly (APRPG) (2). Then, we synthesized stearoyl-APRPG for the modification of liposomes and demonstrated that APRPG is a useful probe for angiogenic vessel-targeting liposomal DDS (2). In fact, liposomes modified with stearoyl-APRPG highly accumulated in the tumor implanted in mice, and the liposomes encapsulating anticancer drugs strongly suppressed the tumor growth (2, 11, 12). Next, we endowed angiogenic vessel-targeted liposomes with long-circulating characteristic by PEGylation. This approach is expected to cause passive targeting of liposomes in tumor tissues in addition to active targeting of them by oligopeptides, since the angiogenic vessels are quite leaky and PEGylated liposomes as well as macromolecules easily accumulate in the interstitial tissues of tumors due to enhanced permeability and retention (EPR) effect (13). For this purpose, we designed a novel conjugate composed of APRPG peptide, PEG and hydrophobic anchor, namely DSPE, and examined the applicability of APRPG–PEG-modified liposomes for cancer treatment (5–7). As a result, it has been demonstrated that APRPG–PEG modification is superior to just APRPG-modification for enhancing antitumor activity of liposomal doxorubicin (14). Furthermore, APRPG–PEG-modified liposomes could deliver an antiangiogenic agent to angiogenic vessels, resulting in suppression of angiogenesis and tumor growth (15).

On the other hand, we developed angiogenic vessel-targeting liposomal 2′-C-cyano-2′-deoxy-1-β-D-*arabino*-pentofuranosylcytosine (CNDAC). CNDAC has a novel anticancer mechanism and induces DNA strand breaks after its incorporation into tumor cell DNA (16). We previously designed 5′-O-dipalmitoylphosphatidyl CNDAC (DPP–CNDAC) to incorporate it into the liposomal bilayer (17), since CNDAC itself is not suitable for the efficient encapsulation in liposomes (see Note 1). APRPG-modified liposomes containing DPP–CNDAC actually caused tumor growth suppression through damaging angiogenic endothelial cells (11). However, in vivo behavior of the liposomes was affected by the presence of the cyano group of DPP–CNDAC on the liposomal surface. It induced aggregation of liposomes, resulting in reduced blood circulation of liposomes. In this case, the potential of APRPG-modification would be attenuated in the blood circulation. Therefore, we masked the CNDAC moiety on the liposomal surface with APRPG–PEG conjugate to erase this undesirable property of DPP–CNDAC in liposomalization (see Note 2). As a result, the improvement of the blood circulation afforded by the use of APRPG–PEG conjugate enhanced the accumulation of the liposomes in the tumor, enabled targeting to the angiogenic endothelial cells, and caused efficient damage to the tumor cells (18). Our studies suggest that PEG-shielding of the liposomal surface should be useful for designing active targeting DDS with oligopeptides as well as passive targeting.

2. Materials

2.1. Lipids, Cells, and other Materials

1. Distearoylphosphatidylcholine (DSPC), DSPE and cholesterol were the products of Nippon Fine Chemical Co., Ltd. (Takasago, Hyogo, Japan).

2. A phage-displayed random peptide library expressing pentadecamer amino acid residues at the N terminus of pIII phage coat protein of M13 phage was kindly provided by Dr. Hideyuki Saya at Keio University.

3. Colon 26 NL-17 colon carcinoma cells were established by Dr. Yamori (Japanese Foundation for Cancer Research, Tokyo, Japan) and kindly provided by Dr. Nakajima (Johnson & Johnson K.K., Tokyo, Japan).

4. A fluorescence dye for labeling liposomes, 1,1′-dioctadecyl-3,3,3′,3′-tetramethylindo-carbocyanine perchlorate

(DiI C18), was purchased from Molecular Probes Inc. (Eugene, OR, USA).

5. Other materials: Reduced Triton X-100 and fetal bovine serum were purchased from Sigma-Aldrich Co. (St. Louis, MO, USA).

2.2. Animals

1. Five-week-old BALB/c, C57BL/6, or BALB/c nu/nu male mice were obtained from Japan SLC Inc. (Shizuoka, Japan). The animals were cared according to the animal facility guideline of the University of Shizuoka.

3. Methods

3.1. Identification of Peptide

1. Angiogenic model mice: Angiogenic vessels were formed on murine dorsal skin for in vivo biopanning (19). Highly metastatic murine B16BL6 melanoma cells (1×10^7 cells/ring) were loaded into a Millipore chamber ring. The chamber rings were dorsally implanted into 5-week-old C57BL/6 male mice. Five days after the implantation, these mice bearing angiogenic vessels on the dorsal skin were used for in vivo biopanning.

2. In vivo biopanning was performed by a modified method as described by Paspualini et al. (20, 21). The phage-displayed peptide library (1×10^{13} cfu) was intravenously injected into angiogenic vessel-bearing mice. Four minutes after the injection, the phages that had accumulated in angiogenic vessels were recovered and titrated. The skin attached to the Millipore chamber ring where the angiogenic vessels had been formed was dissected, minced, and homogenized with ice-cold DMEM containing 1 mM phenyl methyl sulphonyl fluoride. This homogenate was washed three times ($30,000 \times g$ for 10 min) with ice-cold DMEM containing 1% BSA, and the accumulated phages were recovered by infecting *E. coli* K91KAN with them. A part of the phages in the homogenate was used for the titration of the accumulated phages, and the remaining phages were amplified in *E. coli* K91KAN and purified. Then, a second round of biopanning was performed similarly as per the first round. These biopanning steps were repeated for five cycles. At the fifth round of biopanning, the recovery rate of the phage (recovered phage titer to input phage titer) increased about thousand-fold over that of the first round, suggesting that selection of high-affinity phage clones capable of accumulating in the angiogenic site was successful (see Note 3).

3. The selected phages were cloned and the sequence of presented peptides was determined. For in vivo screening, 1.0×10^6 cells of B16BL6 were implanted subcutaneously into the posterior flank of 5-week-old C57BL/6 male mice. Each sample of phage clones (1.0×10^{11} cfu) was injected into tumor-bearing mice via a tail vein when the tumor size had become about 10 mm in diameter. Four minutes after the injection, the phages that had accumulated in the tumor were recovered and titrated. Similar experiment was performed in Meth A sarcoma-bearing mice. As a result, we demonstrated that PRPGAPLAGSWPGTS-presented phage clone highly accumulated in two types of murine tumor.

3.2. Characterization of Peptides

1. Pentadecamer peptides were synthesized by use of Rink amide resin (0.4–0.7 mmol/g) and a peptide synthesizer ACT357, resulting in an amide at the carboxyl terminus.

2. To confirm the capability of the synthetic peptides to accumulate in the tumor, we co-injected 0.25 µmol of synthetic peptide (PRPGAPLAGSWPGTS) and 5×10^8 cfu of corresponding phage clone into B16BL6 melanoma-bearing mice. Four minutes after injection, the titer of phages that had accumulated in the tumor was determined. Tumor accumulation of phage clone was inhibited in the presence of the corresponding synthetic peptide, although a random peptide, GLDLLGDVRIPVVRR, did not affect the phage accumulation.

3. To determine the epitope sequences of the peptides, we synthesized various short peptides based on original 15-mer sequence and examined the inhibitory effect of these peptides against tumor accumulation of the corresponding phage clone. Our results indicated that APRPG in original 15 mer sequences was essential for their affinity.

3.3. Synthesis of DSPE–PEG–APRPG

1. We designed the structure of DSPE–PEG–APRPG (1) as shown in Fig. 1. At first, we synthesized DSPE–PEG–SA (4) (Schemes 1 and 2) and APRPG respectively, and then condensed them to obtain DSPE–PEG–APRPG (1).

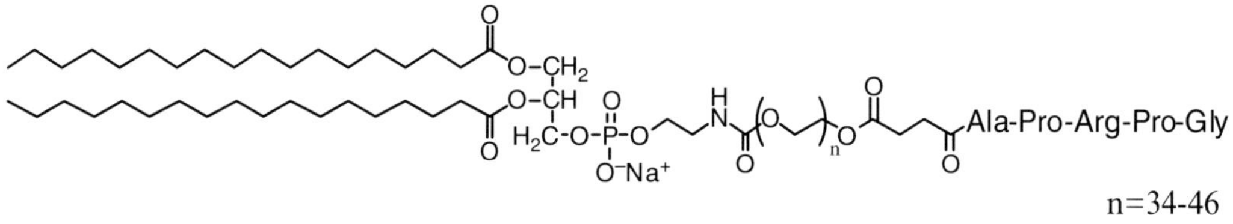

Fig. 1. Structure of DSPE–PEG–APRPG (1). Reproduced with permission from (5)

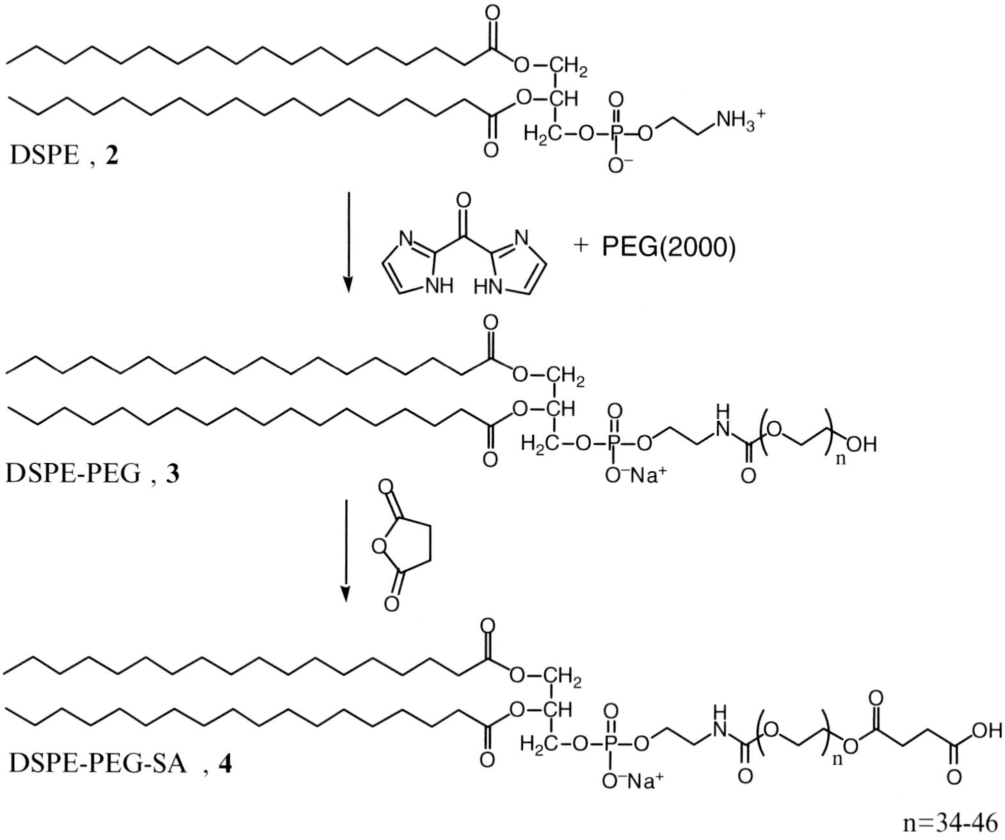

Scheme 1. Pathway for synthesis of DSPE–PEG–SA. Reproduced with permission from (5)

2. To synthesize DSPE–PEG–SA, DSPE (2) 15.0 g and carbonyl diimidazole (CDI) 3.9 g were dissolved in 70 mL of toluene. Reaction was performed at 100 °C for 1 h after addition of triethyl amine 2.0 g. Then, PEG (average molecular weight; 2,000) 40.0 g dissolved in toluene was added dropwise to the solution. The solvent was evaporated in vacuo followed by the reaction, and the product was dissolved in acetone 500 mL and insoluble materials were filtrated and the solvent was evaporated. The reaction mixture was exchanged into Na$^+$ salt with ion exchange resin. Purification by column chromatography on silica gave 11.4 g of the desired product (3) in a 26% yield. In order to use the PEG-end of obtained DSPE–PEG as a carboxylic group (referred to as (4)); it was allowed to react with succinic anhydride 2.1 g in the presence of pyridine 1.7 g in 100 mL of toluene. After powdering with ether, the yield of 4 was 80%.

3. Preparation of APRPG peptide moiety was carried out by the liquid-phase method as shown in Scheme 2. N,N'-dicyclohexylcarbodiimide (DCC, 1.1 equiv. based on peptide) and 1-hydroxybenzotriazol (HOBt, 1.1 equiv. based on peptide)

Angiogenic Vessel-Targeting DDS by Liposomalized Oligopeptides

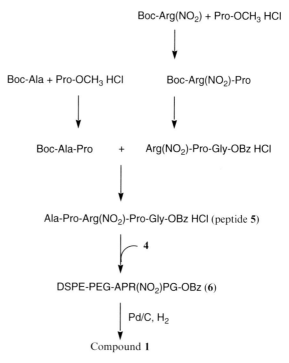

Scheme 2. Pathway for synthesis of DSPE–PEG–APRPG. Reproduced with permission from (5)

were used for peptide coupling in DMF. HCl in 1,4-dioxane was used for deprotection of the Boc group of N-terminal and NaOH was used for deprotection of methyl ester group of C-terminal in water and methanol. In order to avoid racemization, segment condensation was proceeded between Boc-Ala-Pro and Arg(NO_2)-Pro-Gly-OBz to yield 78% of Boc-Ala-Pro-Arg(NO_2)-Pro-Gly-OBz. Next, the Boc protecting group was deprotected by HCl in 1,4-dioxane to obtain peptide (5).

4. Peptide (5) was condensed with (4) (0.93 equiv. based on (5)) in $CHCl_3$ by DCC (1 equiv. based on (5)) and HOBt (1 equiv. based on (5)). The progress of the reaction was monitored by TLC. The reaction was almost complete overnight without any serious side reactions. It was purified by column chromatography on silica. The yield was 83% based on (4). Deprotections of NO_2 group of arginine side chain and benzyl ester group of glycine C-terminal were carried out by 10% palladium–carbon catalytic reduction under hydrogen atmosphere in methanol. It was purified by column chromatography on silica and ion exchange resin. This compound of single spot on TLC was in a 43% yield. This compound (DSPE–PEG–APRPG (1)) was positive for Sakaguchi reagent, while negative for UV lamp on TLC. These showed that NO_2 protecting group and benzyl ester protecting group were deprotected simultaneously (Fig. 1).

3.4. Preparation of PEGylated Liposomal Oligopeptides

1. DSPC and cholesterol with DSPE–PEG or DSPE-PEG–APRPG (10:5:1 as a molar ratio; PEG-lip or APRPG–PEG-lip, respectively) were dissolved in chloroform or chloroform/methanol, dried under reduced pressure, and stored in vacuo for at least 1 h. Liposomes were prepared by hydration of the thin lipid film with 0.3 M glucose, and frozen and thawed for three cycles using liquid nitrogen. Then liposomes were sized by extruding three times through a polycarbonate membrane filter with 100-nm pores (Nucleopore, Maidstone, UK).

2. For an observation of intratumor distribution of liposomes, DiI C18 of the quantity equivalent to 1 mol% of DSPC was added to the liposome. DiI C18-labeled PEG-lip and PEG–APRPG-lip were composed of DSPC, cholesterol, DSPE–PEG and DSPE–PEG–APRPG and DiI C18 (10:5:1:0.1 as molar ratio).

3. For therapeutic experiment, adriamycin (ADM)-encapsulated liposomes were prepared by modification of the remote-loading method as described previously (22). The concentration of ADM was determined by 484 nm absorbance.

4. Particle size and ζ-potential of liposomes diluted with PBS were measured by the use of a Zetasizer Nano ZS (MALVERN, Worcestershire UK, USA).

3.5. Intratumoral Distribution of Liposomal Oligopeptides

DiI C18-labeled liposomes were administered via tail vein of orthotopic tumor model mice (7) on the day 3, 9 and 18 after tumor implantation. Two hours after injection of liposomes, mice were sacrificed and the tumor was dissected. Then, these sections were fluorescently observed by using a microscopic LSM system (Carl Zeiss, Co., Ltd.) (Fig. 2).

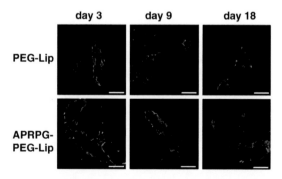

Fig. 2. Intratumoral distribution of DiIC18-labeled liposomes in the orthotopic pancreatic tumors. Mice with orthotopic pancreatic tumor were injected with PEG-Lip or APRPG-PEG-Lip labeled with DiI C18 via a tail vein at the day 3, 9, and 18 after tumor implantation. At 2 h after injection of fluorescence-labeled liposomes, frozen-sections of each tumor were prepared. Green portions indicate CD31-positive regions, red portions liposomal distribution, and yellow portions show the localization of liposomes at the site of vascular endothelial cells. Scale bar represents 100 μm. Reproduced with permission from (7)

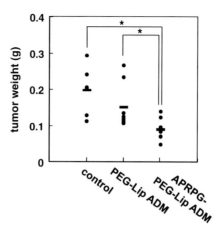

Fig. 3. Therapeutic effect of APRPG–PEG-modified liposome encapsulating ADM on mice with orthotopic pancreatic tumor. Mice with orthotopic pancreatic tumor were injected i.v. with 0.3 M Glucose (control), PEG–LipADM or APRPG–PEG–LipADM for four times at the day 3, 6, 9 and 12 after tumor implantation ($n = 6$–8). Injected dose of liposomal ADM were adjusted to 10 mg/kg as ADM dose in each time. The weight of the tumors was measured at the day 15. Significant differences are shown with asterisks: *$P < 0.05$. Reproduced with permission from (7)

3.6. Therapeutic Efficacy of Adriamycin Encapsulated in Liposomal Oligopeptides

Liposomes encapsulating ADM or 0.3 M glucose solution were administered intravenously into SUIT-2-bearing mice at day 3, 6, 9 and 12 after the implantation of tumor cells. The injected dose of ADM in each administration was 10 mg/kg. The weight of tumor was observed at day 15 (Fig. 3).

3.7. Preparation of DPP–CNDAC Liposomes Modified with Oligopeptides

1. Synthesis of CNDAC and DPP–CNDAC was performed as described previously (16, 17). Briefly, a phosphatidyl group was introduced into CNDAC through transphosphatidylation from 1,2-dipalmitoyl-3-sn-glycerophosphocholine by using phospholipase D.

2. Liposomes were prepared as follows: DPP–CNDAC, DSPC, cholesterol with DSPE–PEG (LipCNDAC/PEG) or DSPE–PEG–APRPG (LipCNDAC/APRPG–PEG) (10/10/5/2 as a molar ratio), or DPP–CNDAC, DSPC, cholesterol without PEG-conjugate (LipCNDAC, 10: 10: 5 as a molar ratio) were dissolved in chloroform/methanol, dried under reduced pressure, and stored in vacuo for at least 1 h. Liposomes were produced by hydration of a thin lipid film with 10 mM phosphate-buffered 0.3 M sucrose (pH 6.8), and frozen and thawed for three cycles by use of liquid nitrogen. Then the liposomes were sized by extrusion thrice through polycarbonate membrane filters with 100-nm-diameter pores. The liposomal solutions were centrifuged at $180,000 \times g$ for 20 min (CS120EX, Hitachi, Japan) to remove the untrapped DPP–CNDAC if present. Then, the liposomes were resuspended in 10 mM phosphate-buffered 0.3 M sucrose.

3. For the determination of the efficacy of trapping DPP–CNDAC in the liposomes, an aliquot of the liposomal solution was solubilized by the addition of reduced Triton X-100, and the amount of DPP–CNDAC was optically determined at 280 nm after the pH of the solution had been adjusted to 1.0. As a result, the encapuslation percent was almost 100%.

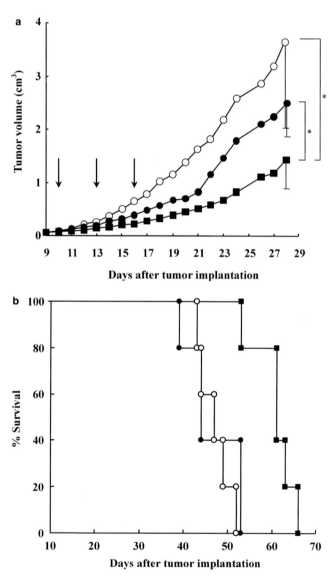

Fig. 4. Therapeutic efficacy of LipCNDAC/APRPG–PEG in tumor-bearing mice. Five-week-old Balb/c male mice (5 or 6 per group) were implanted s.c. with Colon 26 NL-17 carcinoma cells into their left posterior flank. They were injected i.v. with control liposomes (open circle), LipCNDAC/PEG (closed circle) or LipCNDAC/APRPG–PEG (*closed square*) at 15 mg/kg as CNDAC on days 10, 13, and 16 (*arrows*) after tumor implantation. The tumor volume (**a**) and survival time of mice (**b**) were monitored to evaluate the therapeutic efficacy of DPP–CNDAC liposomes. Significant differences from the control liposome-treated group are indicated (*$P < 0.05$*). Reproduced with permission from (18)

4. For the therapeutic study, control liposomes composed of DPPC, DSPC, and cholesterol (10/10/5 as a molar ratio) were prepared similarly as for the other liposomes.

5. Particle size and ζ-potential of liposomes diluted with PBS were measured by use of a Zetasizer Nano ZS. They were 121 ± 4 nm and −29.2 mV for LipCNDAC, 122 ± 6 nm and −6.1 mv for LipCNDAC/PEG, and 102 ± 2 nm and −3.6 mV for LipCNDAC/APRPG–PEG, respectively.

3.8. Therapeutic experiment with APRPG–PEG-Modified Liposomal DPP–CNDAC

LipCNDAC/PEG, LipCNDAC/APRPG–PEG or control liposomes were administered intravenously into colon 26 NL-17 tumor-bearing mice. The injected dose for each administration was 15 mg/kg as CNDAC moiety. The treatment was started when the tumor volume became approxiamtely 0.1 cm^3. The size of the tumor and the body weight of each mouse were monitored daily thereafter (Fig. 4a). Two bisecting diameters of each tumor were measured with slide calipers to determine the tumor volume. Calculation of the tumor volume was performed by using the formula 0.4 (a × b^2), where "a" is the largest and "b" is the smallest diameter. The calculated tumor volume correlated well with the actual tumor weight (r = 0.980) (22). The life spans of tumor-bearing mice were also monitored (Fig. 4b).

4. Notes

1. In general, encapsulation efficiency of drugs into liposomes is dependent on the logP value (octanol/water partition coefficient) of them, when they cannot be liposomalized by special techniques such as a remote-loading method (22). In many cases, it is difficult to encapsulate drugs into liposomes with high encapsulation efficiency since logP value of drugs is not always suit for liposomalization. In addition, it is also difficult to guarantee the quality and the stability of liposomal drugs in such difficult cases. CNDAC is also difficult to be liposomalized with high encapsulation efficiency by a general hydration method. Therefore, phospholipid derivatization of certain drugs to suit liposomal formulations is the useful methodology to develop liposomal drugs.

2. PEG-shielding of the liposomal surface should be useful for designing active targeting DDS with oligopeptides as well as passive targeting. An important aspect of PEGylation is that it serves for not only RES-avoiding but also for the construction of a practical liposomal oligopeptides. In fact, the biodistribution of APRPG-modified DPP–CNDAC liposomes without

PEG was strongly affected by the presence of cyano group of DPP–CNDAC on the liposomal surface. It induced aggregation of the liposomes, resulting in reduced blood circulation of the liposomes. However, the fixed aqueous layer formed by PEG can mask the undesirable surface properties of liposomes, which prevent attenuation of the effect of oligopeptides. The technology used in this study is also applicable to liposomalization of other compounds, DNA or siRNA etc.

3. We indentified oligopeptides specifically bound to tumor angiogenic vessels from a phage-displayed peptide library and applied to a liposomal DDS. The advantage of *in vivo* biopanning the library is that the selected phages have the ability to bind only to angiogenic vessels, not to other tissues. In fact, the amino acid sequences of the phage clones thus obtained were different from any reported sequences. The selected phage clones had high affinity to murine angiogenic vessels.

4. One of the important things to develop liposomal oligopeptides for clinical use, we should investigate whether the peptides selected in the murine model have affinity for angiogenic endothelium in human tumors. In our studies, we demonstrated that our peptides have affinity for human angiogenic endothelium by histochemical staining of the peptides in human cancer samples (2).

Acknowledgements

The authors thank Dr. Noriyuki Maeda and Dr. Yukihiro Namba for the collaboration in the synthesis of APRPG–PEG–DSPE, Dr. Koichi Ogino and Dr. Takao Taki for the phage library project, and Dr. Satoshi Shuto for the synthesis of DPP–CNDAC.

References

1. Asai T, Oku N (2005) Liposomalized oligopeptides in cancer therapy. Methods Enzymol 391:163–176
2. Oku N, Asai T, Watanabe K, Kuromi K, Nagatsuka M, Kurohane K, Kikkawa H, Ogino K, Tanaka M, Ishikawa D, Tsukada H, Momose M, Nakayama J, Taki T (2002) Anti-neovascular therapy using novel peptides homing to angiogenic vessels. Oncogene 21:2662–2669
3. Kondo M, Asai T, Katanasaka Y, Sadzuka Y, Tsukada H, Ogino K, Taki T, Baba K, Oku N (2004) Anti-neovascular therapy by liposomal drug targeted to membrane type-1 matrix metalloproteinase. Int J Cancer 108:301–306
4. Akita N, Maruta F, Seymour LW, Kerr DJ, Parker AL, Asai T, Oku N, Nakayama J, Miyagawa S (2006) Identification of oligopeptides binding to peritoneal tumors of gastric cancer. Cancer Sci 97:1075–1081
5. Maeda N, Takeuchi Y, Takada M, Namba Y, Oku N (2004) Synthesis of angiogenesis-targeted peptides and hydrophobized polyethylene glycol conjugate. Bioorg Med Chem Lett 14:1015–1017

6. Maeda N, Takeuchi Y, Takada M, Sadzuka Y, Namba Y, Oku N (2004) Anti-neovascular therapy by use of tumor neovasculature-targeted long-circulating liposome. J ControlRelease 100:41–52
7. Yonezawa S, Asai T, Oku N (2007) Effective tumor regression by anti-neovascular therapy in hypovascular orthotopic pancreatic tumor model. J Control Release 118:303–309
8. Klibanov AL, Maruyama K, Torchilin VP, Huang L (1990) Amphipathic polyethyleneglycols effectively prolong the circulation time of liposomes. FEBS Lett 268:235–237
9. Sadzuka Y, Nakade A, Hirama R, Miyagishima A, Nozawa Y, Hirota S, Sonobe T (2002) Effects of mixed polyethyleneglycol modification on fixed aqueous layer thickness and anti-tumor activity of doxorubicin containing liposome. Int J Pharm 238:171–180
10. St. Croix B, Rago C, Velculescu V, Traverso G, Romans KE, Montgomery E, Lal A, Riggins GJ, Lengauer C, Vogelstein B, Kinzler KW (2000) Genes expressed in human tumor endothelium. Science 289:1197–1202
11. Asai T, Shimizu K, Kondo M, Kuromi K, Watanabe K, Ogino K, Taki T, Shuto S, Matsuda A, Oku N (2002) Anti-neovascular therapy by liposomal DPP-CNDAC targeted to angiogenic vessels. FEBS Lett 520:167–170
12. Shimizu K, Asai T, Fuse C, Sadzuka Y, Sonobe T, Ogino K, Taki T, Tanaka T, Oku N (2005) Applicability of anti-neovascular therapy to drug-resistant tumor: Suppression of drug-resistant P388 tumor growth with neovessel-targeted liposomal adriamycin. Int J Pharm 296:133–141
13. Matsumura Y, Maeda H (1986) A new concept for macromolecular therapeutics in cancer chemotherapy: Mechanism of tumoritropic accumulation of proteins and the antitumor agent SMANCS. Cancer Res. 46:6387–6392
14. Maeda N, Miyazawa S, Shimizu K, Asai T, Yonezawa S, Kitazawa S, Namba Y, Tsukada H, Oku N (2006) Enhancement of anticancer activity in antineovascular therapy is based on the intratumoral distribution of the active targeting carrier for anticancer drugs. Biol Pharm Bull 29:1936–1940
15. Katanasaka Y, Ida T, Asai T, Maeda N, Oku N (2008) Effective delivery of an angiogenesis inhibitor by neovessel-targeted liposomes. Int J Pharm 360:219–224
16. Matsuda A, Nakajima Y, Azuma A, Tanaka M, Sasaki T (1991) Nucleosides and nucleotides. 100. 2'-C-cyano-2'-deoxy-1-β-D-arabinofuranosyl-cytosine (CNDAC): design of a potential mechanism-based DNA-strand-breaking antineoplastic nucleoside. J Med Chem 34:2917–2919
17. Shuto S, Awano H, Shimazaki N, Hanaoka K, Matsuda A (1996) Nucleosides and nucleotides.150. Enzymatic synthesis of 5'-phosphatidyl derivatives of 1-(2'-C-cyano-2-deoxy-β-D-*arabino*-pentofuranosyl) cytosine (CNDAC) and their notable antitumor effects in mice. Bioorg Med Chem Lett 6:1021–1024
18. Asai T, Miyazawa S, Maeda N, Hatanaka K, Katanasaka Y, Shimizu K, Shuto S, Oku N (2008) Antineovascular therapy with angiogenic vessel-targeted polyethyleneglycol-shielded liposomal DPP–CNDAC. Cancer Sci 99:1029–1033
19. Yonezawa S, Asai T, Oku N (2007) Dorsal air sac model. Angiogenesis assays. Wiley, New York, pp 229–238
20. Pasqualini R, Ruoslahti E (1996) Organ targeting in vivo using phage display peptide libraries. Nature 380:364–366
21. Pasqualini R, Koivunen E, Ruoslahti E (1997) αv integrins as receptors for tumor targeting by circulating ligands. Nat Biotechnol 15:542–546
22. Oku N, Doi K, Namba Y, Okada S (1994) Therapeutic effect of adriamycin encapsulated in long-circulating liposomes on Meth-A-sarcoma-bearing mice. Int J Cancer 58:415–419

Chapter 24

TAT-Peptide Modified Liposomes: Preparation, Characterization, and Cellular Interaction

Marjan M. Fretz and Gert Storm

Abstract

In general, cellular internalization of macromolecular drugs encapsulated in liposomes proceeds via endocytosis. This potentially leads to degradation of the liposome-encapsulated macromolecular content within the endosomal/lysosomal compartment. Therefore, bypassing the endocytic route by conferring a direct plasma membrane translocation property to the liposomes would be very beneficial. Cell penetrating peptides, e.g. TAT-peptide, are exploited in the drug delivery field for their capacity of plasma membrane translocation. Here, we describe the preparation of TAT-peptide modified liposomes and their cellular interaction using live cell flow cytometry and imaging techniques.

Key words: Endocytosis, Cell-penetrating peptides, TAT-peptide, Membrane translocation

1. Introduction

Cellular uptake of (targeted) liposomes is generally mediated by (receptor-mediated) endocytosis. Upon internalization, the liposomes and encapsulated drug will be routed from endosomes to lysosomes. Macromolecules, like proteins, peptides or nucleic acids, encapsulated in liposomes ending up in the endocytic pathway will be degraded, which causes inefficient intracellular delivery. In the recent years, attempts are being made to apply targeted liposomes for cytosolic delivery of macromolecules (1, 2). One approach reported to avoid endocytosis and to achieve direct cytosolic delivery via direct plasma membrane translocation is the use of so-called cell-penetrating peptides (CPP) of which the HIV-1 derived TAT-peptide is an example (3). These CPP have been reported as cytosolic delivery vector for a variety of cargos, like fluorophores, proteins, oligonucleotides and particulate systems (3–9). Torchilin et al. were the first to report on cytosolic delivery of liposomes modified with a CPP on the surface of the liposomes (8). However, the plasma membrane translocation mediated via those CPP was questioned when it was shown that cell fixation could induce rigorous artefacts in the cellular distribution of fluorescently labelled CPP (10, 11). Several studies now pointed out the importance of using live cells to study the route of uptake of CPP and their cargoes (10–12).

TAT-peptide modified liposomes were prepared and cellular association and intracellular distribution of (double) fluorescently labelled particles were assessed by flow cytometry and confocal laser scanning microscopy.

2. Materials

2.1. Preparation and Characterization of Liposomes

1. Egg phosphatidylcholine (EPC), 1,2-distearoyl-glycero-3-phosphoethanolamine-N-(poly(ethylene glycol)2000) (PEG_{2000}–DSPE) (Lipoid GmbH, Ludwigshafen). Lipids can be stored as powder or as stock solution in ethanol at –20°C.

2. Maleimide–PEG_{2000}–DSPE (Shearwater Polymers, Huntsville, AL, USA, currently Nektar Pharmaceuticals). This lipid can be stored as powder or as stock solution in ethanol at –20°C.

3. Stock solutions containing fluorescent labels DiD (1,1′-dioctadecyl-3,3,3′,3′-tetramethylindocarbocyanine 4-chlorobezenesulfonate salt) (Molecular Probes, Eugene, OR, USA) or Lissamine rhodamine B labelled glycerophosphoethanolamine (Avanti Polar Lipids, Alabaster, USA) are made in ethanol and kept at –20°C until use. These lipids are used to label the bilayer of the liposomes.

4. Hepes buffered saline (HBS): 135 mM NaCl and 10 mM Hepes. pH is set to 6.5, 7.0 or 7.4 as stated in the text.

5. FITC–dextran (Mw 70,000 Da, Sigma, St. Louis, MO, USA) is dissolved in HBS pH 7.0 to a concentration of 10 mg/mL. This solution is used for encapsulating a fluorescent marker in the aqueous interior of the liposomes.

6. To remove uncapsulated FITC–dextran or non-coupled TAT-peptide, a column of Sepharose CL-4B (Amersham Pharmacia Biotech, Uppsala, Sweden) is packed in 20% ethanol. Before use the column is equilibrated with HBS pH 6.5 or 7.4.

7. Hydroxylamine solution: 0.5 M Hepes, 0.5 M Hydroxylamine–HCl and 0.25 mM EDTA at pH 7.0. Always prepare a fresh solution before use.
8. Thiol-acetylated TAT-peptide (sequence YGRKKRRQRRRK-S-acetylthiolacetyl) (Ansynth BV, Roosendaal The Netherlands); referred to as TAT-sata is dissolved in HBS, pH 7.4 to a concentration of 5 mg/mL and 1 mg aliquots were kept at −20°C until use.

2.2. Cell Culture

1. Serum free culture medium: Dulbecco's Modified Eagle's Medium (DMEM) containing 3.7 g/L sodium bicarbonate, 4.5 g/L l-glucose supplemented with 2 mM l-glutamine, penicillin (100 IU/mL), streptomycin (100 μg/mL) and amphotericin B (0.25 μg/mL) (Gibco, Grand Island, MY, USA).
2. Complete culture medium: The serum free medium is supplemented with 10% heat-inactivated foetal calf serum (Gibco, Grand Island, MY, USA).
3. Solution of trypsin (0.25% (w/v)) and ethylenediamine tetraacetic acid (EDTA; 0.02% (w/v)) in PBS (Gibco, Grand Island, MY, USA).

2.3. Flow Cytometry

1. Rhodamine–PE labelled liposomes; control (pegylated) and TAT-peptide modified liposomes.
2. Phosphate buffered saline (PBS): 140 mM NaCl, 8.7 mM Na_2HPO_4, and 1.9 mM NaH_2PO_4 pH 7.4 (Braun, Melsungen, Germany).

2.4. Microscopy: Intracellular Distribution

1. Nunc Lab-Tek 16-well chamber slides from Fisher Scientific (Landsmeer, The Netherlands).
2. Phosphate buffered saline (PBS): 140 mM NaCl, 8.7 mM Na_2HPO_4, and 1.9 mM NaH_2PO_4, pH 7.4 (Braun, Melsungen, Germany).
3. Lysostracker Red (Invitrogen, Eugene, OR, USA) solution in PBS. Dilute the stock solution (1 mM) provided by Invitrogen to a final concentration of 75 nM in PBS. This solution can be stored at −20°C.
4. Double labelled TAT-peptide modified liposomes.

2.5. Microscopy: Metabolic and Endocytosis Inhibitors

1. Co-Star 6-well low adherence plates (Corning Life Science BV, Schiphol-Rijk, The Netherlands).
2. Cytochalasin D (Sigma Aldrich, St. Louis, MO, USA) is dissolved in DMSO to a concentration of 5 mg/mL and stored in the fridge until use.
3. For iodoacetamide (Sigma, St. Louis, MO, USA), a fresh stock solution with a concentration of 0.1 M is made in PBS.

4. TAT-peptide modified liposomes labelled with Rhodamine–PE.
5. Phosphate buffered saline (PBS): 140 mM NaCl, 8.7 mM Na_2HPO_4, and 1.9 mM NaH_2PO_4, pH 7.4 (Braun, Melsungen, Germany).

3. Methods

TAT-peptide modified liposomes are prepared by coupling the TAT-peptide to the distal end of PEG-chains on the liposomal surface. For this, maleimide-functionalized PEG-chains are incorporated in the lipid bilayer and the TAT-peptide has been functionalized with a thio-acetyl group at the C-terminus. Those thio-acetyl groups can be converted to sulfhydryl groups. This method has been described in the literature to couple targeting ligands like e.g. antibodies and peptides to liposomes (1, 13). Cellular association, which can include both binding and uptake, can be assessed by flow cytometry. Furthermore, to distinguish between binding and uptake and additionally to evaluate the intracellular localization of the liposomes, live cell confocal laser scanning microscopy can be applied. The use of double fluorescently labelled liposomes, with both the liposomal bilayer and the aqueous compartment labelled, can be used to study the integrity of the liposomes upon incubation with cells and additionally will give information about the possible cytosolic delivery of the encapsulated hydrophilic fluorescent label. Co-localization with markers of the endocytic pathway and the use of metabolic or endocytosis inhibitors will give information about the cellular internalization mechanism.

3.1. Preparation and Characterization of Liposomes

1. EPC, cholesterol, PEG2000–DSPE, and maleimide–PEG2000–DSPE are weighed and dissolved in absolute ethanol in a round bottom flask. The molar ratio of the lipids is 1.85: 1.00: 0.09: 0.06, respectively.
2. The bilayer of the liposomes is fluorescently labelled by adding either DiD or Rho–PE to a final ratio of 0.1 mol percentage.
3. Form a lipid film by evaporation of the ethanol using a rotavapor.
4. Before hydration, flush the lipid film with nitrogen for at least 30 min to obtain a complete dry lipid film.
5. Form liposomes by hydration of the film with either 1 mL of HBS pH 6.5 or 1 mL of HBS (pH 7.0) containing 10 mg/mL FITC dextran in case of rhodamine–PE labelled or DiD labelled liposomes, respectively.
6. Size the liposomes by extrusion to an average size of 150 nm and a polydispersity index below 0.2 (see Note 1). Size distribution should be checked with dynamic light scattering.

7. In case of FITC–dextran containing liposomes: remove the non-encapsulated FITC–dextran using a Sepharose CL-4B column using HBS pH 6.5 as eluent.

8. Split the batch of liposomes into two parts. One part will serve as control liposomes without any TAT-peptide coupled. For the control liposomes, continue with step 11. The second part is used to prepare TAT-peptide modified liposomes by following step 9–11.

9. For the TAT-peptide modified liposomes, deacetylate 1 mg of TAT-sata peptide by adding 20 µL of hydroxylamine solution to 200 µL of peptide solution and leaving it on a rollerbench for 1 h at room temperature to obtain free sulfhydryl groups.

10. Add the deacetylated peptide to the liposomes (1 mg peptide to 21 µmol total lipid) and leave the dispersion overnight at 4°C (see Note 2). The sulfhydryl groups will react with the maleimide-groups present at the distal ends of the PEG-chains resulting in a stable covalent linkage between the peptide and the PEG-chains (see Note 3).

11. Remove non-coupled peptide by size exclusion chromatography using Sepharose CL-4B column but with HBS pH 7.4 as eluent.

12. Characterize the liposomes with respect to size (e.g. dynamic light scattering) and lipid concentration (e.g. determination of phospholipid content using a method described by Rouser et al.) (14).

3.2. Cell Culture

1. The human ovarian carcinoma cell line NIH:OVCAR-3 originates from ATCC (Manassas, USA). OVCAR-3 cells are passaged when confluency is reached to provide new maintenance cultures in 75 cm^2 culture flasks. For maintenance, in general, 1:10 part of the last passage is transferred into a new flask. This procedure is done twice a week.

2. For passaging and seeding, the cells are first washed with PBS and detached from the culture flask by incubating with 3 mL trypsin/EDTA solution for approximately 10 min in the incubator. The trypsin/EDTA solution is inactivated by adding 13 mL of complete culture medium. For maintenance, 1:10 part is transferred into a new culture flask. For seeding or for use in a flow cytometry experiment, the cells are centrifuged (5 min, $300 \times g$), resuspended in the required medium, counted and diluted to the appropriate number of cells/ml.

3.3. Flow Cytometry

1. Prepare a cell suspension of 1×10^6 cells/mL in complete culture medium.

2. Add 100 µL cell suspension (1×10^5 cells) to a FACS tube.

3. For both TAT–liposome and control liposomes, the liposome stock is diluted in complete culture medium: concentrations of liposomes (total lipid) used are 1 mM, 500, 250, 125 and, 62.5 µM.
4. 100 µL of each liposome dispersion made in step 3 is added to the cells in the FACS tubes to obtain final liposome concentrations ranging from 31 to 500 µM total lipid.
5. Incubate the samples for 1 h at 4°C.
6. Centrifuge the samples for 5 min at $300 \times g$.
7. Remove the supernatant and add 400 µL ice-cold PBS.
8. Centrifuge (5 min, $300 \times g$) again and repeat step 7.
9. Resuspend the cells in 400 µL PBS and leave the samples on ice.
10. Analyze the cells by FACS, e.g. a FACScalibur (Becton&Dickinson) by counting at least 5000 viable cells and leave the samples on ice during analysis. For Rhodamine–PE labelled liposomes, the signal is detected in the FL-2 detector of the flow cytometer.
11. Data can be analyzed by flow cytometry programmes like Cell Quest or WinMDI and data is expressed as the mean fluorescence intensity (see Fig. 1).

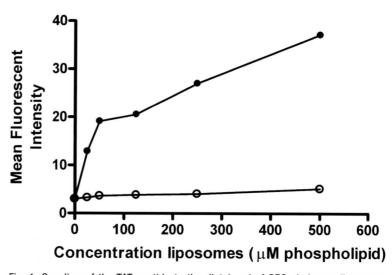

Fig. 1. Coupling of the TAT-peptide to the distal end of PEG-chains on liposomes increases the cellular association with OVCAR-3 cells. OVCAR-3 cells are incubated with various concentrations rhodamine–PE labelled control (*open circle*) or TAT–liposomes (*closed circle*) for 1 h at 4°C, washed and analyzed by flow cytometry. Each data point represents the mean fluorescence intensity of 5,000 cells (mean ± SD; $n=3$). Error bars are within plot symbols when not visible. (Reproduced from (12) with permission from Elsevier Science)

3.4. Microscopy: Intracellular Distribution

1. Cells (10,000 cells/well) are seeded into a 16-well chamber slide and cultured overnight in complete culture medium.
2. Before applying the liposomes in serum-free medium, the cells are washed with 200 µL PBS.
3. 150 nmol of double labelled TAT–liposomes is added to the cells in serum free culture medium. The liposomes are labelled with the bilayer label DiD and FITC–dextran in the aqueous interior of the liposomes. Incubate for 1 h with the labelled TAT–liposomes.
4. Subsequently, incubate the cells for either 1 or 23 h in complete culture medium.
5. Thirty minutes before visualization, incubate with 100 µL Lysotracker Red solution (75 nM) at 37°C.
6. Wash the cells with PBS and mount them in PBS.
7. Cover the sample with a coverslip and seal using transparent nailpolish.
8. Analyze the live cells directly after step 7 with a confocal laser scanning microscope equipped with 488 nm, 568 nm and 633 nm lasers (see Fig. 2) (see Note 4).

3.5. Microscopy: Metabolic or Endocytsosis Inhibitors

1. Cells (300,000 cells/well) are added in a total volume of 5 mL serum free medium in 6-well Co-Star Low adherence plates (see Note 5).
2. In case of low temperature incubation, pre-incubate the cells for 30 min at 4°C prior to the liposome incubation (described in step 4).
3. In case of iodoacetamide or cytochalasin D, incubate the cells with either 1 mM iodoacetamide or 25 µg/mL cytochalasin D for 30 min before continuing with step 4.
4. RhodaminePE labelled liposomes are added (450 nmol total lipid).
5. Cells are incubated for 5 h at either 4°C or 37°C.
6. To remove the non-associated liposomes, the cells are washed twice by centrifugation (5 min, $300 \times g$) with PBS.
7. After the last centrifugation step, the cells are resuspended in approximately 100 µL PBS and an aliquot of the cell suspension is mounted on a glass slide and covered with a coverslip.
8. Seal with transparent nailpolish.
9. Directly visualize the samples using a confocal microscope equipped with a 568 nm laser, suitable for monitoring Rhodamine–PE labelled liposomes (see Fig. 3).

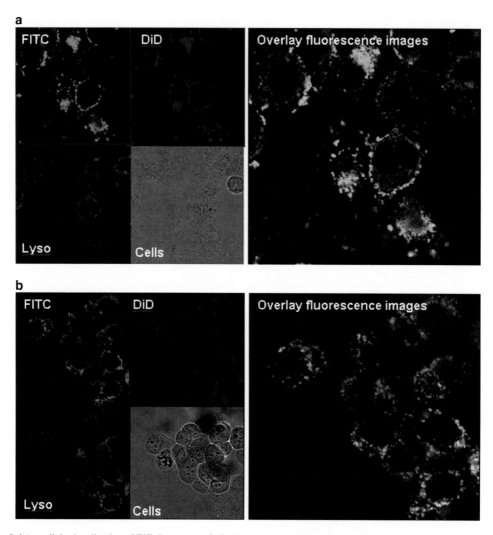

Fig. 2. Intracellular localization of TAT–liposomes. OVCAR-3 cells are incubated with 150 nmol of double fluorescently labelled TAT–liposomes for 1 h and subsequently incubated for 1 h (**a**) or 23 h (**b**) in liposomes-free medium. Thirty minutes before visualization the endocytic pathway is labelled with Lysotracker Red. Live cell imaging is performed with confocal laser scanning microscopy. Double labelled liposomes are used to study the integrity of the liposome during the uptake process: co-localization of both liposomal labels would indicate that the liposomes are intact. 1 h both liposomal labels are localized at the plasma membrane, which represent intact cell-bound TAT–liposomes. The electronically merged image clearly shows lack of co-localization with the endocytic pathway marker, Lysotracker Red. This opposite to the 24 h incubation, both liposomal labels can be seen intracellularly in a punctuate pattern. In the electronically merged image, co-localization with Lysotracker Red is clearly visible. This indicates that the TAT-peptide modified liposomes bind to the plasma membrane and after internalization are present in endocytic vesicles. Therefore, we conclude that the liposomes are internalized by endocytosis. (Reproduced from (12) with permission from Elsevier Science)

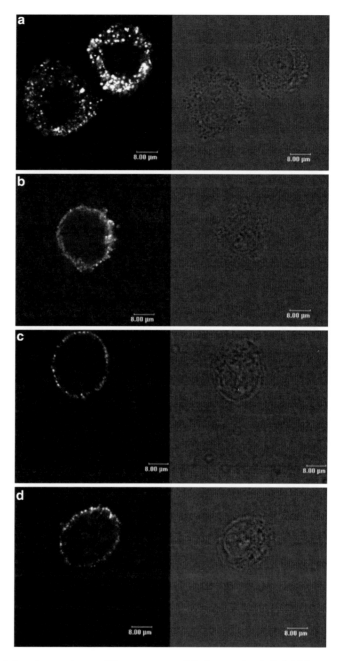

Fig. 3. Low temperature and the presence metabolic or endocytosis inhibitors prevent cellular uptake of TAT–liposomes. OVCAR-3 cells are incubated with Rhodamine–PE labelled TAT–liposomes for 5 h at 37° (**a, c, d**) or at 4°C (**b**). *Left panels* are confocal images, right panels are phase contrast images. Incubation at 37°C without any inhibitor results in intracellular vesicular localization of the TAT–liposomes (**a**). Only plasma membrane binding is observed in case of incubation at 4°C (**b**) and incubation with the metabolic inhibitor iodoactemide (**c**) or endocytosis inhibitor cytochalasin D (**d**). These results indicate that the cellular uptake of TAT–liposomes occurs via endocytosis. (Reproduced from (12) with permission from Elsevier Science)

4. Notes

1. Polydispersity index is an indication of the size distribution of the liposomes. The polydispersity can range from 0.0 for a monodisperse to 1.0 for an entirely heterodisperse suspension.
2. Incubation time, temperature, and lipid concentration can be varied depending on the experimental set-up.
3. Maleimide groups are rather unstable in aqueous solution and, therefore, the peptide should be coupled to the liposomes on the day of lipid hydration.
4. When a triple labelling is used, the use of sequential scanning is preferred. In the normal modus, all the signal are simultaneously acquired in different channels, which in the case of multiple labelling can result in crosstalk, which can ultimately lead to misleading co-localization results.
5. The treatment of the OVCAR-3 cells with cytochalasin D or iodoacetamide resulted in detachment of the cells from normal microscope slides. Therefore, the incubation should be performed in the Low-adherence plates as described.

References

1. Mastrobattista E, Koning GA, Van Bloois L, Filipe AC, Jiskoot W, Storm G (2002) Functional characterization of an endosome-disruptive peptide and its application in cytosolic delivery of immunoliposome-entrapped proteins. J Biol Chem 277:27135–27143
2. Simoes S, Moreira JN, Fonseca C, Duzgunes N, de Lima MC (2004) On the formulation of pH-sensitive liposomes with long circulation times. Adv Drug Deliv Rev 56:947–965
3. Vives E, Brodin P, Lebleu B (1997) A truncated HIV-1 Tat protein basic domain rapidly translocates through the plasma membrane and accumulates in the cell nucleus. J Biol Chem 272:16010–16017
4. Fawell S, Seery J, Daikh Y, Moore C, Chen LL, Pepinsky B et al (1994) Tat-mediated delivery of heterologous proteins into cells. Proc Natl Acad Sci U S A 91:664–668
5. Schwarze SR, Ho A, Vocero-Akbani A, Dowdy SF (1999) In vivo protein transduction: delivery of a biologically active protein into the mouse. Science 285:1569–1572
6. Astriab-Fisher A, Sergueev D, Fisher M, Shaw BR, Juliano RL (2002) Conjugates of antisense oligonucleotides with the Tat and antennapedia cell-penetrating peptides: effects on cellular uptake, binding to target sequences, and biologic actions. Pharm Res 19:744–754
7. Lewin M, Carlesso N, Tung CH, Tang XW, Cory D, Scadden DT et al (2000) Tat peptide-derivatized magnetic nanoparticles allow in vivo tracking and recovery of progenitor cells. Nat Biotechnol 18:410–414
8. Torchilin VP, Rammohan R, Weissig V, Levchenko TS (2001) TAT peptide on the surface of liposomes affords their efficient intracellular delivery even at low temperature and in the presence of metabolic inhibitors. Proc Natl Acad Sci U S A 98:8786–8791
9. Tseng YL, Liu JJ, Hong RL (2002) Translocation of liposomes into cancer cells by cell-penetrating peptides penetratin and tat: a kinetic and efficacy study. Mol Pharmacol 62:864–872

10. Richard JP, Melikov K, Vives E, Ramos C, Verbeure B, Gait MJ et al (2003) Cell-penetrating peptides: a re-evaluation of the mechanism of cellular uptake. J Biol Chem 278(1):585–590
11. Lundberg M, Johansson M (2002) Positively charged DNA-binding proteins cause apparent cell membrane translocation. Biochem Biophys Res Commun 291:367–371
12. Fretz MM, Koning GA, Mastrobattista E, Jiskoot W, Storm G (2004) OVCAR-3 cells internalize TAT-peptide modified liposomes by endocytosis. Biochim Biophys Acta 1665:48–56
13. Koning GA, Morselt HW, Gorter A, Allen TM, Zalipsky S, Scherphof GL et al (2003) Interaction of differently designed immunoliposomes with colon cancer cells and Kupffer cells: an in vitro comparison. Pharm Res 20:1249–1257
14. Rouser G, Fkeischer S, Yamamoto A (1970) Two dimensional then layer chromatographic separation of polar lipids and determination of phospholipids by phosphorus analysis of spots. Lipids 5:494–496

Chapter 25

ATP-Loaded Liposomes for Targeted Treatment in Models of Myocardial Ischemia

Tatyana S. Levchenko, William C. Hartner, Daya D. Verma, Eugene A. Bernstein, and Vladimir P. Torchilin

Abstract

ATP cannot be effectively delivered to most tissues including the ischemic myocardium without protection from degradation by plasma endonucleotidases. However, it has been established that ATP can be delivered to various tissues by its encapsulation within liposomal preparations. We describe here, the materials needed and methods used to optimize the encapsulation of ATP in liposomes, enhance their effectiveness by increasing their circulation time and target injured myocardial cells with liposomal surface anti-myosin antibody. Additionally, we outline methods for ex vivo studies of these ATP liposomal preparations in an isolated ischemic rat heart model and for in vivo studies of rabbits with an induced myocardial infarction. The expectation is that these methods will provide a basis for continued studies of effective ways to deliver energy substrates to the ischemic myocardium.

Key words: ATP, Liposomes, Immunoliposomes, Antimyosin, Ischemia, Isolated rat heart, Rabbit myocardial infarction

1. Introduction

Cardiovascular disease is the leading cause of morbidity, mortality, disability and economic loss in industrialized countries (1). Myocardial ischemia-reperfusion injury (I/R) and acute myocardial infarction (MI) are significant risks during cardiac surgery, progression of coronary artery disease, and following cardiac arrest (2). Acute MI leads to ventricular remodeling, including expansion or aneurysm formation due to cardiomyocyte death in infarcted regions and chamber dilation associated with hypertrophy and fibrosis of non-infarcted regions (3). One of the possible ways to reduce the frequency of heart failure following a period of ischemia is likely to be the timely application of targeted delivery of cardio-protective drugs into the ischemic myocardium that prevents or heals the damage.

In myocardial ischemia, ATP levels in the cardiomyocytes can drop to 20% of their initial value after approximately 15 min (4). During ischemia, the ATP is provided by breakdown of the myocardial glycogen to glucose with subsequent glycolysis to produce a temporary, small supply of ATP. In the continuing absence of an oxygen supply, these ATP sources become depleted, and ATP-dependent ion pumps in the outer membranes of myocytes cease to function, with loss of ion balance. Cells swell and burst, releasing their contents into the circulation as early as 20 min after onset of ischemia in acute myocardial infarction (5). Moreover, a toxic or inflammatory insult to the myocardium also results in the loss of sarcolemmal integrity leading to exposure of intracellular myosin (6, 7) and to extracellular fluids where the concentration of ATP is much lower (8).

Since the key factor responsible for eliciting the decrease in the ATP supply/demand ratio during myocardial ischemia is the relative lack of ATP, delivery of sufficient exogenous ATP should help restore normal cellular levels and could have a cardioprotective effect. However, ATP has a very short half-life in the blood. It is rapidly hydrolyzed to ADP, AMP, and adenosine via a cascade of extracellular ectonucleotidases (9). Additionally, ATP, like other hydrophilic and strongly charged anions, cannot enter cells through the plasma membrane (9, 10). These two restrictions severely limit the direct use of exogenous ATP for use as a therapeutic bioenergetic substrate.

Although current available methods of delivering drugs to an ischemic zone are often limited by impaired myocardial blood flow, accumulation of positively charged liposomes in regions of experimentally infarcted myocardium were reported as early as 1977 by Caride et al. (11). It was later suggested that liposomes actually plug and seal the damaged myocyte membranes, thereby protecting myocytes against permanent ischemia/reperfusion injury (12). Different components of the cardiovascular system have served as targets for the delivery of liposomal drugs including vessel walls, endothelial cells, atherosclerotic lesions, as well as infarcted myocardium (13–16). The accumulation of liposomes and other nanoparticular drug carriers, such as micelles, in ischemic tissues is a general phenomenon and can be explained, at least in part, by the trapping of drug carriers within the ischemic zone (17, 18) via an enhanced permeability and retention (EPR) effect (19, 20). These reports led us to conclude that liposomes could be useful for delivery of ATP to damaged myocytes (17).

The application of liposomes loaded with ATP (ATP-L) has been reported in a variety of other in vitro and in vivo models.

Liposomal ATP protected human endothelial cells from energy failure in a cell culture model of sepsis (21). ATP-L increased the number of ischemic episodes tolerated before electrical silence and brain death in the rat (22, 23). In a hypovolemic shock-reperfusion model in rats, the administration of ATP-L increased hepatic blood flow during shock and reperfusion of the liver (24). The addition of the ATP-L during cold storage preservation of rat liver improved its energy state and metabolism (25, 26). Co-incubation of ATP-L with sperm cells improved their motility (27). Finally, biodistribution studies demonstrated significant accumulation of ATP-L in ischemia-damaged canine myocardium (28).

We describe here, the materials needed and methods used to optimize encapsulation of ATP in liposomes, enhance their effectiveness by increasing their circulation time and target injured myocardial cells with liposomal surface anti-myosin antibody. Additionally, we outline methods for ex vivo studies in an isolated ischemic rat heart model and for in vivo studies of rabbits with an induced myocardial infarction.

2. Materials

2.1. Loading Liposomes with ATP or Preparation of ATP-Loaded Liposomes

1. Egg phosphatidylcholine (PC) (Avanti Polar Lipids, Inc), stored at $-80°C$.
2. Cholesterol (Chol) (Sigma).
3. 1,2-Distearoyl-*sn*-glycero-3-phosphoethanolamine-N-[methoxy(polyethyleneglycol)-2000] (PEG(MW$_{2000}$)) (Avanti Polar Lipids), stored at $-80°C$.
4. 1,2-dioleoyl-3-trimethyl-ammonium-propane (DOTAP), stored at $-80°C$ (Avanti Polar Lipids, Inc).
5. Rhodamine–phosphatidylethanolamine (Rh–PE) (Avanti Polar Lipids, Inc).
6. FITC–Dextran (Fluka Chemicals).
7. Adenosine-5′-triphosphate (ATP) (Sigma).
8. 800, 600, 400 and 200 nm pore size polycarbonate membranes (Whatman).
9. Modified Krebs–Henseleit (K–H) buffer without $CaCl_2$, pH 7.4: 120 mM NaCl, 25 mM $NaHCO_3$, 4 mM KCl, 1.2 mM KH_2PO_4, 1.2 mM $MgSO_4$, 10 mM dextrose.
10. 400 mM ATP in K–H buffer, pH adjusted to 7.4 with 1 M NaOH.
11. Dialysis tubing (Spectra/Por Biotech PVDF membrane, 250 k MWCO) (Spectrum Medical Instruments).

Water is distilled and then deionized. All other chemicals and components of buffer solutions should be analytical grade preparations.

2.2. Determination of ATP Concentration

1. Triton X-100 (Sigma).
2. HPLC 250×4.6 mm stainless steel column Discovery C18 (Supelco, Bellefonte, PA).
3. 0.1 M phosphate buffer, pH 6.0.
4. Methanol (Fluka).

2.3. Visualization of Liposome Accumulation in Ischemic Tissue

1. FITC–dextran (MW 4,000 Da) (Sigma).
2. Rhodamine–phosphatidylethanolamine (Rh–PE) (Avanti Polar Lipids), stored at –80°C.
3. NAP™ columns (Amersham).
4. 0.05% solution of nitro blue tetrazolium (NBT) in PBS.
5. 4% formaldehyde (Sigma).
6. Tissue-Tek OCT 4583 (Sakura Finetek, Torrance, CA).
7. Minotome Plus (TBS, Durham, NC).
8. Fluorescence microscope (Olympus).

2.4. ATP-Loaded Immunoliposomes Specific for Cardiac Myosin

1. Egg phosphatidylcholine (PC) (Avanti Polar Lipids, Inc), stored at –80°C.
2. Cholesterol (Chol) (Sigma).
3. 1,2-Distearoyl-*sn*-Glycero-3-phosphoethanolamine-N-[methoxy(polyethyleneglycol)-2000] (PEG(MW$_{2000}$)) (Avanti Polar Lipids), stored at –80°C.
4. 1,2-dioleoyl-3-trimethyl-ammonium-propane (DOTAP) (Avanti Polar Lipids, Inc Alabaster, Al), stored at –80°C.
5. Adenosine-5'-triphosphate (ATP) (Sigma).
6. 800, 600, 400 and 200 nm pore size polycarbonate membranes (Whatman).
7. Krebs-Henseleit buffer without $CaCl_2$, pH 7.4 (see Subheading 2.2).
8. Dialysis tubing (Spectra/Por Biotech PVDF membrane, 250 k MWCO) (Spectrum Medical Instruments).
9. pNP–PEG–PE. For synthesize, see Chapter by Erdogan & Torchilin.
10. Monoclonal antibody 2G4 (an IgG) obtained, purified and provided for these studies by Dr. B.A. Khaw, Northeastern University (15).

2.5. ELISA

1. ELISA Reader (Labsystems Multiscan CCM 340).
2. 96-well U-bottomed microtiter plates. (Fisher).

3. Pig cardiac myosin (Sigma).
4. PBST: PBS buffer with 0.05% Tween.
5. Heat-inactivated horse serum albumin, 10% solution (Sigma).
6. Horseradish peroxidase-tagged goat anti-mouse IgG (ICN Biomedical).
7. Peroxidase substrate, K-blue (Neogene, Lexington, KY).

2.6. ATP-Loaded Immunoliposomes in Langendorff Model

1. Langendorff apparatus. (Harvard Apparatus Co.).
2. Sprague–Dawley rats (280–330 g).
3. Nembutal (50 mg/mL).
4. K–H buffer (see composition in Subheading 2.1) with 1.7 mM $CaCl_2$, filtrated without pH adjustment (see Note 1).
5. 0.9% NaCl.
6. Latex balloons (see Note 2).
7. Stimulator (PowerLab 4SP, AD Instruments, Colorado Springs, CO).
8. Digital recorder (PowerLab Chart 4, AD Instruments, Colorado Springs, CO).
9. Pressure transducers (Harvard Apparatus Co.)

2.7. ATP-Loaded Immunoliposomes in Rabbit with Experimental Myocardial Infarction

1. New Zealand (NZW) rabbits (2.5–3.5 kg).
2. Ketamine (100 mg/mL).
3. Xylazine (20 mg/mL).
4. Nembutal (50 mg/mL).
5. Harvard rodent positive pressure ventilator (Harvard Apparatus Co.).
6. PageWriter M1700A electrocardiograph (Hewlett-Packard).
7. Unisperse Blue (USB, Ciba).

3. Methods

3.1. Preparation of ATP-Loaded Liposomes

1. Dissolve in chloroform 196 mg of egg phosphatidylcholine, 45 mg of cholesterol, 5 mg of 1,2-distearoyl-sn-glycero-3-phosphoethanolamine-N-[methoxy(polyethylen-eglycol)-2000] and 7. 5 mg of 1,2-dioleoyl-3-trimethyl-ammonium-propane.
2. Mix lipids in 100 mL round-bottomed flask and remove chloroform on rotary evaporator. (Temperature must be higher than transition temperature of the each lipid in the mixture.)

3. Dry lipid film on the surface of the flask to a glassy, clear appearance.
4. Complete removal of chloroform by freeze-drying.
5. Dissolve 220 mg of adenosine-5′-triphosphate in 5 mL of Kreb's–Henseleit buffer and adjust pH to 7.4 with 1 M NaOH.
6. Hydrate lipid film with 5 mL of ATP solution (see Note 3).
7. Freeze the dispersion at −80°C for 30 min followed by thawing at 45°C for 5 min.
8. Repeat freezing and thawing cycle five times.
9. Extrude the liposomes five times through each of 800, 600, 400, and 200 nm pore size polycarbonate membranes.
10. Add the liposomes to a dialysis bag of MWCO 250,000 and separate non-encapsulated ATP by dialysis against the K–H buffer at 4°C overnight (see Note 4).
11. Dilute the liposomal formulations to a final concentration of 4 mg lipids/1 mg ATP/mL with K–H buffer.
12. Add $CaCl_2$ to a concentration 1.7 mM.
13. Prepare control liposomes without added ATP.
14. To prepare Rhodamine-labeled liposomes, add 0.5% Rhodamine–PE in standard formulation (Subheading 3.1, item 1) and hydrate lipid film with 40 mM FITC–dextran in K–H buffer. Remove non-encapsulated FITC–dextran with by gel-filtration on NAP™ column.

3.2. Determination of ATP

The level of liposomal ATP encapsulation is determined by HPLC.

1. Lyse ATP-containing liposomes with 0.5% (v/v) Triton X-100 in distilled water to release ATP from liposomes.
2. Measure the UV absorbance on a Hitachi D-7000 HPLC at 254 nm.
3. Perform chromatography on a 250×4.6 mm stainless steel column packed with 5 µm Discovery C18. The chromatographic conditions are as follows: isocratic elution at room temperature with a mixture of 0.1 M phosphate buffer, pH 6.0, and methanol (96/4 v/v) at a flow-rate of 1 mL/min.

3.3. Visualization of Liposome Accumulation in Ischemic Tissue

1. After perfusion studies, remove heart and slice transversely into 3–4 approximately 3 mm thick heart sections.
2. Incubate sections with a 0.05% solution of NBT at 40–45°C for 20 min.
3. Fix tissues (pink, NBT-negative, infarcted; dark, NBT-positive, non-infarcted) tissue samples in 4% formaldehyde. Rinse with PBS. Immerse in ice-cold isopentane and freeze in liquid nitrogen.

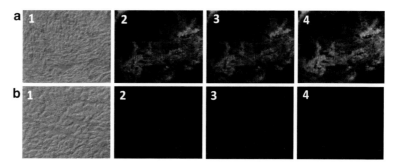

Fig. 1. Microscopy of 7 μM thick heart cryosections fixed with 4% formaldehyde, washed with PBS, and mounted with Fluor mounting media (Trevigen). (**a**) Extensive association of Rh–PE and FITC fluorescence with infarcted (pink, NBT-negative) tissue; (**b**) Lack of fluorescence associated with normal (dark, NBT-positive) tissue. 1 – Transmission microscopy; 2 – Fluorescence microscopy with FITC filter; 3 – Fluorescence microscopy with Rhodamine filter; 4 – Superposition of (2) and (3) (29)

4. Mount frozen samples in Tissue-Tek and prepare 7 μm thick cryosections on the Minotome Plus (Fig. 1).

5. Analyze sections with Olympus microscope with a FITC or a Rhodamine filter.

3.4. Preparation of ATP-Immunoliposomes

ATP–immunoliposomes (ATP–IL) are prepared using the micelle transfer method by modification of the ATP-L with anti-myosin antibody 2G4.

1. Dry 1 mg of pNP–PEG$_{3400}$–PE in chloroform under argon.

2. Freeze-dry under high vacuum overnight.

3. Hydrate the film with 1 mg of 2G4 antibody in PBS, pH 8.0 (see Note 5).

4. Incubate the mixture overnight at room temperature under an argon atmosphere. (PEG–PE-modified 2G4 antibody becomes amphiphilic and spontaneously forms micelles.)

5. Separate micelles from the non-reacted free antibody by overnight dialysis against distilled water at 4°C using a dialysis bag with MWCO of 300 kDa. Store at the same temperature until further use.

6. Incubate an aliquot of PEG–PE-conjugated 2G4 (0.2 mg) antibody with ATP-loaded liposomes at a ratio 10 μg of the modified 2G4 per μmol of total lipid for 2 h at 37°C in K–H buffer, pH 7.4.

7. Separate the immunoliposomes from the non-incorporated PEG–PE–2G4 micelles by the overnight dialysis against K–H buffer, pH 7.4 at 4°C using a dialysis bag with MWCO of 300 kDa.

3.5. ELISA

1. Coat a 96-well U-bottomed microtiter plate with 50 μL of a 10 μg/mL pig cardiac myosin solution by incubation overnight at 4°C.

2. Wash plate three times with PBST.
3. Block wells with 50 µL of 1.0% solution of heat-inactivated horse serum to saturate the non-specific binding sites (see Note 6).
4. Add serial tenfold dilutions of various preparations of liposomes to the wells and incubate for 1 h at room temperature.
5. Wash the plate three times with PBS.
6. Add 50 µL of horseradish peroxidase-tagged goat anti-mouse IgG to each well.
7. Incubate the plate again for 1 h at 37°C and wash with PBST.
8. Add 100 µL of peroxidase substrate, K-blue to each well.
9. Incubate the plate for 15 min at room temperature.
10. Read OD on an ELISA reader at 620 nm with the reference filter at 490 nm.

3.6. Langendorff Model

1. Anesthetize rat with 80 mg/kg Nembutal, i.p.
2. Excise the heart rapidly and place in ice-cold 0.9% NaCl.
3. Attach the heart by the aorta to the cannula of the Langendorff apparatus. The cannula is placed into the cut aortic stump of the isolated heart to perfuse the coronary arteries retrogradely. (The valve separating the aorta and the left ventricle is closed by the pressure in the aorta, and the perfusion fluid is forced through the coronary arteries.)
4. Place a second cannula in the right ventricle via the pulmonary artery to collect coronary venous drainage.
5. Remove thebesian drainage via an apical drain placed through the left ventricular apex.
6. Place a thin-walled latex balloon in the left ventricle via the left atrium, and connect it to a pressure transducer to measure left ventricular developed pressure (LVDP) (systolic–diastolic) and the electronically derived dp/dt.
7. Adjust left ventricular balloon volume to produce an initial LV end-diastolic pressure (LVEDP) of 10 mmHg. Keep this balloon volume constant throughout the experiment.
8. Perfuse heart during normoxia at a coronary perfusion pressure of 80 mmHg.
9. Attach pacing wires to the atria. Pace the heart at 5 Hz.
10. Submerge the heart in the normal saline bath controlled at 37°C.
11. Record LVDP, LVEDP, ±dp/dt, and CPP continuously.
12. Measure the coronary flow rates by timed collections of the coronary venous effluent from the pulmonary artery's cannula. (Discard all perfusates after one passage through the heart.)

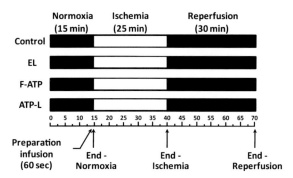

Fig. 2. A schematic diagram of the experimental protocol. Reproduced with permission from (30)

13. After instrumentation is completed, allow the preparation a 30 min stabilization period before starting experiments (Fig. 2).
14. Impose global no-flow ischemia for 25 min by completely stopping coronary perfusion (see Note 7).
15. Record LVDP, LVEDP, ±dp/dt, and CPP continuously during the decrease of perfusion.
16. Stop pacing of the heart.
17. Record LVEDP continuously during 25 min of imposed ischemia.
18. After 25 min of no-flow ischemia, reperfuse the heart for 30 min at a coronary perfusion pressure of 80 mmHg (see Note 8).
19. After 60 s of reperfusion, restore heart pacing at 5 Hz.
20. Record LVDP, LVEDP, ±dp/dt, and CPP continuously during 30 min of reperfusion (Fig. 3).
21. Remeasure the coronary flow rates by timed collections of the coronary venous effluent from the pulmonary artery cannula.

3.7. Measuring Protection of the Systolic and Diastolic Functions of the Myocardium in Isolated Rat Heart Model (Fig. 3)

1. At 15–30 min of stabilization period, infuse the liposomes into coronary circulation over 1 min period.
2. Impose global no-flow ischemia for 25 min (see Subheading 3.5).
3. After 25 min of no-flow ischemia, reperfuse heart for 30 min.

3.8. Experimental Myocardial Infarction in Rabbit

1. Anesthetize rabbits with subcutaneous ketamine (80 mg/kg) and xylazine (8 mg/kg).
2. Intubate via a tracheotomy and ventilate with room air at a tidal volume of 18–22 mL and 46–50 strokes/min using a Harvard rodent positive pressure ventilator. Continuously monitor the ECG (Fig. 4).

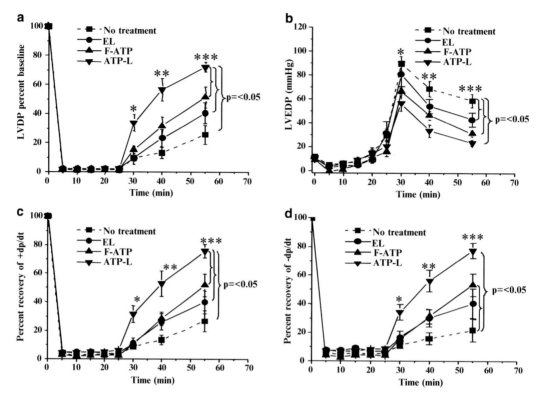

Fig. 3. Effect of various preparations on LVDP (**a**) and LVEDP (**b**) values, as well as on ±dp/dt (**c**, **d**) after global ischemia and reperfusion in isolated rat heart (Mean±SE), $n=7$–10. Reproduced with permission from (30)

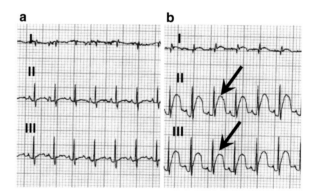

Fig. 4. Typical ECGs of the noninfarcted rabbit (**a**) and ECG with the elevated ST segment after coronary occlusion in the rabbit following an acute experimental infarction. Reproduced with permission from (31)

3. Continue anesthesia to effect with a 1:8 diluted Nembutal infusion via an indwelling marginal ear vein i.v. infusion set at approximately 20–25 mg/h.

4. Make a bilateral parasternal thoracotomy by cutting the ribs beginning at the xyphoid process and retract them to expose the heart.

5. Dissect the pericardium, isolate the aorta from the vena cava and the pulmonary artery. Insert and stabilize a heparinized flexible plastic catheter within the left atrium for rapid infusions into the heart.

6. Isolate an anterior branch of the left coronary artery with a smooth needle on a 3-0 suture for control of flow with PE 50 tubing to form an occlusive snare with the suture.

7. Make test infusions during brief (5–10 s) clamping/occlusion of the aorta to direct flow toward the coronary arteries. Myocardial trapping of infusate should occur as the coronary artery's snare is tightened.

8. Release the snare after 30 min, and reestablish perfusion for at least 3 h.

9. Reocclude the coronary artery, and infuse approximately 3 mL of 1:3 diluted USB via the atrial catheter to demarcate the occlusion-induced ischemic zone (USB stains the normoxic tissue, whereas the ischemic zone termed the "area at risk" remains unstained).

10. Immediately sacrifice (during the occlusion) the anesthetized animal by exsanguination while rapidly dissecting out the heart.

11. Trim and slice the excised left ventricle transversely between apex and base into five to six approximately equal thickness slices with a razor blade.

12. Digitally photograph all slices on both sides to document the site of the occlusion-induced area at risk by the absence of blue staining.

13. Incubate slices with a PBS-buffered 0.05% solution of NBT at 40–45°C for 20 min.

14. Rephotograph the slices and weigh each.

15. Determine the area at risk and irreversibly damaged myocardium from planimetry of both sides of all slices using Adobe Photoshop 7.0 (see Note 9) (Fig. 5).

16. Calculate the total LV weight at risk and infarcted in two independent assessments and average. The infarction size is expressed as the percent of the total LV weight at risk (Fig. 6).

3.9. Measurement of the Protective Effect of ATP-Liposomes

1. Follow the procedure through Subheading 3.8, item 6 and start ATP-L at Subheading 3.8, item 7.

2. Infuse approximately 3 mL of an ATP-L test solution (about 135 mg of lipid, 36 mg of ATP) through the coronary arteries by brief (5–10 s) clamping of the aorta followed by retightening the snare as the infusion ends. Do likewise using control liposomes and PBS buffer.

3. Follow the procedure from Subheading 3.8, item 8.

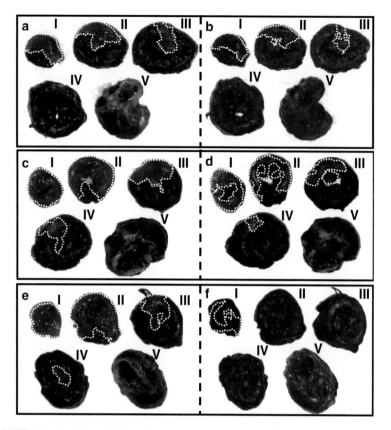

Fig. 5. USB- and NBT-stained sections of infarcted myocardium show the cardioprotective effect of ATP-L after 30 min of coronary occlusion and following 3 h of reperfusion in rabbits with an acute experimental myocardial infarction. (**a, b**) Control K–H buffer-treated animal; (**c, d**) EL treated animal and (**e, f**) ATP-L treated animal. (**a, b, e**) area at risk (USB-unstained red tissue) developed as a result of occlusion; (**b, c, f**) infarcted area at the end of occlusion/reperfusion experiment (NBT-unstained tissue); heart slices I to V represents base-to-apex. Reproduced with permission from (31)

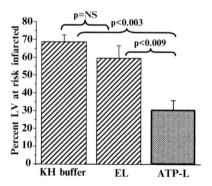

Fig. 6. Summary graph showing the fraction of infarcted area as a percentage of the total area at risk in the KH buffer-treated control group, EL, and ATP-L-treated group. (mean ± SE), $n = 4$–10. Reproduced with permission from (31)

4. Conclusion

This chapter provides background and a brief overview of the literature on the delivery of ATP to various tissues with an emphasis on the myocardium. The practical outline includes efficient methods for the preparation of liposomes containing ATP that can also be readily targeted to the myocardium with anti-myosin. A detailed description of an ex vivo rat and an in vivo rabbit model for testing these formulations is provided. The accompanying summary figures, based on our experience, are provided to show likely results and to illustrate the product of analyses using these methods. These methods provide a practical basis for continued studies of effective ways to deliver energy substrates to the ischemic myocardium.

5. Notes

1. Before and during an experiment buffer must be aerated continuously with a gas mixture (95% oxygen and 5% carbon dioxide) to oxygenate buffer and to stabilize the pH.
2. For instruction how to make the balloon, see: http://www.adinstruments.com/support/knowledge_database/pdf/Balloon_Catheter.pdf
   Instead of a long metal needle, use 20 cm of polyethylene tubing (2.0 mm in diameter) with a Leur connection.
3. During hydration (1 h), lipids must be removed from the surface of the flask by intensive vortexing.
4. Be careful with this dialysis tubing – it is very delicate and easily damaged when closing or opening clips.
5. Keep molar ratio of pNP–PEG3400–PE to antibody at 40:1.
6. In case of many analyses, for better standardization and to decrease the price use 0.5% casein (Sigma) solution in PBST instead of a 1.0% solution of heat-inactivated horse serum.
7. To avoid collapse of coronary arteries, decrease coronary perfusion pressure gradually over a 60 s period from 80 mmHg to 0 mmHg.
8. To avoid damages to coronary arteries, increase coronary perfusion pressure gradually over a 15 s period from 0 mmHg to 80 mmHg.
9. Open picture in Adobe Photoshop and adjust colors to optimize the difference between infarcted and non-infarcted

zones more apparent. Select the infarcted zone (Fig. 5) and measure picsele number. *Ideally*, – two observers should independently assess infarct size to reduce observer bios.

References

1. Carden DL, Granger DN (2000) Pathophysiology of ischaemia-reperfusion injury. J Pathol 190:255–266
2. Karmazyn M (1991) The 1990 Merck Frosst Award: ischemic and reperfusion injury in the heart. Cellular mechanisms and pharmacological interventions. Can J Physiol Pharmacol 69:719–730
3. McKay RG, Pfeffer MA, Pasternak RC, Markis JE, Come PC, Nakao S, Alderman JD, Ferguson JJ, Safian RD, Grossman W (1986) Left ventricular remodeling after myocardial infarction: a corollary to infarct expansion. Circulation 74:693–702
4. Kingsley PB, Sako EY, Yang MQ, Zimmer SD, Ugurbil K, Foker JE, From AH (1991) Ischemic contracture begins when anaerobic glycolysis stops: a 31P-NMR study of isolated rat hearts. Am J Physiol 261:H469–H478
5. Jennings RB, Schaper J, Hill ML, Steenbergen C Jr, Reimer KA (1985) Effect of reperfusion late in the phase of reversible ischemic injury. Changes in cell volume, electrolytes, metabolites, and ultrastructure. Circ Res 56:262–278
6. Khaw BA, Beller GA, Haber E, Smith TW (1976) Localization of cardiac myosin-specific antibody in myocardial infarction. J Clin Invest 58:439–446
7. Khaw BA, Scott J, Fallon JT, Cahill SL, Haber E, Homcy C (1982) Myocardial injury: quantitation by cell sorting initiated with antimyosin fluorescent spheres. Science 217:1050–1053
8. Kuzmin AI, Lakomkin VL, Kapelko VI, Vassort G (1998) Interstitial ATP level and degradation in control and postmyocardial infarcted rats. Am J Physiol 275:C766–C771
9. Gordon JL (1986) Extracellular ATP: effects, sources and fate. Biochem J 233:309–319
10. Puisieux F, Fattal E, Lahiani M, Auger J, Jouannet P, Couvreur P, Delattre J (1994) Liposomes, an interesting tool to deliver a bioenergetic substrate (ATP). in vitro and in vivo studies. J Drug Target 2:443–448
11. Caride VJ, Zaret BL (1977) Liposome accumulation in regions of experimental myocardial infarction. Science 198:735–738
12. Khaw BA, Torchilin VP, Vural I, Narula J (1995) Plug and seal: prevention of hypoxic cardiocyte death by sealing membrane lesions with antimyosin–liposomes. Nat Med 1:1195–1198
13. Lukyanov AN, Hartner WC, Torchilin VP (2004) Increased accumulation of PEG-PE micelles in the area of experimental myocardial infarction in rabbits. J Control Release 94:187–193
14. Torchilin VP, Khaw BA, Smirnov VN, Haber E (1979) Preservation of antimyosin antibody activity after covalent coupling to liposomes. Biochem Biophys Res Commun 89:1114–1119
15. Torchilin VP, Narula J, Halpern E, Khaw BA (1996) Poly(ethylene glycol)-coated anti-cardiac myosin immunoliposomes: factors influencing targeted accumulation in the infarcted myocardium. Biochim Biophys Acta 1279:75–83
16. Trubetskaya OV, Trubetskoy VS, Domogatsky SP, Rudin AV, Popov NV, Danilov SM, Nikolayeva MN, Klibanov AL, Torchilin VP (1988) Monoclonal antibody to human endothelial cell surface internalization and liposome delivery in cell culture. FEBS Lett 228:131–134
17. Palmer TN, Caride VJ, Caldecourt MA, Twickler J, Abdullah V (1984) The mechanism of liposome accumulation in infarction. Biochim Biophys Acta 797:363–368
18. Lukyanov AN, Elbayoumi TA, Chakilam AR, Torchilin VP (2004) Tumor-targeted liposomes: doxorubicin-loaded long-circulating liposomes modified with anti-cancer antibody. J Control Release 100:135–144
19. Maeda H, Wu J, Sawa T, Matsumura Y, Hori K (2000) Tumor vascular permeability and the EPR effect in macromolecular therapeutics: a review. J Control Release 65:271–284
20. Maeda H (2001) The enhanced permeability and retention (EPR) effect in tumor vasculature: the key role of tumor-selective macromolecular drug targeting. Adv Enzyme Regul 41:189–207
21. Han YY, Huang L, Jackson EK, Dubey RK, Gillepsie DG, Carcillo JA (2001) Liposomal ATP or NAD+ protects human endothelial

cells from energy failure in a cell culture model of sepsis. Res Commun Mol Pathol Pharmacol 110:107–116
22. Laham A, Claperon N, Durussel JJ, Fattal E, Delattre J, Puisieux F, Couvreur P, Rossignol P (1988) Intracarotidal administration of liposomally-entrapped ATP: improved efficiency against experimental brain ischemia. Pharmacol Res Commun 20:699–705
23. Laham A, Claperon N, Durussel JJ, Fattal E, Delattre J, Puisieux F, Couvreur P, Rossignol P (1988) Liposomally entrapped adenosine triphosphate: improved efficiency against experimental brain ischaemia in the rat. J Chromatogr 440:455–458
24. Konno H, Matin AF, Maruo Y, Nakamura S, Baba S (1996) Liposomal ATP protects the liver from injury during shock. Eur Surg Res 28:140–145
25. Neveux N, De Bandt JP, Chaumeil JC, Cynober L (2002) Hepatic preservation, liposomally entrapped adenosine triphosphate and nitric oxide production: a study of energy state and protein metabolism in the cold-stored rat liver. Scand J Gastroenterol 37:1057–1063
26. Neveux N, De Bandt JP, Fattal E, Hannoun L, Poupon R, Chaumeil JC, Delattre J, Cynober LA (2000) Cold preservation injury in rat liver: effect of liposomally-entrapped adenosine triphosphate. J Hepatol 33:68–75
27. Skiba-Lahiani M, Auger J, Terribile J, Fattal E, Delattre J, Puisieux F, Jouannet P (1995) Stimulation of movement and acrosome reaction of human spermatozoa by PC12 liposomes encapsulating ATP. Int J Androl 18:287–294
28. Xu GX, Xie XH, Liu FY, Zang DL, Zheng DS, Huang DJ, Huang MX (1990) Adenosine triphosphate liposomes: encapsulation and distribution studies. Pharm Res 7:553–557
29. Hartner WC, Verma DD, Levchenko T, Bernstein EA, Torchilin V (2009) ATP-loaded liposomes for treatment of myocardial ischemia. Wiley Interdisciplinary Reviews, Nanomedicine and Nanobiotechnology, 1(5): 530–539.
30. Verma DD, Levchenko TS, Bernstein EA, Torchilin VP (2005) ATP-loaded liposomes effectively protect mechanical functions of the myocardium from global ischemia in an isolated rat heart model. J Control Release 108:460–471
31. Verma DD, Hartner WC, Levchenko TS, Bernstein EA, Torchilin VP (2005) ATP-loaded liposomes effectively protect the myocardium in rabbits with an acute experimental myocardial infarction. Pharm Res 22:2115–2120

Chapter 26

Intracellular ATP Delivery Using Highly Fusogenic Liposomes

Sufan Chien

Abstract

Healthy cells must maintain a high content of adenosine triphosphate (ATP) because almost all energy-requiring processes in cells are driven, either directly or indirectly, by hydrolysis of ATP. During ischemia or hypoxia, reduced blood flow or disturbed oxygen supply results in the disrupted balance of energy production and utilization, and depletion of high-energy phosphates is the fundamental cause of cell damage. Direct intravenous infusion of high-energy phosphates, such as adenosine triphosphate (ATP), has not produced a consistent result because strongly charged molecules like ATP normally cannot pass the cell membrane in sufficient quantities to satisfy tissue metabolic requirements. Furthermore, the half-life of free ATP in blood circulation is very short, limiting its efficacy as a bioenergetic substrate.

We have developed a new technique for intracellular delivery of high-energy phosphate into normal or ischemic cells by using specially formulated, highly fusogenic, unilamellar lipid vesicles that contain magnesium-ATP. In vitro studies indicated a rapid fusion with the endothelial cells, protection of endothelial cells, and cardiomyocytes during ischemia. In vivo studies have shown enhanced full-thickness skin wound healing in various animal models. This technique has the potential to reduce or eliminate many detrimental effects caused by ischemia or hypoxia.

Key words: Ischemia, Hypoxia, Intracellular delivery, ATP, Liposome, Energy, Endothelial cells, Phospholipids, Wounds

1. Introduction

Tissue ischemia is a universal event in various forms of trauma and life-threatening emergencies. Reduced blood flow or disturbed oxygen supply results in the discrepancy of energy production and utilization. The cessation of blood flow not only halts the influx of oxygen and nutrients, but also stops carrying away toxic wastes. While many of the factors determining the final fate of the cells are still poorly understood, some of these events have been clearly delineated, such as depletion of high-energy phosphate, stimulation of glycolysis, inhibition of cellular energy-dependent processes, generation of oxygen-free radicals, loss of osmotic balance across the membrane, accumulation of metabolic products, intracellular acidosis, and release of lysosomal enzymes (1–4). Among these changes, depletion of high-energy phosphates is the fundamental cause of tissue damage (3, 5–8). Healthy cells must maintain a high content of ATP, and almost all energy-requiring processes in cells are driven, either directly or indirectly, by hydrolysis of ATP (2). Efforts to supplement ischemic cells with ATP have been ongoing for decades with little success. Direct intravenous infusion of ATP would be a simple solution. The membrane phospholipid bilayer of animal cells is selectively permeable only to some small molecules so that the internal composition of the cell is maintained. Most biological molecules are unable to diffuse through the cell membrane unless they have transport proteins (carrier proteins or channel proteins) (9). Larger, especially charged molecules like ATP (10–12), for which no specific transport mechanisms exist, cannot cross cell membrane under normal conditions. Although the presence of equimolar $MgCl_2$ reduces the negative charge of ATP from 3 to 1, this does not alter its permeability simply because it does not change the transporting mechanism. Results from our laboratories have indicated that fructose-1,6-diphosphate can pass through the cell membrane because of its membrane destabilization property (13–15), but only a small amount of ATP can be taken up by the cells, most likely through the membrane pores. Furthermore, the half-life of free ATP in blood circulation is less than 40 s, limiting its efficacy as a bioenergetic substrate (16).

We have developed and used specially formulated, highly fusogenic, unilamellar lipid vesicles that contain magnesium-ATP for intracellular ATP delivery, and preliminary results indicate that this new energy delivery technique can provide a significant protective effect to ischemic tissues. This article reports our encapsulation process and preliminary results with the new ATP delivery technique.

2. Materials

1. The manufacturing of ATP-vesicles is a joint venture between our research team and Avanti Polar Lipids, Inc. (Alabaster, AL).
2. Chemicals: L-α-phosphatidylcholine (Soy PC), 1,2-Dioleoyl-3-trimethylammonium-propane chloride salt (DOTAP), Polyethylene glycol 3350, Mg-ATP, trehalose (Avanti Polar Lipids, Alabaster, AL), Hanks' Salt solution, Na_2HPO_4, KH_2PO_4, and carboxyfluorescein (Sigma-Aldrich, St. Louis, MO).

3. Human umbilical vein endothelial cells (HUEVCs) (American Type Culture Collection, Manassas, VA).
4. TegaDerm™ (3M, Minneapolis, MN).
5. Franz Diffusion Cells (FDC-6, Logans Instrument, Somerset, NJ).
6. Ion Optix Fluorescence and Contractility System (Ion Optix Co. Milton, MA).
7. Adult Sprague–Dawley rats (Harlan Laboratories, Inc., Indianapolis, IN).
8. Adult New Zealand While rabbits (Myrtle's Rabbitry, Thompsons Station, TN).

3. Methods

3.1. Optimizing ATP-Vesicles Formulation and In Vitro Testing

1. Several formulations of ATP-vesicles for efficacy testing were tried. Our goal was to increase the efficacy while maintaining the possibility of cost-effective large-scale manufacturability.
2. The initial formulation consisted of soy derived 95% L-α-phosphatidylcholine (Soy PC) and 1,2-Dioleoyl-3-Trimethylammonium-Propane Chloride Salt (DOTAP) at a 50:1 molar ratio, respectively.
3. The lipid was hydrated with a buffer containing 5-mM Mg-ATP, trehalose at a 1:1 molar ratio with Soy PC, polyethylene glycol (PEG 3350) with a molecular weight of approximately 3,350 at a 4.9:1 molar ratio with Soy PC, Hanks' Salts at a concentration of 9.5-mg/ml, and 10-mM Na_2HPO_4.
4. The lipid was hydrated to a concentration of 20-mg/ml. The formulation was to be lyophilized (freeze-dried) by snap freezing in liquid nitrogen and sublimation of the water, resulting in dry and stable lipid vesicles.
5. Upon close examination of the initial formulation and manufacturing process, several problems became apparent. One problem was associated with pH shifts caused by supercooling. The ice formation drove the buffer salts into the portion that was still liquid. This increase in concentration could cause pH shifts, which could degrade the vesicle membrane lipids. Excessive salts and buffers can cause problems in a lyophilization process. Salts and buffers could lower the glass transition temperature (T_g) of the mixture. In an effective lyophilization process, the product temperature must be held below its glass transition temperature. A lower glass transition temperature resulted in a longer, less efficient lyophilization process (17).

6. The Hanks' Salts were then removed from the formulation. Also, the 10-mM Na_2HPO_4 was replaced with 10-mM KH_2PO_4, because potassium phosphate affects pH shifts less than sodium phosphate (17).

7. A second formulation change included the addition of trehalose. The original formulation called for trehalose at a 1:1 molar ratio with Soy PC. This resulted in an approximate 0.5:1 mass ratio of trehalose to Soy PC. Trehalose was added as a lyoprotectant to minimize lipid vesicle fusion during the lyophilization process (18). The literature reports that a 1:1 mass ratio of sugar to lipid is effective for preserving the lipid bilayer of dry vesicles (19). Therefore, the trehalose amount was increased to a 2:1 molar ratio with respect to Soy PC.

8. The formulation was also optimized by increasing the lipid concentration during the hydration step. The original formulation had a lipid concentration of 20-mg/ml. Experimental data revealed that this lipid concentration resulted in an ATP encapsulation efficiency of approximately 1.5%. In an attempt to increase the amount of deliverable ATP within the vesicle, the ATP concentration in the buffer was increased from 5 to 50 mM. However, the ATP solution was turbid at this elevated concentration at a pH of 7.4. A secondary approach was to increase the lipid concentration, thereby increasing the number of vesicles, and resulting in an increase in encapsulation and deliverable ATP.

9. Based on the basis of our experiments, the 100-mg/ml lipid concentration results in an ATP encapsulation efficiency of 12–16% (see Note 1). This allowed for a larger amount of ATP entrapment in the vesicle, resulting in a larger amount of ATP available for intracellular delivery.

10. The vesicles were processed through a high-pressure emulsifier to reduce the particle size. After the emulsification step, the size was further reduced and tightened by processing the vesicles through extruders containing polycarbonate membranes with pore diameters of 100-nm. According to dynamic light scattering (DLS) data, formulations made with 100-mg/ml lipids generally have a particle size ranging from 90 to 150-nm.

11. The vesicles were lyophilized from individual vials. DLS data has shown that the particle size of the lyophilized vesicles is between 50 and 200-nm.

12. We were successful at identifying a formulation that showed positive results in efficacy studies. Several formulations were manufactured over the past years. The optimal formulation consisted of Soy PC and DOTAP (50:1 mole ratio) at a lipid concentration of 100-mg/ml, 10-mM Mg-ATP, a 2:1 mole ratio of trehalose to lipid, respectively, and 10-mM KH_2PO_4.

This formulation was successfully freeze-dried and retained a particle size less than 200-nm upon rehydration (see Note 2).

13. Before application, ATP-vesicles were dissolved in isotonic solutions such as normal saline for direct application (see Note 3).

14. The process of ATP-vesicle fusion with the cell membrane was visualized using a freeze-fracture electron microscope, conducted by NanoAnalytical Laboratories in San Francisco, CA. An example of one ATP-vesicle beginning to morph with the cell is shown in Fig. 1 (see Note 4).

15. To determine the ability of the ATP-vesicles fusion with the cell membrane, HUVECs were cultured with the lipid vesicles, in which carboxyfluorescein was encapsulated. Within 10 min, the water-soluble carboxyfluorescein filled inside the cells (4).

16. The ATP-vesicles were also tested for their effect on cell survival during various ischemic conditions. Using HUVECs incubated with ATP-vesicles, we tested the effect of 6-h hypoxia (<0.5% O_2) on cell viability as measured by adherence. The experiments were repeated three times for each condition

Fig. 1. A freeze-fracture microscope photo showing one ATP-vesicle beginning to morph with the cell

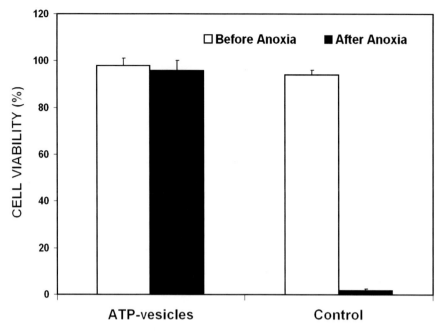

Fig. 2. Comparison of viability of human umbilical vein endothelial cells in culture media treated with and without ATP-vesicles before and after 6 h of anoxia

and the results showed that cells receiving ATP (1 mM)-vesicles remained attached compared to 0% viability in controls (Fig. 2).

17. We further tested this effect by inducing "chemical hypoxia" (20, 21). Potassium cyanide (KCN, 2.5 mM) was used in endothelial culture for 2 h. In the study group ($N=6$), ATP-vesicles were added while no ATP-vesicles were added into the control group ($N=6$). After 2 h, the cells in the study group maintained more than 90% viability as determined by trypan blue exclusion. In the control, no more than 10% of the cells were still alive (Fig. 3).

18. The effect of the ATP-vesicles on cardiomyocyte preservation was also tested during chemical hypoxia induced by KCN. Adult Sprague–Dawley rats (200–250 g) were used and isolation of primary cardiomyocytes was performed according to well-established techniques (22–24). After baseline readings were obtained, the cells were incubated for 30 min with KCN. The group with ATP-vesicles had the lowest amount of enzymes released than that in the other groups (Fig. 4).

19. The effect of ATP-vesicles after chemical hypoxia on myocardial contractility was also determined. After removal of KCN, 50 µL of calcium chloride (2 mM) was added and the cells were stimulated with 0.5–4 Hz, 8-V electric stimulator in an Ion Optix Fluorescence and Contractility System. The contractility data were analyzed with computer software. With 4 Hz stimulation, the velocity of contraction of the

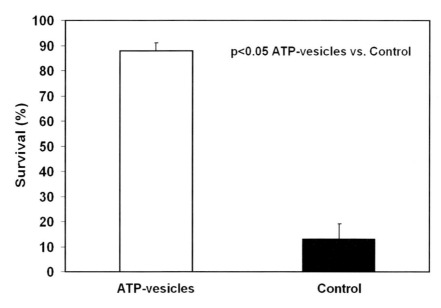

Fig. 3. Comparison of endothelial cell survival under chemical hypoxia

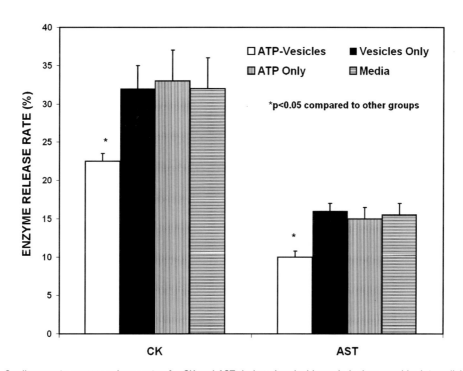

Fig. 4. Cardiomyocyte enzyme release rates for CK and AST during chemical hypoxia is decreased by intracellular ATP delivery. Rat cardiomyocytes were incubated in KCN for a period of 30 min in the presence of either ATP-vesicles, lipid vesicles only, Mg-ATP only, or culture media (M199). ATP-vesicles provided significant protection from CK and AST release during chemical hypoxia. *Significantly different from all groups as tested by ANOVA ($p < 0.05$, $N = 6$)

cardiomyocytes treated with ATP-vesicles was almost twice that of any other group (25) (see Notes 5 and 6).

4. Testing ATP-Vesicles in Wound Care

1. All animal experiments were approved by the IACUC of University of Louisville.
2. Before launching wound study, the ability of ATP-vesicles to penetrate the tissue was tested in skin penetration. The rat skin, which is known to have similar permeability characteristics as those of humans (19), was mounted in the FDC-6 Franz Diffusion Cells (26–28). ATP-vesicles or free ATP solution was placed in the donor dome and the receiving chamber was filled with neutral buffer. ATP, ADP, AMP and their metabolites were measured by HPLC using a modified technique described previously (29–31) in the two chambers at 2, 4, and 24 h and the contents were compared to obtain the permeability ratio. The result indicated that the ATP-vesicles dramatically increased nucleotide penetration through the skin (dermis and epidermis) by 10–20-fold ($N=9$, Fig. 5).

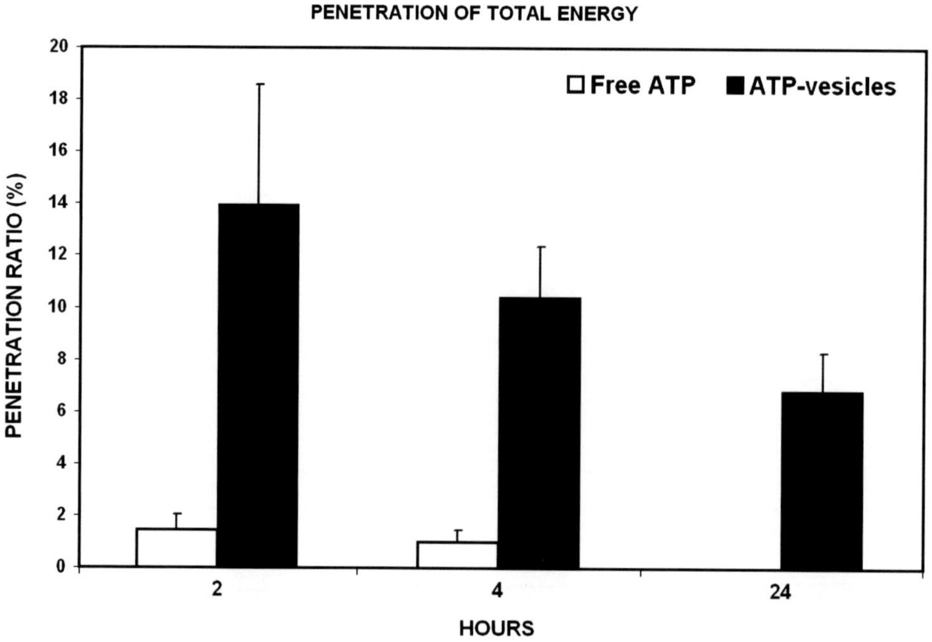

Fig. 5. Comparison of rat skin penetration ratio between ATP-vesicles and free Mg-ATP at 2, 4, and 24 h ($P<0.005$, $N=9$). Encapsulation of ATP dramatically increased nucleotide penetration through the skin. It appeared that the penetration occurred fast and the ATP was gradually metabolized or hydrolyzed with time

3. For wound healing comparison, 16 adult New Zealand white rabbits were used. Five of them were sacrificed at different periods of time for tissue biopsy study before the wounds were healed (some of them used free Mg-ATP as control drug), and wound-healing time was compared in the remaining 11 rabbits between the ATP-vesicles and saline-treated wounds on the non-ischemic ears and ischemic ears (22 wounds in each group).

4. We used a rabbit ischemic ear wound model created by minimally invasive surgery for this study (32). In brief, the rabbits were anesthetized with ketamine hydrochloride 50 mg/kg and xylazine 5 mg/kg, IM. One ear was rendered ischemic by using a minimally invasive surgical technique. The other ear served as a paired normal control. To create ischemic ear, three small vertical incisions were made on the vascular pedicles about 1 cm distal to the base of the ear using a #15 blade. The central artery was ligated, divided and the accompanying nerve was cut as well. The cranial artery and vein were also cut, but the caudal artery and vein were preserved. A circumferential subcutaneous tunnel was made through the three incisions. All the subcutaneous tissues, muscles, nerves, and small vascular branches were discontinued. The skin incisions were closed with 4-0 or 5-0 prolene. Two to four circular full-thickness wounds were created on the ventral side of each ear with a 6-mm stainless steel punch. The distance between the wounds was at least 30 mm. The skin inside the punch wound was removed from the cartilage. The perichondrium was also removed with the skin or separately. The base on which granulation and epithelization took place was the cartilage but the cartilage was not perforated.

5. ATP-vesicles or normal saline was used on two wounds of each ear and healing was compared. An occlusive dressing (TegaDerm™) was used to cover the wound site. This prevented the wounds from becoming desiccated.

6. Dressing changes were made and photos were taken daily until all the wounds were healed.

5. Results and Discussion

The healing times were much shorter in the ATP-vesicles treated wounds than the saline-treated ones. On the ischemic ear, the healing times ranged from 19 to 30 days (mean 22.8 ± 4.1 days) for the saline-treated wounds *vs.* 17–22 days (mean 18.0 ± 1.9 days) for the ATP-vesicles-treated wounds ($p=0.0005$). On the non-ischemic ear, the healing times were 14–17 days (mean 15.5 ± 1.0 days) for the saline-treated wounds and 13–16 days

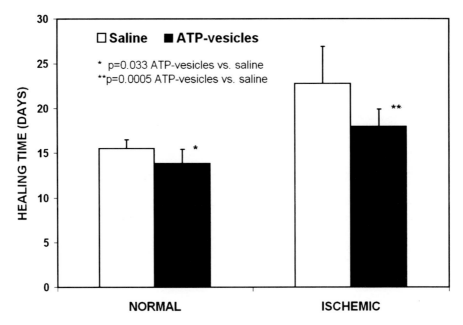

Fig. 6. Comparison of wound healing times between the saline and ATP-vesicles treated wounds on the non-ischemic ears and the ischemic ears in non-diabetic rabbits ($N=22$)

Fig. 7. An example of wound healing comparison between ATP-vesicles-treated and saline-treated wounds on one non-ischemic rabbit ear at day 15

(mean 13.9 ± 1.5 days) for the ATP-vesicles treated wounds ($p=0.033$). There were significant differences in healing times between the two treatments on the ischemic and non-ischemic ears (Fig. 6). A comparison photo of the healing process between the ATP-vesicles-treated and the saline-treated wounds at day 15 is shown in Fig. 7.

ATP-vesicles caused extremely rapid granulation tissue generation. Granulation tissue started to appear only 1 day postoperatively in many of these wounds. By day 5, the granulation tissue almost covered the wound (Fig. 8). Our recent study indicated that such rapid granulation tissue growth also occurred in

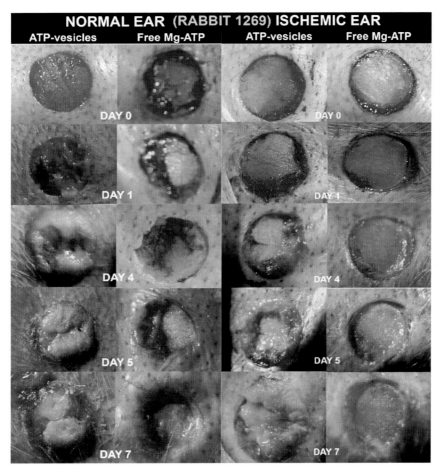

Fig. 8. An example of extremely rapid granulation tissue growth in the wounds treated by ATP-vesicles on rabbit ear. On the non-ischemic ear (*left*), granulation tissue starts to appear only 1 day after surgery. In the Mg-ATP-treated wounds very little granulation tissue is found after 2–3 days. On the ischemic ear (*right*), a similar phenomenon in ATP-vesicles treated wounds occurs but with 2–3 days of delay. Almost no granulation tissue is found in the Mg-ATP-treated wound at 7 days

diabetic animals (Fig. 9). Electron microscope study indicated that a significant cell accumulation as early as day 1 was seen in the ATP-vesicles-treated wounds on the non-ischemic rabbit ear. Scanning electron microscopy indicated numerous inflammatory cells embedded in fibrin-like amorphous matrix (Fig. 10).

The wounds treated by ATP-vesicles showed more CD31 positive cells than those treated by normal saline (33). Although there is currently no anti-rabbit antibody for rabbit CD31 stain, the mouse-anti-human monoclonal antibody has shown a very good cross-species reaction with rabbits in our study. In a separate study, cytokine expression was also tested in the wounds. The results showed that intracellular ATP delivery caused significant upregulation of IL-1β and TNF-α. The expression of TNF-α was significantly higher after only 1-day treatment (34).

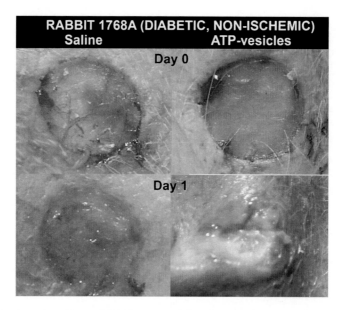

Fig. 9. An example of extremely rapid granulation tissue growth in a diabetic rabbit on the non-ischemic ear. The wound treated by ATP-vesicles shows granulation tissue growth only 1 day after surgery, while the wound treated by saline has no such growth

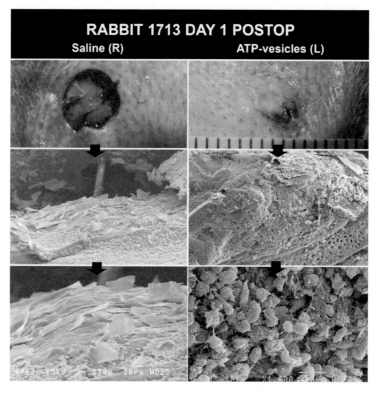

Fig. 10. Scanning electron microscope showing the wounds 1 day after surgery. The wound treated with ATP-vesicles is covered by a layer of granulation tissue which is filled with various cell types. The wound treated by saline only shows some fibrin and red blood cells

6. Notes

1. The current encapsulation rate is still relatively low which may result in high free ATP content in the solution, causing unnecessary side effects. Further increase of encapsulation rate is possible by using a freezing–thawing method (35).

2. If a diameter smaller than 100 nm can be maintained, it may further facilitate the transport of ATP-vesicles through the cell membrane and increase delivery efficiency.

3. Use only isotonic solutions such as normal saline to reconstitute the ATP-vesicles. Using hyper- or hypo-tonic solutions for this purpose may cause membrane leak of the vesicles. The reconstituted ATP-vesicles should be used within 2 h to avoid self-fusion of the vesicles.

4. The effect of the current ATP-vesicles appears to be cell and organ specific. They may work very well for one type of cells or organs, but may not work well for other types of cells or organs. The reason for that is not totally clear, but may be related to many factors such as the maturity of the cells, the structure of the cell membranes, and the existence of specific receptors on the cell membranes. It is also possible that the fusion of lipid vesicles to the cell membrane may have the potential to weaken membrane integrity, which may damage some cells that are already injured by ischemia or reperfusion.

5. Because of the presence of free Mg-ATP in the solution, it may cause purinergic receptor activation if given intravenously. This effect appears less severe if given intra-peritoneally.

6. Due to our continuous modifications, the above results were obtained with ATP-vesicles, but the actual compositions may not be exactly the same in each experiment.

Acknowledgments

This study was supported in part by NIH grants HL64186, DK74566, and AR52984. We thank Drs. Benjamin Chiang and William Ehringer for their experimental contributions in cell culture and preliminary formulation of ATP-vesicles, Dr. Qunwei Zhang for his cytokine studies, and Dr. Jianpu Wang for his wound management in some rabbits. The author also wishes to thank Mr. Drew Chochran of Avanti Polar Lipids for his contribution in the section of Optimizing ATP-vesicles formulation, Mr. Robert Reed for his scanning EM examination, and Ms. Ming Li for her HPLC determinations

References

1. Meisenberg G, Simmons WH (1998) The oxidation of glucose. Glycolysis, the TCA cycle, and oxidative phosphorylation. In: Meisenberg G, Simmons WH (eds) Principles of medical biochemistry. Mosby, St. Louis, MO, pp 297–331
2. Michiels C (2004) Physiological and pathological responses to hypoxia. Am J Pathol 164:1875–1882
3. Connery CP, Hicks GL, Wang T (1990) Positive correlation of functional recovery and tissue ATP levels in the hypothermically stored cardiac explant. Surg Forum 41:282–284
4. Chien S (1997) Metabolic management. In: Toledo-Pereyra LH (ed) Organ procurement and preservation for transplantation, 2nd edn. Springer, New York, pp 83–109
5. Reimer KA, Jennings RB, Hill ML (1981) Total ischemia in dog hearts, in vitro. 2. High energy phosphate depletion and associated defects in energy metabolism, cell volume regulation, and sarcolemmal integrity. Circ Res 49:901–911
6. Whitman G, Kieval R, Wetstein L, Seeholzer S, McDonald G, Harken A (1983) The relationship between global myocardial ischemia, left ventricular function, myocardial redox state, and high energy phosphate profile. A phosphorus-31 nuclear magnetic resonance study. J Surg Res 35:332–339
7. Klein HH, Schaper J, Puschmann S, Neinaber C, Kreuzer H, Schaper W (1981) Loss of canine myocardial nicotinamide adenine dinucleotides determines the transition from reversible to irreversible ischemic damage of myocardial cells. Basic Res Cardiol 76:612–621
8. Stringham JC, Southard JH, Belzer FO (1991) Mechanisms of ATP depletion in the cold-stored heart. Transplant Proc 23:2437–2438
9. Cooper GM (1997) The cell: a molecular approach. ASM Press, Washington, DC
10. Revetto MJ (1985) Myocardial nucleotide transport. Annu Rev Physiol 47:605–616
11. Fedelesova M, Ziegelhoffer A, Krause EG, Wollenberger A (1969) Effect of exogenous adenosine triphosphate on the metabolic state of the excised hypothermic dog heart. Circ Res 24:617–627
12. Parratt JR, Marshall RJ (1974) The response of isolated cardiac muscle to acute anoxia: protective effect of adenosine triphosphate and creatine phosphate. J Pharm Pharmacol 26:427–433
13. Ehringer W, Niu W, Chiang B, Wang OL, Gordon L, Chien S (2000) Membrane permeability of fructose-1, 6-diphosphate in lipid vesicles and endothelial cells. Mol Cell Biochem 210:35–45
14. Ehringer W, Su S, Chiang B, Stillwell W, Chien S (2002) The membrane destabilizing effects of fructose-1, 6-diphosphate on membrane bilayers. Lipids 37:885–892
15. Ehringer WD, Chiang B, Chien S (2001) The uptake and metabolism of fructose-1, 6-diphosphate in rat cardiomyocytes. Mol Cell Biochem 221:33–40
16. Puisieux F, Fattal E, Lahiani M, Auger J, Jouannet P, Couvreur P, Delattre J (1994) Liposomes, an interesting tool to deliver a bioenergetic substrate (ATP) in vitro and in vivo studies. J Drug Target 2:443–448
17. Shim J, Seok KH, Park WS, Han SH, Kim J, Chang IS (2004) Transdermal delivery of mixnoxidil with block copolymer nanoparticles. J Control Release 97:477–484
18. Sinico C, Manconi M, Peppi M, Lai F, Valenti D, Fadda AM (2005) Liposomes as carriers for dermal delivery of tretinoin: in vitro evaluation of drug permeation and vesicle-skin interaction. J Control Release 103:123–136
19. Lee CM, Maibach HI (2006) Deep percutaneous penetration into muscles and joints. J Pharm Sci 95:1405–1413
20. Brown AH, Niles NR, Braimbridge MV, Austen WG (1972) The combination of adenosine-triphosphatase inhibition and provision of high-energy phosphates for the preservation of unperfused myocardium, assessed by ventricular function, histology and birefringence. J Cardiovasc Surg 13:602–616
21. Huang HM, Weng CH, Ou SC, Hwang T (1999) Selective subcellular redistributions of protein kinase C isoforms by chemical hypoxia. J Neurosci Res 56:668–678
22. Altschuld R, Gamelin LM, Kelley RE, Lambert MR, Apel LE, Brierley GP (1987) Degradation and resynthesis of adenine nucleotides in adult rat heart myocytes. J Biol Chem 262:13527–13533
23. Danetz JS, Clemo HF, Davies RD, Embrey RP, Damiano RJ Jr, Baumgarten CM (1999) Age-related effects of St Thomas' Hospital cardioplegic solution on isolated cardiomyocyte cell volume. J Thorac Cardiovasc Surg 118:467–476
24. Delbridge LM, Roos KP (1997) Optical methods to evaluate the contractile function of unloaded isolated cardiac myocytes. J Mol Cell Cardiol 29:11–25
25. Chien S (2008) Metabolic management. In: Toledo-Pereyra LH (ed) Organ procurement and preservation, 3rd edn.

Austin, TX, Landes Bioscience, (http://eurekah.com/chapter/3992), pp 1–42
26. Franz TJ (1975) Percutaneous absorption on the relevance of in vitro data. J Invest Dermatol 64:190–195
27. Bonferoni MC, Rossi S, Ferrari F, Caramella C (1999) A modified Franz diffusion cell for simultaneous assessment of drug release and washability of mucoadhesive gels. Pharm Dev Technol 4:45–53
28. Jacobi U, Taube H, Schafer UF, Sterry W, Lademann J (2005) Comparison of four different in vitro systems to study the reservoir capacity of the stratum corneum. J Control Release 103:61–71
29. Hua D, Zhuang X, Ye J, Wilson D, Chiang B, Chien S (2003) Using fructose-1, 6-diphosphate during hypothermic rabbit heart preservation-a high-energy phosphate study. J Heart Lung Transplant 22:574–582
30. Adams H (1963) Adenosine-5′-triphosphate determination with phosphoglycerate kinase. In: Bergmeyer HU (ed) Methods of enzymatic analysis. Academic Press, New York, pp 539–543
31. Sellevold OFM, Jynge P, Aarstad K (1986) High performance liquid chromatography: a rapid isocratic method for determination of creatine compounds and adenine nucleotides in myocardial tissue. J Mol Cell Cardiol 18:517–527
32. Chien S (2007) Ischemic rabbit ear model created by minimally invasive surgery. Wound Repair Regen 15:928–935
33. Wang J, Zhang Q, Wan R, Mo Y, Li M, Tseng M, Chien S (2009) Intracellular ATP-delivery enhanced skin wound healing in rabbits. Ann Plast Surg 62:180–186
34. Zhang Q, Wan R, Mo Y, Wang J, Chien S (2008) Intracellular ATP delivery accelerates skin wound healing through pro- and anti-inflammatory cytokine expressions (Abstract). Wound Repair Regen 16:A57
35. Liang W, Levchenko TS, Torchilin VP (2004) Encapsulation of ATP into liposomes by different methods: optimization of the procedure. J Microencapsul 21:251–261

Chapter 27

Lipoplex Formation Using Liposomes Prepared by Ethanol Injection

Yoshie Maitani

Abstract

Cationic liposomes composed of 3β-[N-(N′N′–dimethylaminoethane)carbamoyl] cholesterol (DC–Chol) and dioleoylphosphatidylethanolamine (DOPE) (DC–Chol/DOPE liposome, molar ratio, 1:1 or 3:2) prepared by the dry-film method have been often used as non-viral gene delivery vectors. We have shown that a more efficient transfection in medium with serum was achieved using DC–Chol/DOPE liposomes (molar ratio, 1:2) than those (3:2), and preparation method by a modified ethanol injection than the dry-film. The most efficient DC–Chol/DOPE liposome for gene transfer was molar ratio (1:2) and prepared by a modified ethanol injection method. The enhanced transfection is related to an increase in the release of DNA in the cytoplasm by the large lipoplex during incubation in opti-MEM I reduced-serum medium (optiMEM), not to an increased cellular association with the lipoplex. Cationic liposomes rich in DOPE prepared by a modified ethanol injection method will help to improve the efficacy of liposome vector systems for gene delivery.

Key words: Cationic liposome, Gene transfection, DOPE, DC–Chol, Ethanol injection method, Dry-film method, Lipoplex formation

1. Introduction

Liposome-mediated gene delivery is dependent on numerous factors, such as, the formulation of the liposomes including the cationic lipid/neutral lipid ratio, how the liposomes are prepared, the cationic liposome/DNA charge ratio of the complex of cationic liposome and DNA (lipoplex), and the method used to produce the lipoplex. Recently, it was reported that the way in which a liposome was prepared affected transfection efficiency (1), and formation method of lipoplex affected size of lipoplex in which large ones increased the efficiency of transfection (2–7).

Notably, liposomes composed of 3β-[N-(N′,N′-dimethylaminoethane)carbamoyl)cholesterol (DC–Chol) together with dioleoylphosphatidylethanolamine (DOPE) (DC–Chol/DOPE liposome) have been classified as one of the most efficient vectors for the transfection of DNA into cells (8–10) and in clinical trials (11, 12). It has been demonstrated that a 3:2 or 1:1 molar ratio of DC–Chol/DOPE liposome results in high transfection efficiency (10). In these cases, liposomes are mostly prepared by the dry-film method. To further improve the transfection efficiency, it is necessary to evaluate DC–Chol/DOPE liposome from formulation and preparation method of liposome to formation method of their lipoplex.

We reported that greater transfection efficiency in medium with serum was obtained in human cervical carcinoma HeLa cells, using (1) DC–Chol/DOPE liposomes (molar ratio, 1:2) than liposomes (1:1 or 3:2), (2) a modified ethanol injection (MEI) method to prepare liposomes than the dry-film method (13, 14), and (3) a dilution method to form lipoplex than direct mixing. The physicochemical properties of liposomes and lipoplexes can be examined by measuring particle size. Transfection efficiency was evaluated by using plasmid DNA encoding luciferase gene and the cells.

2. Materials

2.1. Materials

1. DOPE (Avanti Polar Lipids Inc. Alabaster, AL), DC–Chol (Sigma, St. Louis, MO).
2. Plasmid DNA encoding firefly luciferase gene under the cytomegalovirus promoter (pCMV-luc) was obtained from Stratagene (pCMV-Tag4, CA).
3. The protein-free preparation of the plasmid was purified following alkaline lysis using maxiprep columns (Qiagen, Hilden, Germany).
4. All other reagents were of analytical grade.

2.2. Preparation of Liposomes by a Modified Ethanol Injection (MEI) Method (MLs)

1. Absolute ethanol, water (see Note 1).
2. A 50-mL round-bottomed spherical Quick-fit flask.
3. A rotary evaporator, water bath.
4. 0.45-μm Millex-HA filters (Millipore, Cork, Ireland).

2.3. Preparation of Liposomes by Dry-Film Method (DLs)

1. Chloroform.
2. N_2 gas to remove the chloroform solvent.
3. Vacuum desiccators to remove solvent from the dried film.

4. Vortex apparatus (Vortex-Genie2, model GT60, Scientific industries, Inc, NY).

5. A bath type sonicator (Honda Electronics, W220R, 200 W, 40 kHz, Tokyo, Japan).

2.4. Formation of Lipoplexes

optiMEM (Invitrogen Corp. Carlsbad, CA), water.

2.5. Determination of Particle Size

1. Dilution of dispersion to an appropriate volume with water.
2. The electrophoresis light-scattering method (Model ELS-800, Ostuka Electronics Co. Ltd., Japan).

2.6. Cryotransmission Electron Microscopy

1. Quantifoil® R1.2/1.3 holey carbon grids at a Leica EM CPC cryo-preparation station.
2. JEOL JEM-3100FFC transmission electron microscope (cryo-TEM).

2.7. Cell Culture and Luciferase Assay

1. For human cervical carcinoma HeLa cells, Dulbecco's Modified Eagle's Medium (DMEM) (Invitrogen) supplemented with 10% fetal bovine serum (FBS) (Invitrogen) and kanamycin sulfate (Wako Pure Chemistry, Osaka, Japan) (100 mg/mL).
2. For human lung adenocarcinoma A549 cells, RPMI-1640 medium (Invitrogen) supplemented with 10% FBS and kanamycin sulfate (100 μg/mL).
3. The plasmid pCMV-luc.
4. Phosphate buffered-saline (pH 7.4, PBS).
5. Cell culture lysis reagent (Toyo Ink, Tokyo, Japan), a picagene luciferase assay kit (Toyo Ink).
6. BCA protein assay reagent (Pierce, Rockford, IL).
7. A chemoluminometer (Wallac ARVO SX 1420 multilabel counter, Perkin Elmer Life Science, Co. Ltd., Japan).

3. Methods

The particle size of the liposomes varied with the volume ratio of ethanol to water. An increase in the ratio of ethanol in the system resulted in an increase in particle size of liposomes at 10–30 mg total lipid/ethanol mL. An increase in the volume of water in the system resulted in a decrease in liposome particle size (15).

MEI method of liposome preparation is similar to the reported ethanol injection method (16, 17), proliposome preparation (17, 18), and coacervation methods (19). The differences between our method and those reported previously are as follows: (a) Water is poured into lipid–ethanol solution at about 70°C although lipid–ethanol solution is injected into the aqueous phase in the ethanol injection method; (b) a high concentration of PC in ethanol (~30 mg/mL ethanol) can be used; (c) after preparation of liposomes, ethanol is removed in a rotary evaporator, not by dialysis; (d) when the solution is diluted with water, small and homogenous liposomes form spontaneously; and (e) a one-step dilution procedure is sufficient for use with drug remote loading.

Morphology of MLs and DLs was evaluated by freeze-fracture and cryotransmission electron microscopy. Both liposomes exhibited single unilamellar vesicles (13).

In regards of formation of lipoplexes, we used direct mixing of liposomes and DNA in water (non-dilution method) or dilution of liposomes and DNA separately in optiMEM (dilution method) for the formation of lipoplexes.

3.1. Preparation of MLs (15)

1. All lipids were dissolved in warm absolute ethanol (1 mL) in a 50-mL round-bottomed spherical Quick-fit flask at about 60°C. 10–30 mg total lipid/ethanol mL.
2. About 70°C water (4 mL) was added rapidly in the lipid–ethanol solution in the flask (see Note 2).
3. The aqueous phase immediately turned milky as the result of the formation of liposomes.
4. The ethanol was then removed by rotary evaporation under the reduced pressure at 40°C while nitrogen gas was passed over the solution.
5. When the water was added to the lipid–ethanol solution, the mix solution was transparent.
6. When reducing ethanol, milky suspension, liposomes came out.
7. The liposome suspension was concentrated to the desired final volume by the removal of water under the same condition (see Note 3).
8. The liposome suspension was immediately sterilized by filtration through a sterile syringe-driven filter unit with a pore size of 450 nm at once (see Note 4).
9. MLs showed smaller size with low polydispersity than DLs as shown in Fig. 1 (see Note 5).

3.2. Preparation of DLs (20)

1. All lipids were dissolved in chloroform and the solution was dried with N_2 gas to remove the chloroform solvent.
2. The dried film was vacuum desiccated for at least 10 min.
3. Water was added, and after sufficient hydration, the film was suspended by vortexing.

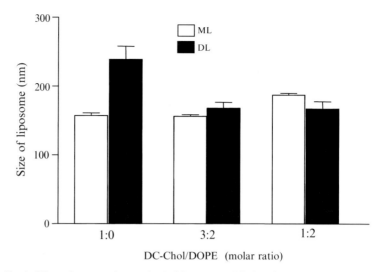

Fig. 1. Effect of preparation method of liposomes (DC–Chol/DOPE) on their size. ML and DL were prepared by the MEI and dry-film method, respectively. The size of the liposome was measured in water. Each result represents the mean ± SD ($n = 3$)

4. The samples were then sonicated for 10 min in a bath type sonicator (see Note 6).
5. The liposome suspension was immediately filtered through 0.45-μm filters for sterilization as shown in Fig. 1 (13) (see Note 7).

3.3. Formation of Lipoplex by Direct Mixing

1. Lipoplexes at charge ratios (+/−) of 1–11 of cationic lipid to DNA were formed by addition of 3.16–11.1 μL of liposome preparation (1–3.75 mg total lipid/ml water) to 1 μg of DNA in 5 μL of water with 10 rounds of pipetting with gentle shaking and leaving at room temperature for 10–15 min.
2. The optimal lipoplexes (1 μg of DNA) were as follows: for the liposome composed of DC–Chol/DOPE (molar ratio, 1:0): 11.1 μL of liposome suspension (1 mg total lipid/ml water) at a charge ratio of (+/−) of 7, for the liposome composed of DC–Chol/DOPE (3:2): 3.16 μL of liposome suspension (1.92 mg total lipid/ml water) at a charge ratio of (+/−) of 2, and for the liposome composed of DC–Chol/DOPE (1:2): 3.16 μL of liposome suspension (3.75 mg total lipid/mL water) at a charge ratio of (+/−) of 2.
3. The direct mixing yield a smaller lipoplex compared with dilution method as shown in Fig. 2 (see Note 8).

3.4. Formation of Lipoplex by Dilution Method

The liposomes and DNA were diluted separately to 125 μL in optiMEM, allowed to stand for 5 min, mixed, and incubated at room temperature for a further 20 min.

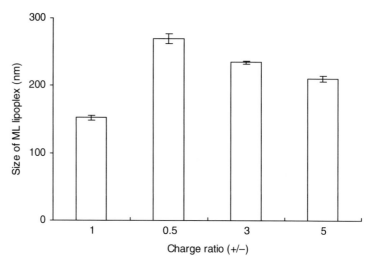

Fig. 2. Effect of the charge ratio (+/−) of ML (DC–Chol/DOPE = 1:2) to plasmid DNA on size of lipoplexes formed by direct mixing which was left at room temperature for 15 min and incubated with RPMI-1640 medium for further 15 min. The lipoplex at various charge ratios (+/−) of DC–Chol to DNA were prepared by addition of each liposome preparation contained 0.9 mM DC–Chol concentration (1.67, 10, 16.7 µL for the charge ratio (+/−) of 0.5, 3 and 5) to 1 µg of DNA in 5 µL of water

3.5. Transfection of Lipoplexes into Cells

1. HeLa cells were grown in DMEM supplemented with 10% FBS at 37°C in a humidified 5% CO_2 atmosphere.
2. A549 cells were grown in RPMI-1640 medium supplemented with 10% FBS in a humidified 5% CO_2 atmosphere.
3. The lipoplex was diluted with DMEM for HeLa cells or RPMI-1640 medium for A549 cells containing 10% FBS to a final concentration of 1 µg of DNA per 0.5 mL of medium per well in 12 well plates, and incubated with the cells for 24 h in the medium.
4. The optimal molar ratio of DC–Chol to DOPE of MLs was 1:2 in A549 and HeLa cells, showed the highest transfection efficiency as shown in Figs. 3 and 4.
5. The optimal charge ratios of DC–Chol/DNA(+/−) in ML lipoplexes were decided in HeLa cells as shown in Fig. 4.

3.6. Determination of Particle Size

1. After by diluting of dispersion to an appropriate volume with water, the cumulant particle size of liposomes was determined using a dynamic light scattering instrument as shown in Figs. 1 and 2.
2. In the process of formation of lipoplex, the size of liposomes and lipoplexes was determined to be diluted in water within 5 min and 20 min after incubating liposomes and lipoplexes in optiMEM, respectively as shown in Fig. 5a, b (see Note 9).

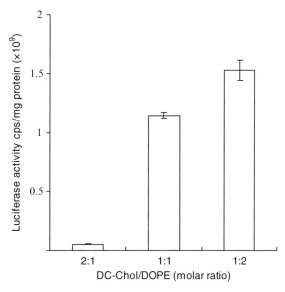

Fig. 3. Transfection efficiency of ML (DC–Chol/DOPE = 1:2, 1:1, 2:1) lipoplexes formed by direct mixing at a charge ratio (+/−) of 3 in A549 cells. Lipoplexes were diluted with RPMI-1640 containing FBS to a final concentration of 1 μg of DNA in 0.5 mL of medium per well, and the cells were incubated for 24 h in the medium

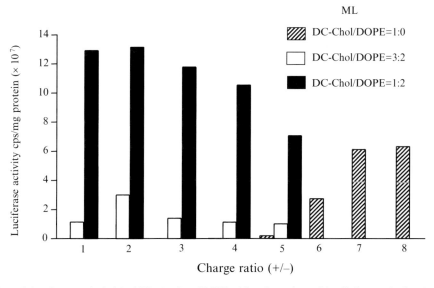

Fig. 4. Effect of the charge ratio (+/−) of MLs to plasmid DNA of lipoplexes formed by dilution method on transfection efficiency in HeLa cells. Lipoplexes were diluted with DMEM containing FBS to a final concentration of 2 μg of DNA in 1 mL of medium per well, and the cells were incubated for 24 h in the medium. Each result represents the mean ($n = 2$)

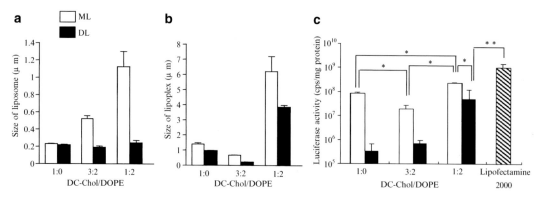

Fig. 5. Change of size and transfection efficiency of lipoplexes formed by dilution method. Size of ML and DL after dilution in optiMEM within 5 min (**a**), size of their lipoplex after incubation in optiMEM within further 20 min (**b**), and transfection efficiency of lipoplexes in HeLa cells (**c**). Lipoplexes at a charge ratio (+/−) of 2 for ML and DL (DC–Chol:DOPE = 3:2, 1:2), and of 7 for ML and DL (DC–Chol) from the results in Fig. 4. Transfection condition was the same with Fig. 4. Each result represents the mean ± SD ($n=3$). *$P < 0.05$. **$P < 0.01$ by ANOVA test

3.7. Luciferase Assay

1. Luciferase expression was measured according to the luciferase assay system.
2. Incubation was terminated by washing the plates three times with cold PBS.
3. Cell lysis solution was added to the cell monolayers and subjected to one cycle of freezing (−70°C) and thawing at 37°C, followed by centrifugation at 20,400 g for 5 s.
4. The supernatants were stored at −70°C until the assays. Aliquots of 20 µL of the supernatants were mixed with 100 µL of luciferase assay system (Pica gene) and counts per sec (cps) were measured with a chemoluminometer.
5. The protein concentration of the supernatants was determined with BCA reagent, using bovine serum albumin as a standard and cps/µg protein was calculated (see Note 10).
6. The dilution method using each ML and DL, yielded a larger lipoplex and higher transfection efficiency in HeLa cells as shown in Fig. 5c.

3.8. Cryotransmission Electron Microscopy

1. A drop of 2.5 µL of the lipid dispersion (more than 1 mM lipid concentration) was applied to the grid, excess liquid was blotted by touching with a piece of filter paper and the grid immediately plunged into liquid ethane kept at −165°C.
2. The grid was then transferred to the microscope by a cryo-transfer device.
3. Cryo-electron microscopy was performed on a JEOL JEM-3100FFC transmission electron microscope (cryo-TEM) (see Note 11).

4. Notes

1. Unless stated otherwise, water and all solution should be used in Milli Q water.
2. When the ethanol ratio to water was 20% (v/v), the particle size of liposomes was less than 200 nm with a small value of polydisperse index (PI) (<0.1). However, at ethanol ratios of over 50% (v/v) the particle size of liposomes was more than 1000 nm (15).
3. Even the large size of liposomes in 60% (v/v) ethanol could be decreased by filtration through polycarbonate membranes with a pore diameter of 0.2 μm at room temperature, or by sonication in bath type sonicater for 10 min.
4. The size of liposomes (140 – 160 nm) was not markedly affected by pH of the medium, and the various ion strengths of the medium did not change the size of liposomes up to ion strength 2 (15).
5. ML produced by the MEI method had a homogeneous size distribution ranging from 150 to 180 nm obtained without sonication. MEI is an easy method leading spontaneously to smaller particles with low polydispersity. Thus, use of the MEI method is suggested to obtain small liposomes (13).
6. The size of liposomes can be also adjusted by Extruder (LIPEX™ Extruder, Northern Lipids Inc. Canada).
7. DL was prepared by briefly sonicating liposomes until a homogeneous size distribution ranging from 150 to 230 nm was obtained (Fig. 1).
8. Lipoplexes prepared by dilution method with ML(1:2) at a charge ratio (+/−) of 3 were significantly larger (2247±352 nm) than those prepared by direct mixing (1427±83 nm), showing more than two-times greater transfection activity in HeLa cells in the presence of 10% FBS. This finding suggested that the dilution method yielded a larger lipoplex and higher transfection efficiency than direct mixing.
9. In in vivo transfection, the size of lilpoplexes should be small. In these cases, direct mixing is preferred to dilution method.
10. Lipofectamine™ 2000 (Invitrogen) was used positive control in transfection of luciferase in the cells. 5 μL of Lipofectamine™ 2000 was used for 2 μg of DNA to form a DNA complex in Opti-MEM, according to the manufacturer's protocol.
11. It is equipped with field emission gun (FEG), helium temperature specimen stage, omega-type energy filter and Gatan MegaScan 795 2Kx2K CCD camera. For improved contrast of ice-embedded specimens we employed a novel Zernike-type

phase plate at the back focal plane of the objective lens (21). It provides a true phase contrast regime revealing details in the image which are hidden in the conventional defocus phase contrast mode. All images were taken by the CCD camera with the TEM operated at 300 kV acceleration voltage, zero-loss energy filter mode, ×60,000 indicated magnification and employing the phase plate. At that magnification, the specimen resolution at the CCD is 3.0 Å/pix. To minimize electron beam damage, we employed a minimum dose protocol which irradiates the area of interest only during the image exposure. The total dose to the specimen was about 6 $e^-/Å^2$.

References

1. Tranchant I, Thompson B, Nicolazzi C, Mignet N, Scherman D (2004) Physicochemical optimisation of plasmid delivery by cationic lipids. J Gene Med 6:S24–S35
2. Felgner JH, Kumar R, Sridhar CN, Wheeler CJ, Tsai YJ, Border R, Ramsey P, Martin M, Felgner PL (1994) Enhanced gene delivery and mechanism studies with a novel series of cationic lipid formulations. J Biol Chem 269:2550–2561
3. Zhang YP, Reimer DL, Zhang G, Lee PH, Bally MB (1997) Self-assembling DNA-lipid particles for gene transfer. Pharm Res 14:190–196
4. Ross PC, Hui SW (1999) Lipoplex size is a major determinant of in vitro lipofection efficiency. Gene Ther 6:651–659
5. Turek J, Dubertret C, Jaslin G, Antonakis K, Scherman D, Pitard B (2000) Formulations which increase the size of lipoplexes prevent serum-associated inhibition of transfection. J Gene Med 2:32–40
6. Almofti MR, Harashima H, Shinohara Y, Almofti A, Li W, Kiwada H (2003) Lipoplex size determines lipofection efficiency with or without serum. Mol Membr Biol 20:35–43
7. Hattori Y, Maitani Y (2005) Folate-linked nanoparticle-mediated suicide gene therapy in human prostate cancer and nasopharyngeal cancer with herpes simplex virus thymidine kinase. Cancer Gene Ther 12:796–809
8. Zhou X, Huang L (1994) DNA transfection mediated by cationic liposomes containing lipopolylysine: characterization and mechanism of action. Biochim Biophys Acta 1189:195–203
9. Farhood H, Gao X, Son K, Yang YY, Lazo JS, Huang L, Barsoum J, Bottega R, Epand RM (1994) Cationic liposomes for direct gene transfer in therapy of cancer and other diseases. Ann N Y Acad Sci 716:23–34 discussion 34–35
10. Farhood H, Serbina N, Huang L (1995) The role of dioleoyl phosphatidylethanolamine in cationic liposome mediated gene transfer. Biochim Biophys Acta 1235:289–295
11. Nabel GJ, Nabel EG, Yang ZY, Fox BA, Plautz GE, Gao X, Huang L, Shu S, Gordon D, Chang AE (1993) Direct gene transfer with DNA-liposome complexes in melanoma: expression, biologic activity, and lack of toxicity in humans. Proc Natl Acad Sci U S A 90:11307–11311
12. Nabel EG, Yang Z, Muller D, Chang AE, Gao X, Huang L, Cho KJ, Nabel GJ (1994) Safety and toxicity of catheter gene delivery to the pulmonary vasculature in a patient with metastatic melanoma. Hum Gene Ther 5:1089–1094
13. Maitani Y, Igarashi S, Sato M, Hattori Y (2007) Cationic liposome (DC–Chol/DOPE=1:2) and a modified ethanol injection method to prepare liposomes, increased gene expression. Int J Pharm 342:33–39
14. Ding W, Hattori Y, Higashiyama K, Maitani Y (2008) Hydroxyethylated cationic cholesterol derivatives in liposome vectors promote gene expression in the lung. Int J Pharm. 354:196–203 doi:10.1016/j.ijpharm.2007.10.051
15. Maitani Y, Soeda H, Wang J, Takayama K (2001) Modified ethanol injection method for liposomes containing β-sitosterol β-D-glucoside. J Liposome Res 11:115–125
16. Kremer JMH, van der Esker MW, Pathmamanoharan C, Wiersems PH (1977) Vesicles of variable diameter prepared by a modified injection method. Biochemistry 16:3932–3935
17. Perrett S, Golding M, Williams WP (1991) A simple method for the preparation of liposomes

for pharmaceutical applications: characterization of the liposomes. J Pharm Phaarmacol 43: 154–161

18. Park K, Lee M, Hwang K, Kim C (1999) Phospholipid based microemulsions of flurbiprofen by the spontaneous emulsification process. Int J Pharm 183:145–154

19. Ishii F, Takamura A, Ishigami Y (1995) Procedure for preparation of lipid vesicles (liposomes) using the coacervation (phase separation) technique. Langmuir 11:483–486

20. Bangham AD, Standish MM, Watkins JC (1965) Diffusion of univalent ions across the lamellae of swollen phospholipids. J Mol Biol 13:238–252

21. Danev R, Nagayama K (2006) Applicability of thin film phase plates in biological electron microscopy. Biophysics 2:35–43

Chapter 28

Acid-Labile Liposome/pDNA Complexes

Michel Bessodes and Daniel Scherman

Abstract

One of the bottlenecks to achieve gene or drug delivery to cells is the spatio-temporal release of the cargo to effect the correct task whereever and whenever it is supposed to. The aim of this chapter is to describe the synthesis and properties of lipids, which can form liposome complexes with DNA and to fall apart in slight acidic condition such as those encountered in the late endosome, thus destabilizing the liposome membrane and releasing their content.

Key words: Gene delivery, Acid labile liposome, DNA, Transfection

1. Introduction

Attempts to correct pathological disorders in cell by the delivery of the correct gene that would either produce the missing or underexpressed protein, or induce the death of (i.e.) a tumor cell, has been the aim of numerous research teams in the past decades. Among the different technologies tentitatively used to reach this goal, cationic liposome complexes of DNA, so-called lipoplexes, have been extensively studied (1–4). However, many challenging problems should be resolved before an efficient gene delivery could be obtained with these systems. One of these tricky steps is the escape of DNA from the complex, at the right time and in the right place. DNA release and traffic to the nucleus is indeed a recognized major barrier to efficient transfection (5), even if the other steps occurring between endosome internalization of lipoplexes and DNA expression are still rather unclear today. Thus, it is generally assumed that a triggered system that could help DNA escape from the lipoplexes could improve gene transfer efficacy. The lipid compounds whose synthesis will be described here have been designed to address this particular outcome. Their chemical structure results in a pH controlled lability of the liposomes obtained from their aqueous formulation, which is obtained by the chemical disruption of the lipid molecule. This should be opposed to pH sensitivity of other liposomal systems that respond to pH change by physicochemical collapse of the liposomal structure (6–15).

2. Materials

2.1. Solvents and Chemicals

1. All the solvents used were ACS reagent grade from SDS ("Solvants, documentation, synthèse"; France) and were used without further purification unless specifically notified.
2. The reagents and catalysts were obtained from Sigma/Aldrich.
3. The silica gel used in purifications was from SDS (silica 60, 200μ), glass columns for low-pressure flash chromatography were obtained from Aldrich glassware.
4. Washing solutions were made from saturated sodium chloride (35%, NaClsat), saturated sodium hydrogen carbonate (10%, $NaHCO_3$ sat) and 0.5 M potassium hydrogen sulfate (68 g/l, $KHSO_4$), in regular deionized water.

2.2. Biological Reagents

1. Plasmid DNA used contained the luciferase reporter gene (for in vitro experiments) including the CMV sequence. Plasmids present the normalized criteria of quality: endotoxin level <20 EU/mg, supercoiled DNA >90%, E. coli-derived DNA contaminant <5%, RNA contaminant <5%, and protein contamination <1%. Such a plasmid preparation has been described (16).
2. 1,2-Dioleoyl-sn-glycero-3-phosphoethanolamine (DOPE) (Avanti Polar Lipids, Birmingham, AL, USA).
3. The different buffers were prepared using milliQ™ water and filtered with a 0.2 μm syringe filter prior to use.

3. Methods

We have included the orthoester linkage in the lipid structure for its unique properties of hydrolysis in acidic media which are closely related to structure and substitution pattern (17). Also, this linker is expected to respond to the lower pH observed in the endosome, thus leading to pH degradable lipoplexes (see Note 1). We have synthesized two cationic lipids (**2** and **4**) differentiated

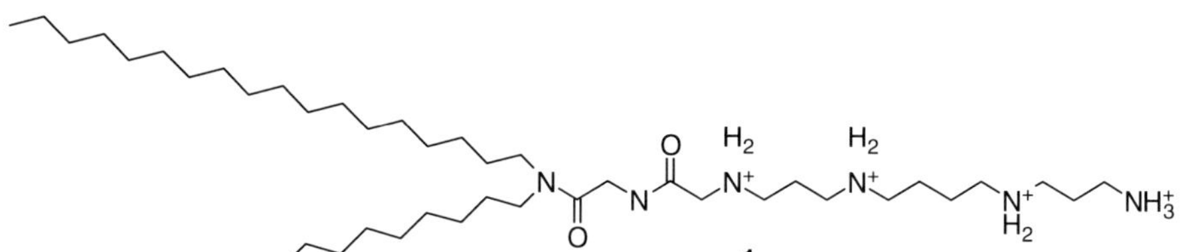

Scheme 1. pH labile cationic lipids (**2** and **4**) and their stable analogs (**1** and **3**)

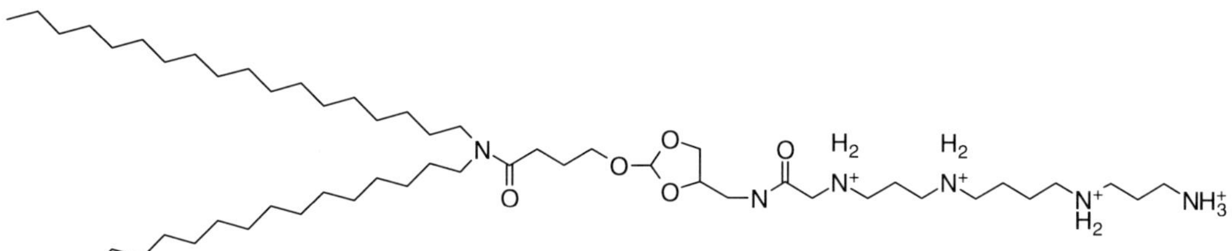

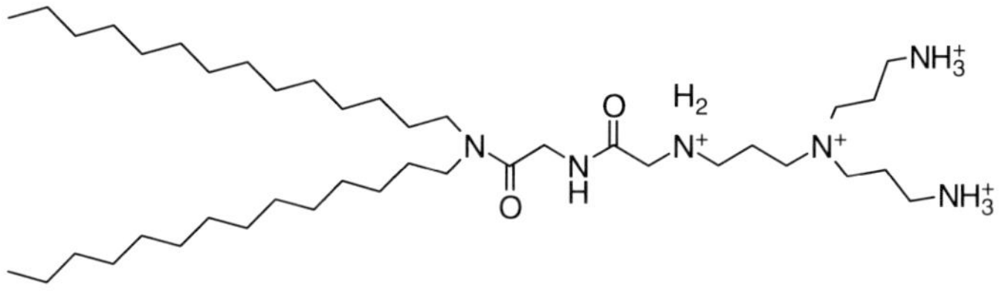

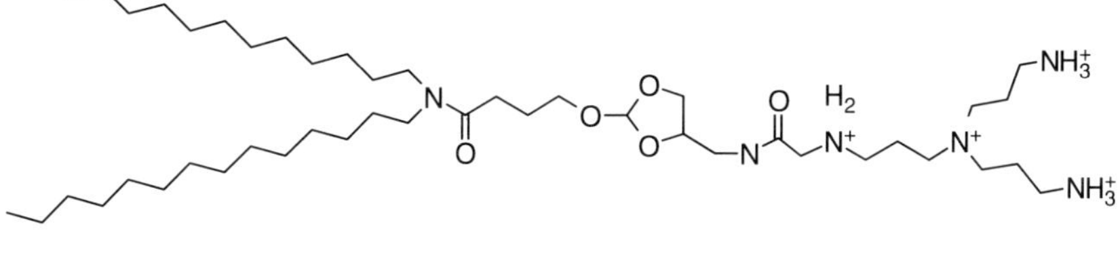

by the structure of the cationic head, as degradable analogs of compounds **1** and **3** commonly used in our group (Scheme 1). Both compounds bore the orthoester hydrolysable linker between the double alkyl chains and the polar cationic head (18). Lipoplexes formulated with this type of compounds are expected to collapse following acid hydrolysis, hopefully inducing endosomal escape

3.1. Chemical Synthesis

3.1.1. Synthesis of the pH Labile Lipids (Scheme 2)

3.1.1.1. N-(2,3-dihydroxypropyl)-2,2,2-trifluoroacetamide (5)

and release of their content in the cytoplasm. They were compared to the lipoplexes obtained with the nonlabile cationic lipids analogs those synthesis has been described elsewhere (19).

3-aminopropan-1,2-diol (15 g, 164.6 mmol, 1 eq) was placed in a round-bottom flask equipped with magnetic stirrer and solubilized in THF (100 ml). The solution was cooled to 0°C in an ice bath and ethyl trifluoroacetate (21.5 ml, 181.1 mmol, 1.1 eq) was slowly added (see Note 2). The reaction medium was stirred 2 h at room temperature, then evaporated to give 29 g (95% yield) of the title compound as a colorless oil.

NMR ^{1}H (300 MHz, CDCl$_3$, δ ppm): 2.12 (t, J = 5.5 Hz: 1H); 2.61 (d, J = 5 Hz: 1H); 3.25–3.85 (m: 4H); 3.92 (m: 1H); 6.88 (m: 1H).
M/z (D/CI) = 188 (MH$^+$).
IR (KBr): 3108; 2943; 1710; 1564; 1214; 1187; 1158; 1050 and 727 cm^{-1}.

3.1.1.2. 2,2,2-Trifluoro-N-(2-methoxy-[1,3]dioxolan-4-ylmethyl)-acetamide (6)

Compound **5** (29 g; 155 mmol, 1 eq) and trimethylorthoformate (685 mmol, 4.4 eq) were dissolved in dichloromethane (75 ml). Paratoluenesulfonic acid (300 mg, 1.7 mmol, 0.01 eq) was added, and the mixture was stirred at room temperature during **2** h. The reaction medium was diluted with dichloromethane (500 ml) and washed successively with NaHCO$_3$ sat (3 × 200 ml), and NaClsat

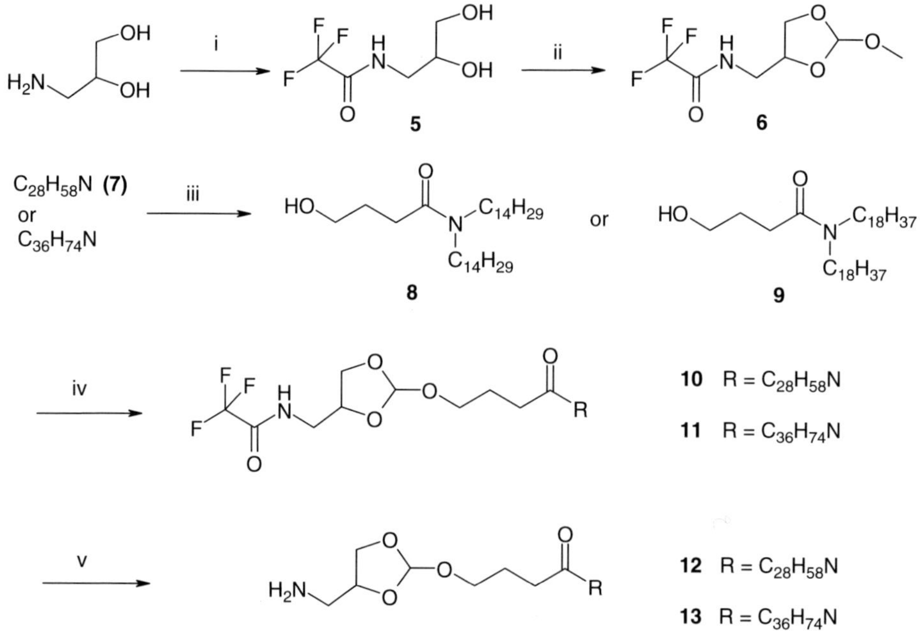

Scheme 2. (i) ethyl trifluoroacetate, 0°C, THF; (ii) trimethyl orthoformate, p-toluene sulfonic acid; (iii) butyrolactone, AlCl$_3$; (iv) p-toluene sulfonic acid, neat, 80°C; (v) NaOH 4%, THF, 50:50, 20°C

(3 × 200 ml). The organic phase was dried on magnesium sulfate, filtered and evaporated to dryness. Product **2** was obtained as a pure colorless oil (30 g, 85%)

¹H NMR (300 MHz, CDCl$_3$, δ ppm, (see Note 3): 3.33 and 3.37 (2s: 3H); de 3.35–3.80 (m: 3H); 4.10–4.25 (m: 1H); 4.50 (m: 1H); 5.73 and 5.78 (2s: 1H); 6.66 and 7.55 (2m: 1H).
M/z (D/CI) = 247 (MNH4$^+$).
IR (CH$_2$Cl$_2$): 3427; 3312; 2945; 1728; 1548; 1213; 1176; 1073 and 982 cm^{-1}.

3.1.1.3. Ditetradecylamine chlorohydrate (7)

Bromotetradecane (74 g, 267.1 mmol, 1 eq) and tetradecylamine (57 g, 267.1 mmol, 1 eq) were dissolved in absolute ethanol (400 ml). Sodium carbonate (70.8 g, 667 mmol, 2.5 eq) was added and the mixture was refluxed overnight. The reaction medium was then evaporated, taken up in dichloromethane (1,500 ml) and washed successively with water (3 × 200 ml) and NaClsat (1 × 400 ml). The organic phase was dried over calcium chloride and concentrated.

TLC (Rf = 0.55; DCM/MeOH 9/1; Ninhydrine, I$_2$/H$_2$SO$_4$). The salification of the amines was obtained by dissolution of the residue in a solution of hydrochloric acid in isopropanol and crystallization (48.4 g; 41%) (see Note 4).
¹H NMR (300 MHz, CDCl$_3$, δ ppm): 0.88 (t, J = 7 Hz: 6H); 1.15–1.45 (m: 44H); 1.90 (m: 4H); 2.90 (m: 4H); 9.48 (m: 2H).
M/z (D/CI) = 410 (MH$^+$).
IR (KBr): 2953; 2920; 2851; 2794; 2746; 2531; 2441; 1468 and 721 cm^{-1}.

3.1.1.4. 4-Hydroxy-N,N-ditetradecyl-butyramide: (8)

A solution of triethylamine (39 ml; 280.1 mmol; 5 eq) in chloroform (100 ml) was added dropwise to a cooled (10°C) solution of aluminum chloride (22.4 g; 168.1 mmol; 3 eq) in chloroform (75 ml), the mixture was allowed to warm at room temperature. Ditetradecylamine chlorhydrate (7) (25 g; 56 mmol, 1 eq) and butyrolactone (5.2 ml; 67.2 mmol, 1.2 eq) in chloroform (350 ml) were then slowly added to the stirred mixture. After 2 h, the reaction medium was diluted with water (200 ml) and stirred for an additional 30 min. It was filtered over Celite, washing with chloroform. The organic phase was decanted and washed with NaClsat (3 × 150 ml), then dried over magnesium sulfate, filtrated and concentrated. Column chromatography on silica eluted with cyclohexane/ethyl acetate (1/1) gave 21.2 g (76%) of a white powder.

TLC (Rf = 0.25; C$_6$H$_{12}$/AcOEt 1/1; I$_2$/H$_2$SO$_4$).
¹H NMR (300 MHz, CDCl$_3$, δ ppm): 0.88 (t, J = 7 Hz: 6H); 1.15–1.40 (m: 44H); 1.54 (m: 4H); 1.90 (m, J = 6.5 Hz: 2H); 2.50 (t, J = 6.5 Hz: 2H); 3.22 (m: 2H); 3.30 (m: 2H); 3.41 (t broad, J = 5.5 Hz: 1H); 3.70 (m: 2H).

M/z (D/CI) = 496 (MH⁺).
IR (KBr): 3444; 2917; 2850; 1626; 1471; 1053 and 717 cm⁻¹.

3.1.1.5. 4-Hydroxy-N,N-dioctadecyl-butyramide: (9)

An identical procedure was applied to dioctadecylamine (6 g, 11.5 mmol) which gave 4.9 g (70%) of a white powder (9) after column chromatography (C_6H_{12}/AcOEt 1/1).

TLC (Rf = 0.3; C_6H_{12}/AcOEt 1/1; I_2/H_2SO_4).
¹H NMR (400 MHz, CDCl₃, δ ppm): 0.90 (t, J = 7 Hz: 6H); 1.20–1.40 (m: 60H); 1.50–1.65 (m: 4H); 1.92 (m: 2H); 2.51 (t, J = 6.5 Hz: 2H); 3.23 (t broad, J = 8 Hz: 2H); 3.31 (t broad, J = 8 Hz: 2H); 3.71 (t broad, J = 5 Hz: 2H).
M/z (EI) = 607 (MH⁺).
IR (CH_2Cl_2): 3281; 2933; 2855; 1616 and 1467 cm⁻¹.

3.1.1.6. N,N-ditetradecyl-4-{4-[(2,2,2-Trifluoro-acetylamino)-methyl]-[1,3]dioxolan-2-yloxy}-butyramide (10)

4-hydroxy-N,N-ditetradecyl-butyramide (8) (3.5 g; 7.1 mmol) and 2,2,2-trifluoro-N-(2-methoxy-[1,3]dioxolan-4-ylmethyl)-acetamide (6) (1.8 g; 7.8 mmol, 1.1 eq) were mixed without solvent. A catalytic amount of PPTS (18 mg; 0.071 mmol, 0.01 eq) was added and the neat mixture was heated at 80°C during 3 h (see Note 9). The reaction medium was dissolved in heptane (200 ml), washed with NaHCO₃sat (3 × 50 ml), acetonitrile (3 × 50 ml) and concentrated to dryness.

TLC (Rf = 0.4; C_6H_{12}/AcOEt 1/1; I_2/H_2SO_4).
M/z (D/CI) = 693 (MH⁺).

3.1.1.7. N,N-dioctadecyl-4-{4-[(2,2,2-Trifluoro-acetylamino)-methyl]-[1,3]dioxolan-2-yloxy}-butyramide (11)

The same reaction conditions were applied to the dioctadecyl derivative **9** (2.8 g; 4.6 mmol) leading to compound **11**.

TLC (Rf = 0.4; C_6H_{12}/AcOEt 1/1; I_2/H_2SO_4).
M/z (D/CI) = 805 (MH⁺).

3.1.1.8. 4-(4-Aminomethyl-[1,3]dioxolan-2-yloxy)-N,N-ditetradecyl-butyramide (12)

To a solution of compound 10 in THF (20 ml) was added sodium hydroxide (20 ml; 4%). The reaction medium was stirred overnight at room temperature. It was concentrated and the residue extracted with diethyloxide (3 × 200 ml). The organic phase was dried over calcium chloride, filtered and evaporated. Chromatography on silica (DCM/MeOH 9/1) led to a colorless oil (1.6 g; 38% two steps **10 and 12**).

TLC (Rf = 0.2; DCM/MeOH 9/1; Ninhydrine, I_2/H_2SO_4).
¹H NMR (300 MHz, CDCl₃, δ ppm) (see Note 3): 0.89 (t, J = 7 Hz: 6H); 1.15–1.45 (m: 44H); 1.52 (m: 4H); 1.58 (m: 2H); 1.95 (m, 2H); 2.39 (t, J = 6.5 Hz: 2H); 2.75–3.00 (m: 2H); 3.21 (m: 2H); 3.29 (m: 2H); 3.60 (m: 2H); 3.71 and 3.80 (dd, J = 7.5 Hz; 6 Hz t, J = 7.5 Hz: 1H total); 4.06 and 4.14 (t, J = 7.5 Hz: 1H total); 4.21 and 4.33 (m: 1H total); 5.82 and 5.85 (s: 1H total).

M/z (D/CI) = 597 (MH+).

IR (CCl$_4$): 2927; 2855; 1645; 1467; 1139; 1091 and 1067 cm^{-1}.

3.1.1.9. 4-(4-Aminomethyl-[1,3]dioxolan-2-yloxy)-N,N-dioctadecyl-butyramide (13)

The same procedure as above was applied to compound **11**, leading to **13**.

TLC (Rf = 0.2; DCM/MeOH 9/1; Ninhydrine, I$_2$/H$_2$SO$_4$).
^{1}H NMR (300 MHz, CDCl3, δ ppm) (see Note 3): 0.90 (t, J = 7 Hz: 6H); 1.15–1.40 (m: 60H); 1.65 (m: 4H); 1.95 (m: 2H); 2.40 (t, J = 7.5 Hz: 2H); 2.75–3.00 (m: 2H); 3.21 (t broad, J = 8 Hz: 2H); 3.29 (t broad, J = 8 Hz: 2H); 3.61 (m: 2H); 3.65–3.85 (m: 1H); 4.00–4.40 (m: 2H); 5.82 and 5.84 (2s: 1H total).
M/z (D/CI) = 709 (MH+).
IR (CCl4): 2927; 2855; 1644; 1467; 1139 and 1067 cm^{-1}.

3.1.2. Synthesis of the Cationic Heads (Scheme 3)

3.1.2.1. 2,2,2-Trifluoro-N-[3-(2,2,2-trifluoro-acetylamino)-propyl]-N-(4-{trifluoroacetyl-[3-(2,2,2-trifluoro-acetylamino)- propyl]-amino}-butyl)- acetamide): (14)

A mixture of spermine (8 g, 39.5 mmol, 1 eq) and triethylamine (33 ml, 237 mmol, 6 eq) in dichloromethane (75 ml) was cooled to 0°C in an ice bath. A solution of trifluoroacetic anhydride (41.5 g, 198 mmol) in dichloromethane (100 ml) was added dropwise over 1 h. The reaction medium was stirred overnight at room temperature, then neutralized by addition of sodium hydrogen carbonate solution (75 ml, 5% ww) (see Note 5). The aqueous phase was extracted with dichloromethane (3 × 150 ml) and the combined organic phases washed with potassium hydrogen sulfate (3 × 100 ml, 0.5 M), brine and dried over magnesium sulfate. Concentration yielded (14) as a yellowish powder (22.5 g, 97%).

TLC (Rf = 0.5; DCM/MeOH 95/5; Ninhydrine, I$_2$/H$_2$SO$_4$).
^{1}H NMR (300 MHz, CDCl3, δ ppm): 1.67 (m: 4H); 1.80–2.05 (m: 4H); 3.30–3.55 (m: 12H); 6.64 and 6.80 (2m: 1H); 7.12 and 7.32 (2m: 1H).
M/z (D/CI) = 604 (MNH4+).
IR (CH2Cl$_2$): 3429; 3339; 2954; 1726; 1686; 1548; 1447; 1206; 1174 and 1151 cm^{-1}.

3.1.2.2. (Trifluoroacetyl-{3-[trifluoroacetyl-(4-{trifluoroacetyl-[3-(2,2,2-trifluoro-acetylamino)-propyl]-amino}-butyl)-amino]-propyl}-amino)-acetic acid: (15)

Compound **14** (10 g, 17.0 mmol, 1 eq) in DMF (40 ml) was added dropwise to a cooled mixture of 60% sodium hydride (1 g, 25.6 mmol, 1.5 eq) in DMF (60 ml), under an argon atmosphere. After 1 h at ambient temperature, it was again cooled to 0°C before *tert*-butyl bromoacetate (3.66 g, 18.7 mmol, 1.1 eq) was added. The reaction medium was stirred overnight at room temperature, then diluted with ethyl acetate (500 ml), washed with NaHCO$_3$sat (3 × 100 ml), and dried over magnesium sulfate. Concentration gave a yellow oil as the tert-butyl ester. TLC (Rf = 0.7; DCM/MeOH 95/5; I$_2$/H$_2$SO$_4$).

Deprotection was performed by dissolution in trifluoroacetic acid/dichloromethane (50:50, 100 ml) after 3 h at room

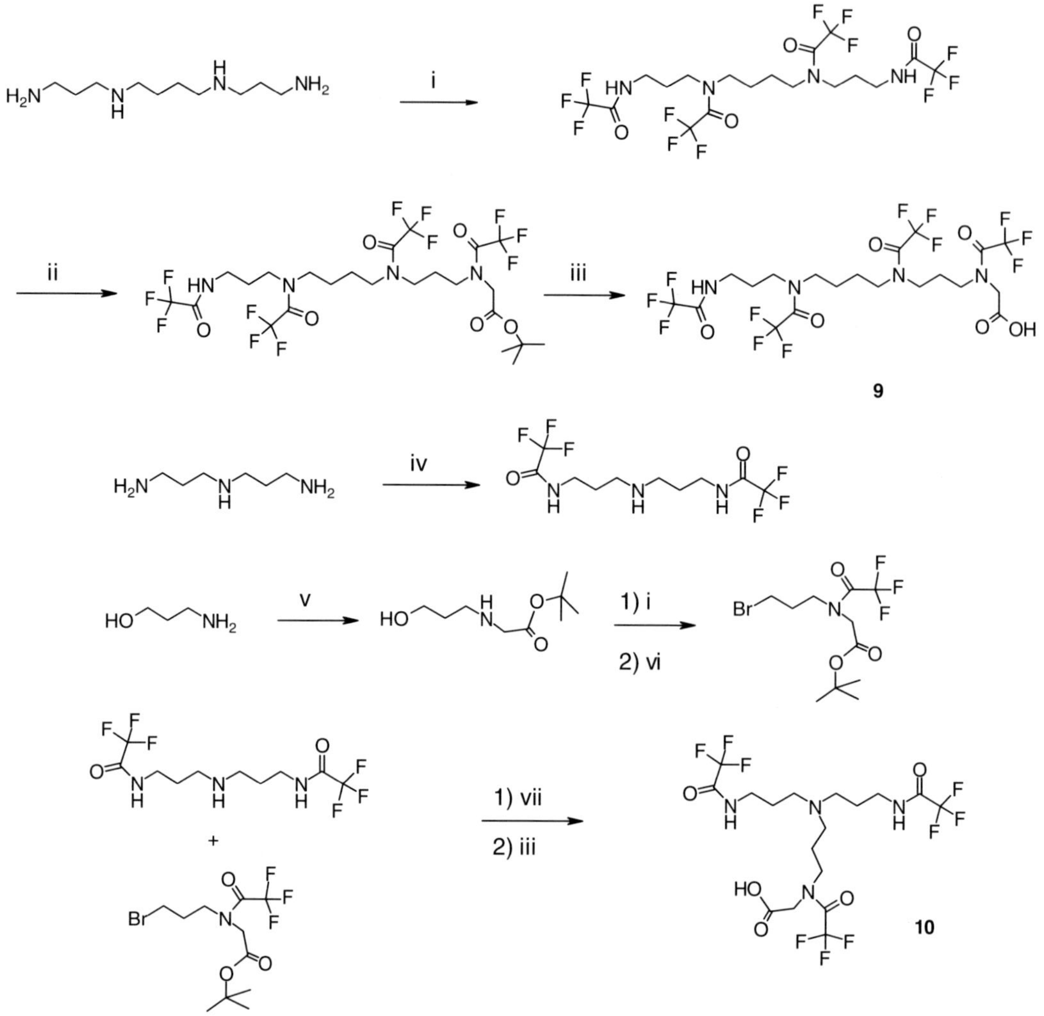

Scheme 3. (i) trifluoroacetic anhydride, Et$_3$N, CH$_2$Cl$_2$, 0°C → 20°C; (ii) *tert*-butyl bromoacetate, NaH, DMF, 20°C; (iii) trifluoroacetic acid, CH$_2$Cl$_2$, 20°C; (iv) ethyl trifluoroacetate, THF, 0°C; (v) *tert*-butyl bromoacetate, CH$_2$Cl$_2$, 0°C; (vi) triphenylphosphine, carbon tetrabromide, THF, 20°C; (vii) K$_2$CO$_3$, CH$_3$CN, reflux

temperature. The reaction medium was evaporated to dryness then diluted with dichloromethane (50 ml), the product was extracted with NaHCO$_3$sat (3×150 ml) then acidified with concentrated hydrochloric acid. The product was extracted with dichloromethane (3×300 ml), the organic phase was dried (magnesium sulfate), filtrated and concentrated. Chromatography on silica (DCM/MeOH 8/2) yielded **15** as a yellow powder (3.5 g, 32%).

TLC (Rf = 0.2; DCM/MeOH 8/2; I$_2$/H$_2$SO$_4$).
^{1}H NMR (400 MHz, (CD$_3$)$_2$SO d6, δ ppm): 1.62 (m: 4H); 1.80–2.00 (m: 4H); 3.28 (m: 2H); 3.0–3.60 (m: 10H); 4.01 (s: 2H); 9.15–9.35 (m: 1H).
M/z (D/CI) = 662 (MNH$_4^+$).

IR (KBr): 3370; 2957; 1688; 1561; 1470; 1446; 1205; 1144; 760 and 693 cm^{-1}.

3.1.2.3. 2,2,2-Trifluoro-N-{3-[3-(2,2,2-trifluoro-acetylamino)-propylamino]-propyl}-acetamide: (16)

3,3′-imino-bispropylamine (35 g, 266.7 mmol, 1 eq) was dissolved in anhydrous THF (150 ml) under an argon atmosphere. The solution was cooled to 0°C in an ice bath and ethyl trifluoroacetate (65 ml, 546.8 mmol, 2.05 eq) was added dropwise (see Note 2). After 3 h, the reaction medium was warmed to room temperature and was left 2 more hours under argon. The insoluble was discarded on paper filter and the solution concentrated. After drying (see Note 6), a white powder was obtained (85.3 g, 99%).

TLC (Rf = 0.7; EtOH/NH$_3$ 8/2; Ninhydrine).
^{1}H NMR (300 MHz, CDCl3, δ ppm): 1.74 (m, J = 6 Hz: 4H); 2.73 (t, J = 6 Hz: 4H); 3.46 (m: 4H); 8.18 (m: 2H).
M/z (D/CI) = 324 (MH+).
IR (KBr): 3293; 3098; 1704; 1564; 1211; 1184; 1163 and 723 cm–^{1}.

3.1.2.4. tert-Butyl (3-Hydroxy-propylamino)-acetate: (17)

To a cooled (0°C) solution of 3-aminopropanol (196 ml, 2.56 mol, 25 eq) in DCM (250 ml), tert-butyl bromoacetate (20 g, 102.5 mmol, 1 eq) in DCM (200 ml) was added dropwise. After 2 h, the reaction medium was warmed to room temperature and kept three more hours. The solution was then washed with NaHCO$_3$sat (3 × 150 ml), NaClsat (150 ml). After drying over magnesium sulfate, filtration, and evaporation, a colorless oil was obtained (17.8 g, 92%).

TLC (Rf = 0.55; DCM/MeOH 8/2; Ninhydrine, I$_2$/H$_2$SO$_4$).
^{1}H NMR (300 MHz, (CD3)2SO d6, δ ppm): 1.44 (s: 9H); 1.54 (m, J = 6.5 Hz: 2H); 2.56 (t, J = 6.5 Hz: 2H); 3.19 (s: 2H); 3.46 (t, J = 6.5 Hz: 2H); 3.70–4.70 (m: 1H).
M/z (D/CI) = 190 (MH+).
IR (CH2Cl2): 3616; 3329; 2982; 1731; 1394; 1369; 1234; 1157; 1075 and 846 cm^{-1}.

3.1.2.5. tert-Butyl-[(3-Hydroxy-propyl)-trifluoroacetyl-amino]-acetate: (18)

tert-Butyl-(3-hydroxy-propylamino)-acetate (**17**) (17.65 g, 93.3 mmol, 1 eq) and triethylamine (26 ml, 186.6 mmol, 2 eq) were dissolved in dichloromethane (100 ml) and cooled to 0°C, trifluoroacetic anhydride (21.5 g, 102.6 mmol, 1.1 eq) was added dropwise. The reaction medium was then stirred overnight at room temperature. The solution was washed with NaHCO$_3$sat (3 × 50 ml), KHSO$_4$ 0.5 M (3 × 50 ml) and NaClsat (50 ml). The organic phase was dried over magnesium sulfate, filtrated and concentrated to a pale yellow oil (24.7 g, 93%).

TLC (Rf = 0.25; C$_6$H$_{12}$/AcOEt 1/1; Ninhydrine, I$_2$/H$_2$SO$_4$, UV).
^{1}H NMR (300 MHz, (CD$_3$)$_2$SO d6, δ ppm) (see Note 7): 1.44, 1.46 (2s: 9H); 1.60–1.80 (m: 2H); 3.35–3.55 (m: 4H); 4.09, 4.25 (m: 2H); 4.57, 4.61 (t, J = 5 Hz: 1H).

M/z (D/CI) = 303 (MNH$_4^+$).
IR (CH$_2$Cl$_2$): 3618; 2983; 1747; 1694; 1371; 1235; 1205; 1150 and 845 cm^{-1}.

3.1.2.6. *tert*-Butyl [(3-Bromo-propyl)-trifluoroacetyl-amino]-acetate: (18)

To a stirred solution of **17** (10 g, 35.0 mmol, 1 eq) and triphenylphosphine (12.4 g, 47.3 mmol, 1.35 eq) in THF (150 ml), carbon tetrabromide (15.1 g, 45.6 mmol, 1.3 eq) in acetonitrile (60 ml) was added dropwise. After 4 h the reaction medium was concentrated, taken up in ethyl acetate and filtrated on paper filter. The filtrate was concentrated to dryness, taken up in cyclohexane and filtrated on fritted glass (n°3). After a final concentration and purification on silica column (C$_6$H$_{12}$/AcOEt 8/2), a lightly yellow colored oil was obtained (10.4 g, 85%).

TLC (Rf = 0.6; C$_6$H$_{12}$/AcOEt 1/1; Ninhydrine, I$_2$/H$_2$SO$_4$, UV).
^{1}H NMR (300 MHz, (CD$_3$)$_2$SO d6, δ ppm) (see Note 7): 1.43, 1.45 (s: 9H); 2.05–2.20 (m: 2H); 3.50–3.65 (m: 4H); 4.11, 4.28 (m: 2H).
M/z (D/CI) = 365 (MNH$_4^+$).
IR (CH$_2$Cl$_2$): 2983; 1746; 1697; 1371; 1235; 1207; 1145; 846 and 675 cm^{-1}.

3.1.2.7. *tert*-Butyl[(3-{Bis-[3-(2,2,2-trifluoro-acetylamino)-propyl]-amino}-propyl)-trifluoroacetyl-amino]-acetate: (19)

To a solution of compound **18** (26 g, 74.7 mmol, 1 eq) and the amine (**16**) (24.1 g, 74.7 mmol, 1 eq) in acetonitrile (130 ml), potassium carbonate (30 g, 224 mmol, 3 eq) was added and the mixture heated to reflux during 6 h. The reaction medium was then filtrated and evaporated to dryness. Chromatography on silica (C$_6$H$_{12}$/AcOEt 2/8) yielded (**19**) as a pale yellow oil (16.6 g; 3 8%).

TLC (Rf = 0.35; AcOEt; I$_2$/H$_2$SO$_4$, UV).
^{1}H NMR (300 MHz, CDCl$_3$, δ ppm) (see Note 7): 1.47, 1.48 (2s: 9H); 1.65–1.85 (m: 6H); 2.40–2.60 (m: 6H); 3.40–3.55 (m: 6H); 3.97, 4.07 (m: 2H); 7.45–7.65 (m: 2H).
M/z (D/CI) = 591 (MH$^+$).
IR (CH$_2$Cl$_2$): 3431; 3343; 2957; 1721; 1696; 1547; 1371; 1208; 1168 and 844 cm^{-1}.

3.1.2.8. [(3-{Bis-[3-(2,2,2-trifluoro-acetylamino)-propyl]-amino}-propyl)-trifluoroacetyl-amino]-acetic acid: (20)

Trifluoroacetic acid (50 ml) was added to a solution of compound **19** (15.8 g, 28.76 mmol, 1 eq) in DCM (50 ml). The mixture was stirred at room temperature during 2 h, then concentrated. A faint yellow gum was obtained (18.7 g, 100%).

^{1}H NMR (300 MHz, CDCl$_3$, δ ppm) (see Note 7): 1.80–2.10 (m: 6H); 3.12 (m: 6H); 3.29 (m: 4H); 3.50 (m: 2H); 4.13, 4.29 (m: 2H); 9.50–9.75 (m: 2H).
M/z (D/CI) = 535 (MH$^+$).
IR (neat): 3317; 3093; 2954; 2734; 2625; 1715; 1561; 1187 and 722 cm^{-1}.

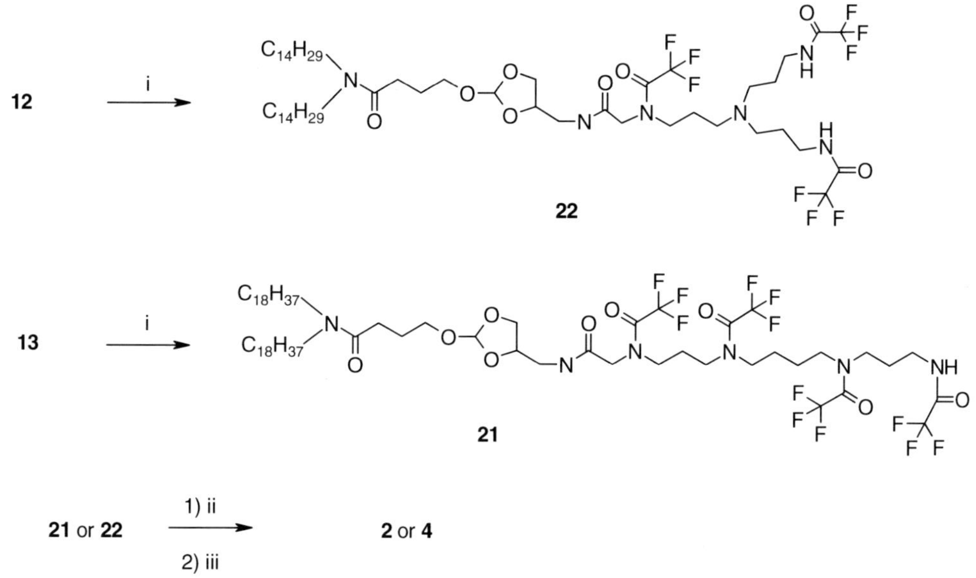

Scheme 4. (i) BOP, triethyl amine, CH$_2$Cl$_2$; (ii) NaOH 4%, THF, 50:50, 20°C; (iii) Dowex 21K – Cl$^-$, MeOH

3.1.3. Assembly of the Acid-Labile Cationic Lipid (Scheme 4)

3.1.3.1. N,N-Dioctadecyl-4-(4-{[2-(trifluoroacetyl-{3-[trifluoroacetyl-(4-{trifluoroacetyl-[3-(2,2,2-trifluoro-acetylamino)-propyl]-amino}-butyl)-amino]-propyl}-amino)-acetylamino]-methyl}-[1,3]dioxolan-2-yloxy)-butyramide: (21)

4-(4-Aminomethyl-[1,3]dioxolan-2-yloxy)-N,N-dioctadecyl-butyramide (**9**) (800 mg, 1.13 mmol, 1 eq), triethylamine (390 µl, 2.8 mmol, 2.5 eq), and the acid **15** (800 mg, 1.24 mmol, 1.1 eq) were dissolved in dichloromethane (10 ml). BOP (600 mg, 1.36 mmol, 1.1 eq) was added and the mixture was stirred 2 h at ambient temperature. The reaction medium was concentrated, taken up in ethyl acetate (150 ml), and washed with NaHCO$_3$sat (3×40 ml), NaClsat (40 ml). It was dried over magnesium sulfate, filtrated and evaporated. Column chromatography on silica (EtOAc) gave 1.1 g of a white powder (73%) (see Note 8).

TLC (Rf = 0.35 and 0.45; EtOAc; I$_2$/H$_2$SO4).
^{1}H NMR (400 MHz, (CD$_3$)$_2$SO d6, δ ppm): 0.87 (t, J = 7 Hz: 6H); 1.00–1.65 (m: 68H); 1.65–1.95 (m: 6H); 2.31 (t, J = 7 Hz: 2H); 3.10–3.55 (m: 20H); 3.64 (m: 1H); 4.00–4.25 (m: 4H); 5.81 (s: 1H); 8.29, 8.34 (2m: 1H); 9.44, 9.50 (m: 1H).
M/z (LSIMS) = 1357 (MNa$^+$).

3.1.3.2. 4-{4-[(2-{3-[4-(3-Amino-propylamino)-butylamino]-propylamino}-acetylamino)-methyl]-[1,3]dioxolan-2-yloxy}-N,N-dioctadecyl-butyramide; hydrochloride salt: (2)

To a solution of (**21**) (290 mg, 0.22 mmol) in THF (3 ml), 4% sodium hydroxide was added under vigorous stirring. The mixture was left overnight at room temperature, then concentrated. Column chromatography on silica (DCM/MeOH/NH$_3$, 45/45/10) gave the pure product which was then lyophilized to a white powder (180 mg, 87%).

TLC (Rf = 0.2; DCM/MeOH/NH$_3$ 42.5/42.5/15; Ninhydrine, I$_2$/H$_2$SO$_4$).

The lyophilisate was then dissolved in a 1:1 mixture of ethanol water and eluted over an ion exchange resin column (DOWEX 21K; Cl⁻ form).

¹H NMR (400 MHz, CDCl$_3$, δ ppm): 0.89 (t, J = 7 Hz: 6H); 1.20–1.40 (m: 60H); 1.52 (m: 4H); 1.65–1.90 (m: 8H); 1.93 (m: 2H); 1.97 (s: 3H); 2.37 (t, J = 7 Hz: 2H); 2.73 (m: 4H); 2.81 (t, J = 6.5 Hz: 2H); 2.89 (m: 4H); 2.95 (t, J = 6.5 Hz: 2H); 3.15–3.30 (m: 4H); 3.32 (dd, J = 17 Hz: 2H); 3.45 (m, J = 14 and 6.5 Hz: 1H); 3.55–3.65 (m: 1H); 3.61 (m, 2H); 3.74 (t, J = 8 Hz: 1H); 4.06 (t, J = 8 Hz: 1H); 4.31 (m: 1H); 5.79 (s: 1H); 7.81 (t, J = 5.5 Hz: 1H).
M/z (D/CI) = 951 (MH⁺).
IR (KBr): 2956; 2918; 2850; 1637; 1560; 1468; 1410; 1140; 1063 and 721 cm⁻¹.

3.1.3.3. 4-[4-({2-[(3-{Bis-[3-(2,2,2-trifluoro-acetylamino)-propyl]-amino}-propyl)-trifluoroacetyl-amino]-acetylamino}-methyl)-[1,3]dioxolan-2-yloxy]-N,N-ditetradecyl-butyramide: (22)

4-(4-aminomethyl-[1,3]dioxolan-2-yloxy)-N,N-ditetradecyl-butyramide (8) (1.65 g, 2.76 mmol, 1 eq), triethylamine (1.9 ml, 13.8 mmol, 5 eq), and carboxylic acid (20) (2 g, 3.04 mmol, 1.1 eq) were dissolved in DCM (30 ml), BOP (1.8 g, 4.14 mmol, 1.5 eq) was added and the mixture was stirred 1 h at room temperature. The reaction medium was concentrated, taken up in ethyl acetate (200 ml) and washed with NaHCO$_3$sat (3 × 40 ml), NaClsat (40 ml). After evaporation, it was purified on silica column (AcOEt) to give a light yellow oil (2.2 g, 72%).

TLC (Rf = 0.4; DCM/MeOH 9/1; I$_2$/H$_2$SO$_4$).
¹H NMR (300 MHz, CDCl$_3$, δ ppm): 0.88 (t, J = 7 Hz: 6H); 1.15–1.40 (m: 44H); 1.51 (m: 4H); 1.70–2.00 (m: 8H); 2.37 (m: 2H); 2.40–2.60 (m: 6H); 3.15–3.35 (m: 4H); 3.45 (m: 4H); 3.45–3.85; 3.95–4.50 (m: 11H); 5.79, 5.81 (s: 1H); 7.82, 7.90 (m: 2H).
M/z (ES) = 1136 (MNa⁺).
IR (CCl$_4$): 3306; 3085; 2928; 2855; 1706; 1626; 1552; 1467; 1208; 1164; 840 and 558 cm⁻¹.

3.1.3.4. 4-{4-[(2-{3-[Bis-(3-amino-propyl)-amino]-propylamino}-acetylamino)-methyl]-[1,3]dioxolan-2-yloxy}-N,N-ditetradecyl-butyramide; hydrochloride salt: (4)

Compound **22** (2.1 g, 1.88 mmol) was dissolved in THF (30 ml), sodium hydroxide was added (30 ml, 1 N) with vigorous stirring. The mixture was left overnight at ambient temperature, then it was concentrated and purified on silica column (DCM/MeOH/NH$_3$, 45/45/10). TLC (Rf = 0.2; DCM/MeOH/NH$_3$ 42.5/42.5/15; Ninhydrine, I$_2$/H$_2$SO$_4$). After concentration, an aqueous solution of the product was eluted from an ion exchange resin (FLUKA; DOWEX 21K; Cl⁻ form) to give after lyophilization 1.3 g of a white powder (84%).

¹H NMR (300 MHz, (CD$_3$)$_2$SO d6, δ ppm): 0.89 (t, J = 7 Hz: 6H); 1.15–1.40 (m: 44H); 1.40–1.60 (m: 4H); 1.72 (m: 8H); 2.31 (m: 2H); 2.40–2.55 (m: 6H); 2.70–2.90 (m: 6H); 3.15–3.75; 4.00–4.40 (m: 13H); 5.83, 5.86 (s:).

M/z (D/CI) = 825 (MH⁺).
IR (CH$_2$Cl$_2$): 3384; 3240; 2927; 2854; 1675; 1632; 1529; 1468; 1143 and 1065 cm^{-1}.

3.2. Analytical and Purification Methods

1. CCM were performed on silica-coated aluminum plates (Merck silica gel 60 F$_{254}$; 0.2 mm).
2. Detection was obtained under a UV lamp (254 nm), and/or by spraying an ethanolic solution of ninhydrine with subsequent heating, or a 1:1 mixture of 1 M iodine and 1 M sulfuric acid.
3. Column chromatography were performed on silica gel (SDS; 0.063–0.200 mm) using low pressure columns with a flow rate in the range 3–10 ml/min.
4. Analytical HPLC was performed using a Merck–Hitachi instrument equipped with an automatic injector (AS-2000A), an intelligent pump (L6200A), and a UV detector (L-4000). Lipids were detected at 220 nm. Columns used were small stainless steel Browlee columns, "Aquapore Butyl" 7 µm (C$_4$) (Applied Biosystem; 30/4.6 mm). Mobile phases were gradient mixtures of water and acetonitrile (SDS; HPLC grade), degassed under vacuum by filtration on 0.2 µm filter, and modified by the addition of 0.1% and 0.08%, respectively of trifluoroacetic acid (obviously, except in the case of acid sensitive compounds). Injected volumes were 30 µl of 1 mg/ml concentration. Gradient conditions were: 0–3 min; 20% acetonitrile, 3–20 min; 20–100% acetonitrile; 20–35 min; 100% acetonitrile, with a 1 ml/min constant flow rate (180 bar pressure).
5. Preparative HPLC was performed on a Gilson instrument equipped with two pumps (305/303 model), a magnetic mixer (811C model, 23 ml), an injection loop (5 ml), a UV detector (119 model, variable length), and a fraction collector (202 model) with 10 ml glass tubes.
6. The preparative column was a C4 Vydac (214TP1022 model; 250/22 mm). Mobile phases were the same as for analytical HPLC. Gradient used was: 0–10 min; 10% acetonitrile, 10–110 min; 10–100% acetonitrile, 110–120 min; 100% acetonitrile with a 18 ml/min flow rate, 9 ml fractions were collected (two tubes/min).

3.3. Structural Analysis

1. NMR spectra were recorded with a Bruker spectrometer at 400 MHz (^{1}H) and 100 MHz (^{13}C). Chemical shifts are expressed in ppm from TMS as internal standard. Multiplicity was designed by usual abbreviations: s (singlet); d (doublet); t (triplet); m (multiplet).
2. IR spectra were performed with Fourier transform spectrometer Nicolet 510 and Perkin-Elmer 2000. Samples were prepared by the KBr pellet technique unless otherwise stated.

3. Mass spectra were obtained with the following conditions: ES (Electrospray) on an LCTOF MICROMASS; EI (Electronic Impact) and D/CI (Desorption/Chemical Ionization; ammonia) on a SSQ7000 FINNIGAN; LSIMS (Liquid Secondary Ion Mass Spectrometry) with a SCIEX PERKIN ELMER (Cs; 35 KeV; 3-nitrobenzyl alcohol matrix).

3.4. Physicochemical Analysis and General Methods

3.4.1. DNA Compaction at Neutral pH

1. Lipoplexes were prepared in a Hepes/mes/pipes buffer at pH = 8.0, at different charge ratios from liposomes obtained with compounds **2** and **4** described above, in association with the co-lipid DOPE (cationic lipid/DOPE 1:1), and were compared to those obtained from stable cationic lipids **1** and **3**.

2. Plasmid DNA used contained the luciferase (Luc) reporter gene under the cytomegalovirus (CMV) promoter.

3. Compaction of DNA in the liposomes was verified by addition of ethidium bromide (EtBr; 3 µL of a 1 g/L solution for 8 µg of DNA) and measurement of the fluorescence (λ_{em} 590 nm), which decreases when DNA is compacted.

4. The DNA compaction was also evidenced by the loss of DNA electrophoretical mobility on agarose gel.

5. The size of the complexes was measured as a function of the cationic lipid/DNA charge ratio by dynamic light scattering with a Coulter N4+ particle sizer at 632 nm. The results, comparable to those obtained with the non-pH labile cationic lipids, are shown in Fig. 1a–c (see Note 10).

6. For cationic lipid/DNA charge ratio up to 1, small particles were obtained but DNA was not compacted as shown by the high level of fluorescence obtained. Between 1 and 4–6 nmol cationic lipid/µg DNA (which in the case of the cationic lipids described here is also equivalent to charge ratio), large aggregated particles were observed with low fluorescence indicating compaction of DNA.

7. Above 4–6 nmol cationic lipid/µg DNA, small particles were obtained containing compacted DNA. The formation of complexes between DNA and pH labile cationic lipids thus obeys similar rules as those observed with classical nonlabile cationic lipids.

8. These lipids could then be used as gene vectors according to their stability and DNA compaction properties.

3.4.2. Degradation Studies at pH 5

1. The degradation of lipoplexes obtained with pH labile cationic lipids **2** and **4** was evaluated by measuring the fluorescence percentage of EtBr intercalated with DNA. Acidic degradation of pH labile cationic lipids should result in the liberation of DNA and thus to an increase in the observed fluorescence, since EtBr will intercalate in uncompacted DNA.

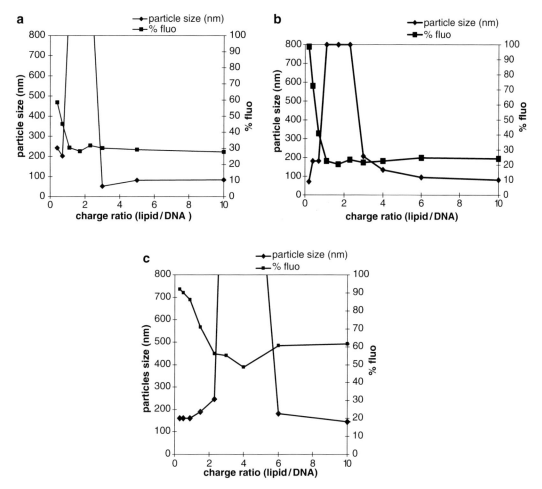

Fig. 1. Effect of the lipid/DNA charge ratio on particle size and DNA compaction

2. Liposomes containing either pH labile or stable cationic lipids were mixed with different amounts of DNA in order to obtain three different charge ratios: 0.4 nmol cationic lipid/µg DNA corresponding to the uncompacted zone, 1.7 nmol cationic lipid/µg DNA corresponding to the aggregated zone, and 6 nmol cationic lipid/µg DNA corresponding to small lipoplexes with compacted DNA.

3. Preparations were incubated at 37°C in 0.1 M acetic acid/sodium acetate buffer (pH 5) and the fluorescence of EtBr was measured (λ_{em} 590 nm) as a function of time. Complexes prepared at 0.4 charge ratio showed high fluorescence levels, confirming the accessibility of noncompacted DNA to EtBr (data not shown).

4. Fluorescence level was low and stable over time when considering lipoplexes prepared with stable cationic lipid **3** (or **1**) at charge ratios 1.7 and 6.

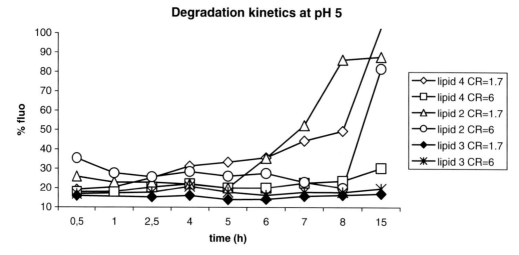

Fig. 2. Degradation kinetic of lipoplexes obtained with stable lipid **3** and pH labile lipids **2** and **4** at pH 5, and charge ratios 1.7 and 6. DNA release induced by degradation is measured by the % of fluorescence of intercalated ethidium bromide at 590 nm^{Em}; 260 nm^{Ex} (100% refers to naked DNA)

5. Lipoplexes containing pH labile cationic lipid **2** or **4** showed an increase of fluorescence at pH 5, after 4–6 h with a charge ratio of 1.7 and after 8–12 h with a charge ratio of 6 (Fig. 2) (see Note 10).

6. On the other hand, no fluorescence increase was observed at pH 7 for all the lipids tested (data not shown).

7. It is noteworthy that the charge ratio influences strongly the release of DNA; in the case of an excess of cationic lipid, it is necessary to hydrolyze a larger number of molecules to destabilize the complexes, thus delaying the degradation of lipoplexes. As a matter of fact, as the cationic lipid/DNA reaches the limit charge ratio insuring colloidal stability, the kinetics of release increases. This behavior could advantageously be exploited to optimize the time needed for content release after exposition of the complexes to acidic medium, thus offering a temporal controlled delivery.

3.4.3. In Vitro Transfection Experiments with pH Labile Lipoplexes

1. Formulations containing pH labile cationic lipids **2** and **4** were evaluated in vitro in transfection experiments with Hela cells with three different charge ratios (1.5, 6.0, and 10), using plasmid DNA containing the luciferase (Luc) reporter gene under the cytomegalovirus (CMV) promoter.

2. The results were compared to those obtained with non-pH labile formulations.

3. Micelles formulations were preferred to liposome, as they gave better results in preliminary experiments.

4. The complexes were prepared in 20 mM Hepes buffer at pH 7.4 and diluted to a final concentration of 1 mM, before incubation with the appropriate amount of DNA.

5. Experiments were performed with or without addition of calf serum (10%).
6. In order to evidence the influence of endosome acidification on transfection, experiments were also performed in the presence of bafilomycin A1, an inhibitor of endosome acidification (20).
7. The results of transfection with pH labile lipid **4** and stable lipid **3** in the absence of serum, with or without bafilomycin A, are shown in Fig. 3. Formulations containing lipids **1** and **2** gave results similar to those obtained with **3** and **4** and for the sake of clarity are not represented in the figure.
8. Size measurements were performed for all the formulations, and aggregation was observed at 1.5 charge ratio, while small particles (85 nm) were obtained for the other ratios.
9. For a charge ratio = 10 which is the charge ratio commonly used in systemic injection, significant enhancement (x12) of transfection was obtained with the pH labile formulation versus stable formulation.
10. No significant differences were observed with bafilomycin-treated cells at ratios 1.5 and 6. However, at charge ratio = 10, the transfection level was decreased to one-fourth with bafilomycin-treated cells in the case of pH labile complexes compared to untreated cells, whereas no decrease was observed with the stable complexes in the same conditions.
11. Serum decreased dramatically the transfection as expected, especially at high charge ratios. Nevertheless, a better transfection was still observed with the pH labile formulations (data not shown).

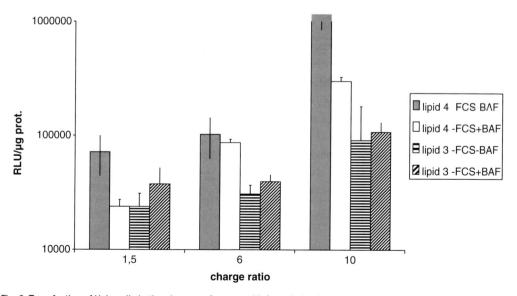

Fig. 3. Transfection of Hela cells in the absence of serum with formulation including stable lipid **3** or pH labile lipid **4**, with or without bafilomycin A

12. In conclusion, pH labile formulations gave a higher transfection level than their stable analogs. Results obtained in the presence of bafilomycin suggest that this effect could be attributed to a destabilization of the complexes in the endosome where acidification occurs. The dramatic loss of transfection observed in the presence of serum, especially at high charge ratios, was an expected phenomenon occurring with nonlabile cationic lipids as well.

4. Notes

1. It should be noted that the acid-labile synthon described here could also be introduced into other molecules for different delivery purposes, prodrugs, etc.
2. Partial hydrolysis of the trifluoroacetate ester releases trifluoroacetic acid. It is essential that the pH of the ester be neutral, otherwise the amine would protonate and fail to react. Therefore, the solution of the ester is advantageously treated with solid sodium or potassium carbonate prior to use (CAUTION: gas evolvement).
3. 50/50 mixture of diastereoisomers.
4. The crude concentrate was dissolved in a warm mixture of isopropanol (600 ml) and 5 M HCl in isopropanol (300 ml) which induced crystallization of the product as a white flaky powder. This was then thoroughly washed with isopropanol and dichloromethane.
5. The neutralization with sodium hydrogen carbonate generates a lot of gas (CO_2). Care should be taken as to avoid overflow and/or pressure building.
6. The residue was further dried 16 h in a vacuum oven at 40°C under high vacuum.
7. A mixture of rotamers was obtained; thus, some signal doubling was observed.
8. Two diastereoisomers were obtained and could be separated; however, since there was no significant difference in physicochemical or biological behavior, only one is described.
9. Since the reaction is theoretically an equilibrium, the methanol evolved in the reaction is best driven out by a gentle flush of nitrogen or argon.
10. Results obtained with the two stable lipids were very much comparable; therefore, only comparison with lipid **3** is represented.

References

1. Behr JP et al (1989) Efficient gene transfer into mammalian primary endocrine cells with lipopolyamine-coated DNA. Proc Natl Acad Sci USA 86:6982–6986
2. Lee ER et al (1996) Detailed analysis of structures and formulations of cationic lipids for efficient gene transfer to the lung. Hum Gene Ther 7:1701–1717
3. Bragonzi A, Conese M (2002) Non-viral approach toward gene therapy of cystic fibrosis lung disease. Curr Gene Ther 2:295–305
4. Anderson DM et al (2003) Stability of mRNA/cationic lipid lipoplexes in human and rat cerebrospinal fluid: methods and evidence for nonviral mRNA gene delivery to the central nervous system. Hum Gene Ther 14:191–202
5. Lechardeur D et al (1999) Metabolic instability of plasmid DNA in the cytosol: a potential barrier to gene transfer. Gene Ther 6:482–497
6. Leroux J-C (ed) (2004) pH-responsive carriers for enhancing the cytoplasmic delivery of macromolecular drugs. Adv Drug Deliv Rev 56(7):925–1050
7. Yatvin MB et al (1980) pH sensitive liposomes: possible clinical implications. Science 210:1253–1255
8. Chu C-J et al (1990) Efficiency of cytoplasmic delivery by pH-sensitive liposomes to cells in culture. Pharm Res 7:824–834
9. Lee RJ et al (1998) The effects of pH and intraliposomal buffer strength on the rate of liposome content release and intracellular drug delivery. Biosci Rep 18:69–78
10. Subbarao NK et al (1987) pH-dependent bilayer destabilization by an amphipathic peptide. Biochemistry 26:2964–2972
11. Sudimack JJ et al (2002) A novel pH-sensitive liposome formulation containing oleyl alcohol. Biochim Biophys Acta 1564:31–37
12. Shi G et al (2002) Efficient intracellular drug and gene delivery using folate receptor-targeted pH-sensitive liposomes composed of cationic/anionic lipid combinations. J Control Release 80:309–319
13. Singh RS et al (2004) On the gene delivery efficacies of pH-sensitive cationic lipids via endosomal protonation. A chemical biology investigation. Chem Biol 11:713–723
14. Bell PC et al (2003) Transfection mediated by gemini surfactants: engineered escape from the endosomal compartment. J Am Chem Soc 125:1551–1558
15. Jennings KH et al (2002) Aggregation properties of a novel class of cationic gemini surfactants correlate with their efficiency as gene transfection agents. Langmuir 18:2426
16. Soubrier F, Cameron B, Manse B, Somarriba S, Dubertret C, Jaslin G, Jung G, Le Caer C, Dang D, Mouvault J-M, Scherman D, Mayaux J-F, Crouzet J (1999) pCOR: a new design of plasmid vectors for nonviral gene therapy. Gene Ther 6:1482–1488
17. Ahmad M et al (1977) Ortho ester hydrolysis. The complete reaction mechanism. J Am Chem Soc 99:4827–4828
18. Bessodes M et al (2000) Acid sensitive compounds for delivering drugs to the cells. US 60/239,116, October 2000, PCT Int.Appl. WO 02/20510 (March 2002) 73.
19. Byk G, Dubertret C, Escriou V, Frederic M, Jaslin G, Rangara R et al (1998) Synthesis, activity and structure-activity relationship studies of novel cationic lipids for DNA transfer. J Med Chem 41(2):229–235
20. Yoshimori T et al (1991) Bafilomycin A1, a specific inhibitor of vacuolar-type H(+)-ATPase, inhibits acidification and protein degradation in lysosomes of cultured cells. J Biol Chem 266:17707–17712

Chapter 29

Serum-Resistant Lipoplexes in the Presence of Asialofetuin

Conchita Tros de ILarduya

Abstract

Vectors proposed for gene delivery generally fall into two categories: viral and nonviral. They differ primarily in their assembling process. A viral vector is assembled in a cell, whereas a nonviral vector is constructed in a test tube. While vectors based on viral-based delivery systems are related to safety concerns, immune response, and formulation issues, the problem of nonviral ones is related to their low efficiency for encapsulating large DNA molecules, which has been an important technical obstacle to their utilization. Moreover, for most nonviral vectors, high efficiency in vitro *transfection* correlates with a global excess of cationic charges. This excess can in vivo facilitate nonspecific interactions with many undesired elements such as extracellular matrix and negatively charged serum components. Scientists have been using liposomes for gene delivery since the late 1970s. However, it was only after the introduction of cationic liposomes, which were shown to complex DNA and form the termed "lipoplexes," which offered some promise for an easy and efficient liposomal gene delivery. In this protocol, we describe the preparation of serum-resistant lipoplexes in the presence of the ligand asialofetuin (AF), in order to design efficient gene therapy carriers to deliver genes to the liver. It is also interesting to note, that although most of the current protocols imply covalent binding of the ligand, our complexes have been formulated by simple mixing of the three components in a studied and established order of addition. Lipoplexes containing the optimal amount of AF (1 µg/µg DNA) showed 16-fold higher transfection activity in HepG2 cells than nontargeted (plain) complexes.

Key words: Cationic liposomes, Lipoplexes, Asialoglycoprotein receptor, Asialofetuin, Liver gene therapy, Gene delivery

1. Introduction

The liver possesses a variety of characteristics that make this organ very attractive for gene therapy. The proportion of administered macromolecules internalized by hepatocytes depends on their particle size and biochemical characteristics. Only relatively small molecules can pass the fenestrae of sinusoidal endothelial cells in the liver. On the other hand, the basic mechanism underlying targeted delivery is ligand–receptor interactions. Specific targeting to the liver has been achieved by using ligands that bind the asialoglycoprotein receptor (ASGPr), which is uniquely present on hepatocytes in large numbers with high-affinity binding (1). Asialofetuin (AF), a glycoprotein having triantennary galactose terminal sugar chains, is known as an excellent ligand molecule selectively recognized by ASGPr (2).

For a successful gene therapy, effective and safe delivery of genes into appropriate cells is required. For liver gene delivery, both virus-mediated (3–5) and nonviral systems have been considered. Although some of the virus-mediated gene transfer systems have been found to be quite effective, their usefulness is limited, given that they induce an immune response, leading to the rapid rejection of transduced cells. On the other hand, the main problem of nonviral systems is mostly the low efficiency of transfection and gene expression as well. A decisive advance in nonviral-mediated gene transfer was made by Felgner and associates when they first reported the use of cationic lipids with high efficiency of DNA delivery into cells (6). The transfection protocol using the cationic liposomes is very simple. The lipid and the DNA are mixed to form a complex called "lipoplex" (7) by condensation of the DNA through electrostatic charge–charge interactions, usually in a ratio with a little excess of cationic lipid. This ensures an overall positive charge on the lipoplex and improves the interaction with the negatively charged plasma membrane of the cell. Furthermore, by carefully controlling the complexation conditions, relatively homogeneous and physically stable suspensions can be obtained. This quantitative complexation eliminates the need for a separate step to remove unencapsulated material, and all of the polynucleotide is utilized for each experiment. These complexes are efficient in transfecting cells both in vitro and in vivo (8–11). Because of its convenience and efficacy, cationic lipid-mediated gene delivery technology has become a promising system for gene delivery as an alternative to viral-based vectors.

It is interesting also to note that the ability of serum to inhibit lipofection is an often described phenomenon (12). In cell culture systems, liposome-mediated gene transfection is usually carried out in serum-free medium or in at most 10–20% serum. The inhibitory effect of serum on transfection mediated by lipoplexes to hepatocytes (13) has been also reported. Thus, during the in vitro assessment of transfection reagent, it is important to emulate in vivo conditions by using high concentration of serum, as is has been done in this protocol.

2. Materials

2.1. Preparation and Characterization of Asialofetuin-Lipoplexes

1. 1,2-Dioleoyl-3-(trimethylammonium) propane (DOTAP) and cholesterol (Chol) (Avanti Polar Lipids, Alabaster, AL, USA). Lipids are dissolved in chloroform and stored at −80°C.
2. Asialofetuin Type I (AF) (Sigma, Madrid, Spain). It is dissolved in distilled water and stored in aliquots at −80°C.
3. Plasmid, encoding for luciferase (pCMVLuc, VR-1216) (Clontech, Palo Alto, CA, USA).
4. N-(2-hydroxyethyl) piperazine-N'-[2-ethanesulfonic acid] (HEPES) (Sigma, Madrid, Spain). It is stored at room temperature.
5. Buffer Hepes Glucose (BHG): 10-mM HEPES, pH 7.4, 10% D-(+)-glucose. This buffer is stored refrigerated.
6. Polycarbonate membranes (Avestin, Toronto, Canada) (100 nm).
7. Sterilization membranes (Millex 0.22 μm, Millipore, Bedford, MA, USA).
8. Restriction enzymes: *Pst* I (10 U/μl), *Xho* I (10 U/μl), and *Blg* II (10 U/μl), (Invitrogen Life Technologies, USA). All were stored at −20°C.

2.2. Agarose Gel Electrophoresis

1. DNase I (140 U/μl, stored at −20°C) and ethidium bromide (GibcoBRL Life Technologies, Barcelona, Spain).
2. Sodium dodecyl sulfate (SDS).
3. Glycerol (>99%).
4. Stacking gel (0.8% agarose in TBE Buffer 1×) agarose D-1 Low EEO (Hispanlab S.A. Pronadisa™, Madrid, Spain).
5. Running buffer, Tris–boric–EDTA: 100-mM Tris, pH 8.4, 90-mM boric acid, and 1-mM ethylenediamine tetraacetic acid (EDTA).
6. Loading buffer (Invitrogen Life Technologies, Barcelona, Spain), stored at 4°C.
7. Bromophenol blue (Sigma, Madrid, Spain).
8. Molecular weight markers: 1-Kb DNA ladder (Invitrogen Life Technologies, Barcelona, Spain).

2.3. Cell Culture Transfection and Lysis

1. HepG2 (human hepatoblastoma) cells (American Type Culture Collection, Rockville, MD, USA).

2. Dulbecco's modified Eagle's medium (DMEM) with Glutamax, supplemented with 10% (v/v) heat-inactivated fetal bovine serum (FBS), penicillin (100 units/ml) and streptomycin (100 μg/ml) (all GibcoBRL Life Technologies, Barcelona, Spain).

3. Phosphate-Buffered Saline (PBS): 0.15-M NaCl, pH 7.4 (Invitrogen Life Technologies, Spain).

4. Solution of Trypsin-EDTA 10× (Gibco BRL, Life Technologies, Barcelona, Spain) (stored at −20°C).

5. Teflon cell scrapers (TPP, Switzerland).

6. Flask 75 cm^3 for cell culture (TPP, Switzerland).

7. 48-well plates (Iwaki Microplate 48-well, Japan).

8. Reporter lysis buffer (RLB) (Promega, USA) (stored at room temperature).

9. Luciferase Assay Kit (Promega, USA) was stored at −20°C. Luciferase is reconstituted and stored at −80°C.

10. Kit Bio-Rad DC Protein Assay and bovine serum albumin (BSA) standard (Bio-Rad, Hercules, CA, USA).

3. Methods

Cationic lipid:pDNA complexes (lipoplexes) generally are prepared by the simple mixing together of the two components; however, it is also important to consider that the applied protocols for complex formation and subsequent modifications strongly influence the properties of the transfection particle. Also, the order of addition of components to form the lipoplex affects considerably lipofection activity.

3.1. Preparation of Asialofetuin-Lipoplexes

3.1.1. Preparation of Liposomes

1. DOTAP/Chol (1:0.9 molar ratio) liposomes were prepared by drying a chloroform solution of the lipids by rotary evaporation under reduced pressure.

2. Take 0.28-ml DOTAP (25-mg/ml) and 0.35 ml Chol (10 mg/ml) and dry by rotary evaporation.

3. Rehydrate the film with 1 ml of Buffer Hepes Glucose, to give a final concentration of 10-mM DOTAP/9-mM Chol.

4. The resulting multilamellar vesicles were extruded five times through polycarbonate membranes with 100-nm pore diameter using a Liposofast device (Avestin, Toronto, Canada), to obtain a uniform size distribution. For that, pass the liposome emulsion back and forth through the membrane (usually five to ten passes is sufficient) (see Notes 5 and 7).

3.1.2. Preparation of Asialofetuin-Lipoplexes

5. Liposomes were filter-sterilized through 0.22-μm membranes (see Subheading 2).
6. Liposomes were stored at 4°C under nitrogen.

1. Lipoplexes were prepared at 4/1 (±) charge ratio by sequentially mixing 100 μl of a solution of 10-mM HEPES, 10% (w/v) glucose buffer (pH 7.4) without (plain lipoplexes) or with a variable amount of AF (0.01, 0.1, 1, 4.5, 9, 18, and 36 μg) (AF-lipoplexes), and 22.8 nmoles of the DOTAP/Chol liposome suspension (see Note 2).
2. After incubating for 15 min at room temperature, 100 μl of water containing 1 μg of pCMVLuc were added and gently mixed (see Note 3).

3.2. Characterization of Asialofetuin-Lipoplexes

The particle size of complexes was measured by dynamic light scattering, and the overall charge by zeta potential measurements, using a particle analyzer (Zetamaster, Malvern Instruments, Spain).

3.2.1. Particle Size and Zeta Potential Measurements

1. Samples of the prepared complexes (2.5 ml) were measured three times for 60 s at 1,000 Hz and an electric current of 3 mA with zero field correction (results are shown in Fig. 1) (see Notes 8 and 9).

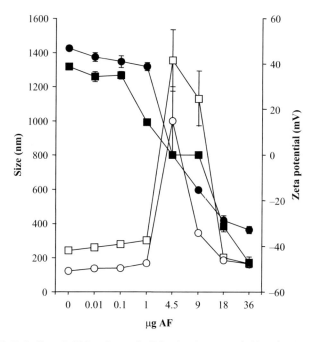

Fig. 1. Optimization of AF-lipoplexes. Particle size (*open symbols*) and zeta potential (*closed symbols*) of DOTAP/Chol liposomes (*circles*) and lipoplexes (*squares*) in the absence or presence of different amounts of asialofetuin. Each value represents the mean ± SD ($n = 3$) (10).

3.3. Nuclease Resistance of Asialofetuin-Lipoplexes by Agarose Gel Electrophoresis

To evaluate the role of our complexes in the protection of DNA, naked plasmid and lipoplexes with different amounts of AF were incubated in the presence of the enzyme DNase I. Samples were analyzed by 0.8% agarose gel electrophoresis and the integrity of the plasmid in each formulation was compared with untreated DNA as a control.

1. First, lipoplexes were prepared at 4/1 (±) charge ratio containing 2.5 µg of pCMVLuc.
2. DNase I (1 unit per µg of DNA) was added to 2.5 µg (DNA) of each sample and the mixtures were incubated at 37°C for 30 min (see Note 10).
3. Two microliters of EDTA (0.5 M) were immediately added to stop DNase degradation.
4. Sodium dodecyl sulfate (SDS) was included to a final concentration of 1% to release DNA from the complexes.

3.3.1. Agarose Gel Electrophoresis

1. Add the loading buffer (30% glycerol, 25-mg bromophenol blue, and 10-ml water) to the prepared samples.
2. Prepare the running buffer (Tris–boric–EDTA, pH 8.4: 100-mM Tris, 90-mM boric acid and 1 mM EDTA).
3. Prepare agarose 0.8 % in TBE 1× (dissolve 0.44-g agarose in 55 ml of buffer 1×).
4. Add ethidium bromide to the previous solution to a concentration of 0.5 µg/ml.
5. Insert the comb in the electrophoresis cuvette.
6. Once the stacking gel has set, carefully remove the comb.
7. Add the running buffer to the chambers of the gel unit and load each sample in a well. Include one well for prestained molecular weight markers.
8. Complete the assembly of the gel unit and connect to a power supply. The gel can be run at 80 mV, 140 mA during 2 h and visualized under UV illumination after ethidium bromide staining, using a camera Gel (doc 2000, Bio-Rad USA) (Results are shown in Fig. 2) (see Note 11).

3.4. In vitro Transfection Studies

3.4.1. Protocol of Transfection

1. For transfection, 10^5 HepG2 cells were seeded in 1 ml of medium in 48-well culture plates 24 h before addition of the complexes and used at approximately 80% confluency (see Notes 12 and 14).
2. Cells were washed twice with 1 ml of DMEM without antibiotics and then 0.3 ml of fetal bovine serum and 0.2 ml of complexes were added gently to each well.
3. After a 4-h incubation in 60% FBS (at 37°C in 5% CO_2) the medium was replaced by 1 ml of DMEM containing 10% FBS.

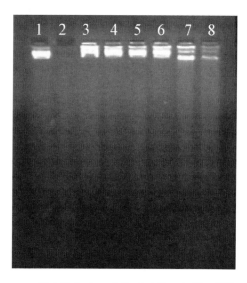

Fig. 2. Protection assay. Stability to degradation by DNase I of free DNA or DNA formulated in AF-lipoplexes with different amounts of the ligand. Untreated DNA (*lane 1*), DNA treated with DNase: naked plasmid (*lane 2*), DNA inside lipoplexes prepared with zero (*lane 3*), 0.1 (*lane 4*), 1 (*lane 5*), 4.5 (*lane 6*), 9 (*lane 7*), and 18 (*lane 8*) μg AF/μg DNA (10)

4. Cells were further incubated for 48 h in medium containing 10% FBS.

3.4.2. Luciferase Activity Measurement

1. After 48 h, cells were washed with phosphate-buffered saline (PBS).

2. Cells were lysed using 100 μl of reporter lysis buffer (RLB 1×) at room temperature for 10 min, followed by two alternating freeze–thaw cycles to open pores in the cells.

3. Cell lysate was centrifuged for 2 min at $12,000 \times g$ to pellet debris.

4. Twenty microliters of the supernatant was assayed for total luciferase activity using the luciferase assay reagent, according to the manufacturer's protocol. For that, reconstitute the Luciferase Assay Substrate (luciferin, lyophilized) with 10 ml of Luciferase assay buffer (mix well before use). We will need 100 μl of substrate per 20 μl of sample. Blank (20 μl lysis buffer + 100 μl substrate). Recombinant luciferase (Promega, USA) is used as standard (see Notes 15 and 16).

5. A luminometer (Sirius-2, Berthold Detection Systems, Innogenetics, Diagnóstica y Terapéutica, S.A., Barcelona) was used to measure luciferase activity (measurement parameters: delay time 3 s and measuring time 10 s).

3.4.3. Protein Content Determination

The protein content of the lysates was measured by the DC Protein Assay reagent using bovine serum albumin as the standard.

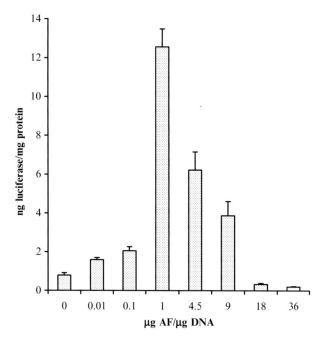

Fig. 3. In vitro transfection activity by AF-lipoplexes. Transfection of HepG2 cells by plain (nontargeted) and AF-lipoplexes as a function of the amount of the ligand. The data represent the mean ± SD of three wells and are representative of three independent experiments (10)

1. Put in each well of a 96-well plate 10 µl of sample.
2. Add 25 µl of reagent A and 200 µl of reagent B.
3. Wait for 15 min at room temperature until a blue color appears.
4. Measure absorbance at 650–750 nm.

Results of transfection activity, expressed as ng luciferase/mg protein, are shown in Fig. 3 (see Note 17).

4. Notes

1. Plasmid was amplified in *Escherichia coli*, isolated and purified using a QIAGEN EndoFree® Plasmid Giga Kit (QIAGEN GmbH, Hilden, Germany). Purity was confirmed by 0.8% agarose gel electrophoresis followed by ethidium bromide staining and the DNA concentration was measured by UV absorption at 260 nm.
2. For the preparation of lipoplexes, the lipid-to-DNA charge ratio was calculated as the mole ratio of DOTAP (one charge per molecule) to nucleotide residue (average MW 330).

3. It is essential to prepare the complexes by following the order: AF + cationic liposomes + plasmid DNA.
4. The lipoplexes resulted to be of easy preparation and very stable, without having problems of aggregation.
5. The extrusion of a lipid solution through a polycarbonate membrane provides liposomes with the desired and uniformity of size. The LiposoFast™-Basic used for extrusion of liposomes has virtually zero dead volume allowing for almost complete sample recovery.
6. It is better to prepare freshly lipoplexes to get optimal in vitro transfection results.
7. It is important to consider that the size and overall charge of starting liposomes was 120 nm and 47 mV, respectively.
8. By increasing the amount of AF to 1 μg (per microgram of DNA), the particle size of AF-liposomes increased slightly to approximately 130 nm. The same behavior was observed for AF-lipoplexes, which reached a size of 302 nm (Fig. 1).
9. The zeta potential of AF-liposomes and lipoplexes showed clearly positive values by using amounts of AF lower than 1 μg. AF-complexes aggregated at 4.5 and 9 μg AF/μg DNA, which corresponds to a value of the zeta potential close to the electroneutrality (Fig. 1).
10. Before addition of DNAse I, incubate the complexes at room temperature (30 min).
11. Naked DNA was degraded quickly (Fig. 2, lane 2), while the plasmid inside AF-lipoplexes containing an amount of AF below 4.5 μg was protected completely (Fig. 2, lanes 3–8).
12. The cells were maintained at 37°C under 5% CO_2 in Dulbecco's modified Eagle's medium-high glucose (DMEM), supplemented with 10% (v/v) heat-inactivated fetal bovine serum (FBS), penicillin (100 units/ml), streptomycin (100 μg/ml), and l-glutamine (4 mM).
13. DMEM should be stored refrigerated at 0–4°C.
14. Cells were passaged twice a week.
15. Luciferase substrate (luciferin) should be reconstituted at room temperature or lower. Do not reconstitute at elevated temperatures.
16. Equilibrate the reagents at room temperature before use.
17. Complexes containing the ligand in the range from 0.01 to 9 μg AF/μg DNA gave always higher values of gene expression compared to plain lipoplexes (without ligand). Complexes prepared with amounts of the ligand above 9 μg decreased gene expression dramatically. No gene expression was detected with the naked plasmid (Fig. 3).

Acknowledgments

This work was supported by a grant from the Government of Navarra (Department of Education) and the University of Navarra Foundation.

References

1. Ashwell G, Morell AG (1974) The role of surface carbohydrates in the hepatic recognition and transport of circulating glycoproteins. Adv Enzymol Relat Areas Mol Biol 41:99–128
2. Wu J, Liu P, Zhu JL, Maddukuri S, Zern MA (1998) Increased liver uptake of liposomes and improved targeting efficacy by labelling with asialofetuin in rodents. Hepatology 27:772–778
3. Kitten O, Cosset FL, Ferry N (1997) Highly efficient retrovirus-mediated gene transfer into rat hepatocytes in vivo. Hum Gene Ther 8:1491–1494
4. Schiedner G, Morral N, Parks RJ, Wu Y, Koopmans SC, Langston C, Graham FL, Beaudet AL, Kochanek S (1998) Genomic DNA transfer with a high-capacity adenovirus vector results in improved in vivo gene expression and decreased toxicity. Nat Genet 18:180–183
5. Nakai H, Herzog RW, Hagstrom JN, Walter J, Kung SH, Yang EY, Tai SJ, Iwaki Y, Kurtzman GJ, Fisher KJ, Colosi P, Couto LB, High KA (1998) Adeno-associated viral vector-mediated gene transfer of human blood coagulation factor IX into mouse liver. Blood 91:4600–4607
6. Felgner PL, Gadek TR, Holm M, Roman R, Chan HW, Wenz M, Northrop JP, Ringold GM, Danielsen M (1987) Lipofection: a highly efficient, lipid-mediated DNA-transfection procedure. Proc Natl Acad Sci USA 84:7413–7417
7. Felgner PL, Barenholz Y, Behr JP, Cheng SH, Cullis P, Huang L, Jessee JA, Seymour L, Szoka F, Thierry AR, Wagner E, Wu G (1997) Nomenclature for synthetic gene delivery systems. Hum Gene Ther 8:511–512
8. Tros de Ilarduya C, Düzgünes N (2000) Efficient gene transfer by transferrin lipoplexes in the presence of serum. Biochim Biophys Acta 1463:333–342
9. Tros de Ilarduya C, Arangoa MA, Moreno-Aliaga MJ, Düzgünes N (2002) Enhanced gene delivery in vitro and in vivo by improved transferrin-lipoplexes. Biochim Biophys Acta 1561:209–221
10. Arangoa MA, Düzgünes N, Tros de Ilarduya C (2003) Increased receptor-mediated gene delivery to the liver by protamine-enhanced-asialofetuin-lipoplexes. Gene Ther 10:5–14
11. Tros De Ilarduya C, Buñuales M, Qian C, Düzgünes N (2006) Antitumoral activity of transferrin-lipoplexes carrying the IL-12 gene in the treatment of colon cancer. J Drug Target 14:527–535
12. Zelphati O, Uyechi LS, Barron LG, Szoka FC Jr (1998) Effect of serum components on the physico-chemical properties of cationic lipid/oligonucleotide complexes and on their interactions with cells. Biochim Biophys Acta 1390:119–133
13. Jarnagin WR, Debs RJ, Wang SS, Bissell DM (1992) Cationic lipid-mediated transfection of liver cells in primary culture. Nucleic Acids Res 20:4205–4211

Chapter 30

Anionic pH Sensitive Lipoplexes

Nathalie Mignet and Daniel Scherman

Abstract

To provide long circulating nanoparticles which can carry a gene to tumors, we have designed anionic pegylated lipoplexes that are pH sensitive. Anionic pegylated lipoplexes have been prepared from the combined formulation of cationic lipoplexes and pegylated anionic liposomes. The neutralization of the particle surface charge as a function of the pH was monitored by light scattering, in order to determine the ratio between anionic and cationic lipids that would give pH sensitive complexes. This ratio has been optimized to form particles sensitive to pH change in the range 5.5–6.5. Compaction of DNA into these newly formed anionic complexes was checked by DNA accessibility to picogreen. The transfection efficiency and pH sensitive property of these formulations were shown in vitro using bafilomycin, a vacuolar H^+-ATPase inhibitor.

Key words: Anionic lipoplexes, Pegylated lipoplexes, pH sensitive lipoplexes, Gene delivery to tumor, Anionic cholesterol

1. Introduction

Non viral gene delivery still suffers from limited in vivo applications due to a poor vector efficacy to deliver the gene at the target. Works on DNA delivery involve different types of cationic lipids (1, 2) or polymers (3) and degradable cationic lipid/DNA complexes (4, 5) or polymer/DNA complexes (6).

In the present work, we have focused our studies on the pH sensitive strategy. This might involve the use of pH labile components into the formulation, such as pH degradable lipid (7–9) or polymer (10). The pH sensitivity might also be provided by a mixture of titrable anionic and cationic lipids in the formulation. Numerous works have been reported on such pH sensitive lipoplexes, mostly to increase gene delivery into the cells upon endosome pH drop (11, 12). These are mostly cationic lipoplexes, as they allow a high DNA compaction and an efficient cell uptake. For instance, in the LPDII particles, DNA/polylysine complexes were mixed with acid-sensitive CHEMS/DOPE/DOPE-Peg-folate liposomes to form acid-sensitive lipoplexes (13). The pH sensitivity is given by the CHEMS and DOPE lipids which are able to form a hexagonal phase at acidic pH, which might induce a lipoplexes destabilization in the endosomal compartment (14, 15). This destabilization might also be brought by fusogenic peptides (16). In our case, the idea is different. The concept is to form tunable pH sensitive lipoplexes (17), meaning that only the ratio between the anionic and the cationic lipid would be responsible for the pH sensitivity, whatever the anionic and cationic lipid chosen.

Moreover, delivering DNA to tumor using non viral vectors requires two main necessities:

- a long circulation time to obtain the highest possible vector accumulation in the tumor vascularization; Since this requires poorly charged lipoplexes, in particular, a limited amount of cationic charges (18), we have chosen to develop anionic lipoplexes.

- a reversal of these anionic charges or an amplification of the cationic charges in the tumor environment upon pH drop, in order to obtain cationic lipoplexes that will efficiently enter the cells by binding to the anionic plasma membrane and deliver its DNA content; for this last step, acid sensitive PEG had been developed (19–21); a fusogenic lipid might also be incorporated into the lipoplex.

To reach this goal, we have designed tunable anionic pH sensitive lipoplexes. These complexes are anionic at physiological pH and pegylated to improve their circulation time, as compared to cationic lipoplexes (22). Moreover, they become cationic at pH under 6, in order to promote efficient tumor cell internalization, since it is widely recognized that extracellular pH is acidic in ischemic tumor tissue (see Scheme 1).

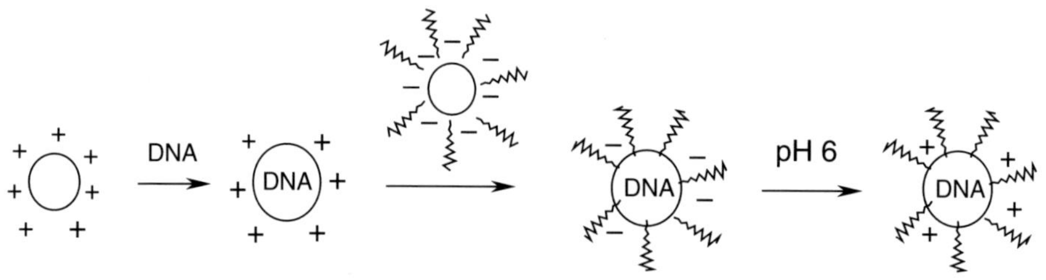

Scheme 1. Schematized representation of the sequential process used for the formation of anionic lipoplexes and charge reversibility upon reduction of the pH. It does not take into account the structure of the complexesor the position of DNA in the particle lipidic bilayer

2. Materials

2.1. Abbreviations of the Lipids Used

PEG: Polyethylene glycol; DOPE: Dioleoylphosphatidylethanolamine; CHEMS: Cholesteryl hemisuccinate; DPPC: Dipalmytoylphosphatidylcholine. The names of the different lipids were generated with AutoNom 2000 software which is based on IUPAC rules. The cationic lipid whose name, according to the nomenclature is 2-{3-[Bis-(3-amino-propyl)-amino]-propylamino}-*N*-ditetradecylcarbamoyl methyl-acetamide or RPR209120, that we called DMAPAP, was previously described in the supporting information of Thompson et al. (23). The tetracarboxylated derivative, *[(2-{Cholesteryloxycarbonyl-[2-(bis-carboxymethyl-carbamoyloxy)-ethyl]-amino}-ethoxycarbonyl)-carboxymethyl-amino]-acetic acid*, that we named CCTC, is described in Mignet et al. (22).

2.2. Chemicals and DNA Provided or Synthesized

L-α-Dioleoyl Phosphatidylethanolamine (DOPE), and 1,2-Dipalmitoyl-*sn*-Glycero-3-Phosphocholine (DPPC) were purchased from Avanti Polar Lipids. The chol-PEG$_{110}$ was obtained in one step from the reaction of cholesteryl chloroformate and α-amino-ω-methoxy-PEG. The luciferase encoding gene was obtained as reported (22). Picogreen® was purchased from Molecular Probes, USA, Bafilomycin from Sigma, France, the BCA kit from Pierce and the luciferase kit from Promega.

2.3. Equipment

Anionic liposome was prepared by the film method on a rotary evaporator Heidolph, VWR, equipped with a vacuubrand CVC2 to control the pressure. Sonication was performed on sonicator branson 1210. Size and zeta potentials measurements were performed on a Zeta Sizer NanoSeries from Malvern Instruments equipped with a MPT2 autotitrator. Fluorescence was measured on a multilabel plate reader Wallac Victor2 1420 Multilabel Counter, Perkin Elmer, France, equipped with excitation and emission filters (350 ± 10 nm, 450 ± 10 nm).

2.4. Buffer

1. 50 mM Tris(hydroxymethyl)aminolethane-maleate, pH 7:

   Prepare a solution of tris acid maleate 50 mM by mixing 6 g of tris(hydroxymethyl)aminomethane and 5.8 g of maleic acid in 1 L H$_2$O. Prepare a 50 mM NaOH solution. Mix the 50 mL of the tris acid maleate solution and 48 mL of NaOH solution. Verify and adjust the pH by adding one of the Tris acid maleate or NaOH solutions.

2. 5% glucose (5 g in 100 mL H$_2$O).

3. Tris-maleate/glucose buffer: Mix the tris-maleate and the glucose solutions (1:1, vol:vol) obtain 25 mM Tris-maleate 25 mM, 2.5% glucose.

4. 40 mM Hépès and 20% glucose: Mix the Hépès and the glucose solutions (1:1, vol:vol) obtain 20 mM Hépès, 10% glucose.

3. Methods

Obtaining anionic self-associated lipoplexes is not obvious, since a competition occurs between the anionic charge of the particles and the anionic charges of the DNA. The easiest way to proceed is to obtain cationic lipoplexes core by mixing cationic lipid and DNA. Then an anionic pegylated liposome is added to the mixture to form an anionic pegylated lipoplex (Scheme 1). The PEG lipid was used to avoid aggregation that would occur by mixing the cationic lipoplexes and the anionic liposome, and will also be useful to limit protein interaction upon systemic injection (24). The cationic lipid (23) we used exhibiting two primary amines, one secondary and one tertiary amine, we designed original negatively charged cholesterol bearing four carboxylate moieties (22) in order to limit the amount of cholesterol in the liposome.

Thanks to the presence of the carboxylate moieties, the anionic particle charge should be reversed to a cationic one at a determined appropriate pH. To reach this goal, the ratio between the two lipids has to be optimized, taking into account the DNA negative charges. The optimal ratio was determined by zeta potential measurements combined with titration experiments, as will be described (Data obtained represented Fig. 1). Light scattering was also used to insure that the particle size remained in the right size range (around 100–200 nm), which is fundamental to maintain the particle circulating into the blood stream (25). To evaluate if plasmid DNA was well confined into the structure formed, picogreen™ was used, and the fluorescence associated to free DNA or compacted DNA was measured. Compacted DNA does not allow picogreen™ to intercalate into the DNA base pair and gives a low fluorescence level. The intracellular pH sensitivity of the anionic pegylated lipoplexes was shown in vitro using bafilomycin, an ATPase inhibitor, which reduces endosome acidification (Fig. 2).

3.1. Preparation of Cationic Liposomes

1. Dissolve separately the lipids DMAPAP (10 µmol, 10 mg) and DOPE (10 µmol, 7.3 mg) in chloroform (500 µL each). Take care that the lipids are well dissolved separately before mixing them (see Note 1).
2. Mix them into a round bottom flask (10 mL) (see Note 2).
3. Put the flask at the evaporary evaporator to remove the solvent in a pressure controlled manner. First, reduce the pres-

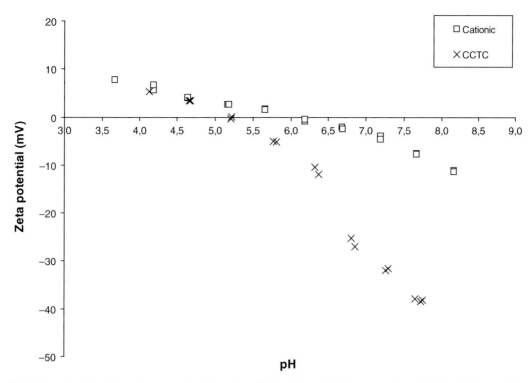

Fig. 1. Zeta potential of the particles as a function of the pH. The pH was initially measured at pH 8.25 and decreased by addition of a solution of HCl at 0.1 M, in order to obtain steps with a pH decrease of 0.5. A measure of the electrophoretic mobility and dynamic diameter was measured every variation of 0.5 in pH after the pH was stabilized. The measures were taken at room temperature

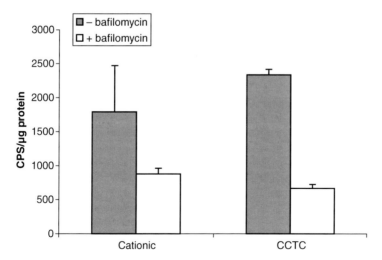

Fig. 2. Transfection of the cationic and anionic formulations using a plasmid encoding for the luciferase reporter gene in presence and absence of bafilomycin. The Y axis represents the level of luciferase expression per protein (μg), mean values + SD. The background level of the untreated cells was removed for each sample

sure from 1,000 to 200 mbar in approximately 15 min with a middle rotation speed. When the drop forms, increase the rotation speed at its maximum level to drag the drop into the film. Then, reduce the pressure from 200 to 5 mbar in 30 min, and leave the film under reduced pressure for an additional hour (see Note 3).

4. The film being dry, add 1 mL (to afford a final concentration of 20 mM) milliQ filtered (0,22 μm) H_2O and leave the flask under gentle rotation overnight at room temperature (see Note 4).

5. Mix gently the mixture on a vortex, if the film is not fully detached from the wall. Sonicate the particles during 5 min to afford a rather homogenous size distribution of approximately 150–200 nm.

6. Control the size by dynamic light scattering (see Note 5). For measurements on a nanoZS (Malvern Instruments), dilute 5 μL of the particles obtained in a 500 μL cuve, start the measure in the automatic mode.

3.2. Preparation of Cationic Lipoplexes

Lipoplexes were prepared in tris-maleate 25 mM, glucose 2.5% with a charge ratio cationic lipid/anionic lipid = 6, which corresponds to a ratio total lipid to DNA = 12.

1. Dilute the DMAPAP/DOPE suspension initially at 20 mM to 5 mM in H_2O.

2. Dilute 6 μg pDNA in 100 μL tris-maleate 25 mM, glucose 2.5%

3. Dilute 15 μL of the 5 mM DMAPAP/DOPE suspension in 100 μL tris-maleate 25 mM, glucose 2.5%.

4. Add the plasmid DNA to the cationic liposome dropwise in few seconds, with constant vortexing (see Note 6).

5. Leave the sample 1 h at room temperature to incubate before using it or adding it to the anionic liposomes.

3.3. Preparation of Anionic Pegylated Liposomes

The anionic liposomes were prepared by the film method as described for the cationic liposome in Subheading 3.1.

1. Dissolve separately the lipids DPPC (5 μmol, 3.7 mg), CCTC (15 μmol, 7.4 mg) and Chol-PEG_{110} (0.5 μmol, 2.5 mg) in chloroform (200 μL, 600 μL and 200 μL respectively). Take care that the lipids are well dissolved separately before mixing them (see Note 1).

2. Mix them into a round bottom flask (10 mL) and evaporate the choloroform under reduced pressure as described in Subheading 3.1.3 (see Notes 2 and 3).

3. The film being dry, add 1 mL (to afford a final concentration of 20.5 mM) milliQ filtered (0,22 μm) H_2O and leave the flask under gentle rotation overnight at room temperature.

4. Mix gently the mixture on a vortex, if the film is not fully detached from the wall.

5. Filter the particles successively on 0.45 and 0.22 μm polyethylsulfonate filters.

6. Control the size by dynamic light scattering (see Note 5). For measurements on a nanoZS (Malvern Instruments), dilute 5 μL of the particles obtained in a 500 μL cuve, start the measure in the automatic mode.

3.4. Preparation of Anionic Pegylated Lipoplexes

The preformed cationic lipoplexes were added to the anionic liposomes according to the charge lipid ratio DMAPAP/CCTC (±) = 1.3.

1. Prepare a suspension of the anionic liposomes at 0.55 mM from the 20.5 mM suspension prepared in Subheading 3.3.

2. Dilute 100 μL of this anionic liposome suspension in Hépès 20 mM, glucose 10%

3. Add the preformed lipoplexes described in Subheading 3.2 to the suspension of the anionic liposomes.

3.5. Titration Experiments

1. Prepare the HCl 0.1 M buffer and fix it to the titrator (see Note 7).

2. Rinse the autotitrator cables with H_2O. Prime the HCl buffer and rinse again.

3. Dilute the anionic pegylated lipoplexes prepared in Subheading 3.4 in 10 mL H_2O and fix the flask at the autotitrator, including the pH electrode.

4. Enter the protocol: zeta potential measurements with an initial pH point taken at the pH of the solution, in this case, pH 8.25, and points taken every 0.5 pH change until pH 3.7 is reached. Stir between each measure to insure the solution is homogeneous. Electrophoretic mobility is converted automatically to the ζ potential, according to the Smoluchowski equation, by the system.

3.6. DNA Complexation Checked by Fluorescence

1. Prepare the picogreen® solution as described by the provider (1/200 in tris–EDTA buffer).

2. Load into a 96-well plate free DNA or complexed DNA (40 ng) in tripliquets.

3. Add 200 μL of the picogreen solution (Subheading 3.6.1) to each well filled with DNA and three more to obtain the picogreen background level.

4. Read the emission at 450 nm under an excitation at 350 nm on a multiplate reader able to measure fluorescence.

5. For the calculation, calculate the mean and the standard error on each tripliquets. Remove the picogreen background from

the sample data. Calculate the percentage of fluorescence of each sample by dividing the sample data by the value of the free DNA taken as 100% fluorescence.

3.7. In Vitro Experiments

1. B16 murine cells were grown into DMEM supplemented with L-glutamine (29.2 mg/mL), penicillin (50 units/mL), streptomycin (50 units/mL), and 10% fetal bovine serum.

2. The day before the experiment, seed B16 cells into 24-well culture plates at a density of 50,000 cells per well and incubate at 37°C, under 5% CO_2 for 24 h.

3. One hour before transfection, wash the cells with fresh medium, with or without bafilomycin.

4. Add 100 μL of cationic or anionic lipoplexes containing 0.5 μg DNA onto each well in tripliquet, and incubate the plates at 37°C for 6 h in the presence of 5% CO_2, then replace by fresh medium for 18 h.

5. Wash the cells twice with PBS and treat with 200 μL of a passive lysis buffer (Promega). After 15 min, centrifuge the cells for 5 min at 12,000 tr/min.

6. Add 10 μL of the supernatant and 10 μL of iodoacetamide to a 96-well plate, and incubate at 37°C for 1 h. Quantify the protein content with the BCA protein assay KIT (PIERCE) and report to BSA taken as a reference curve (see Note 8).

7. Quantify the luciferase activity using a commercial kit Luciferase assay system (PROMEGA): Load 10 μL of the lysed cells into a 96-well plate and place it into the lecture plate reader. Load the luciferase substrate to the injector. 50 μL of the luciferase substrate is injected via an injector, and the absorbance is read immediately at 563 nm on a Wallac Victor2 1420 Multilabel Counter (Perkin Elmer).

8. For the calculation, background of the untreated cells, taken as negative controls, was removed from the sample data. The relative counts obtained for luciferase quantification were divided by the protein content in each well to normalize the results per μg of protein. The cationic formulation was taken as the positive reference formulation.

4. Notes

1. Solubility of the lipids should be checked with intensive care, since presence of non-soluble entities will appear in the film and reduce particle homogeneity after hydration.

2. The ratio between the volume to be reduced (or the amount of lipids) and the round bottom flask is important, since the

film should occupy as much flask wall as possible. From the surface of the flask occupied by the film will depend the number of layers in the liposomes.

3. Make sure that the film is not crackled by a too rapid pressure reduction. If so, dissolve again the lipids in 1 mL $CHCl_3$ and start again part 3. It is always preferable to obtain an homogeneous film along the flask wall, it will provide more homogeneous liposome size after the hydration step.

4. Evaporation and hydration time are usually reported as shorter, but we have found that taking time to do these steps are required to form homogeneous liposome sizes.

5. All buffers and water used should be filtered on 0.22 μm filters, since any dust might interfere with light scattering experiments.

6. In order to maintain an excess of cationic charges and, hence, avoid precipitation by going through a charge ratio (+/−) equal to 1, we add DNA on the cationic lipid and not the opposite order.

7. All buffers should be degazed prior titration to avoid any volume error.

8. The concentration range of BSA should be done for each experiment, since the values are fully dependent on the incubation time. It should be performed in the same buffer as the buffer used for the cells, in this case passive lysis buffer.

Acknowledgments

The author would like to thank Caroline Richard for her dedicated work during her fellowship and Michel Bessodes for providing the anionic cholesterol derivatives.

References

1. Miller A (1998) Cationic liposomes for gene therapy. Angew Chem 41:1768–1785
2. Nicolazzi C, Garinot M, Mignet N, Scherman D, Bessodes M (2003) Cationic lipids for transfection. Curr Med Chem 10:1263–1277
3. Eliyahu H, Barenholz Y, Domb A (2005) Polymers for DNA delivery. Molecules 10: 34–64
4. Hirko A, Tang F, Hughes J (2003) Cationic lipid vectors for plasmid DNA delivery. Curr Med Chem 10:1185–1193
5. Byk G, Wetzer B, Frederic M, Dubertret C, Pitard B, Jaslin G, Scherman D (2000) Reduction-sensitive lipopolyamines as a novel nonviral gene delivery system for modulated release of DNA with improved transgene expression. J Med Chem 43:4377–4387
6. Luten J, Van Nostrum C, De Smedt S, Hennink W (2008) Biodegradable polymers as non-viral carriers for plasmid DNA delivery. J Control Release 126:97–110
7. Garinot M, Masson C, Mignet N, Bessodes M, Scherman D (2007) Synthesis and advantages of acid-labile formulations for lipoplexes. In: Gregoriadis G (ed) Liposome technology, vol 1, 3rd edn. CRC Press, Boca Raton, pp 139–163

8. Martin B, Sainlos M, Aissaoui A, Oudrhiri N, Hauchecorne M, Vigneron JP, Lehn JM, Lehn P (2005) The design of cationic lipids for gene delivery. Curr Pharm Des 3:375–394
9. Roux E, Stomp R, Giasson S, Pezolet M, Moreau P, Leroux JC (2002) Steric stabilization of liposomes by pH-responsive N-isopropylacrylamide copolymer. J Pharm Sci 91:1795–1802
10. Kim Y, Park J, Lee M, Kim Y, Park T, Kim S (2005) Polyethylenimine with acid-labile linkages as a biodegradable gene carrier. J Control Release 103:209–219
11. Shi G, Guo W, Stephenson S, Lee R (2002) Efficient intracellular drug and gene delivery using folate receptor-targeted pH-sensitive liposomes composed of cationic/anionic lipid combinations. J Control Release 80:309–319
12. Guo X, Szoka FC (2003) Chemical approaches to triggerable lipid vesicles for drug and gene delivery. Acc Chem Res 36:335–341
13. Lee R, Huang L (1996) Folate-targeted, anionic liposome-entrapped polylysine-condensed DNA for tumor cell-specific gene transfer. J Biol Chem 271:8481–8487
14. Hafez I, Cullis P (2000) Cholesteryl hemisuccinate exhibits pH sensitive polymorphic phase behavior. Biochim Biophys Acta 1463:107–114
15. Koynova R, Wang L, Tarahovsky Y, MacDonald R (2005) Lipid phase control of DNA delivery. Bioconjug Chem 16:1335–1339
16. Hafez I, Ansell S, Cullis P (2000) Tunable pH-sensitive liposomes composed of mixtures of cationic and anionic lipids. Biophys J 79:1438–1446
17. Simões S, Slepushkin V, Gaspar R, de Lima M, Düzgüneş N (1998) Gene delivery by negatively charged ternary complexes of DNA, cationic liposomes and transferrin or fusigenic peptides. Gene Ther 7:955–964
18. Zalipsky S, Brandeis E, Newman MS, Woodle MC (1994) Long circulating, cationic liposomes containing amino-PEG-phosphatidylethanolamine. FEBS Lett 353:71–74
19. Masson C, Garinot M, Mignet N, Wetzer B, Mailhe P, Scherman D, Bessodes M (2004) pH sensitive PEG lipids containing orthoester linkers: new potential tools for nonviral gene delivery. J Control Release 99:423–434
20. Choi J, MacKay J, Szoka F (2003) Low-pH-sensitive PEG-stabilized plasmid-lipid nanoparticles: preparation and characterization. Bioconjugate Chem 14:420
21. Kale A, Torchilin V (2007) Enhanced transfection of tumor cells in vivo using "Smart" pH-sensitive TAT-modified pegylated liposomes. J Drug Target 15:537–548
22. Mignet N, Richard C, Seguin J, Largeau C, Bessodes M, Scherman D (2008) Anionic pH-sensitive pegylated lipoplexes to deliver DNA to the tumors. Int J Pharm 361(1–2):194–201
23. Thompson B, Mignet N, Hofland H, Lamons D, Seguin J, Nicolazzi C, de la Figuera N, Kuen R, Meng Y, Scherman D, Bessodes M (2005) Neutral post-grafted colloidal particles for gene delivery. Bioconjug Chem 16:608–614
24. Mignet N, Cadet M, Bessodes M, Scherman D (2007) Incorporation of PEG lipid into lipoplexes: on-line incorporation assessment, and pharmacokinetics advantages. In: Gregoriadis G (ed) Liposome technology, vol 2, 3rd edn. CRC Press, Boca Raton, pp 273–292
25. Tranchant I, Thompson B, Nicolazzi C, Mignet N, Scherman D (2004) Physicochemical optimization of plasmid delivery by cationic lipids. J Gene Med 1(Suppl 1):S24–S35

Chapter 31

Liposomal siRNA Delivery

Jeffrey Hughes, Preeti Yadava, and Ryan Mesaros

Abstract

With the recent discovery of small interfering RNA (siRNA), to silence the expression of genes in vitro and in vivo, there has been a need to deliver these molecules to the cell. Forming a lipid/nucleic acid complex has become a solution and is explored here. Certain methods and ideas are used, such as: the positive/negative electrostatic interaction with a cationic lipid and an anionic RNA molecule, the size of the lipid vesicle aiding the uptake target tissues, targeted lipoplexes which can increase efficiency, and the protection of the siRNA molecule from the natural defenses of the immune system. Many lipid formulations exist and can be experimented with to achieve varying results depending on the application.

Key words: Lipoplex, siRNA, Liposome, Gene delivery, Transfection

1. Introduction

1.1. RNAi

Gene silencing or "antisense therapy" is the form of gene therapy that involves the use of "antisense" (DNA, RNA or a chemical analogue) technology. A number of technologies offer the potential for specific therapeutic gene silencing. One such technology, that was discovered relatively recently and is believed to be very potent and specific, is mediated by double-stranded RNA (dsRNA). This technology, called RNA interference, (RNAi) is being viewed as an improved system and a successor to antisense techniques.

RNAi is the natural process of sequence-specific, posttranscriptional gene silencing by dsRNA, homologous in sequence to the target gene (1). In 1998, Fire and Mello first demonstrated that dsRNA led to inheritable downregulation 10 times greater than either the sense, or the antisense strand (2). This led to the

Table 1
Commercial transfection lipids

Name	Company
Lipofectin	Invitrogen
Oligofectamine	Qiagen
Lipofectamine	Invitrogen
TransIT-TKO	Mirus
RNAifect	Mirus
Lipofect ACE	Life Technologies
Lipofection	Life Technologies
DMRIE-C	Invitrogen
Transfectam	Promega
TransFast	Promega
Trojene	Avanti Polar Lipids

recognition of RNA as being responsible for the "homology dependent gene silencing" (HDGS) (3) observed earlier. RNAi has been demonstrated to be involved in at least three distinct roles: antiviral defense (4), control of chromatin structure and function (5), and gene regulation (6).

The mechanism of RNAi-induced silencing is known to proceed via a two step process (7). First is the cleavage of long dsRNAs by the ribonuclease "Dicer" generating small interfering RNAs (siRNAs), 21–23 nucleotides in length (8), with 3′ dinucleotides overhangs. The second is messenger RNA (mRNA) degradation involving the formation of a "RNA-induced silencing complex" (RISC) (9). The RISC unwinds the two strands of the siRNA, and based on the relative stability of base-pairing at the two ends of the siRNA duplex, one strand is selected and incorporated into the RISC to form the "mature RISC." The strand incorporated into the RISC (guide or antisense strand) identifies a complementary target mRNA while the "sense" or "passenger" strand is degraded. RISC is reported to be a ~500 kDa complex containing the Argonaute-2 (Ago-2) protein, a DEAD-box helicase (Gemin-3), a protein of unknown function (Gemin-4) and dicer (a family of RNase III proteins containing dual RNase III domains and dsRNA-binding motifs). Dicer processes dsRNA into siRNAs and binds to Ago-2 via the PAZ domains (10).

RNAi technology has been widely used to sequence-selectively inhibit the expression of targeted genes in both cell culture and animal models (11). However, the introduction of dsRNAs longer than 30 base pairs in mammalian cells elicits a nonspecific interferon response (through protein kinase (PKR) and 2′,5′-oligoadenylate synthetase) leading to nonspecific gene silencing. This can be avoided by the use of siRNAs that resemble the natural products of Dicer.

1.2. Small Interfering RNA

In 2001, Elbashir et al. used synthetic siRNAs to demonstrate that they could mediate RNAi without requiring cleavage by Dicer, while avoiding the immune response generated against long dsRNA (12). siRNA mediated gene silencing results in highly efficient gene silencing in cell culture for up to about 96 h in HeLa cells (13). Variables such as siRNA efficacy, transfection efficiency, cell type and protein stability all contribute to observed differences in the effectiveness of siRNA-mediated RNAi.

siRNA 19–23 base pairs in length can bypass Dicer and induce RNAi by incorporating within RISC, thereby eliminating the PKR/interferon response (14). siRNAs may activate a sequence-independent interferon response (4). However, all siRNA may not be immunogenic under all conditions. Heidel et al. reported that chemically synthesized siRNAs do not elicit an immune response in mice, even when administered under conditions that permit them to be taken up by cells and perform detectable down-regulation of a target gene (15).

siRNAs are several-fold more potent inhibitors of gene expression than antisense oligonucleotides, DNAzymes, or ribozymes (16). Also siRNA provides exquisite specificity and allows allele-specific silencing in dominant familial disease where the disorder is caused by a point mutation. Despite rapid progress in the RNAi field and the undoubted therapeutic promise of this technology, there remains important questions to be resolved before clinical application is possible. The possibility of nonspecific effects on gene expression is a major concern. Also, given that the RNAi pathway can be saturated, it is conceivable that the natural functions of this pathway such as regulation of development and protection of the genome (transposon mobility) (17) could be usurped. siRNA may also be unable to distinguish between single point mutations.

The study of siRNA biochemistry has led researchers to believe that:

- RISC-mediated mRNA cleavage is a catalytic process with reactions in vitro showing the cleavage of ten target mRNAs per RISC complex (7).
- siRNAs are assembled asymmetrically into RISC complexes with destruction of one strand.
- The siRNA strand with most unstable duplex at 5′ terminus is preferentially assembled into RISC (18).

1.3. Synthetic siRNA

For gene down regulation, siRNA can be delivered either as a plasmid or viral vector expressing short hairpin RNA (shRNA) (19) or as in vitro-synthesized duplexes. Exogenously produced or intracellularly expressed siRNAs have recently been developed to trigger RNAi against specific targets in mammalian cells (12). siRNA prepared in vitro have the ability to be labeled (20) for analysis, while DNA-based vectors and cassettes that express siRNA within the cells are preferred for long term studies and where selectability (antibody) is required (21). These methods have proven to be quick, inexpensive and effective for knockdown experiments in vitro and in vivo (22). Another advantage that synthetic siRNA-based system has is that their site of action is the cytosol. Thus, they avoid the slow and inefficient process of uptake by the nucleus.

Xu et al. found duplex siRNAs to be more potent than single-stranded antisense siRNAs. They found that sequential transfection with the sense siRNAs lead to gene silencing while single-stranded antisense siRNAs were inefficient when used alone. Thus, they concluded that the structural character of siRNA molecules might be a more important determinant of siRNA efficiency than their cellular persistence (23). The synthetic siRNA first used by Elbashir et al. were 21-mer duplexes with phosphodiester bonds and 3′ dinucleotide overhangs (21). Since then, several chemical modifications of siRNA and/or complexation with a carrier/delivery particle have been researched to delay or avoid degradation (24). It has been demonstrated that the 5′ phosphate group on the antisense strand but not on the sense strand (25), and 5′ phosphodiester linkage on the antisense strand (26) are required for activity (mRNA cleavage).

Some of the chemical modifications that have been reported to be tolerated include:

- Modifications at the 3′ end of antisense strand (27, 28).
- Limited number of phosphorothioate ($P=S$) and Boranophosphonate ($P=B$) linkages (instead of phosphodiester ($P=O$)) (29, 30).
- Blunt 19-mer duplexes with extensive 2′-O-methyl modifications; blunt 25-mer duplexes with sense-strand 2′-O-methyl modifications (31); asymmetric 27-mer duplexes (32).
- Modification of the 2′ position of the ribose including 2′-O-methyl and 2′-deoxy-2′-fluoro have been shown to confer serum stability, while 2′-deoxy-2′-fluoro (2′-F) and locked nucleic acid (LNA where a methylene bridge connects the 2′-O with the 4′-C of the ribose) have been shown to improve target binding affinity.
- Conjugation of cholesterol to siRNA has been demonstrated to improve serum protein binding, improve pharmacokinetics,

and increase delivery to hepatocytes (33, 34). A dinitrophenol end modification has been shown to improve transfection and increase intracellular stability in tissue culture (35).

- siRNA terminals are most sensitive to nucleases; thus it may be preferable to limit the use of several modifications to the terminal positions (or 3′ overhangs).

1.4. microRNA

Similar to siRNA, microRNAs (miRNAs) were also discovered in *C. elegans* and are endogenous RNAs that mediate sequence specific gene silencing (36, 37). Despite their functional similarity to siRNAs, miRNAs differ from siRNAs in regard to their origin, evolutionary conservation, and the types of genes they silence. miRNAs have contributed significantly to the understanding of RNAi and its role in cellular and developmental biology.

Another mediator of gene downregulation is 29 nucleotide stem loop analogue of pri-miRNAs called short hairpin RNAs (shRNAs) (38). siRNA may be generated through in vitro transcription (IVT) methods generally done using bacteriophage promoters with (linearized) DNA templates and is cheaper than chemical synthesis. Research has demonstrated that plasmids that express siRNAs and shRNAs are suitable for RNAi. But synthetic siRNAs are still preferred for most common, transient applications.

siRNA holds therapeutic promise for silencing dominantly acting disease genes, particularly if mutant alleles can be targeted selectively. Allele-specific silencing of disease genes with siRNA can be achieved by targeting either a linked single-nucleotide polymorph (SNP) or the disease mutation directly. The ability to accomplish selective gene silencing has led to the hypothesis that siRNAs might be used to suppress gene expression for therapeutic benefit (39). Cancer is an attractive target for siRNA therapy due to the specificity of the mutated gene target within cancer cells and the relative ease of targeting these cells (40). RNAi has been demonstrated to also be effective against influenza virus (41), ocular angiogenesis (42) and hepatic C (43).

The delivery of preformed siRNA duplexes offers an attractive alternative to the introduction of large DNA or RNA which requires processing to become active siRNA. The delivery of siRNA into the cell faces similar problems to the delivery of most macromolecules intracellulary. Delivery vectors used to address larger nucleic acids have been adapted for the siRNA delivery and will be discussed below.

1.5. Lipoplexes

Lipoplexes are the formulation of a cationic lipid/nucleic acid mixture to form a nonviral vector for gene delivery. An advantage of a nonviral delivery pathway is the lowered chance of immunological response (44). Electrostatic interactions aid in the formation of these complexes and come from the amine groups of the lipid and the phosphate groups of the nucleic acid. A ratio of positive to negative charge exists, which can impact transfection efficiency. This ratio can vary from 1 to 10, or higher, and is named the nitrogen/phosphate ratio (N/P ratio). A slight positive charge is desirable for interaction with the negative cell membrane. Different ratios should be used in trials to determine the most effective formulation based on the application. The exact ratio is cell line and experiment specific, and requires optimization. N/P ratio can be determined by taking nmol of nitrogen in the lipid, divided by the nmol of phosphate in the RNA.

This type of delivery is used to aid in the internalization of the RNA molecules, usually by endocytosis at the cell membrane, with the release of the RNA component once inside of the endosome. However, currently, the exact mechanism is not known. Often, these vectors also include additional components such as polyethylene glycol conjugated lipid for increased in vivo circulation and cholesterol for added stability, in addition to a cationic lipid. One common cationic lipid used today is DOTAP (1,2-dioleoyl-3-trimethylammonium-propane) which provides a positive charge to interact with the negative phosphate groups of dsRNA. In addition to this, a helper lipid, such as DOPE (dioleoylphosphatidylethanolamine) is common. This helper lipid improves transfection efficiency, although the mechanism is not well understood. DOPE is currently the most common helper lipid in practice, but new combinations are being explored. Some other investigated helper lipids include:

- 1,2-Dilauroyl-*sn*-Glycero-3-Phosphoethanolamine (DLPE).
- 1,2-Dipalmitoyl-*sn*-Glycero-3-Phosphoethanolamine (DPPE).
- 1,2-Dimyristoyl-*sn*-Glycero-3-Phosphoethanolamine (DMPE).
- Diphytanoyl Phosphoethanolamine.
- 1,2-Distearoyl-*sn*-Glycero-3-Phosphoethanolamine (DSPE).
- Dilinoleoyl Phosphoethanolamine.
- Dielaidoyl Phosphoethanolamine.

2. Materials

2.1. Cell Culture

1. 10% fetal bovine serum (FBS, HyClone) in Dulbecco's Modified Eagle's Medium (DMEM).
2. 6–96 well plates.
3. Trypsin solution: 0.25% in sterile water.
4. HEPES buffered saline: 10 mM HEPES, 150 mM NaCl, pH 7.4.

5. Phosphate buffered saline (PBS): 10 mM Phosphate, 150 mM NaCl, 2.7 mM KCl, pH 7.4.
6. Cell culture flasks (BD Falcon).

2.2. Liposome Formation

1. Round bottom flasks of appropriate size (50–100 ml usually) or glass tubes, inert gas (nitrogen or argon), vacuum desiccator (see Note 1).
2. 1:1 molar ratio DOPE:DOTAP (1,2-Dioleoyl-*sn*-Glycero-3-Phosphoethanolamine:1,2-Dioleoyl-3-Trimethylammonium-Propane) liposomes are popular, but other lipids can be substituted.
3. Rehydration medium: PBS or HEPES which may contain dissolved material to be entrapped within the liposome.
4. Probe sonicator (Mandel Scientific Co., Inc.) or bath sonicator (Laboratory Supplies Co., Inc.)
5. Pressure extruder (Northern Lipids).
6. Size or charge characterization equipment such as Zetasizer (Malvern).

3. Methods

3.1. Preparation of Liposomes

1:1 molar DOTAP/DOPE liposomes are popular due to the simplicity of the mixture and the fact that they will spontaneously form lipoplexes with a negatively charged nucleic acid. Different molar ratios and different cationic lipids can be tested experimentally, depending on the application. A small molar fraction of PEG conjugated lipid can also be incorporated to aid with in vivo work (45). In addition to this, a targeting ligand may also be attached to the liposomal membrane to enhance cell specific binding. In most cases, the liposomes are prepared separately and mixed fresh with the nucleic acid to avoid aggregation. There are numerous methods for liposome preparation, which are amendable for the creation of transfection reagents. One of the simplest protocols is provided below.

3.2. DOTAP/DOPE Liposome Formation

1. Add 1:1 w/w ratio of DOTAP/DOPE to a round bottomed flask or tube. Dissolve the lipids in chloroform. For example, 10 mg of DOTAP and 10 mg of DOPE in an excess of chloroform (15–25 ml) (see Note 1).
2. Evaporate the organic solvent (usually chloroform) with inert gas or under vacuum. The process can be accelerated with the addition of mild heat.
3. Further removal of trace solvent can be performed for 4 h to overnight with a vacuum desiccator.

4. Rehydrate the lipid film with an aqueous solution to result in a concentration of 1 mg DOTAP per 10 ml, rehydration media (see Note 2).

4.1. The osmolarity should be equal to that of the medium you will be using to transfect cells, physiological fluids, or any other in vitro/in vivo mediums.

5. Size reduction can then be performed by either extrusion or sonication. A size of about 100 nm is usually preferred.

6. Allow the formulation to rehydrate overnight at 4°C before introducing the siRNA.

3.3. Ethanol–Calcium Method of Forming Neutral Lipoplexes

At times, charge can be an issue, since it can cause the vesicles that may have a high in vivo clearance to interact with proteins in the blood. Neutrally charged lipoplexes can be implemented to lower the toxicity, increase circulation time, and decrease interaction with proteins (46). Since the neutral charge allows these liposomes to circulate longer, they will be more likely to accumulate in tumor tissue where the vasculature lends itself to particle uptake and retention. However, the charge of a liposome is an important factor in complexing a negatively charged nucleic acid. Often, positively charged lipids must be used to obtain adequate levels of transfection (45).

The problem with using this type of approach is that smaller amounts of DNA will be entrapped in the liposome and the size of the liposome could increase dramatically, both unwanted effects. A way to combat this is to condense the DNA using a divalent cation as previously described by Bailey and Sullivan. In this case, Ca^{2+} is used because of the weaker binding to DNA. Also, this phenomenon is dependent on the dielectric constant of the buffer; therefore, ethanol is added to allow a lower concentration of Ca^{2+} to be used. DOPC liposomes are also used due to the higher stability when forming small unilamellar vesicles (SUVs) and their tendency to be in the liquid crystalline state at normal physiological temperatures. Other formulations that can be tested are DOPC:DOPE:Cholesterol (1,2-Dioleoyl-sn-Glycero-3-Phosphocholine, L-alpha-Dioleoyl Phosphatidylethanolamine) 1:1:1 or DOPC:DOPE 1:1, which are both neutral as well. Entrapment efficiency of 80% in SUVs has been reported using this method (47).

1. Form SUVs, as described in the protocol above, by thin film rehydration in 1–5 ml of 10 mM Tris, leaving a final concentration of 80 mM lipid (see Note 3).

    1.1. DOPC, DOPC:DOPE (1:1), DOPC:DOPE:Chol (1:1:1), or another formulation which yields a neutral charge can also be tried here. Roughly 300 mg of lipid will yield an 80 mM concentration when rehydrated with 5 ml media.

2. Probe sonicate at 40 W for 0.5–1 min, then centrifuge at between 18,000–20,000 ×g for 5 min to remove larger particles (see Note 4).

3. Mix 250 μl/20 μmol SUVs, 0.1 mg siRNA, and 10 mM Tris buffer to a final volume of 400 μl.

4. Add 600 μl of a calcium/100% ethanol/Tris 7.4 mixture dropwise, with high vortexing, to form aggregated complexes.

   4.1. Some experimental concentrations of these mixtures used for each lipid formulation are:

   4.1.1. DOPC, 35–50% ethanol, 0–5 mM Ca^{2+}

   4.1.2. DOPC/DOPE, 20–40% ethanol, 0–10 mM Ca^{2+}

   4.1.3. DOPC/DOPE/Chol, 35–45% ethanol, 5–15 mM Ca^{2+}

5. Then dialyze for 24 h with 500× volume or Tris 7.4 pH buffer, with two changes of buffer.

6. If osmolarity within a physiological range is needed, dialyze against 500× volume of PBS for 24 h (see Note 5).

3.4. Sonication

Several methods can be performed after the above protocol to obtain a certain size of liposomes for complexation with siRNA; probe sonication can be used to reduce the size of liposomes by introducing a mechanical rupture of membranes; however, probe sonication can create lipid breakdown or introduce pieces of titanium from the sonicator itself into the sample.

An alternative to titanium probe sonication is a bath sonicator (see Note 6). SUVs can be formed by sonication for 10 min to an hour. Instead of inserting a probe directly into the sample, a water bath is used to transfer the energy to a tube placed in the center. Glass or plastic tubes can be used and the sample can also be covered with an inert gas such as argon or nitrogen (48). This closed system cannot be accomplished with a probe sonicator. The liquid should go from cloudy to clear after being sonicated. It is an affordable way (under $1,000) to introduce the energy needed to reduce the size of liposomal formulations. These sonicators are available from Laboratory Supplies Company, New York (516-681-7711).

3.5. Extrusion

Another useful method of obtaining liposomes of a desired size is to pass them through membranes that contain pores of a defined size. These pores can range from more than 1 μm down to 50 nm. The size distribution of the liposomes can also be tightly controlled with several passes through the polycarbonate membrane (49). A simple extruder is sold by Avanti Polar Lipids (http://www.avantipolarlipids.com) and is a small handheld model (Fig. 1). This model allows the user to pass the solution back and forth through a membrane by using pressure applied by hand on a syringe. Extrusion should also be done above the phase transition temperature of the lipids you are using. This can also be accomplished by using a heating block here.

After about 10–15 passes through the membrane, the solution should appear clearer and have a size distribution close to the membrane pore size (50). Another type of extruder can be attached to a tank of inert gas and which will push the lipid solution through a chamber containing polycarbonate membranes (Fig. 2). Northern Lipids (http://www.northernlipids.com) sells

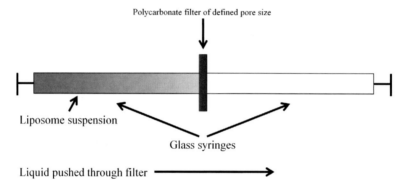

Fig. 1. A schematic of a mini hand extruder. The liquid is pushed back and forth through the filter. The apparatus can also be placed in a heating block

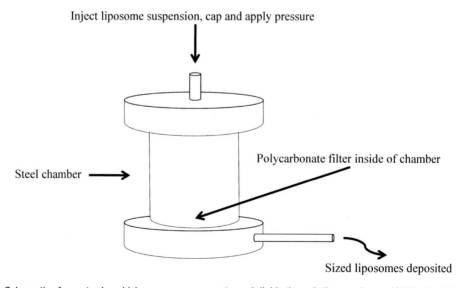

Fig. 2. Schematic of an extruder which uses gas pressure to push lipids through the membrane. Lipids placed in the top and pressure from the gas tank forces lipids through the membrane and out the bottom of the cylinder

the LIPEX Extruder for this purpose. This type of extruder is desirable due to the fact that you can keep constant pressures on your liposomes, there are no moving parts, and you can also cycle temperature controlled water through the unit (see Note 7). More detailed protocols and setup are included from the manufacturer for this type of equipment.

3.6. Lipoplex Formation

1. Suspend liposomes in serum free media at a ratio of 8:1 (w/w) to the oligonucleotide.
2. Determine the amount of RNA needed to obtain a 10–50 nM final concentration in the well for transfection. It is also possible to experimentally test several concentrations or N/P ratios in this step. The liposome concentration can be kept constant while varying the RNA amounts of visa versa (51).
   2.1. A control of naked siRNA can be used to determine the increase in transfection.
   2.2. 1 µg RNA per 10 µg lipid can be used for transfections.
3. Add siRNA to the liposomal suspension and incubate at room temperature for 10–15 min.
4. Additionally, the formulation may be passed through a 0.22 µm filter to sterilize.
5. Transfect cell by incubating in HEPES buffered saline, 3 ml mixture/100 mm culture plate at 37°C.

3.7. Transfection

1. Seed cell a day prior to the transfection in a multiwell plate.
   1.1. For a 6-well plate, seed cells in FBS supplemented wgrowth media at 2×10^5 cell/well.
2. Check cells the following day to make sure they have grown to about 40–70% confluency, as well as checking to ensure the cells are healthy.
3. Remove the growth medium and optionally wash the cells with PBS.
4. Add 1 ml transfection medium with the lipoplex to each well, ensuring that serum is not present at this step (adjust for different well sizes). Some lipids are maybe serum resistant which allows for use with serum containing media.
   4.1. The concentration of siRNA should be 20 µM, which is about 0.25 µg/µl.
5. Place back in incubator for 4–5 h. Transfection times can vary based on the cell line being used, proteins targeted, or formulation being tested.
6. Remove transfection medium and add normal growth medium.
7. Incubate for another 1–2 days and assay for gene silencing (see Note 8).

Several commercially available lipid formulations can be purchased. A few examples are given below in Table 1. They will contain their own protocol that can vary from the one found here.

4. Notes

1. Lipids and liposomes should be stored under inert gas, especially if double bonds are present in the lipid structure. The inert gas can also be used to evaporate the organic solvent from stock lipid solutions.
2. Rehydration can take quite a long time depending on the lipids involved. At times, gentle heating and agitation can aid in this process. Times from 1 h to overnight have been used to achieve good rehydration.
3. Saline can hinder the complexation of nucleic acids. Care should be taken to use buffers which will not interfere with this process.
4. Size of the liposomes should be between 100 and 200 nm for best circulation times in vivo, this can be checked by a dynamic light scattering method. The size dependency has to do with accumulation in the liver, ability to extravasate into tissues, and any possible immune reactions.
5. siRNA can be purchased commercially from companies such as Invitrogen, Sigma-Aldrich, Applied Biosystems, Ambion, and Open Biosystems. In many cases, specific sequences can be custom made or simply purchased premade. Many universities will also be able to provide custom made oligonucleotides.
6. Bath sonication is desirable because the high energy input of a probe sonication could disrupt the lipid structure because of the high heat. If probe sonication is performed, an ice bath should be used to keep the formulation from being heated excessively. Long periods of probe sonication should also be avoided, using a pulsing timer can help with this.
7. Lipids should be extruded above their phase transition temperature. Many extruders have a port for water to warm the unit. This is essential for the ability of the lipids to pass through the polycarbonate membrane. Extrusion also produces a much narrower size distribution which will become more uniform with multiple passes through the membrane. Membranes should also be used two at a time to produce the best results.
8. Luciferase gene expression is common as well as green fluorescent protein expression. These are some standard examples used to evaluate gene silencing.

References

1. Hannon GJ (2002) RNA interference. Nature 418(6894):244–251
2. Fire A, Xu S, Montgomery MK, Kostas SA, Driver SE, Mello CC (1998) Potent and specific genetic interference by double-stranded RNA in Caenorhabditis elegans. Nature 391(6669):806–811
3. Ruiz F, Vayssie L, Klotz C, Sperling L, Madeddu L (1998) Homology-dependent gene silencing in Paramecium. Mol Biol Cell 9(4):931–943
4. Hammond SM, Bernstein E, Beach D, Hannon GJ (2000) An RNA-directed nuclease mediates post-transcriptional gene silencing in Drosophila cells. Nature 404(6775):293–296
5. Elbashir SM, Lendeckel W, Tuschl T (2001) RNA interference is mediated by 21- and 22-nucleotide RNAs. Genes Dev 15(2):188–200
6. Hamilton AJ, Baulcombe DC (1999) A species of small antisense RNA in posttranscriptional gene silencing in plants. Science 286(5441):950–952
7. Zamore PD, Tuschl T, Sharp PA, Bartel DP (2000) RNAi: double-stranded RNA directs the ATP-dependent cleavage of mRNA at 21 to 23 nucleotide intervals. Cell 101(1):25–33
8. Bernstein E, Caudy AA, Hammond SM, Hannon GJ (2001) Role for a bidentate ribonuclease in the initiation step of RNA interference. Nature 409(6818):363–366
9. Martinez J, Patkaniowska A, Urlaub H, Luhrmann R, Tuschl T (2002) Single-stranded antisense siRNAs guide target RNA cleavage in RNAi. Cell 110(5):563–574
10. Hammond SM, Boettcher S, Caudy AA, Kobayashi R, Hannon GJ (2001) Argonaute2, a link between genetic and biochemical analyses of RNAi. Science 293(5532):1146–1150
11. Xia H, Mao Q, Paulson HL, Davidson BL (2002) siRNA-mediated gene silencing in vitro and in vivo. Nat Biotechnol 20(10):1006–1010
12. Elbashir SM, Harborth J, Lendeckel W, Yalcin A, Weber K, Tuschl T (2001) Duplexes of 21-nucleotide RNAs mediate RNA interference in cultured mammalian cells. Nature 411(6836):494–498
13. Brown D, Jarvis R, Pallotta V, Byrom M, Ford L (2002) RNA interference in mammalian cell culture: design, execution and analysis of the siRNA effect. Ambion TechNotes 9(1):3–5
14. Jorgensen R (1990) Altered gene expression in plants due to trans interactions between homologous genes. Trends Biotechnol 8(12):340–344
15. Heidel JD, Hu S, Liu XF, Triche TJ, Davis ME (2004) Lack of interferon response in animals to naked siRNAs. Nat Biotechnol 22(12):1579–1582
16. Khan A, Benboubetra M, Sayyed PZ, Ng KW, Fox S, Beck G, Benter IF, Akhtar S (2004) Sustained polymeric delivery of gene silencing antisense ODNs, siRNA, DNAzymes and ribozymes: in vitro and in vivo studies. J Drug Target 12(6):393–404
17. Kamath RS, Martinez-Campos M, Zipperlen P, Fraser AG, Ahringer J (2001) Effectiveness of specific RNA-mediated interference through ingested double-stranded RNA in Caenorhabditis elegans. Genome Biol 2(1):RESEARCH0002
18. Schwarz DS, Hutvagner G, Du T, Xu Z, Aronin N, Zamore PD (2003) Asymmetry in the assembly of the RNAi enzyme complex. Cell 115(2):199–208
19. Gilmore IR, Fox SP, Hollins AJ, Sohail M, Akhtar S (2004) The design and exogenous delivery of siRNA for post-transcriptional gene silencing. J Drug Target 12(6):315–340
20. Grunweller A, Gillen C, Erdmann VA, Kurreck J (2003) Cellular uptake and localization of a Cy3-labeled siRNA specific for the serine/threonine kinase Pim-1. Oligonucleotides 13(5):345–352
21. Kawasaki H, Miyagishi M, Taira K (2003) Development of siRNA libraries by in vitro dicing and optimized efficient expression vectors for siRNAs in mammalian cells. Tanpakushitsu Kakusan Koso 48(Suppl 11):1638–1645
22. Brummelkamp TR, Bernards R, Agami R (2002) A system for stable expression of short interfering RNAs in mammalian cells. Science 296(5567):550–553
23. Xu Y, Linde A, Larsson O, Thormeyer D, Elmen J, Wahlestedt C, Liang Z (2004) Functional comparison of single- and double-stranded siRNAs in mammalian cells. Biochem Biophys Res Commun 316(3):680–687
24. Morrissey DV, Lockridge JA, Shaw L, Blanchard K, Jensen K, Breen W, Hartsough K, Machemer L, Radka S, Jadhav V et al (2005) Potent and persistent in vivo anti-HBV activity of chemically modified siRNAs. Nat Biotechnol 23(8):1002–1007

25. Schwarz DS, Hutvagner G, Haley B, Zamore PD (2002) Evidence that siRNAs function as guides, not primers, in the Drosophila and human RNAi pathways. Mol Cell 10(3):537–548
26. Harborth J, Elbashir SM, Vandenburgh K, Manninga H, Scaringe SA, Weber K, Tuschl T (2003) Sequence, chemical, and structural variation of small interfering RNAs and short hairpin RNAs and the effect on mammalian gene silencing. Antisense Nucleic Acid Drug Dev 13(2):83–105
27. Chiu YL, Rana TM (2003) siRNA function in RNAi: a chemical modification analysis. RNA 9(9):1034–1048
28. Czauderna F, Fechtner M, Dames S, Aygun H, Klippel A, Pronk GJ, Giese K, Kaufmann J (2003) Structural variations and stabilising modifications of synthetic siRNAs in mammalian cells. Nucleic Acids Res 31(11):2705–2716
29. Hall AH, Wan J, Shaughnessy EE, Ramsay SB, Alexander KA (2004) RNA interference using boranophosphate siRNAs: structure-activity relationships. Nucleic Acids Res 32(20):5991–6000
30. Amarzguioui M, Holen T, Babaie E, Prydz H (2003) Tolerance for mutations and chemical modifications in a siRNA. Nucleic Acids Res 31(2):589–595
31. Chen D, Texada DE, Duggan C, Liang C, Reden TB, Kooragayala LM, Langford MP (2005) Surface calreticulin mediates muramyl dipeptide-induced apoptosis in RK13 cells. J Biol Chem 280(23):22425–22436
32. Rose SD, Kim DH, Amarzguioui M, Heidel JD, Collingwood MA, Davis ME, Rossi JJ, Behlke MA (2005) Functional polarity is introduced by Dicer processing of short substrate RNAs. Nucleic Acids Res 33(13):4140–4156
33. Lorenz C, Hadwiger P, John M, Vornlocher HP, Unverzagt C (2004) Steroid and lipid conjugates of siRNAs to enhance cellular uptake and gene silencing in liver cells. Bioorg Med Chem Lett 14(19):4975–4977
34. Soutschek J, Akinc A, Bramlage B, Charisse K, Constien R, Donoghue M, Elbashir S, Geick A, Hadwiger P, Harborth J et al (2004) Therapeutic silencing of an endogenous gene by systemic administration of modified siRNAs. Nature 432(7014):173–178
35. Liao H, Wang JH (2005) Biomembrane-permeable and ribonuclease-resistant siRNA with enhanced activity. Oligonucleotides 15(3):196–205
36. Basyuk E, Suavet F, Doglio A, Bordonne R, Bertrand E (2003) Human let-7 stem-loop precursors harbor features of RNase III cleavage products. Nucleic Acids Res 31(22):6593–6597
37. Lee Y, Ahn C, Han J, Choi H, Kim J, Yim J, Lee J, Provost P, Radmark O, Kim S et al (2003) The nuclear RNase III Drosha initiates microRNA processing. Nature 425(6956):415–419
38. Siolas D, Lerner C, Burchard J, Ge W, Linsley PS, Paddison PJ, Hannon GJ, Cleary MA (2005) Synthetic shRNAs as potent RNAi triggers. Nat Biotechnol 23(2):227–231
39. Brummelkamp TR, Bernards R, Agami R (2002) Stable suppression of tumorigenicity by virus-mediated RNA interference. Cancer Cell 2(3):243–247
40. Urban-Klein B, Werth S, Abuharbeid S, Czubayko F, Aigner A (2005) RNAi-mediated gene-targeting through systemic application of polyethylenimine (PEI)-complexed siRNA in vivo. Gene Ther 12(5):461–466
41. Ge Q, Filip L, Bai A, Nguyen T, Eisen HN, Chen J (2004) Inhibition of influenza virus production in virus-infected mice by RNA interference. Proc Natl Acad Sci USA 101(23):8676–8681
42. Kim B, Tang Q, Biswas PS, Xu J, Schiffelers RM, Xie FY, Ansari AM, Scaria PV, Woodle MC, Lu P et al (2004) Inhibition of ocular angiogenesis by siRNA targeting vascular endothelial growth factor pathway genes: therapeutic strategy for herpetic stromal keratitis. Am J Pathol 165(6):2177–2185
43. Wilson JA, Jayasena S, Khvorova A, Sabatinos S, Rodrigue-Gervais IG, Arya S, Sarangi F, Harris-Brandts M, Beaulieu S, Richardson CD (2003) RNA interference blocks gene expression and RNA synthesis from hepatitis C replicons propagated in human liver cells. Proc Natl Acad Sci USA 100(5):2783–2788
44. Daan JA, Hennink WE, Storm G (2002) Drug targeting system: fundamentals and applications to parenteral drug delivery In: Hillery AM, Lloyd AW, Swarbrick J (eds) Drug delivery and targeting for pharmacists and pharmaceutical scientists. Routledge, UK
45. Hillery AM, Lloyd AW, Swarbrick J (2002) Drug delivery and targeting for pharmacists and pharmaceutical scientists. Taylor & Francis, London, pp 118–143
46. Daan JA, Hennink WE, Storm G (2002) Drug targeting system: fundamentals and

applications to parenteral drug delivery In: Lloyd AW, Hillery AM, Swarbrick J (eds) drug delivery and targeting for pharmacists and pharmaceutical scientists. Routledge, UK

47. Bailey AL, Cullis PR (1997) Membrane fusion with cationic liposomes: effects of target membrane lipid composition. Biochemistry 36:1628–1634

48. Bailey A, Sullivan S (2000) Efficient encapsulation of DNA plasmids in small neutral liposomes induced by ethanol and calcium. Biochim Biophys Acta 1–2:239–252

49. Sims RPA, Larose JAG (1962) Use of iodine vapor as a general detecting agent in thin layer chromatography. J Am Oil Chem Soc 39:232

50. Magotoshi M, Samir SA, Noriaki T (1983) Size and permeability of liposomes extruded through polycarbonate membranes. Int J Pharm 17:215–224

51. Preeti SA, Wei W, Fuxing T, King MA, Meyer EM, Hughes JA (2001) Transgene delivery with a cationic lipid in the presence of amyloid β (βAP) peptide. Neurochem Res 23(3):195–202

Chapter 32

Complexation of siRNA and pDNA with Cationic Liposomes: The Important Aspects in Lipoplex Preparation

José Mario Barichello, Tatsuhiro Ishida, and Hiroshi Kiwada

Abstract

In the last two decades, cationic liposomes have been investigated as vehicles for nucleic acids [plasmid DNA (pDNA) and small interfering RNA (siRNA)] delivery in vitro and in vivo. The formation of cationic liposomes–nucleic acids complexes, termed lipoplexes, depends on a number of experimental variables. The quality of the nucleic acid and the cationic liposome as well as the selection of diluents for diluting the concentrated stocks strongly affect the resulting lipoplexes and their efficiency of gene-expression or gene-silencing effect following transfection. In addition, the molar ratio of cationic lipid nitrogen (N) to siRNA or pDNA phosphate (P) (N/P ratio) influences the final characteristics of the lipoplexes, such as size, surface zeta potential, and reproducibility, thereby reflecting their efficiency following transfection. The methods presented in this chapter could be helpful to obtain reliable and reproducible lipoplexes and experimental results.

Key words: Cationic liposome, siRNA, pDNA, Complex formation, siRNA-lipoplex, pDNA-lipoplex, N/P ratio, Lipoplex formation under vortex-mixing, Spontaneous lipoplex formation

1. Introduction

Gene therapy has gained rapid momentum as a new modality for treating several diseases such as cancer, infection, and hereditary disorders (1). More recently, the discovery of the RNA interference (RNAi) mechanism in mammalian cells revolutionized the field of functional genomics (2). The ability to simply, effectively, and specifically downregulate the expression of genes in mammalian cells by small interfering RNA (siRNA) holds enormous scientific, commercial, and therapeutic potential. Nevertheless, the success of gene therapies is predicated on the development of sufficient and safe delivery vector.

Cationic liposome is one of most used nonviral vectors for pDNA, ODNs, and siRNA in vitro and in vivo (3–5). Cationic liposomes are constituted of cationic lipids and helper lipids, such as DOPE and cholesterol, and have a unilamellar structure with a positive surface charge at neutral pH (pH 7.4). The cationic liposomes offer the attractive ability of complexing with nucleic acid molecules such as pDNA, ODNs, and siRNA. Cationic liposomes–nucleic acids complexes, termed lipoplexes, are basically formed through the spontaneous electrostatic interaction between the positively charged liposome with the negatively charged phosphate backbone of the nucleic acid (6). The resulting lipoplex can prevent nucleic acid molecules from degradation by metabolic enzymes, such as DNase or RNase, and overcome the electrostatic repulsion of the cell membrane, resulting in enhanced uptake by the cell (1, 6–12).

The molar ratio of cationic liposome to nucleic acid determines the proportion of electrostatic neutralization, which reflects the entire surface charge and the size of resulting lipoplexes (13). Therefore, lipoplex formation should be affected by experimental variables such as nucleic acid/cationic liposome concentration, time and medium for the complexation, the number and/or order of addition steps, and the presence of serum during lipoplex formation. In this section, we will present instructions to form lipoplexes and discuss the most important aspects to be considered in siRNA- or pDNA-lipoplex formation.

2. Materials

2.1. siRNA and pDNA Stocks and Dilutions

1. siRNA and/or pDNA of interest.
2. TE buffer, pH 8.0: 10 mM Tris–HCl, 1 mM EDTA. Store at room temperature. (See Note 1).
3. Phosphate-buffered saline (PBS): 137 mM NaCl, 2.7 mM KCl, 4.3 mM Na_2HPO_4 and 1.47 mM KH_2PO_4. The final pH is adjusted to 7.4. (See Note 1).
4. Doubled distilled water. (See Note 1).
5. Culture medium (Table 1). (See Note 1).
6. Microtubes of flat top. (See Note 2).
7. BD Falcon™ Conical Tubes (Franklin Lakes, NJ, USA). (See Note 2).

2.2. Cationic Liposome Stock and Dilution

1. Cationic liposome of interest.
2. Doubled distilled water. (See Note 1).
3. 150 nm Sodium chloride solution (See Note 1).
4. Culture medium (Table 1). (See Note 1).

Table 1
Diluents routinely used in siRNA- and pDNA-lipoplex preparation

Description

Dulbecco's modified eagle medium (D-MEM)[a,b]
Opti-MEM I reduced serum medium[a,b]
Minimum essential medium (MEM)[a,b]
Minimum essential medium eagle (S-MEM)[a,b]
Endothelial cell basal medium (EBM-2)[a,b]
F-12 Ham's nutrient mixture (F-12)[a,b]
DMEM/Ham's F12[a,b]
RPMI-1640[a,b]
Phosphate buffer saline (PBS), pH 7.4: 137 mM NaCl, 2.7 mM KCl, 4.3 mM Na_2HPO_4 and 1.47 mM KH_2PO_4[b]
TE buffer, pH 8.0: 10 mM Tris–HCl, 1 mM EDTA[b]
Sodium chloride solution (0.150 M)[b]
9% sucrose solution[b]
Potassium chloride solution, 0.075 M[b]

[a]Media are composed of a mixture of essential salts, nutrients, and buffering agents. Sterile media are usually purchased in solution. Alternatively, packaged premixed powders are available. Powdered media and concentrated formulations usually do not contain sodium bicarbonate
[b]It must be sterile and the storage condition of solutions after opened is at 2°C to 8°C (+4°C)

5. Microtubes of flat top. (See Note 2).
6. BD Falcon™ Conical Tubes. (See Note 2).

2.3. Complex Formation

1. Culture medium (Table 1). (See Note 1).
2. 9% Sucrose solution. (See Note 1).
3. Microtubes of flat top. (See Note 2).
4. BD Falcon™ Conical Tubes. (See Note 2).

2.4. Gel Electrophoresis

1. Powdered agarose for routine use. Store at room temperature.
2. Running Tris–Borate–EDTA (TBE) buffer (1×): 45 nM Tris-Borate, 1 mM Na_2EDTA; pH 8.2. Store at room temperature.
3. Ethidium bromide (EtBr) solution (10 mg/l in water). Store at 2–8°C (+4°C). EtBr is sensitive to light, therefore uses an aluminum foil to protect the amber glass flask.
4. Doubled distilled water. (See Note 1).

3. Methods

During lipoplex formation, the positively charged head group of lipid interacts with the phosphate backbone of the nucleic acid (6, 7). Like any association process, the thermodynamic driving force for spontaneous lipoplex formation is the lowering in total free energy of the lipoplexes when compared with those of the liposomes and the nucleic acid (6, 14). Controlling the parameters contributing to lipoplex formation is important to obtain reliable and reproducible lipoplexes and experimental results. In this regard, the quality of the nucleic acid and cationic liposome as well as the selection of diluents for diluting the concentrated stocks strongly affects the resulting lipoplexes. In addition, the molar ratio of cationic lipid nitrogen (N) to siRNA or pDNA phosphate (P) (N/P ratio) influences the final characteristics of the lipoplexes, such as size, surface zeta potential, and reproducibility, and thereby reflects the efficiency of gene-expression or gene-silencing effect following transfection (6, 7, 13–15).

For preparation of lipoplexes, first of all, concentrated stocks of nucleic acid and cationic liposome are prepared and stored at optimal condition (−80°C to 4°C). Then, the concentrated stocks are diluted with required volume of diluent according to desired N/P ratio. An example calculation of N/P ratio is presented. Then, the diluted nucleic acids and liposomes are mixed and stand at room temperature for desired time to allow the lipoplex formation. The formation and stability of complexes can be assessed with the gel retardation assay.

3.1. siRNA or pDNA Stocks

1. If siRNA or pDNA are lyophilized, those must be hydrated prior to use. To hydrate those, investigators should follow the instructions provided with the siRNA or pDNA. (See Note 3).

2. If no instructions were provided, siRNA or pDNA should be hydrated with PBS buffer (pH 7.4), doubled distilled water (for short-term storage), culture medium, or various buffers to a convenient stock concentration (i.e., 20–100 µM) (Table 1). (See Notes 1 and 4).

3. Determine concentration of siRNA or pDNA after dilution as follows: Dilute the sample containing siRNA or pDNA properly and measure the A_{260} against blank (diluent used). One A_{260} unit of siRNA corresponds to ~40 µg/ml, while one A_{260} unit of pDNA corresponds to ~50 µg/ml. (See Note 5).

4. As an example for determining the concentration, siRNA is used. Use 100 mM NaCl, TE buffer (pH 8.0) for blank and for siRNA dilutions, respectively. The A_{260} of a 1:10 dilution of concentrated siRNA stock = 1.7509. Therefore, the concentration is $1.7509 \times 40\,\mu g/ml \times 10$ (Dilution Factor) = 700.36 µg/ml.

5. Determine molar concentration (µg/nmol) of siRNA stock. The average molecular weight of a nucleotide is 333.5 or 0.3335 µg/nmol.

    (a) Multiply the number of nucleotides (nt)×2 strands=total number nt×0.3335 µg/nmol = x µg/nmol. Example of 21 nt siRNA: 21×2=42×0.3335 µg/nmol=14.007 µg/nmol.

    (b) Divide the concentration of siRNA (µg/ml) by the number of nucleotides (µg/nmol) to determine siRNA molar concentration (µM). Example: 700.36 (µg/ml)/14.007 (µg/nmol) = 50 (nmol/ml) or 50 µM (0.665 µg/µl).

6. Split aliquot of the siRNA or pDNA solution (100 µl) into small volume tubes (0.2 to 0.5 ml) and store them as concentrated stock at −20°C to −80°C. For best results, limit freeze-thaw events of each tube to no more than three.

3.2. Cationic Liposome Stock

1. If cationic liposomes are lyophilized, those must be hydrated prior to use. To hydrate, investigators should follow the instructions provided with the cationic liposome. Store at 2–8°C (+4°C). (See Notes 3 and 4).

2. If cationic liposomes are freshly prepared. Adjust it to a convenient stock concentration using PBS buffer (pH 7.4), sucrose solution, potassium chloride solution, sodium chloride solution, or double water (for short-term storage) (Table 1). (See Note 6).

3. These instructions use a cationic liposome stock concentration of 10 mM, which corresponds to 4 µmol/ml of cationic lipid. Always mix the cationic liposome with an appropriate diluent before adding siRNA and/or pDNA solutions.

3.3. siRNA or pDNA Dilutions

1. Calculate the required volume of concentrated siRNA or pDNA stock according to desired N/P ratio, and the total volume required before preparing the dilutions. The example calculation uses a final concentration of 75 nM of siRNA. The required volume of siRNA concentrated stock is calculated using the total volume of siRNA-lipoplex required (1,000 µl). This volume represents the siRNA dilution plus the cationic liposome dilution. Therefore, the final volume of siRNA dilution is 500 µl (Table 2) (See Note 7).

2. To determine the required volume (µl) of siRNA concentrated stock, multiply the required volume of lipoplex (µl) by the final concentration of siRNA (µM) and divide it by the concentration of siRNA stock (µM).

$$\text{Example:} \frac{1{,}000\,(\mu l) \times 0.075\,(\mu M)}{50\,(\mu M)} = 1.5\,\mu l$$

Table 2
Volume of concentrated stocks (cationic liposome and siRNA) calculated according to desired N/P ratio for preparing their dilutions and the siRNA-lipoplex at a desired N/P ratio

N/P ratio	Cationic liposome[a]		siRNA[a]	
	Stock (μl)	Diluent volume (μl)	Stock (μl)	Diluent volume (μl)
20.0	15.0	485.0	1.5	498.5
6.0	4.5	495.5	1.5	498.5
0.8	0.6	499.4	1.5	498.5

[a]In this example of calculation, the amount of cationic liposome was varied, while maintaining the same concentration of siRNA

The diluting procedure can be varied and separated in two or three steps when the final concentration of siRNA required is much lower than that of the concentrated stocks

3. Prepare the siRNA or pDNA dilution in an appropriate diluent suggested in Table 1. (See Notes 1 and 8). Add the required volume of diluent into appropriated microtubes or BD Falcon™ conical tubes. (See Note 9).

4. Mix the siRNA or pDNA concentrated stock gently before use and add the required volume to those microtubes or BD Falcon™ conical tubes containing diluent. Mix thoroughly by gentle pipetting up and down few times, or by gentle vortexing for 10 s.

3.4. Calculation of N/P Ratio

1. The N/P ratio represents the number of equivalents that forms the lipoplex, wherein 1 μg of the siRNA or pDNA equals 3 nmol of phosphate (15–17), and one equivalent is the amount of cationic liposome (nitrogen resides) required to neutralize the negative charges of siRNA or pDNA phosphate groups.

2. The volume of cationic liposome stock necessary to neutralize the negative charges of 1 μg of nucleic acid will depend on the amount of cationic lipid and the concentration of lipids in cationic liposome stock (See Note 10).

3. Therefore, the following calculation can be used to determine the required volume of cationic lipid from any stock cationic liposome suspension (adapted from (15, 16)):

$$\text{Volume cationic liposome stock } (\mu l) = \frac{(\mu g \text{ of siRNA (or pDNA)} \times \text{nmol of phosphate per } \mu g \times \text{desired N/P ratio})}{\text{nmol of cationic lipid per } \mu l \text{ of cationic liposome stock}}$$

4. In the example calculation, siRNA-lipoplexes are prepared at various N/P ratios (Table 2), wherein 1 µl of a 10 mM cationic liposome stock contains 4 nmol of cationic lipid (nitrogen residues), the desired N/P ratio is 20 and the siRNA amount is 1 µg. Thus, the calculation is:

$$\text{Volume cationic liposome stock } (\mu l) = \frac{(1 \times 3 \times 20)}{4} = 15\,\mu l$$

3.5. Cationic Liposome Dilutions

1. Calculate the required volume of cationic liposome concentrated stock according to desired N/P ratio, and the total volume required before preparing the dilutions.

2. In the example calculation, the total volume of lipoplex (1,000 µl) represents the siRNA dilution plus the cationic liposome dilution. Therefore, the final volume of cationic liposome dilution is 500 µl. (See Note 7).

3. Prepare the cationic liposome dilution in an appropriate diluent suggested in Table 1. (See Notes 1 and 8). Add the required volume of diluent into appropriated microtubes or BD Falcon™ conical tubes. (See Note 9).

4. Mix the cationic liposome concentrated stock gently before use and add the required volume of cationic liposome concentrated stock (Table 2) to those microtubes or BD Falcon™ conical tubes containing diluent. Do not let concentrated stock cationic liposome suspension touch any plastic other than the pipette tip. Mix gently by pippeting few times, or by gentle vortexing for 10 s, and let it stand for no more than 20 min before preparing the complex.

3.6. Lipoplex Formation

3.6.1. Spontaneous

1. This is a general procedure for lipoplex formation based on the electrostatic interaction between the positively charged lipid in the liposome and the negatively charged phosphate backbone of the nucleic acid (pDNA or siRNA).

2. Combine the diluted nucleic acid solution with the diluted cationic liposome prepared in separated tubes (in the example calculation, the total volume is 1,000 µl). Combine the dilutions in the prescribed order of protocol, since the order of dilution addition is important to achieve the optimal results (See Note 11). Mix by pipetting carefully up and down few times, or by gentle vortexing for 10 s to avoid precipitation, and let it stand for 10–45 min at room temperature to allow the nucleic acid-lipoplex formation. Depending on the concentration of nucleic acid and cationic liposome, the solution may appear cloudy. (See Note 12).

3. Lipoplexes are usually stable for up to 6 h, but it is strongly recommended its use immediately after preparation. Longer incubation times may decrease activity of nucleic acid.

3.6.2. Under Vortex-Mixing

1. This method is suggested as an alternative method for siRNA-lipoplex formation, when the spontaneous formation of lipoplex did not produce an effective gene silencing effect (See Note 13). It is important to note that application of vortex-mixing to form pDNA-lipoplex is not practicable, since it may damage the structure of pDNA.

2. Combine the diluted siRNA solution with the diluted cationic liposome prepared in separated tubes (in the example calculation, the total volume is 1,000 μl). Combine the dilutions in the prescribed order of protocol, since the order of dilution addition is important to achieve the optimal results (See Note 11). Mix immediately by applying a strong vortex-mixing (>2,500 rpm) for 10 min to allow siRNA-lipoplex formation. (See Note 12).

3. Lipoplexes are usually stable for more than 6 h, but it is strongly recommended its use immediately after preparation. Longer incubation times may decrease activity of siRNA.

3.7. Gel Electrophoresis

1. Gel retardation is a method widely used to assess the formation and stability of lipoplexes between siRNA or pDNA and cationic liposomes (18). Their complexation leads to the formation of a large complex which is unable to migrate toward the anode during electrophoresis in an agarose gel.

2. Prepare a 1.0-mm thick, 2% gel by dissolving the needed amount of powdered agarose in TBE buffer in a glass flask (Becker or Erlenmeyer). Heat the agarose solution in a microwave oven or in water bath to allow all of the grains of agarose to dissolve. (See Note 14).

3. Cool down the solution to 50°C. EtBr (final concentration of 0.5 μg/ml) can be added to the agarose solution in this step or you can submerge the gel in an EtBr solution once it solidifies.

4. The open borders of a clean and dry glass or acrylic plate should be sealed with tape for forming a mold. A comb of 0.5–1.0 mm above the plate is positioned in order to permit the formation of a complete well when the agarose solidifies. (See Note 15). Using a Pasteur pipette seal the glass plate with small amounts of agarose solution. Once the seals are set, pour the gel in the glass plate.

5. Remove the comb and the tape when the gel has completely hardened (20–40 min at room temperature). Place the gel in the electrophoresis tank and add enough TBE buffer to the tank to cover the gel (about 1 mm of depth). The top of the wells should be submerged in TBE buffer.

6. If dilution is necessary, mix the lipoplexes prepared as described above and standards with TBE buffer. Slowly load

the mixture or the original lipoplex suspension into the wells with a micropipette. (See Note 16).

7. Close the lid of the tank and be sure that samples are correctly positioned with respect to the anode and the cathode (naked siRNA and pDNA will migrate toward the anode). Apply the desired voltage (1–5 V/cm) to the gel to begin the electrophoresis. (See Note 17).

8. Visualize and photograph the agarose gel stained with ethidium bromide under transillumination at 300 nm (UV light). The nucleic acid-lipoplex should remain inside the well, while the free or weakly bound nucleic acid should run in the gel. An example of gel retardation assay is shown in Fig. 2.

4. Notes

1. Solutions (including doubled distilled water, buffer and culture medium) should be sterile and DNase- and RNase-free grade and stored at 2–8°C (+4°C) after opened. Unless stated otherwise, all solutions should be prepared in water that has a resistivity of 18.2 MΩ cm and total organic content of less than five parts per billion. This standard is referred to as "doubled distilled water" in this text.

2. Ensure that all glassware and plasticware are DNase- and RNase-free grade. The use of DNase- and RNase-free sterile, disposable, plastic tubes is recommended throughout the procedure.

3. When investigators work with chemicals and nucleic acids, they always wear a suitable lab coat, disposable gloves, and protective goggles. Perform all steps of protocol in a laminar flow cell culture hood using sterile techniques.

4. siRNA or pDNA solution might be dissolved in buffer before lyophilized. In this case, only double distilled water should be used as diluent.

5. Quality of siRNA and pDNA strongly influences the lipoplex formation and characteristics of resulting lipoplexes such as size, zeta potential, and reproducibility. It is recommended that OD_{260}/OD_{280} ratio is 1.8 or greater. Therefore, for executing the procedures described above, only siRNA and pDNA with the higher quality should be used. Purification with specific kits is highly recommended.

6. It is not recommended to use culture medium in cationic liposome concentrated stock. They are composed of a mixture of essential salts, nutrients, and buffering agents that will inhibit the complex formation ability of siRNA and pDNA.

7. It is recommended the preparation of 10% more dilution than is required to allow for loss during pipetting, i.e., for a dilution of 500 µl prepare enough dilution for 550 µl.
8. For in vivo experiments, the buffer should be adjusted to the condition.
9. It is not recommended using media containing serum, antibiotics, or proteins during either dilutions or lipoplex formation, as they may inhibit the complexation process.
10. During calculation of the N/P ratio. The number of equivalents required to prepare nucleic acid-lipoplexes of adequate properties depends on the cationic lipid and the cationic liposome formulation. Therefore, optimization of the number of equivalents may further increase the gene-expression or gene-silencing in your particular application.
11. A major influence on the lipofection level was found when the mode of lipoplex preparation was varied [lamellarity of the vesicles (multilamellar large vesicles or large unilamellar vesicles), mixing order, and number of mixing steps]. Mixing plasmid DNA and cationic large unilamellar vesicles (1:1) in two steps instead of one step resulted in a higher lipofection, when at the first step the DNA/cationic lipid mole ratio was 0.5 than when it was 2.0. Only static light-scattering measurement, which is related to particle size and particle size instability, revealed differences between the lipoplexes as a function of (19).
12. If precipitate forms after adding transfection reagent, certify that the cationic liposome/pDNA or siRNA concentration is not too high or serum is present during formation of complex. Solve these problems by increasing the volume of serum-free water, buffer, or medium and by using only serum-free medium during formation of complex.
13. This siRNA-lipoplex formation method was developed for improving the gene silencing efficiency of a cationic liposome, LipoTrust™. Figures 1 and 2 illustrate how the application of vortex-mixing during lipoplex formation affected the gene silencing efficiency of siRNA.
14. The buffer should not occupy more than 50% of the volume of the flask. If part of the buffer evaporated during the heating, bring the solution back to the original volume through the addition of double distilled water.
15. It is of crucial importance to avoid air bubbles under or between the teeth of the comb.
16. The volume (5–20 µl) of lipoplex sample should be adequate to the well size in order to avoid any mix of the samples between wells. Dye can be added to lipoplex and standard samples in order to readily determine the running and the ending point.

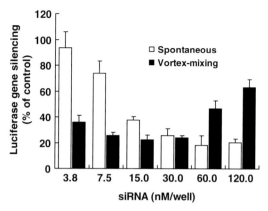

Fig. 1. Effect of vortex speed (2,500 rpm for 10 min) on the luciferase gene silencing efficiency of a siRNA-LipoTrust™ lipoplex in HeLa cells. The amount of cationic lipid was 9.6 μM. The siRNA doses correspond to the cationic lipid$^+$/siRNA$^-$ charge ratio of 30.45, 15.24, 7.62, 3.81, 1.90 and 0.95, respectively. LipoTrust™ is constituted of dioleoylphosphatidylethanolamine, cholesterol and the cationic lipid O,O'-ditetradecanoyl-N-(α-trimethyl ammonioacetyl) diethanolamine chloride (DC-6-14) in the molar ratio of 0.75/0.75/1.00. For this cationic liposome, an expressive gene silencing effect in vitro was obtained at lower siRNA dose with application of a higher vortex-mixing during complex formation

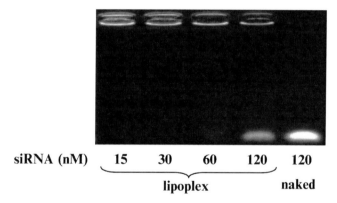

Fig. 2. Gel retardation assay of naked siRNA and siRNA-LipoTrust™ lipoplexes formed spontaneously in a 2% agarose gel. The amount of cationic lipid was 9.6 μM. The siRNA doses correspond to the cationic lipid$^+$/siRNA$^-$ charge ratio of 0.95, 1.90, 3.81 and 7.62, respectively. LipoTrust™ is constituted of dioleoylphosphatidylethanolamine, cholesterol and the cationic lipid O,O'-ditetradecanoyl-N-(α-trimethyl ammonioacetyl) diethanolamine chloride (DC-6-14) in the molar ratio of 0.75/0.75/1.00

17. If the apparatus is working, bubbles should form in the buffer when the electric field is applied. Besides, dyes should run with time after electrophoresis is started. Endurance and voltage of the electrophoresis depends on the nature of the sample run and the desired resolution.

Acknowledgments

The research was supported in part by the Health and Labour Sciences Research Grants for Research on Advanced Medical Technology from The Ministry of Health, Labour and Welfare of Japan. We thank the Japan Association for the Advancement of Medical Equipment for supporting the Postdoctoral Fellowship for Dr. Jose Mario Barichello.

References

1. Templeton NS, Lasic DD, Frederik PM, Strey HH, Roberts DD, Pavlakis GN (1997) Improved DNA: liposome complexes for increased systemic delivery and gene expression. Nat Biotechnol 15:647–652
2. Elbashir SM, Harborth J, Lendeckel W, Yalcin A, Weber K, Tuschl T (2001) Duplexes of 21-nucleotide RNAs mediate RNA interference in cultured mammalian cells. Nature 411:494–498
3. Dass CR, Choong PF (2006) Selective gene delivery for cancer therapy using cationic liposomes: in vivo proof of applicability. J Control Release 113:155–163
4. Zhang S, Zhao B, Jiang H, Wang B, Ma B (2007) Cationic lipids and polymers mediated vectors for delivery of siRNA. J Control Release 123:1–10
5. Garcia-Chaumont C, Seksek O, Grzybowska J, Borowski E, Bolard J (2000) Delivery systems for antisense oligonucleotides. Pharmacol Ther 87:255–277
6. Lasic DD, Templeton NS (1996) Liposome in gene therapy. Adv Drug Deliv Rev 20:221–266
7. Lima MCP, Simoes S, Pires P, Faneca H, Duzgunes N (2001) Cationic lipid-DNA complexes in gene delivery: from biophysics to biological applications. Adv Drug Deliv Rev 47:277–294
8. Felgner JH, Kumar R, Sridhar CN, Wheeler CJ, Tsai YJ, Border R et al (1994) Enhanced gene delivery and mechanism studies with a novel series of cationic lipid formulations. J Biol Chem 269:2550–2561
9. Sternberg B, Hong K, Zheng W, Papahadjopoulos D (1998) Ultrastructural characterization of cationic liposome-DNA complexes showing enhanced stability in serum and high transfection activity in vivo. Biochim Biophys Acta 1375:23–35
10. Li CX, Parker A, Menocal E, Xiang S, Borodyansky L, Fruehauf JH (2006) Delivery of RNA interference. Cell Cycle 5:2103–2109
11. Spagnou S, Miller AD, Keller M (2004) Lipidic carriers of siRNA: differences in the formulation, cellular uptake, and delivery with plasmid DNA. Biochemistry 43:13348–13356
12. Li W, Ishida T, Tachibana R, Almofti MR, Wang X, Kiwada H (2004) Cell type-specific gene expression, mediated by TFL-3, a cationic liposomal vector, is controlled by a post-transcription process of delivered plasmid DNA. Int J Pharm 276:67–74
13. Almofti MR, Harashima H, Shinohara Y, Almofti A, Li W, Kiwada H (2003) Lipoplex size determines lipofection efficiency with or without serum. Mol Membr Biol 20:35–43
14. Hirsch-Lerner D, Zhang M, Eliyahu H, Ferrari ME, Wheeler CJ, Barenholz Y (2005) Effect of "helper lipid" on lipoplex electrostatics. Biochim Biophys Acta 1714:71–84
15. Hofland HEJ, Shephard L, Sullivan SM (1996) Formation of stable cationic lipid/DNA complexes for gene transfer. Proc Natl Acad Sci USA 93:7305–7309
16. Boussif O, Lezoualc'h F, Zanta MA, Mergny MD, Scherman D, Demeneix B, Behr J-P (1995) A versatile vector for gene and oligonucleotide transfer into cells in culture and in vivo: Polyethylenimide. Proc Natl Acad Sci USA 92:7297–7301
17. Doyle SR, Chan CK (2007) Differential intracellular distribution of DNA complexed with polyethylenimine (PEI) and PEI-polyarginine PTD influences exogenous gene expression within live COS-7 cells. Genet Vaccines Ther 5:11
18. Lleres D, Weibel JM, Heissler D, Zuber G, Duportail G, Mely Y (2004) Dependence of the cellular internalization and transfection efficiency on the structure and physicochemical properties of cationic detergent/DNA/liposomes. J Gene Med 6:415–428
19. Zuidam NJ, Hirsch-Lerner D, Margulies S, Barenholz Y (1999) Lamellarity of cationic liposomes and mode of preparation of lipoplexes affect transfection efficiency. Biochim Biophys Acta 1419:207–220

Chapter 33

Effective In Vitro and In Vivo Gene Delivery by the Combination of Liposomal Bubbles (Bubble Liposomes) and Ultrasound Exposure

Ryo Suzuki and Kazuo Maruyama

Abstract

Gene delivery with a physical mechanism using ultrasound (US) and nano/microbubbles is expected as an ideal system in terms of delivering plasmid DNA noninvasively into a specific target site. We developed novel liposomal bubbles (Bubble liposomes (BLs)) containing the lipid nanobubbles of perfluoropropane which were utilized for contrast enhancement in ultrasonography. BLs were smaller in diameter than conventional microbubbles and induced cavitation upon exposure ultrasound. In addition, when coupled with US exposure, BLs could deliver plasmid DNA into various types of cells in vitro and in vivo. The transfection efficiency with BLs and US was higher than that with conventional lipofection method. Therefore, the combination of BLs and US might be an efficient and novel nonviral gene delivery system.

Key words: Liposomes, Nanobubbles, Gene delivery, Ultrasound, Noninvasive, Nonviral vector

1. Introduction

Ultrasound (US) has been utilized as a useful tool for in vivo imaging, destruction of renal calculus and treatment for fibroid of the uterus. It was reported that US was proved to increase permeability of the plasma membrane and reduce the thickness of the unstirred layer of the cell surface, which encourages the DNA entry into cells (1, 2). The first studies applying ultrasound for gene delivery used frequencies in the range of 20–50 kHz (1, 3). However, these frequencies, along with cavitation, are also known to induce tissue damage if not properly controlled (4–6). To improve this problem, many studies using therapeutic ultrasound for gene delivery, which operates at frequencies of

1–3 MHz, intensities of 0.5–2.5 W/cm², and pulse-mode have emerged (7–9). In addition, it was reported that the combination of therapeutic US and microbubble echo contrast agents could enhance gene transfection efficiency (10–14). In the sonoporation with microbubbles, it was reported that estimates of pore size based on the physical diameter of maker compounds were most commonly in the range of 30–100 nm, and estimates of membrane recovery time ranged from a few seconds to a few minutes (15). Therefore, it is thought that plasmid DNA is effectively and directly transferred into the cytosol via these pores. Conventional microbubbles including US contrast agents based on protein microspheres and sugar microbubbles are commercially available, the size of these bubbles being about 1–6 μm (16). For example, although the mean diameter of Optison microbubbles is about 2.0–4.5 μm, and they contain bubbles of up to 32 μm in diameter. Tsunoda et al. reported that some mice died immediately after the i.v. injection of Optison without ultrasound exposure due to lethal embolisms in vital organs (17). The same problem has not been reported in humans, but there is the possibility that Optison can not pass through capillary vessels. Therefore, microbubbles should generally be smaller than red blood cells. From this stand point of view, it is necessary to develop novel bubbles which are smaller than conventional microbubbles. Using liposome technology, we developed novel liposomal bubbles containing perfluoropropane gas. We called these bubbles "Bubble liposomes (BLs)." BLs were smaller than Optison (18–21). In addition, BLs could effectively deliver plasmid DNA by the combination with US exposure in vitro and in vivo.

2. Materials

2.1. Preparation of BLs (18)

1. 1,2-distearoyl-sn-glycero-phosphatidylcholone (DSPC) and 1,2-distearoyl-sn-glycero-3-phosphatidyl-ethanolamine-methoxypolyethyleneglycol (DSPE-PEG(2 k)-OMe) (NOF corporation, Tokyo, Japan).
2. Chloroform.
3. Diisopropyl ether.
4. Phosphate buffered saline (pH 7.4) (PBS): 137 mM NaCl, 8.10 mM Na_2HPO_4, 2.68 mM KCl, 1.47 mM KH_2PO_4 (Wako Pure Chemical Industries).
5. Perfluoropropane (Takachiho Chemical Industries, Tokyo, Japan).
6. Rotary evaporator (TOKYO RIKAKIKAI, Co. Ltd. (EYELA), Tokyo, Japan).

7. Extruding apparatus (Northern Lipids Inc., Vancouver, BC).
8. Bath-type sonicator (42 kHz, 100 W) (Branson Ultrasonics Co., Danbury, CT).
9. Liposome sizing filters (pore sizes: 100 and 200 nm) (Nuclepore Track-Etch Membrane, Whatman plc, UK).
10. 0.45 μm pore size filter (MILLEX HV filter unit, Durapore PVDF membrane) (Millipore Corporation, MA).
11. Dynamic light scattering (ELS-800) (Otsuka Electronics Co., Ltd., Osaka, Japan).
12. Phospholipid C-test wako (Wako Pure Chemical Industries).

2.2. Transmission Electron Microscopy of BLs (20)

1. Sodium alginate (500-600cP).
2. Calcium chloride.
3. Glutaraldehyde.
4. Cacodylate buffer.
5. Osmiumtetroxide.
6. Ethanol.
7. Epan812.
8. Uranyl acetate.
9. Electron microscope: JEOL JEM12000EX at 100 kV.

2.3. In Vitro Ultrasonography with BLs (19)

1. Ultrasound imaging equipment: UF-750XT (Fukuda Denshi Co. Ltd., Tokyo, Japan).
2. 9 MHz linear probe (9 MHz, Fukuda Denshi Co. Ltd.)

2.4. Gene Delivery with BLs and US In Vitro and In Vivo

1. Cells: COS-7 cells (the African green monkey kidney fibroblast cell line), S-180 cells (mouse sarcoma), Meth-A fibrosarcoma cells (mouse fibrosarcoma), Jurkat cells (human T cell line), Colon 26 cells (mouse colon adenocarcinoma), B16BL6 cells (mouse meranoma), Human umbilical vein endothelial cells (HUVEC) (Kurabo Industries, Osaka, Japan).
2. Culture media: Dulbecco's modified Eagle's medium (DMEM), RPMI-1640, Eagle's medium (MEM) and medium 199 (Sigma Chemical Co., St. Louis, MO), Supplements: Fetal bovine serum (FBS, GIBCO, Invitrogen Co., Carlsbad, CA), HEPES and heparin (Wako Pure Chemical Industries), endothelial cell growth supplement (ECGS) (Sigma Chemical Co.), Antibiotics: Penicillin and Streptomycin (Wako Pure Chemical Industries).
3. COS-7 cells and S-180 cells were cultured in DMEM supplemented with 10% heat-inactivated FBS. Meth-A fibrosarcoma cells and Jurkat cells were cultured with RPMI-1640 supplemented with 10% heat inactivated FBS. Colon 26 cells

were cultured with RPMI-1640 supplemented with 10% heat-inactivated FBS and 2.5% HEPES. B16BL6 cells were cultured with MEM supplemented with 10% heat-inactivated FBS. HUVECs were cultured in a DMEM and medium 199 mixture with 15% heat-inactivated FBS, heparin (3.25 U/mL) and ECGS. All culture media contained 100 U/ml penicillin and 100 μg/ml streptomycin.

4. Animals: ddY mice (4–6 weeks age, male), Anesthetic agent: NEMBUTAL (Dainippon Sumitomo Pharma Co., Ltd., Osaka, Japan), Adhesive agent (Aron Alpha) (Daiichi Sankyo Co., Ltd., Tokyo, Japan).

5. Ultrasound equipments and probes for gene delivery – Ultrasound equipments: Sonopore 3000 and Sonopore 4000 (NEPAGENE Co. Ltd.), Probe: KP-T6 (diameter: 6 mm) and KP-T8 (diameter: 8 mm), KP-T20 (diameter: 20 mm) (NEPAGENE Co., Ltd.)

6. Assessment of cytotoxicity: MTT [3-(4,5-s-dimethylthiazol-2-yl)-2,5-diphenyl tetrazolium bromide] (Dojindo, Kumamoto, Japan), Sodium dodecyl sulfate (SDS) (Wako Pure Chemical Industries), Microplate reader (POWERSCAN HT; Dainippon Pharmaceutical, Osaka, Japan).

7. Luciferase assay: Cell lysis buffer (0.1 M Tris–HCl (pH 7.8), 0.1% Triton X-100, 2 mM EDTA), Luciferase assay system (Promega, Madison, WI), Luminometer (TD-20/20) (Turner Designs, Sunnyvale, CA).

8. In vivo luciferase imaging: Escain (Mylan Inc., Tokyo, Japan), D-luciferin and In vivo luciferase imaging system (IVIS) (Caliper Life Sciences, MA).

3. Methods

3.1. Preparation of BLs (18)

1. DSPC and DSPE-PEG(2 k)-OMe were dissolved in 8 mL of 1:1 (v/v) chloroform/diisopropyl ether.

2. Four milliliter of PBS (pH 7.4) was added into the lipid solution. The mixture was sonicated to make suspension, and evaporated at 65° (water bath) to remove solvent.

3. After evaporation, liposome suspension was passed through sizing filters (pore sizes: 100 and 200 nm) using an extruding apparatus. And the size of liposomes was adjusted to less than 200 nm.

4. The liposomes suspension was sterilized by passing them through a 0.45 μm pore size filter. (see Fig. 1a, c)

5. Finally, size of the sterilized liposomes was measured with dynamic light scattering (ELS-800). The average diameter of

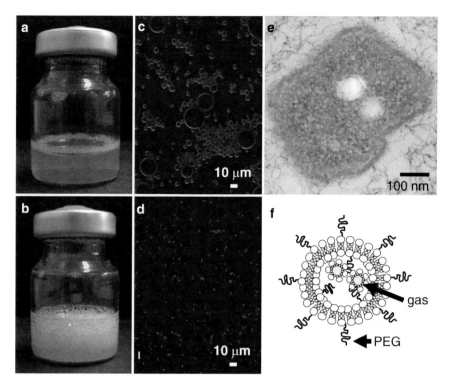

Fig. 1. Aspect and structure of BLs. PEG-liposomes (**a**) were sonicated with supercharged perfluoropropane gas. After that, they became to BLs (**b**). Optison (**c**) and BLs (**d**) were observed with microscope using the darklite illuminator (NEPAGENE, Co., Ltd). (**e**): Transmission electron microscopy (TEM) of BLs. (**f**): Scheme of structure of BLs

these liposomes were about 150–200 nm. In addition, lipid concentration was measured with the Phospholipid C-test wako.

6. The lipid concentration of liposomes suspension was adjusted to 1 mg/mL with PBS.
7. Two milliliter of the liposomes suspension (lipid conc. 1 mg/mL) was entered into sterilized vial (vial size: 5 mL).
8. The vial was filled with perfluoropropane, capped and then supercharged with 7.5 mL of perfluoropropane.
9. The vial was placed in a bath-type sonicator (42 kHz, 100 W) for 5 min to form BLs (see Fig. 1b, d and Note 1).

3.2. Transmission Electron Microscopy of BLs (20)

1. BLs were suspended into sodium alginate (500-600cP) solution (0.2% (w/v) in PBS).
2. This suspension was dropped into calcium chloride solution (100 mM in PBS) to hold BLs within calcium alginate gel.
3. The beads of calcium alginate gel containing BLs were prefixed with 2% glutaraldehyde solution in 0.1 M Cacodylate buffer.
4. The beads were postfixed with 2% OsO_4, dehydrated with an ethanol series, and then embedded in Epan812 (polymerized at 60°).

5. Ultrathin sections were made with an ultramicrotome at a thickness of 60–80 nm.
6. Ultrathin sections were mounted on 200 mesh copper grids.
7. They were stained with 2% uranyl acetate for 5 min and Pb for 5 min.
8. The samples were observed with JEOL JEM12000EX at 100 kV (see Fig. 1e; Notes 2 and 3).

3.3. In Vitro Ultrasonography with BLs (19)

1. BLs were placed into latex tube filled with degassed PBS (10 mL) in a water bath (See Fig. 2a, c).
2. The probe (9 MHz) of an ultrasound imaging equipment was positioned under the water bath.
3. BLs in the tube were imaged (see Fig. 2 b, e, g).

3.4. In Vitro Gene Delivery with BLs and US

3.4.1. Transfection of Plasmid DNA into Cells with BLs and US (21)

1. Plasmid DNA, cells and BLs were suspended in culture medium with 10% FBS (final volume; 500 µL) in 2 mL polypropylene tubes.
2. The probe (KP-T6) (2 MHz, diameter: 6 mm) of US was placed into the suspension.
3. US was exposed to the suspensions with Sonopore 3000 or 4000 under the condition of various US parameters (Duty, Intensity, Exposure time, Burst rate) (see Fig. 2c).

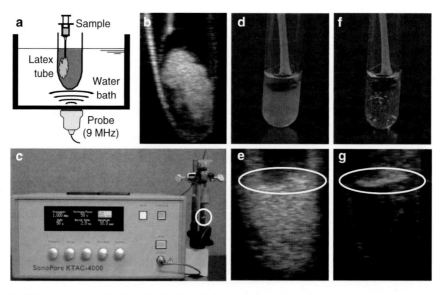

Fig. 2. In vitro Ultrasonography with BLs. The Method of ultrasonography for observation of BLs was shown in (a). BLs were injected into PBS filled latex tube in the water bath. Then, the samples were observed with ultrasonography (b). To confirm the disruption of BLs by US exposure using Sonopore 4000 (c), BLs were observed with naked image (d, f) and ultrasonography (e, g) before (d, e) and after (f, g) US exposure (2 MHz, 2.5 W/cm^2, 10 s). *Circle* in (c, e, g) shows US probe

4. After US exposure, the cells were washed twice with PBS and then resuspended in fresh culture medium.

5. The cells were cultured in culture plate or wells.

6. After 2 days culture of cells, the expression of transgene was measured (see Fig. 3; Notes 4 and 5).

3.4.2. Assessment of Cytotoxicity by the Treatment of BLs and US to Cells (18)

1. Cells (1×10^5) and BLs were suspended in culture medium with 10% FBS (final volume; 500 µL) in 2 mL polypropylene tubes.

2. US was exposed to cells using Sonopore 3000 or 4000 with a probe (KP-T6) (2 MHz, diameter: 6 mm).

3. After US exposure, the cells were washed twice with PBS and then resuspended in fresh culture medium.

4. One hundred microliter of the cells suspension were cultured in 96 well plates for 24 h.

5. Cell viability was assayed using MTT, as described by Mosmann, with minor modifications (22). Briefly, MTT (5 mg/mL, 10 µL) was added to each well, and the cells were incubated at 37°C for 4 h. The formazan product was dissolved in 100 µL of 10% SDS containing 15 mM HCl. Color intensity was measured using a microplate reader at test and reference wavelengths of 595 and 655 nm, respectively.

3.5. In Vivo Gene Delivery with BLs and US

3.5.1. Gene Delivery for Femoral Artery (18)

1. The femoral artery was exposed by operation.

2. BLs (250 µg) and plasmid DNA (10 µg) suspension (300 µL) was slowly injected into the femoral artery of ddY mice (6 weeks age, male) using 30-gauge needle (M-S Surgical MFG. Co. Ltd., Tokyo, Japan).

3. In the same time, US (frequency: 1 MHz, duty: 50%, intensity: 1 W/cm^2, time: 2 min) was transdermally exposed to downstream of injection site using Sonopore 3000 or 4000 with a probe (KP-T8) (diameter: 8 mm).

4. After 2 days of injection, the mice were sacrificed and the femoral artery of US exposure area was collected. Then, gene expression in the artery was measured (see Fig. 4; Notes 6 and 7).

3.5.2. Gene Delivery for Ascites Tumor (20)

1. S-180 cells (1×10^6 cells) were i.p. injected into ddY mice (4 weeks age, male) on day 0.

2. When S-180 cells grew as the ascites tumor in mice after 8 days of the injection, the mice were anaesthetized with NEMBUTAL Injection (50 mg/kg), then injected with 510 µL of plasmid DNA and BLs (500 µg) in PBS.

3. US (frequency: 1 MHz, duty: 50%, intensity: 1 W/cm^2, time: 1 min) was transdermally exposed to the abdominal area using Sonopore 300 or 4000 with a probe (KP-S20) (diameter: 20 mm).

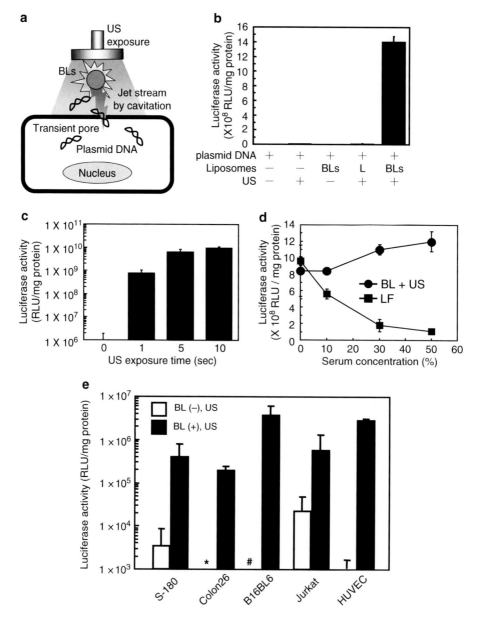

Fig. 3. Property of gene delivery with BLs and US exposure (**a**) Schema of transfection mechanism by BLs and US. The mechanical effect based on the disruption of BLs by US exposure, which results in generation of some pores on plasma membrane, is associated with direct delivery of extracellular plasmid DNA into cytosol. (**b**) Luciferase expression in COS-7 cells transfected by BLs and US. COS-7 cells (1×10^5 cells/500 μL/tube) were mixed with pCMV-Luc (5 μg) and BLs (60 μg). The cell mixture was exposed with US (Frequency: 2 MHz, Duty: 50%, Burst rate: 2 Hz, Intensity: 2.5 W/cm^2, Time: 10 s). The cells were washed and cultured for 2 days. After that, luciferase activity was measured. (**c**) Effect of US condition on transfection efficiency with BLs. COS-7 cells were exposed with US (Frequency: 2 MHz, Duty: 50%, Burst rate: 2 Hz, Intensity: 2.5 W/cm^2, Time: 0, 1, 5, 10 s) in the presence of pCMV-Luc (0.25 μg) and BLs (60 μg). Luciferase activity was measured as above. (**d**) Effect of serum on transfection efficiency of BLs. COS-7 cells in the medium containing FBS (0, 10, 30, 50% (v/v)) were treated with US (Frequency: 2 MHz, Duty: 50%, Burst rate: 2 Hz, Intensity: 2.5 W/cm^2, Time: 10 s), pCMV-Luc (0.25 μg) and BLs (60 μg) or transfected with lipoplex of pCMV-Luc (0.25 μg) and lipofectin (1.25 μg). (**e**) In vitro gene delivery to various types of cell using BLs and US. The method of gene delivery was same as above. S-180: mouse sarcoma cells, Colon26: mouse colon adenocarcinoma cells, B16BL6: mouse melanoma cells, Jurkat: human T cell line, HUVEC: human umbilical endothelial cells. Luciferase activity was measured as above. * $<10^3$ RLU/mg protein, # $<10^0$ RLU/mg protein Each data represents the mean ± S.D. ($n=3$). L: PEG-liposomes, LF: Lipofectin

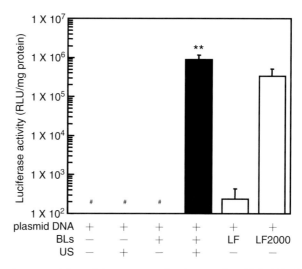

Fig. 4. In vivo gene delivery into mouse ascites tumor cells with Bubble liposomes. S-180 cells (1×10^6 cells) were i.p. injected into ddY mice. After 8 days, the mice were anaesthetized, then injected with 510 μL of pCMV-Luc (10 μg) and Bubble liposomes (500 μg) in PBS. Ultrasound (frequency: 1 MHz, duty: 50%; intensity: 1.0 W/cm^2, time: 1 min) was transdermally applied to the abdominal area. In another experiment, pCMV-Luc (10 μg) – Lipofectin (50 μg) or Lipofectamine 2000 (50 μg) complex was suspended in PBS (510 μL) and injected into the peritoneal cavity of mice. After 2 days, S-180 cells were recovered from the abdomens of the mice. Luciferase activity was determined, as described in Materials and Methods. Each bar represents the mean ± S.D. ($n = 3$–6). **$P < 0.01$ compared to the group treated with plasmid DNA, Bubble liposomes, ultrasound exposure or lipofection with Lipofectin or Lipofectamine 2000. LF, Lipofectin. LF2000, Lipofectamine 2000. # < 10^2 RLU/mg protein

4. After 2 days of US exposure, ascites tumor cells were recovered from the abdomen of the mice. Then, the gene expression in the recovered cells was measured (see Fig. 4).

3.5.3. Gene Delivery for Solid Tumor (20)

1. S-180 cells (1×10^6 cells) were inoculated into the left footpad of ddY mice (5 weeks age, male).

2. At day 4, when the thickness of the footpad was over 3.5 mm (normal thickness was about 2 mm), the left femoral artery was exposed by operation.

3. BLs (100 μg) and plasmid DNA suspension (100 μL) were injected into the femoral artery using 30-gauge needle.

4. In the same time, US (frequency: 0.7 MHz, duty: 50%, intensity: 1.2 W/cm^2, time: 2 min) was transdermally exposed to the tumor tissue using Sonopore 3000 or 4000 with a probe (KP-T8) (diameter: 8 mm).

5. The needle hole was then closed with an adhesive agent and skin was put in a suture.

6. After 2 days of US exposure, the mice were sacrificed and the tumor tissues were collected. Then, the gene expression of the tumor tissue was measured (see Fig. 6 and Note 8).

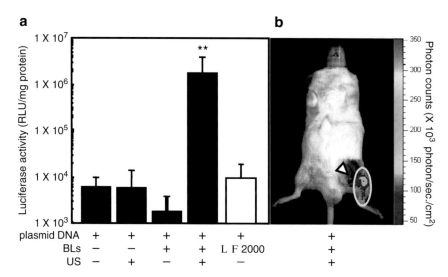

Fig. 5. Gene delivery to femoral artery with Bubble liposomes Each sample containing plasmid DNA 10 μg was injected into femoral artery. At the same time, ultrasound (frequency, 1 MHz; duty, 50%; burst rate, 2 Hz; intensity, 1 W/cm^2; time 2 min) was exposed to the downstream area of injection site. (**a**) Luciferase expression in femoral artery of the ultrasound exposure area at 2 days after transfection, Luciferase expression was determined as described in Materials and Methods. Data are shown as means ± S.D. ($n=5$). (LF2000: Lipofectamine 2000) **$P < 0.01$ compared to the group treated with plasmid DNA, ultrasound exposure, Bubble liposomes or Lipofectamine 2000. (**b**) In vivo luciferase imaging at 2 days after transfection in the mouse treated with plasmid DNA, Bubble liposomes and ultrasound exposure. The photon counts are indicated by the pseudocolor scales. Arrow head shows injection site and circle shows ultrasound exposure area.

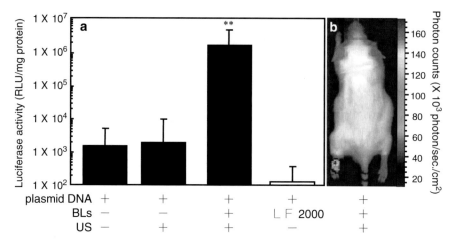

Fig. 6. In vivo gene delivery into mouse solid tumor with Bubble liposomes. S-180 cells (1×10^6 cells) were inoculated into left footpad of ddY mice. After 4 days, the mice were anaesthetized, then injected with 100 μL of pCMV-Luc (10 μg) in absence or presence of Bubble liposomes (100 μg) in PBS. Ultrasound (frequency: 0.7 MHz, duty: 50%; intensity: 1.2 W/cm^2, time: 1 min) was transdermally exposed to tumor tissue. In another experiment, pCMV-Luc (10 μg) – Lipofectamine 2000 (25 μg) complex was suspended in PBS (100 μL) and injected into the left femoral artery. After 2 days, tumor tissue was recovered from the mice. Luciferase activity was determined as described in Materials and Methods. (**a**) Luciferase activity in solid tumor. Each bar represents the mean ± S.D. for five mice/group. **$P<0.01$ compared to the group treated with plasmid DNA, ultrasound exposure or Lipofectamine 2000. (**b**) In vivo luciferase imaging in the solid tumor bearing mice. The photon counts are indicated by the pseudocolor scales. LF 2000, Lipofectamine 2000

3.6. Measurement of Reporter Gene Expression

3.6.1. Luciferase Assay

1. The lysis buffer (0.1 M Tris–HCl (pH 7.8), 0.1% Triton X-100, 2 mM EDTA) was added to the sample cells in vitro or tissues in vivo. In the case of the tissues in vivo, they were homogenized before next step.
2. The cells or the homogenized tissues in lysis buffer were repeatedly frozen and thawed three times to completely disrupt the cell membranes.
3. After that, the lysate of the cells or tissues was centrifuged and the supernatant was collected in other tube.
4. Luciferase activity in the supernatant was measured using a luciferase assay system and a luminometer. The activity is reported in relative light units (RLU) per mg protein of cells or tissue.

3.6.2. In Vivo Luciferase Imaging

1. The mice were anaesthetized with Escain and i.p. injected with D-luciferin (150 mg/kg).
2. After 10 min, luciferase expression was observed with in vivo luciferase imaging system (IVIS).

4. Notes

1. There are some important points to prepare BLs. The air in the vial containing the liposome suspension is completely replaced with perfluoropropane. After that, it needs to be supercharged in the vial with perfluoropropane. And the vial is sonicated with a bath-type sonicator (42 KHz, 100 W) (BRANSONIC 2510 J-DTH, Branson Ultrasonics). In this step, sonication power and the vial position in the bath are very important. Because we have experimented that BLs were not prepared using other type of bath sonicator (UC-1 (38 KHz, 80 W), IKEDA RIKA, Japan) with low intensity of ultrasound exposure. In addition, BLs were not prepared using other gas such as air, nitrogen gas or carbonic dioxide gas. Therefore, it thought that it is important for the preparation of BL to use hydrophobic gas such as perfluoropropane.
2. To fix BLs as a sample for transmission electron microscope, BLs were held within calcium alginate gel. The handling of BLs was improved by holding within the gel. The advantage for using this gel is to make the gel even at low temperature. Because BLs became unstable according to increasing temperature. Therefore, it is thought that the gel, such as agarose, which has gel point at high temperature is inappropriate for this purpose.
3. It was thought that liposomes were reconstituted by sonication under the condition of supercharge with perfluoropropane.

Then, perfluoropropane was entrapped within lipids like micelles. In addition, the lipid nanobubbles were encapsulated within liposomes. To confirm the structure of BLs, we observed BLs with transmission electron microscope. Interestingly, BLs had nanobubbles into lipid bilayer. Therefore, we called this "Bubble liposome" because of this structure. This structure of BLs was different from that of conventional microbubbles and nanobubbles which had lipid monolayer.

4. This protocol can be adapted for many other types of cell. In the gene transfection for adherent cells, the transfection efficiency in the condition of suspension was higher than that in the condition of adhesion on the culture plate. Although this result is unclear, it is thought that the distance between BLs and cells is important. Because BLs entrapping gas is easy to flow and result in getting away from the adherent cells on the plate.

5. In in vitro gene delivery, it is very important to fix the location of it, in order to reduce the experimental error of each data. The efficiency of this gene delivery was not affected even in the presence of serum. Moreover, the gene expression was observed even under the condition of US exposure for 1 s. From these results, it was suggested that this system could immediately deliver plasmid DNA into cells.

6. In in vivo gene delivery, echo jelly is necessary for US exposure to mice. Gene expression was observed in the arrested area of US exposure. Because it is thought that the mechanical effect based on the disruption of BLs by US exposure results in generation of some pores on plasma membrane of the cells in the area of US exposure.

7. This system is thought that there is not a serious damage for the cells in blood such as red blood cells by the disruption of BLs in blood stream by US exposure.

8. The transfection efficiency with the gene delivery system by sonoporation mechanism using BLs and US was higher than conventional lipofection method with Lipofectin and Lipofectamine 2000. Therefore, it is expected that this system might be an effective nonviral gene delivery system.

Acknowledgements

We are grateful to Dr. Katsuro Tachibana (Department of Anatomy, School of Medicine, Fukuoka University) for technical advice regarding the induction of cavitation with ultrasound, to Dr. Naoki Utoguchi, Mr. Yusuke Oda, Mr. Eisuke Namai,

Ms. Tomoko Takizawa, Ms. Kaori Sawamura and Ms. Kumiko Tanaka (Department of Pharmaceutics, School of Pharmaceutical Sciences, Teikyo University), Yoichi, Negishi (School of Pharmacy, Tokyo University of Pharmacy and Life Science), and Dr. Kosuke Hagisawa (Department of Medical Engineering, National Defense Medical College) for excellent technical advice and assistance, to Mr. Yasuhiko Hayakawa, Mr. Takahiro Yamauchi, and Mr. Kosho Suzuki (NEPAGENE Co., Ltd.) for technical advice regarding ultrasound exposure using Sonopore 3000 and 4000, and Sonitron 2000.

This study was supported by an Industrial Technology Research Grant in 2004 from NEDO, JSPS KAKENHI (16650126), MEXT KAKENHI (160700392, 19700423), a Research on Advanced Medical Technology (17070301) in Health and Labour Sciences Research Grants from Ministry of Health, Labour and Welfare, and the Program for Promotion of Fundamental Studies(07-24) in Health Sciences of the National Institute of Biomedical Innovation (NIBIO).

References

1. Fechheimer M, Boylan JF, Parker S, Sisken JE, Patel GL, Zimmer SG (1987) Transfection of mammalian cells with plasmid DNA by scrape loading and sonication loading. Proc Natl Acad Sci U S A 84:8463–8467
2. Miller MW, Miller DL, Brayman AA (1996) A review of in vitro bioeffects of inertial ultrasonic cavitation from a mechanistic perspective. Ultrasound Med Biol 22:1131–1154
3. Joersbo M, Brunstedt J (1990) Protein synthesis stimulated in sonicated sugar beet cells and protoplasts. Ultrasound Med Biol 16:719–724
4. Miller DL, Pislaru SV, Greenleaf JE (2002) Sonoporation: mechanical DNA delivery by ultrasonic cavitation. Somat Cell Mol Genet 27:115–134
5. Guzman HR, McNamara AJ, Nguyen DX, Prausnitz MR (2003) Bioeffects caused by changes in acoustic cavitation bubble density and cell concentration: a unified explanation based on cell-to-bubble ratio and blast radius. Ultrasound Med Biol 29:1211–1222
6. Wei W, Zheng-zhong B, Yong-jie W, Qing-wu Z, Ya-lin M (2004) Bioeffects of low-frequency ultrasonic gene delivery and safety on cell membrane permeability control. J Ultrasound Med 23:1569–1582
7. Duvshani-Eshet M, Machluf M (2005) Therapeutic ultrasound optimization for gene delivery: a key factor achieving nuclear DNA localization. J Control Release 108:513–528
8. Tata DB, Dunn F, Tindall DJ (1997) Selective clinical ultrasound signals mediate differential gene transfer and expression in two human prostate cancer cell lines: LnCap and PC-3. Biochem Biophys Res Commun 234:64–67
9. Kim HJ, Greenleaf JF, Kinnick RR, Bronk JT, Bolander ME (1996) Ultrasound-mediated transfection of mammalian cells. Hum Gene Ther 7:1339–1346
10. Greenleaf WJ, Bolander ME, Sarkar G, Goldring MB, Greenleaf JF (1998) Artificial cavitation nuclei significantly enhance acoustically induced cell transfection. Ultrasound Med Biol 24:587–595
11. Shohet RV, Chen S, Zhou YT, Wang Z, Meidell RS, Unger RH, Grayburn PA (2000) Echocardiographic destruction of albumin microbubbles directs gene delivery to the myocardium. Circulation 101:2554–2556
12. Taniyama Y, Tachibana K, Hiraoka K, Namba T, Yamasaki K, Hashiya N, Aoki M, Ogihara T, Yasufumi K, Morishita R (2002) Local delivery of plasmid DNA into rat carotid artery using ultrasound. Circulation 105:1233–1239
13. Taniyama Y, Tachibana K, Hiraoka K, Aoki M, Yamamoto S, Matsumoto K, Nakamura T, Ogihara T, Kaneda Y, Morishita R (2002) Development of safe and efficient novel nonviral gene transfer using ultrasound: enhancement of transfection efficiency of naked plasmid DNA in skeletal muscle. Gene Ther 9:372–380

14. Sonoda S, Tachibana K, Uchino E, Okubo A, Yamamoto M, Sakoda K, Hisatomi T, Sonoda KH, Negishi Y, Izumi Y, Takao S, Sakamoto T (2006) Gene transfer to corneal epithelium and keratocytes mediated by ultrasound with microbubbles. Invest Ophthalmol Vis Sci 47:558–564
15. Newman CM, Bettinger T (2007) Gene therapy progress and prospects: ultrasound for gene transfer. Gene Ther 14:465–475
16. Lindner JR (2004) Microbubbles in medical imaging: current applications and future directions. Nat Rev Drug Discov 3:527–532
17. Tsunoda S, Mazda O, Oda Y, Iida Y, Akabame S, Kishida, T, Shin-Ya M, Asada H, Gojo S, Imanishi J, Matsubara H, Yoshikawa T (2005) Sonoporation using microbubble BR14 promotes pDNA/siRNA transduction to murine heart. Biochem Biophys Res Commun 336:118–127
18. Suzuki R, Takizawa T, Negishi Y, Hagisawa K, Tanaka K, Sawamura K, Utoguchi N, Nishioka T, Maruyama K (2007) Gene delivery by combination of novel liposomal bubbles with perfluoropropane and ultrasound. J Control Release 117:130–136
19. Suzuki R, Takizawa T, Negishi Y, Utoguchi N, Maruyama K (2007) Effective gene delivery with liposomal bubbles and ultrasound as novel non-viral system. J Drug Target 15:531–537
20. Suzuki R, Takizawa T, Negishi Y, Utoguchi N, Sawamura K, Tanaka K, Namai E, Oda Y, Matsumura Y, Maruyama K (2008) Tumor specific ultrasound enhanced gene transfer in vivo with novel liposomal bubbles. J Control Release 125:137–144
21. Suzuki R, Takizawa T, Negishi Y, Utoguchi N, Maruyama K (2008) Effective gene delivery with novel liposomal bubbles and ultrasonic destruction technology. Int J Pharm 354:49–55
22. Mosmann T (1983) Rapid colorimetric assay for cellular growth and survival: application to proliferation and cytotoxicity assays. J Immunol Methods 65:55–63

Chapter 34

Liposomal Magnetofection

Olga Mykhaylyk, Yolanda Sánchez-Antequera, Dialekti Vlaskou, Edelburga Hammerschmid, Martina Anton, Olivier Zelphati, and Christian Plank

Abstract

In a magnetofection procedure, self-assembling complexes of enhancers like cationic lipids with plasmid DNA or small interfering RNA (siRNA) are associated with magnetic nanoparticles and are then concentrated at the surface of cultured cells by applying a permanent inhomogeneous magnetic field. This process results in a considerable improvement in transfection efficiency compared to transfection carried out with nonmagnetic gene vectors. This article describes how to synthesize magnetic nanoparticles suitable for nucleic acid delivery by liposomal magnetofection and how to test the plasmid DNA and siRNA association with the magnetic components of the transfection complex. Protocols are provided for preparing magnetic lipoplexes, performing magnetofection in adherent and suspension cells, estimating the association/internalization of vectors with cells, performing reporter gene analysis, and assessing cell viability. The methods described here can be used to screen magnetic nanoparticles and formulations for the delivery of nucleic acids by liposomal magnetofection in any cell type.

Key words: Nucleic acid delivery to cultured cells, Magnetic nanoparticles, Magnetic lipoplexes, Magnetofection

1. Introduction

Since the first reports on magnetically enhanced nucleic acid delivery in the year 2000 (1, 2), magnetofection has become a well-established method and has been predominantly used for in vitro applications. It has been shown to potentiate viral (3, 4) and non-viral nucleic acid delivery, including plasmids or small constructs such as antisense oligonucleotides, and synthetic siRNA and PCR products (5–10). The nucleic acids can be directly associated with magnetic nanoparticles in naked form or can be incorporated into a complex composed of magnetic particles and other components such as cationic lipids or polymers, thus forming magnetic lipoplexes or polyplexes.

Many enhancers (11) known to be efficient in the transfection of a particular cell line can be combined with magnetic nanoparticles to construct magnetic vectors. This protocol uses either the commercially available transfection reagent DreamFect™-Gold or SM4-31 as enhancers (4 μl/μg DNA, OZ Biosciences). To perform magnetofection *in vitro*, magnetic vectors are added to cell culture supernatants. To concentrate the applied vector dose at the cell surface, cell culture plates are placed on magnetic plates consisting of an array of suitably positioned permanent magnets that generate an inhomogeneous magnetic field. The diffusion limitation to delivery is overcome, and transfection/transduction is synchronized and greatly accelerated. The vector dose required for efficient transfection/transduction is therefore considerably reduced. Together, these features constitute a substantial improvement in transfection/transduction kinetics and efficiency. Magnetic devices and magnetic nano- or microparticles are commercially available, along with standardized application protocols for various vector types and cell culture formats (OZ Biosciences, Marseille, France, http://www.ozbiosciences.com; Chemicell, Berlin, Germany, http://www.chemicell.com). The commercially available magnet array for magnetofection produces high-gradient magnetic fields (70–250 mT and a field gradient of 50–130 T/m) in the vicinity of the cells, and sediments the full vector dose on the cells within minutes.

The development of new magnetic nanoparticles is expected to lead to further improvements of the technique (12), because the biophysical properties of the particles have a major impact on their formulations with vectors and on their function in biological systems in vitro and in vivo. A large variety of coating compounds is useful in magnetofection (13, 14), and further improvements can be expected. Therefore, this article provides protocols for every step from magnetic nanoparticle synthesis and characterization to their use in liposomal magnetofection of plasmid DNA or siRNA. The magnetic nanoparticles described here differ in their coating material; are stable enough to be stored over extended periods; and are sufficiently biocompatible for application in living cells. These particles achieve efficient nucleic acid delivery to adherent cells in vitro by magnetofection and can be associated with nucleic acids alone or with nucleic acids and an enhancer to form nonviral lipoplexes (15). Subsequently, we describe how the binding of nucleic acids to magnetic nanoparticles in combination with a third agent that enhances transfection (known as an enhancer) can be characterized using radioactively labelled nucleic acids prepared according to the modified Terebesi procedure (16). This labeling can be used to determine suitable ratios and mixing orders of magnetic nanoparticles, nucleic acids, and third components, in order to choose formulations that are potentially useful for magnetofection. For the screening purposes presented here, it is most useful to use reporter genes such as eGFP and luciferase reporters, which allow rapid and sensitive result evaluation in cell lysates and even in living cells. Using the eGFP reporter gene, the percentage of transfected cells can be easily determined. Here we focus on a 96-well screening format for magnetic nanoparticles to be used in nonviral liposomal magnetofection. Vectors are prepared in a serum- and supplement-free medium and transferred to the cells in triplicate in a volume of 50 μl per well. To obtain dose-response data, we recommend performing serial dilutions of a given vector composition, such that the highest plasmid or siRNA dose transferred to the cell culture plate is 500 ng plasmid or 200 ng siRNA per well, respectively.

According to our results, for most of the tested cells (HeLa cells, H441, M1, Jurkat cells), the optimum nanomaterial-to-nucleic acid ratio for magnetic vectors containing enhancers described in this protocol is between 0.5 and 1 iron-to-nucleic acid wt/wt ratio. We describe procedures for determining the level of association of transfection complex with cells, quantifying the internalization of complexes, and evaluating transfection efficiency in cell lysates with respect to the toxicity determined using an MTT-based cell viability test. We illustrate these protocols with DNA and siRNA magnetofection results obtained in adherent HeLa and H441 and difficult-to-transfect suspension Jurkat cells. Presenting the results in terms of absolute units of reporter gene expression normalized per weight of total protein in the examined cell lysate, as described in the protocol, is especially important in siRNA transfection experiments and makes it possible to distinguish between gene down-regulation and toxicity effects. These protocols should enable the skilled experimentalist to practice the method independently and to contribute further to the field.

2. Materials

2.1. Synthesis of Magnetic Nanoparticles Suitable as Components of Magnetofection Complexes

1. Iron(II) chloride tetrahydrate (Sigma-Aldrich).
2. Iron(III) chloride hexahydrate (Sigma-Aldrich).
3. Argon.
4. 10% hydroxylamine hydrochloride solution in water (see Note 1).
5. SO-Mag1 precipitation solution: 15 ml of 28–30% ammonium hydroxide.
6. SO-Mag1 coating component 1: 0.2 g of Tetraethyl orthosilicate (Sigma-Aldrich).

7. SO-Mag1 coating component 2: 0.3 g of 3-(trihydroxysilyl) propylmethylphosphonate (Sigma-Aldrich).
8. NDT-Mag1 precipitation/coating solution: 12.5 ml of 28–30% ammonium hydroxide solution (Sigma-Aldrich) plus 1.25 ml Lithium 3-[2-(perfluoroalkyl)ethylthio]propionate (ZONYL®FSA; Sigma-Aldrich) filled with water to a total volume of 25 ml and degassed with argon/helium. NDT-Mag1 coating solution: 1 ml of 1,9-Nonanedithiol (Sigma-Aldrich) dissolved in 24 ml toluene.
9. Polyethylenimine 25 kDa, branched (PEI-25$_{Br}$ Sigma-Aldrich).
10. Tetramethylrhodamine isothiocyanate (TRITC) solution: 2.5 mg TRITC (Sigma-Aldrich)/ml DMSO. Store in dark at 4°C.
11. 0.1 N Sodium-Carbonate buffer, pH 9.0.
12. PEI-Mag2 precipitation/coating solution: 2.5 g polyethylenimine 25 kD, branched (PEI-25$_{Br}$; Sigma-Aldrich) plus 12.5 ml 28–30% ammonium hydroxide solution (Sigma-Aldrich) plus 1.25 ml Lithium 3-[2-(perfluoroalkyl)ethylthio]propionate (ZONYL®FSA; Sigma-Aldrich) filled with water to a total volume of 50 ml and degassed with argon/helium.
13. PalD1-Mag1 precipitation/coating solution: 2 g palmitoyl dextran PalD1 (see Note 2) plus 15 ml of 28–30% ammonium hydroxide solution filled with water to a total volume of 50 ml and degassed with argon/helium.
14. PL-Mag1 coating solution 1: 2 g Pluronic F-127 (Sigma-Aldrich) filled with water to a total volume of 25 ml and degassed with argon/helium. PL-Mag1 precipitation/coating solution: 15 ml of 28–30% ammonium hydroxide solution plus 7.5 ml of ammonium bis[2-(perfluoroalkyl)ethyl] phosphate solution (ZONYL®FSE; Sigma-Aldrich) filled with water to a total volume of 25 ml and degassed with argon/helium.

2.2. Determination of Magnetic Nanoparticle Concentration in terms of Dry Weight and Iron Content

1. Ammonium acetate buffer for iron determination: Dissolve 25 g ammonium acetate (Sigma) in 10 ml water, add 70 ml glacial acetic acid and adjust volume to 100 ml with water.
2. 10% Hydroxylamine hydrochloride (Sigma-Aldrich) in water.
3. 0.1% Phenanthroline solution: Dissolve 100 mg 1,10-phenanthroline monohydrate (Sigma) in 100 ml water, add 2 drops of concentrated hydrochloric acid (Fluka). If necessary, warm to obtain a clear solution.
4. Iron stock solution: Dissolve 392.8 mg ammonium iron(II) sulfate hexahydrate (Sigma) in a mixture of 2 ml concentrated sulfuric acid and 10 ml water, add 0.05N KMnO$_4$ dropwise

until pink color persists and adjust the volume to 100 ml with water.

5. Standard iron solution (make fresh as required): Dilute iron stock solution 1 to 25 with water just before calibration measurements.

6. 0.05N $KMnO_4$ solution: Dissolve 0.790 g $KMnO_4$ in 100 ml water.

2.3. Radiolabeling (Iodination) of Nucleic Acids

1. DNA solution: Luciferase gene plasmid p55pCMV-IVS-luc⁺ containing the firefly luciferase cDNA under the control of the cytomegalovirus (CMV) promoter (Plasmid Factory, Bielefeld, Germany) (pBLuc 5 mg/ml); see also Note 3.

2. siRNA solution: Reconstitute 5 nmol (78.3 µg) GFP-22 siRNA (Qiagen) with 39.1 µl of the siRNA suspension buffer at 2 µg siRNA/µl and store in aliquots at −20°C.

3. Sodium ^{125}iodide in 40 mM NaOH, activity: 2 mCi in 20 µl. Caution: Radioactive material! Store at ambient temperature, 15–20°C. Retains iodination efficiency for over 2 months in storage.

4. 250 µM potassium iodide in water. Prepare on the day of DNA labeling, from 25 mM potassium iodide.

5. 1 M sodium hydroxide in water.

6. 30 mM Thallium trichloride tetrahydrate (Sigma-Aldrich) solution in water. To obtain a clear solution, heat the tube to 70°C using a water bath. Solution is stable, and can be stored for at least a year.

7. 1 M sodium sulfite in water. Prepare on the day of siRNA labeling.

8. 1 M ammonium acetate buffer, pH 7.

9. Disposable Sephadex G25 PD-10 desalting columns (GE Healthcare).

2.4. Testing the Nucleic Acid Association and Magnetic Sedimentation in Transfection Complexes with Magnetic Nanoparticles

1. Suspension of magnetic nanoparticles: Dilute stock suspension of magnetic nanoparticles in water at a concentration of 720 µg iron/ml or 288 µg iron/ml for testing DNA or siRNA association with and magnetic sedimentation of magnetic nanoparticles, respectively. Prepare just before the experiment.

2. DreamFect™Gold (DF-Gold) liposomal transfection reagent (OZ Biosciences).

3. SM4-31 liposomal transfection reagent (OZ Biosceinces).

4. ^{125}I-labeled DNA solution: 12 µg/ml total DNA (pBLuc, Plasmid Factory) comprising 2×10^5 CPM/ml ^{125}I-labeled DNA from Subheading 3.3, in RPMI medium without supplements, or in whatever solvent of interest.

5. ^{125}I-labelled siRNA solution: 4.8 μg/ml total siRNA (GFP-22 siRNA, Qiagen) comprising 2×10^5 CPM/ml ^{125}I-labelled siRNA from Step 3.3 in RPMI medium without supplements, or whatever solvent of interest.

6. 96-Magnets magnetic plate (magnetic plate; OZ Biosciences).

2.5. Cell Culture and Plating for Transfection

1. NCI-H441 human pulmonary epithelial (H441) cells derived from papillary carcinoma of the lungs (ATCC).
2. NCI-H441 cells stably expressing eGFP (H441-GFP cells).
3. Human cervical epithelial adenocarcinoma (HeLa) cells (ATCC).
4. HeLa cells stably expressing eGFP (HeLa-GFP cells).
5. Jurkat human T cell leukemia cells (Jurkat cells, DSMZ Cat no. ACC 282).
6. H441 culture medium: Modified RPMI 1640 medium with 2 mM l-glutamine, 10 mM HEPES, 1 mM sodium pyruvate, 4.5 g/l glucose, 1.5 g/l sodium bicarbonate supplemented with 10% heat-inactivated FCS, 100 U/ml penicillin, 100 μg/ml streptomycin and 2 mM l-Glutamin. Split the cells 1 to 4–5 when they are about 80–90% confluent.
7. HeLa culture medium: DMEM supplemented with 2 mM l-Glutamin, 1 mM sodium pyruvate supplemented with 10% heat-inactivated FCS, 100 U/ml penicillin, and 100 μg/ml streptomycin. Split the cells 1 to 5–7 when they are about 80–90% confluent.
8. Jurkat culture medium: Modified RPMI 1640 medium with 2 mM l-Glutamin, 100 U/ml penicillin, 100 μg/ml streptomycin supplemented with 10% heat-inactivated FCS.
9. Trypsin/EDTA solution, 0.25%/0.02% (wt/vol).
10. Dulbecco's PBS w\o Ca^{2+}, Mg^{2+} solution (PBS).

2.6. Preparation of Magnetic Nanoparticle-nucleic acid Magnetic Lipoplexes

1. Magnetic nanoparticles synthesized according to Subheading 3.1, or commercially available magnetic nanoparticles to be tested suspended in water at 36 μg/ml for siRNA delivery and at 90 μg/ml for DNA delivery with magnetic transfection complexes just before the experiment (concentration refers to iron content), see also Note 4.

2. DreamFect™ Gold (DF-Gold) liposomal transfection reagent (OZ Biosceinces) or SM4-31 liposomal transfection reagent (SM4-31, OZ Biosceinces) as an enhancer for siRNA delivery: mix 5.8 μl of the liposomal transfection reagent with 34.2 μl of water in a tube for each siRNA transfection complex to be tested. To prepare enhancer for DNA delivery, mix 14.4 μl liposomal transfection reagent with 25.6 μl of water. This results in a liposomal transfection reagent to nucleic

acid v/w ratio of 4 to 1, when used according to the protocol of Subheading 3.6.

3. siRNA stock solution (100×): reconstitute 5 nmol siRNA e.g., GFP-22 siRNA (Qiagen), at 480 μg siRNA/ml with 162.9 μl of the siRNA suspension buffer, and store in aliquots at −20°C.

4. siRNA solution: prepare by 1 to 100 dilution of the 100× siRNA stock solution with a serum- and supplement-free medium (e.g., RPMI 1640).

5. Plasmid DNA solution: prepare DNA solution e.g. luciferase reporter plasmid or eGFP plasmid, at a concentration of 12 μg DNA/ml by dilution of the stock solution with a serum- and supplement-free medium (e.g., RPMI 1640).

6. Luciferase reporter plasmid p55pCMV-IVS-luc+ containing the firefly luciferase cDNA under the control of the cytomegalovirus (CMV) promoter.

7. eGFP plasmid containing eGFP under the control of the EF-1 promoter (BD Biosciences, Clontech, Heidelberg).

2.7. Magnetofection

1. Adherent or suspension type cells plated for transfection according to Subheading 3.5 steps 1–6 or 7, respectively.

2. Magnetic transfection complexes, and appropriate controls if necessary, prepared according to Subheading 3.6, just before magnetofection.

3. 96-Magnets magnetic plate (magnetic plate; OZ Biosciences).

2.8. Evaluation of Transfection Complex Association with Cells and Internalization into Cells by Microscopy and Fluorescence-activated Cell Sorting

1. GFP-22 siRNA labeled with rhodamine (siRNA-Rho, Qiagen).

2. Hoechst 33342 stock solution: Hoechst 33342, trihydrochloride trihydrate (Invitrogen) in 1 mg/ml of water. Store in the dark at 4°C.

3. YOYO-1 iodide (491/509) stock: 1 mM solution in DMSO (Invitrogen). Store in aliquots in the dark at −20°C.

4. Rhodamine-labeled magnetic nanoparticles synthesized according to Subheading 3.1. Step 10 of the protocol.

5. FACS buffer: PBS supplemented with 1% FCS.

6. Propidium iodide (PI, Sigma) stock (1 mg/ml).

7. Trypan Blue (TB) 0.4% (TB, Sigma).

2.9. Quantification of the Internalization of the Transfection Complex into Cells using Radioactively Labeled siRNA

1. ^{125}I-labeled DNA solution: 12 μg/ml total DNA (pBLuc, Plasmid Factory) comprising 1×10^6 CPM/ml ^{125}I-labeled DNA (from Subheading 3.3) in RPMI medium without supplements, or in whatever solvent of interest.

2. ^{125}I-labeled siRNA solution: 4.8 μg/ml total siRNA (GFP-22 siRNA, Qiagen) comprising 1×10^6 CPM/ml ^{125}I-labeled

siRNA (from Subheading 3.3) in RPMI medium without supplements, or in whatever solvent of interest.

3. Other reagents as in Subheading 2.6.

2.10. Quantification of Luciferase Reporter Gene Expression in Cell Lysate

1. Lysis buffer: 0.1% Triton X-100 in 250 mM Tris, pH 7.8.
2. Luciferin buffer: 35 µM D-luciferin (Roche Diagnostics), 60 mM DTT (Sigma-Aldrich), 10 mM magnesium sulfate, 1 mM ATP, in 25 mM glycyl-glycin-NaOH buffer, pH 7.8.
3. Luciferase standard stock: 0.1 mg luciferase per ml (Roche Diagnostics) and 1 mg BSA per ml (Sigma) in 0.5 M Tris-acetate buffer, pH 7.5. Store in aliquots at −70°C.
4. BioRad protein assay reagent.
5. BSA stock solution: 1.5 mg/ml BSA (Sigma) in PBS. Store at 4°C.

2.11. Evaluation of eGFP Gene Expression

1. Purified recombinant SuperGlo GFP (GFP; Qiagen). GFP stock solution: 500 ng GFP per µl PBS. Store in small portions at −70°C.
2. Clear bottom black-walled plate, 96-well (Greiner Bio-One).

2.12. MTT-based Test for Toxicity of Transfection Complexes

1. MTT solution: 1 mg thiazolyl blue tetrazolium bromide (MTT; Sigma-Aldrich) per ml and 5 mg/ml glucose in Dulbecco's PBS solution (solution must be stored at −20°C).
2. MTT solubilization solution: 10% Triton X-100 in 0.1N hydrochloric acid in anhydrous isopropanol (solution can be stored at room temperature: 15–25°C).

3. Methods

3.1. Synthesis of Magnetic Nanoparticles Suitable as Components of Complexes for Liposomal Magnetofection

1. To synthesize the SO-Mag1 nanomaterial, dissolve 0.025 mol (6.8 g) of ferric chloride hexahydrate and 0.0125 mol (2.5 g) of ferrous chloride tetrahydrate in 200 ml double-distilled water and filter using a 0.2 µm filter flask or bottle-top filter; transfer the solution to a 500 ml round bottom flask (make fresh as required). Remove dissolved oxygen by continuous argon or helium bubbling through the solution for ~10 min. Cool to 2–4°C and continue bubbling argon/helium.
2. To obtain a primary precipitate, rapidly add SO-Mag1 precipitation solution, heat the material to 90°C over a 15 min interval, and stir at this temperature for 30 min. Add coating component 1 and stir at 90°C for the next 30 min.
3. Add SO-Mag1 coating component 2 and stir at 90°C for 30 min.

4. Cool the mixture to 25°C (no more inert gas bubbling is needed) and incubate for 24 h with continuous stirring.

5. Release the particles by adding ethanol, separate from the mixture by exposure to a gradient magnetic field and wash two times with ethanol and once with water.

6. Sonicate the product for 10 min using a resonance frequency of about 20 kHz, 75 mW, impulses 60 s/30 s interval and dialyze extensively against water using Spectra/Por® 6 50 kD cut-off dialysis membrane to remove excess unbound stabilizer. Sterilize the suspension using a 25 kGy dose of ^{60}Co gamma-irradiation; see also Note 5.

7. Similarly, synthesize the NDT-Mag1 nanomaterial. Perform step 1 of the synthesis protocol. At step 2, add NDT-Mag1 precipitation/coating solution followed by addition of NDT-Mag1 coating solution. Stir at 90°C for 60 min. Perform further synthesis according to steps 4–6 of this section.

8. PL-Mag2 and PEI-Mag2 and PalD1-Mag1 nanomaterials (used in examples shown in figures) were synthesized with precipitation/coating solutions as described in Subheading 2.1, using steps 1, 2, 4, and 6 of the synthesis protocol of this Subheading 3.1. Steps 3 and 5 are omitted.

9. To obtain the SO-Mag2 nanomaterial by surface decoration of the SO-Mag1 nanomaterial via spontaneous adsorbtion of branched polyethylenimine 25 kD (PEI-25$_{Br}$), mix 5 ml of aqueous SO-Mag1 suspension (20 mg Fe/ml) with 0.5 ml of aqueous PEI-25$_{Br}$ solution (10 mg/ml), incubate for 1 h, and dialyze extensively against water.

10. To obtain tetramethylrhodamine-labeled SO-Mag2 nanomaterial (SO-Mag2-Rho), mix a 2.4 ml suspension of SO-Mag2 nanoparticles containing 2.5 mg Fe/ml of 0.1 M Na-Carbonate buffer pH 9 with 100 µl TRITC solution from Subheading 2.1 step 10 containing 2.5 mg TRITC/ml of DMSO, and incubate for 24 h in the dark. Wash the nanoparticles with a 0.01 M Na-carbonate buffer, pH 9, using magnetic separation, resuspend in the buffer, determine iron concentration according to steps from Subheading 3.2, of the protocol and adjust the volume to obtain the desired concentration of the material. Rhodamine-labeled PEI-Mag2 nanomaterial can be prepared in a similar way.

The characteristics of selected nanomaterials synthesized according to Subheading 3.1 of the protocol are given in Table 1 and shown in Fig. 1; see also Note 6.

3.2. Determination of Magnetic Nanoparticle Concentration in terms of Dry Weight and Iron Content

1. To determine the magnetic nanoparticle concentration in suspension in terms of iron content, take 20 µl aliquots of the magnetic nanoyparticle suspension and add 200 µl of concentrated hydrochloric acid and 50 µl of water. Wait until

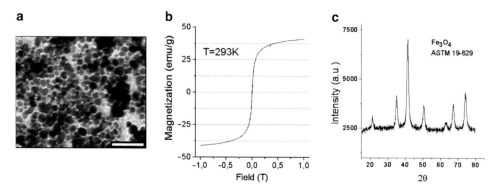

Fig. 1. Characteristics of PL-Mag1 nanoparticles typical for core-shell magnetic nanoparticles synthesized according to Subheading 3.1 of the protocol. (a) TEM image. The scale bar equals 50 nm, and the magnetite core size is of 5–12 nm. (b) Magnetization curve. Saturation magnetization of 34 emu g^{-1} or 82 emu (g Fe)$^{-1}$. (c) X-ray diffraction pattern. The average crystallite size <d> determined from the broadening of the X-ray diffraction peak is of 10.6 nm

the magnetic nanoparticles are completely dissolved, and then adjust the volume to 5 ml with water.

2. Transfer 20 μl of the solution from step one to a microcentrifuge tube, add 20 μl of concentrated hydrochloric acid, 20 μl of hydroxylamine hydrochloride solution, 200 μl of ammonium acetate buffer, 80 μl of 1,10-phenanthroline solution, and 860 μl of water. Mix well and allow to stand for 20 min.

3. Prepare a blank sample by mixing 20 μl of concentrated hydrochloric acid, 20 μl of hydroxylamine hydrochloride solution, 200 μl of ammonium acetate buffer, 80 μl of 1,10-phenanthroline solution, and 880 μl of water (see Note 7).

4. Measure the absorbance of the samples from step 2 at 510 nm against the blank (step 3) using a spectrophotometer (e.g., Beckman DU 640).

5. To construct a calibration curve for determining the iron concentration, add increasing amounts of iron standard solution to microcentrifuge tubes (e.g., 50, 70, 90 up to 150 μl) and adjust the volume to 150 μl with water. Use 150 μl of water instead of iron solution to prepare a blank sample. To each tube, add 20 μl of concentrated hydrochloric acid, 20 μl of 10% hydroxylamine hydrochloride solution, 200 μl of ammonium acetate buffer, 80 μl of 0.1% 1,10-phenanthroline solution, and 730 μl of water. Mix well and allow to stand for 20 min. Measure the absorbance at 510 nm against the blank. Plot the absorbance at 510 nm as a function of the iron concentration in the standard samples. Use linear regression as an approximation function to calculate the iron concentration in the magnetic nanoparticle samples.

6. To determine the iron concentration per dry weight of magnetic nanoparticles, freeze-dry under high vacuum as follows: transfer 1 ml aliquots of magnetic nanoparticle suspensions into pre-weighed glass vials, freeze the samples (at −80°C or in liquid nitrogen) and dry overnight under high vacuum using a lyophilizer. Weigh the vials again to calculate the dry weight. Add 1 ml of concentrated hydrochloric acid, wait until the magnetic nanoparticles are completely dissolved, and then transfer 20 μl of the resultant solution to a microcentrifuge tube and determine the iron content by following steps 1–5 from this section. Calculate the iron concentration per dry weight of magnetic nanoparticles (see Note 8). Sample results are given in Table 1.

3.3. Radiolabeling (Iodination) of Nucleic Acids (see Note 9)

1. *To label DNA* radioactively, prepare in vial 1 (ideally a conical screw cap microcentrifuge tube), a mixture of 20 μl DNA solution (5 μg DNA/μl) and 80 μl 0.1 M ammonium acetate buffer, pH 5. *To label siRNA*, prepare in vial 1, a mixture

Table 1
Characteristics of the magnetic nanoparticles synthesized according to steps from Subheading 3.1 (see Note 6)

Parameter	Nanoparticles					
	SO-Mag1	SO-Mag2	NDT-Mag1	PalD1-Mag1	PL-Mag1	PEI-Mag2
Mean magnetite crystallite size $\langle d \rangle$ (nm)[a]	11	11	11.6	8.5	10.6	9
Mean hydrated diameter D_N (nm)[b]	250 ± 86	427 ± 90	92 ± 49	55 ± 10	223 ± 2	63 ± 36
Iron content (g Fe/g dry weight)	0.75	0.73	0.50	0.526	0.41	0.56
Saturation magnetization of the "core" M_s (A·m²/kgFe)[c]	118	118	74	55	82	62
ξ-Potential in water (mV)[c]	−47.8 ± 8.6	+37.4 ± 1.6	−14.6 ± 0.7	−15.6 ± 1.6	−13.3 ± 1.6	+55.4 ± 1.6

[a]Determined from broadening of the X-ray diffraction peak (shown in Fig. 1c)
[b]Assemblies of magnetic nanoparticles
[c]Determined using magnetization data shown in Fig. 1b

of 15 μl siRNA solution (2 μg siRNA/ml) and 15 μl 0.1 M ammonium acetate buffer, pH 5. (See Note 10).

2. In vial 2, prepare a mixture of 5 μl 250 μM potassium iodide, 5 μl sodium ^{125}iodide (0.5 mCi), and 30 μl 0.1 M sodium hydroxide.

3. Add 50 μl 30 mM thallium trichloride solution to vial 2 (see Note 11), quickly mix and immediately transfer the contents of vial 2 to vial 1, incubate the vial at 60°C for 45 min, and then cool on ice.

4. Add 50 μl of 0.1 M sodium sulfite, then 150 μl of 1 M ammonium acetate buffer, pH 7, incubate for 60 min at 60°C, and then cool on ice.

5. During the incubation of the previous step, equilibrate a Sephadex G25 PD10 desalting column with water according to the manufacturer's instructions. Apply the reaction mixture to the column and let it penetrate the column bed. Position a rack with 20 aligned microcentrifuge tubes under the column for fraction collection. Add 5 ml of water twice for elution, collecting 11 drops in each (≈400–500 μl) of the microcentrifuge tubes aligned in the rack.

6. Using a handheld radiation monitor, determine the early eluting product fractions with the highest radioactivity, which will likely be between fractions 6 and 10 (see Figs. 2a and a' for radiolabeling of DNA and siRNA, respectively).

7. Transfer a 20 μl aliquot of the product fraction to a scintillation vial, and determine the radioactivity (CPM) using a gamma counter (e.g. Wallac 1480 Wizard 3″ automatic gamma counter). In another aliquot of the product fraction, determine the DNA concentration (see Fig. 2) by measuring the absorbance D at 260 nm and using the following formula: DNA(siRNA) concentration (μg/ml) = (D_{260}) × (dilution factor) × (50 μg nucleic acid/ml).

Figures 2a and a' show the efficiency of the procedure described in Subheading 3.3, of the protocol for radiolabelling and isolation for both siRNA and DNA, allowing the specific activity to be obtained up to 4.2×10^5 CPM/μg DNA and 9.8×10^5 CPM/μg siRNA. Some conversion of the supercoiled DNA into circular DNA is also observed, as shown in the electrophoresis data in Fig. 2b.

3.4. Testing the Nucleic Acid Association with and Magnetic Sedimentation of Transfection Complexes with Magnetic Nanoparticles

This procedure can be accomplished in 2 h.

1. For use as a transfection enhancer for DNA lipo(magneto)fection, mix 20.2 μl DF-Gold (or SM4-31) and 119.8 μl of water (prepared fresh before the experiment). This results in an enhancer-to-nucleic acid volume/weight ratio of 4 to 1, if testing is performed according to this protocol. In general,

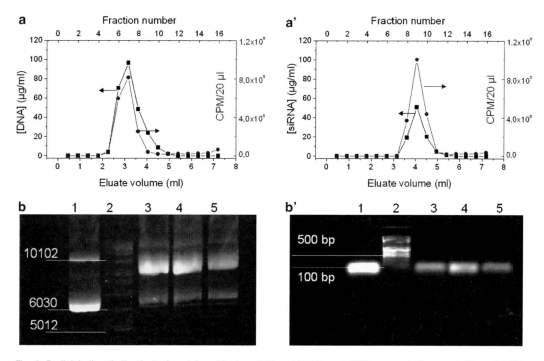

Fig. 2. Radiolabeling (iodination) of nucleic acids. (**a**, **a′**) Plasmid DNA and siRNA concentrations and ^{125}I-radioactivity (CPM/20 μl aliquots) measured in the fractions after purification of the labeled nucleic acids on a Sephadex column G25 PD10. (**b**, **b′**) Fraction probes (ca. 1 μg nucleic acid) were electrophoresed (100 V; 30 min for siRNA and 90 min for plasmid DNA) on an EtBr/1% agarose gel in TBE buffer. Lane 1. Unlabeled nucleic acid. Lane 2. (**b**) Supercoiled DNA Ladder and (**b′**) peqGold DNA Ladder. Lanes 3, 4, and 5. Fractions of the ^{125}J-labeled nucleic acids

any other transfection reagent can be tested as an enhancer instead.

2. In a 96-well round bottom plate (Techno Plastic Products), add 20 μl of magnetic nanoparticle suspension (step 1) into well A1 (corresponding to 5.76 μg iron of magnetic nanoparticles). Add 10 μl of water to each well from A2 to A6.

3. Transfer 10 μl from A1 into A2, mix, transfer 10 μl from A2 into A3, etc., down to A6. Discard the excess 10 μl from A6. Well A7 is a reference.

4. Add 20 μl of enhancer dilution to each well from A1 to A7; mix well with a pipette. To measure the nucleic acid association with magnetic nanoparticles in the absence of enhancer, add 20 μl of water to each well.

5. Add 150 μl ^{125}I-labeled DNA or siRNA solution comprising 2×10^5 CPM/ml ^{125}I-labeled nucleic acid (from Subheading 3.3) to each well from A1 to A7; mix well with a pipette (see Note 12). Incubate for 15 min to allow complex formation.

6. To sediment magnetic transfection complexes, place the plate on the magnetic plate for 30 min.

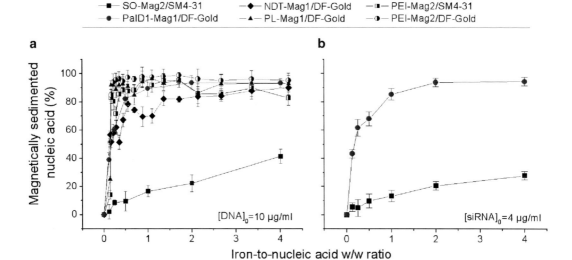

Fig. 3. DNA and siRNA association and magnetic sedimentation with magnetic nanoparticles. Steps from Subheading 3.4 were perfomed in triplicate to form triplexes of nucleic acids with SO-Mag2, PalD1-Mag1, PL-Mag1, NDT-Mag1 and PEI-Mag2 magnetic nanoparticles in the presence of DF-Gold or SM4-31 lipid transfection reagents as enchancers (4 μl enchancer preparation per 1 μg nucleic acid). This figure shows (**a**) DNA and (**b** siRNA associated and magnetically sedimented with magnetic nanoparticles plotted against nanoparticle concentration in terms of iron-to-nucleic acid weight/weight ratio. Starting DNA and siRNA concentrations of 10 and 4 μg ml^{-1}, respectively. A range of iron-to-nucleic acid ratios (w/w) from 0.25 to 4 has been examined. Most of the formulations shown here exhibit high association and magnetic sedimentation of nucleic acids with magnetic nanomaterials in a wide range of iron-to-nucleic acid w/w ratios. Relatively lower associations of magnetic nanoparticles with SM4-31 lipid transfection agent as an enhancer, are nevertheless enough to considerably improve plasmid DNA delivery and siRNA delivery efficiency compared to similar lipoplex efficiency (illustrative examples are shown in Figs. 9 and 12 for adherent HeLa and H441 cells)

7. Carefully sample 50 μl supernatant from each well using a pipette. Transfer each sample together with the pipette tip into the scintillation vial, taking care to avoid disturbing magnetically sedimented complexes.

8. Measure the radioactivity (CPM) in every vial using the gamma counter.

Calculate magnetic sedimentation of the nucleic acids associated with the magnetic nanoparticles (%) as follows:

Magnetically sedimented DNA (siRNA) (%) = $[1 - CPM_{sample}/CPM_{ref}] \times 100$, where CPM_{ref} is the radioactivity measured in the reference well A7, if the assay is carried out following the above protocol. Sample results are shown in Figs. 3a and b, for different DNA and siRNA lipoplexes comprising magnetic nanoparticles (see also Note 13).

3.5. Cell Culture and Plating for Transfection

1. Culture H441 cells (human adenocarcinoma bronchial epithelial cells) at 37°C in a 5% CO_2 atmosphere. Split the cells at a ratio of 1:4 to 1:5 every 4–5 days before reaching 100% confluence. Seed plates 24 h before transfection (see Note 14). H441 cells are used as an example, but other cell lines could be used instead.

2. For plating, wash the cells with PBS, aspirate the supernatant and add 2 ml trypsin-EDTA (0.25%) solution per 75-cm² cultivation flask. Shake gently so that the solution can cover the entire cell are, and then remove all the trypsin with a Pasteur pipette and incubate the flask at 37°C for 2–3 min. Observe the cells under a microscope; when the cells are detached, immediately add 10 ml of H441 culture medium to arrest the trypsin action.

3. Count the cells using a microscope counting chamber (hemocytometer) and resuspend in H441 culture medium at a density of 1.67×10^5 cells per ml before transferring to a reagent reservoir.

4. Transfer 150 μl of the cell suspension per well to the 96-well flat bottom plate (Techno Plastic Products) or to a clear bottom black-walled plate, 96-well (Greiner Bio-One) using a multichannel pipette (see Note 15). 25,000 cells per well reach 50% confluence before magnetofection 24 h later.

5. Store the plate in a cell culture incubator at 37°C in a 5% CO_2 atmosphere until transfection, usually 24 h later. The cells should be approximately 50% confluent at the time of transfection.

6. Adherent cells that divide more rapidly than H441 cells (such as NIH-3T3 or HeLa cells) should be plated at a density of 5,000–10,000 cells per well.

7. Cultivate suspension-type Jurkat cells at a density of 0.5–1.5×10^6 cells/ml. Split the cells at a ratio of about 1:2 to 1:3 every 2–3 days. For transfection experiments, use cells up to passage 12–13. Just prior to transfection, count the cells using a hemocytometer, sediment them by centrifugation at $300 \times g$ (1,200 rpm on a Heraeus Megafuge 2.0) and resuspend at a density of 1.33×10^5 cells/ml before transferring to a reagent reservoir. Transfer 150 μl of the cell suspension per well to the 96-well plate (U-bottom) using a multichannel pipette (20,000 cells per well). Perform the transfection just after plating the cells.

3.6. Preparation of Magnetic Nanoparticle-DNA (siRNA) Lipoplexes

1. To form lipoplexes containing magnetic nanoparticles, add 20 μl of a suspension of magnetic nanoparticles to be tested (from Subheading 2.6 step 1) into wells A4, A7, A10 and E1, E4, E7 and E10 of flat-bottom 96-well plates. This will result in a magnetic nanoparticle iron-to-DNA(siRNA) ratio of 0.5:1 (wt/wt) (see Notes 16 and 17). Add 40 μl of the enhancer solution (from Subheading 2.6 step 2 or 3 for DNA or siRNA transfection, respectively) to each of the wells and mix using a pipette.

2. Add 300 μl of the DNA (or siRNA) solution from Subheading 2.6 step 4 (12 μg DNA/ml or 4.8 μg siRNA/ml in a serum- and

Table 2
Characteristics of selected DNA and siRNA lipoplexes generated at DNA and siRNA concentrations of 10 and 4 µg ml^{-1}, respectively

Complex	Iron-to-nucleic acid w/w ratio	ξ-Potential (mV)	Mean hydrated diameter D (nm)
Short interfering RNA lipoplexes			
DF-Gold/pBLuc	–	+16.9 ± 4.7	742 ± 111
PalD1-Mag1/DF-Gold/siRNA	0.5:1	+12 ± 6.3	968 ± 289
SM4-31/siRNA	–	–	1,052 ± 503
SO-Mag2/SM4-31/siRNA	0.5:1	–	728 ± 128
Plasmid DNA lipoplexes			
DF-Gold/pBLuc	–	+16.9 ± 4.7	742 ± 111
NDT-Ma1/Df-Gold/pBLuc	0.5:1	+16.5 ± 3.2	1,730 ± 172
PL-Mag1/Df-Gold/pBLuc	0.5:1	-2.5 ± 3.3	1,509 ± 374
PalD1-Mag1/DF-Gold/pBLuc	0.5:1	–	529 ± 214
SM4-31/pBLuc	–	–	528 ± 214
SO-Mag2/SM4-31/pBLuc	0.5:1	–	687 ± 159

supplement-free medium such as RPMI 1640, which delivers 3.6 µg DNA or 1.44 µg siRNA per well) to each of the same wells and mix well using a pipette. This results in a final volume of 360 µl in wells A4, A7 and A10 (see also Note 18). The characteristics of selected nucleic acid magnetic lipoplexes prepared according to this protocol are given in Table 2.

3. For the untransfected control setup, add 300 µl of serum- and supplement-free medium and 60 µl of water to well A1. For other controls and references (e.g. magnetic nanoparticle-DNA or siRNA duplexes without enhancer, enhancer-DNA or siRNA complexes without magnetic nanoparticles), substitute the omitted component(s) with medium and/or water. Incubate for 15 min at RT.

4. During the incubation time, fill each of the remaining wells of columns 1, 4, 7, and 10 with 180 µl of serum- and supplement-free medium (RPMI 1640). Prepare a 1:1 dilution series, when the 15 min incubation time is over, as follows: transfer 180 µl, from each of A1, A4, A7, and A10 to B1, B4, B7, and B10, respectively, using a multichannel pipette, then mix, transfer 180 µl from the respective wells in row B to row C and so on down to row D or further to cover the desired nucleic acid dose range per well.

3.7. Magnetofection

Timing: ~30–40 min plus 48–72 h to allow reporter gene transfection for DNA delivery or down-regulation for siRNA delivery. Magnetofection should be carried out under sterile conditions.

1. Check the plates prepared for transfection according to Subheading 3.5 steps 1–5, under the microscope for cell state and confluence. A confluence of ~40–50% before transfection is preferable for H441 cells.

2. Cells that divide more rapidly than H441 cells (NIH-3T3 or HeLa cells) can be transfected at a lower confluence of ~30–40%. Aspirate the medium from the wells and add 150 µl of fresh cultivation medium per well.

3. Transfer 50 µl each of the transfection complex dilutions prepared according to steps from Subheading 3.6, into the culture plates with the seeded cells, as follows: Using a multichannel pipette, mix the dilutions of transfection complex prepared in column 1 of the complex preparation plate (from Subheading 3.6 step 10) by pipetting up and down, then transfer 50 µl to the wells of columns 1, 2, and 3 (to test each composition and dilution of transfection complex in triplicate) of the cell culture plate (Subheading 3.5 step 5). Transfer 50 µl, from each well of column 4 of the complex preparation plate to columns 4, 5, and 6 of the cell culture plate. Transfer 50 µl from each well of column 7 of the complex preparation plate to columns 7, 8 and 9 of the cell culture plate. Transfer 50 µl from each well of column 10 of the complex preparation plate to columns 10, 11, and 12 of the cell culture plate. This results in delivery of 500, 250, 125, and 62.5 ng DNA or 200, 100, 50, and 25 ng siRNA per well in rows A (E), B (F), C (G), and D (H) and so on.

4. Place the cell culture plate on a magnetic plate for 15–30 min to create a permanent magnetic field at the cell layer with a field strength and gradient of 70–250 mT and 50–130 T/m, respectively.

5. Remove the magnetic plates after 20–30 min exposure of the cells to the magnetic field (see Note 19), and incubate the plate containing the transfected cells in a cell culture incubator at 37°C in a 5% CO_2 atmosphere until evaluation.

6. To transfect suspension cells (e.g., Jurkat cells), just after plating the cells, centrifuge the 96-well round-bottom plate with suspension cells from Subheading 3.5 step 7 at 300×*g* for 5 min to sediment the cells. Do not remove the supernatant (see Note 20). Transfer 50 µl of each of the transfection complexes (Subheading 3.6 step 10) to the cell culture plate from Subheading 3.5 step 7, as described in Subheading 3.7 step 3. Take care to avoid pellet dispersion. Place the plate on the magnetic plate for 30 min. Remove the magnetic plate and

place the culture plate with transfected cells into a cell culture incubator and incubate the plate at 37°C in a 5% CO_2 atmosphere until evaluation.

7. To allow reporter gene expression or down-regulation, plates must usually be incubated for 24–72 h after transfection. See also Note 21.

3.8. Evaluation of the Association of Transfection Complexes with Cells and their Internalization into Cells by Microscopy and Fluorescence-activated Cell Sorting

1. To evaluate nucleic acid transfection complex association with cells, prepare transfection complexes as described in Subheading 3.6 steps 1–3. Then, add 1 µl of the 1 mM solution in DMSO of the cell-impermeable intercalating nucleic acid stain YOYO-1 iodide per 360 µl complex (corresponds to one dye molecule per 5.5 bp), incubate for 15 min in the dark, and perform dilutions as described in Subheading 3.6 step 5. Perform transfections according to steps of Subheading 3.7. To visualize the association/localization of YOYO-1-labeled transfection complexes with cells, after incubation at 37°C in a 5% CO_2 atmosphere, use a fluorescence microscope and observe with 490/509 nm green fluorescence. Sample results are shown in Fig. 4a.

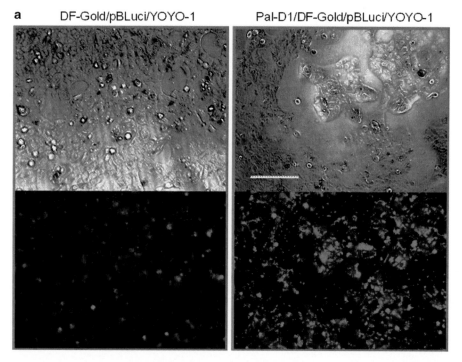

Fig. 4. Association/internalization of plasmid DNA lipoplexes and magnetic lipoplexes with H441 human pulmonary epithelial cells by microscopy and flow cytometry. (a) H441 cells were incubated for 30 min at the magnetic plate with DF-Gold/pBLuc lipoplexes and PalD1-Mag1/DF-Gold/pBLuc magnetic lipoplexes labeled with YOYO-1 intercalating DNA stain at a DNA concentration of 250 ng/25,000 cells/0.33 cm^2 and an iron-to-siRNA w/w ratio of 0.5 as described in Subheading 3.7 of this protocol, and were observed after 10 h with a fluorescence microscope. Images were obtained at an original magnification of 10; scale bar = 200 µm. Pictures show phase contrast images (at the top) and fluorescence images taken at 530/30 nm for YOYO-1 labeled plasmid DNA (at the bottom). Fluorescence microscopy data prove the association of the magnetic

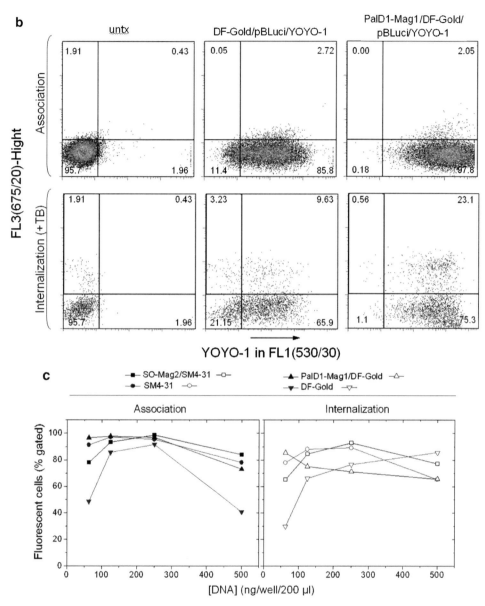

Fig. 4. (continued) transfection complexes with a majority of the cells and considerably higher cell association of magnetic complexes versus non-magnetic lipoplexes. (**b**) Twenty-four hours post-transfection, the cells were trypsinized, washed and resuspended in 1% FCS in PBS. Vector cell association and internalization were analyzed using a FACS Vantage microflow cytometer. The figures show density plots of untransfected H441 cells (untx); cells transfected with lipoplexes containing DreamFect-Gold (DF-Gold/pBLuci/YOYO-1) and magnetic triplexes containing PalD1-Mag1 magnetic nanoparticles (PalD1-Mag1/DF-Gold/pBLuc/YOYO-1) for vector *association* analysis with YOYO-1-labeled luciferase plasmid. For analysis of vector *internalization*, the cells were additionally incubated with Trypan Blue (TB) at a final TB concentration of 1 mg ml^{-1} to quench fluorescence of the complexes that were associated with the cells but not internalized into the cells. The numbers in squares indicate the percentages of gated cells with untreated cells as a reference. (**c**) Percentage of YOYO-1 positive H441 cells (*Association*) and of H441 cells that only have internalized complexes (*Internalization*) versus DNA concentration at transfection for plasmid lipoplexes with Dreamfect-Gold or SM4-31 lipid reagents or magnetic lipoplexes made up of these lipids with PalD1-Mag1 or SO-Mag2 magnetic nanoparticles. The results given here clearly show that most of the cells are associated with transfection complexes. There is no considerable difference between the percentages of cells associated with magnetic and non-magnetic transfection complexes made with SM4-31. However, DreamFect-Gold lipoplexes exhibit less cell association/internalization compared to DreamFect-Gold magnetic triplexes

2. To evaluate the association of siRNA transfection complexes with cells, and their internalization into cells, prepare the transfection complexes with rhodamine-labeled GFP-siRNA (siRNA-Rho), according to steps of Subheading 3.6, and perform transfection of the cells as described in Subheading 3.7. Alternatively, the rhodamine-labeled magnetic nanoparticles synthesized according to step 8 from Subheading 3.1 of the protocol can be used.

3. After incubation at 37°C in a 5% CO_2 atmosphere, observe the plate using a fluorescence microscope at 510/650 nm (red fluorescence) to visualize localization of the siRNA-Rho complexes or rhodamine-labeled magnetic nanoparticles (for an example, see Fig. 5a and Fig. 6).

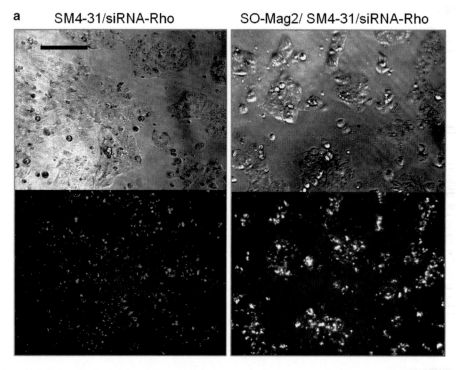

Fig. 5. Association/internalization of the siRNA lipoplexes and magnetic lipoplexes with H441 human pulmonary epithelial cells by microscopy and flow cytometry. (a) H441 cells were incubated for 30 min at the magnetic plate with SM4-31/siRNA-Rho lipoplexes and SO-Mag2/SM4-31/siRNA-Rho magnetic lipoplexes with rhodamin-labeled GFP-siRNA (siRNA-Rho) at a siRNA concentration of 32 nM (100 ng) siRNA/25,000 cells/200 µl/0.33 cm^2 and an iron-to-siRNA w/w ratio of 1, and were observed with a fluorescence microscope after 10 h. Images were obtained at an original magnification of 10×; scale bar = 200 µm. The pictures show phase contrast images (*at the top*) and fluorescence images taken at 510/650 nm for siRNA-Rho (*at the bottom*). The results indicate a higher level of cell association of siRNA when delivered with magnetic complexes compared to non-magnetic lipoplexes. (b) H441-cells were transfected in a 96-well plate as described in Subheading 3.7. Sixty hours post-transfection, the cells were trypsinized, washed and resuspended in 1% FCS in PBS. Vector cell/association/internalization was analyzed using a FACS Vantage microflow cytometer. Figures show density plots of untransfected cells (untx); cells transfected with lipoplexes Df-Gold/siRNA-Rho comprising

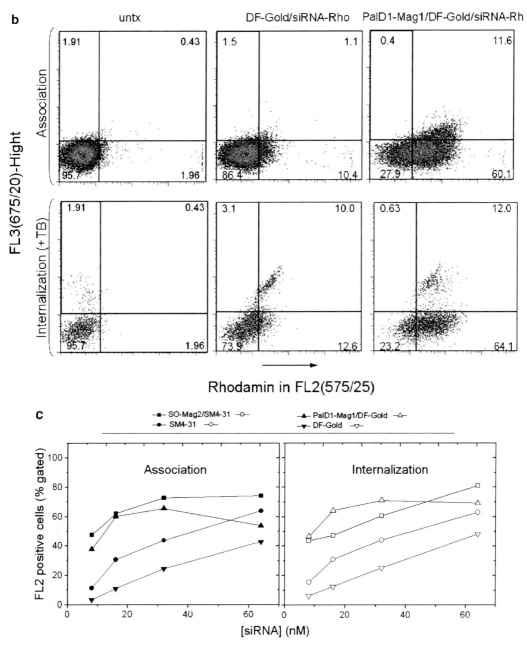

Fig. 5. (continued) DreamFect-Gold and siRNA-Rho and cells transfected with magnetic triplexes made up of PalD1-Mag1 magnetic nanoparticles, Dreamfect-Gold and siRNA-Rho (PalD1/Df-Gold/siRNA-Rho) for vector association analysis. For analysis of vector internalization, the cells were additionally incubated with Trypan Blue (TB) at a final TB concentration of 1 mg ml^{-1} to quench the fluorescence of the complexes associated with the cells but not internalized into the cells. The siRNA dose was 100 ng per well in the examples in Fig. **b**. The numbers in squares indicate the percentages of gated cells with untreated cells as a reference. (**c**) The percentage of rhodamine-positive H441 cells associated with siRNA (*Association*) and the percentage of H441 cells that have internalized complexes (*Internalization*) are shown versus siRNA concentration for cells and complexes as in Fig. **b**. These data are given for cells transfected with lipoplexes SM4-31/siRNA-Rho or magnetic triplexes comprising SO-Mag2 magnetic nanoparticles, SM4-31 and siRNA-Rho (SO-Mag2/SM4-31/siRNA-Rho). The results show that the magnetic complexes are characterized by higher cell association/internalization upon magnetofection compared to siRNA lipofection

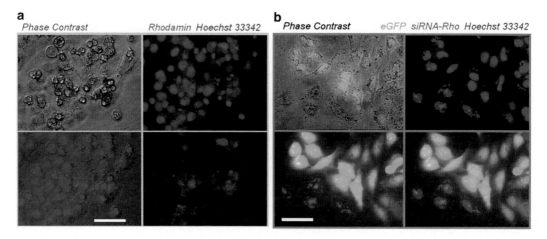

Fig. 6. Internalization and perinuclear localization of magnetic DNA and siRNA lipoplexes post-liposomal magnetofection in HeLa cells detected by microscopy. (**a**) HeLa human cervical carcinoma cells and (**b**) HeLa cells stably transfected with eGFP protein were incubated for 30 min at the magnetic plate with SO-Mag2-Rho/SM4-31/pBLuc and PalD1-Mag1/DF-Gold/GFP-siRNA-Rho triplexes, at plasmid DNA and siRNA concentrations of 125 ng plasmid/10,000 cells/0.33 cm^2 or 100 ng siRNA/10,000 cells/0.33 cm^2, respectively, and an iron-to-nucleic acid wt/wt ratio of 0.5, and a SM4-31(DF-Gold)-to-siRNA v/w ratio of 4 to 1, and were observed after 24 h with a fluorescence microscope. Images were obtained at an original magnification of 40×, scale bar=50 µm. Hoechst 33342 was used as a nuclear counterstain. The pictures show fluorescence images taken at 490/509 nm (*green fluorescence*) for eGFP fluorescence, 510/650 nm (*red fluorescence*) for rhodamine-labeled magnetic nanoparticles SO-Mag2-Rho, synthesized according to steps from Subheading 3.1, and rhodamine-labeled siRNA, and at 350/461 nm (*blue fluorescence*) for Hoechst 33342 nuclear staining, or overlays thereof. Fluorescence microscopy data prove the association of the magnetic transfection complexes with a majority of the cells, and are indicative of internalization into cells and perinuclear localization of the complexes

4. To allow visualization of the location of the internalized complexes in relation to the cell nuclei, add 1 µl per well of the cell-permeable nuclear counterstain Hoechst 33342 (1 mg ml^{-1} stock solution; this results in a final Hoechst concentration of 5–10 µg/ml). Incubate for 15–20 min and observe with 350/461 nm blue fluorescence filters for Hoechst dyes (results are shown in Figs. 6a and b).

5. To quantify the percentage of cells that are associated with or have taken up (internalized) transfection complexes, perform FACS analysis on cells from Subheading 3.8 step 1 or 2.

6. Wash the adherent cells with 150 µl PBS per well, aspirate the supernatant with a Pasteur pipette, add 10 µl Trypsin-EDTA (0.25%) solution per well and incubate the flask at 37°C for 2–3 min. Observe the cells under a microscope. When the cells are detached, immediately add 200 µl of complete cell culture medium to arrest trypsinization. This step is omitted for suspension type cells such as Jurkat cells.

7. Combine cells from triplicate wells of cell culture plates in a fluorescence-activated cell sorting (FACS) tube. Centrifuge, at 300×*g* (1,200 r.p.m. on a Heraeus Megafuge 2.0) for 5 min, remove supernatants carefully and add 1 ml PBS supplemented with 1% FCS (FACS buffer). Centrifuge, again at

300×g for 5 min, discard the supernatants carefully, and resuspend the cells in 0.5 ml FACS buffer.

8. Analyze the cells on a flow cytometer: excite fluorescence with an argon laser >488 nm and detect green YOYO-1 fluorescence using a 530/30 nm bandpass filter and rhodamine fluorescence using a 575/26 nm bandpass filter. Analyze a minimum of 10,000 events per sample (see Note 22).

9. The percentages of cells with associated transfection complexes are determined as a percentage of gated fluorescent events detected with the appropriate filter, using untreated cells as a reference.

10. To quench fluorescence from the complexes that are associated with cells but are not internalized into cells and to determine the percentage of cells that have internalized the transfection complexes, add trypan blue (TB) stock solution to the cell suspension for FACS analysis to a final TB concentration of 1 mg/ml, and analyze the cells again on a flow cytometer according to Subheading 3.8 steps 8 and 9.

11. Alternatively, to stain the YOYO-1 labeled complexes that are associated with cells but are not internalized into cells, add 1 µl/ml cell suspension of the cell-impermeable nucleic acid stain propidium iodide (PI) stock solution diluted 1 to 10 (100 µg/ml) to obtain a final PI concentration of 1 µg/ml, and incubate for 10 min. Analyze the cells on a flow cytometer and detect YOYO-1 fluorescence using a 530/30 nm bandpass filter and propidium iodide fluorescence using a 575/26 nm bandpass filter. Examples of results for a semi-quantitative evaluation of the complex association/internalization are given for adherent and for suspension-type Jurkat cells in Figs. 4b and c, Figs. 5b and c and in Figs. 7a and b, respectively.

3.9. Quantification of the Internalization of Transfection Complexes Into Cells using Radioactively Labeled Nucleic Acids

1. To quantify transfection complex internalization into cells, prepare the transfection according to 3.6 using ^{125}I-labelled DNA or siRNA solution from Subheading 2.9 and perform transfection of the cells according to steps of Subheading 3.7. Reserve 50 µl of each of the transfection complexes as a reference.

2. After incubation at 37°C in a 5% CO_2 atmosphere, wash the cells with 150 µl PBS per well at different time points post-transfection, and aspirate the supernatant with a Pasteur pipette. To remove extracellularly bound complexes, add 100 µl per well 100 U/ml heparin solution containing 75 mM sodium azide to inhibit endocytosis (17).

3. After incubation at 37°C in a 5% CO_2 atmosphere for 30 min, wash the cells with 150 µl PBS per well, aspirate the supernatant, add 10 µl Trypsin-EDTA (0.25%) solution per well and incubate the flask at 37°C for 2–5 min.

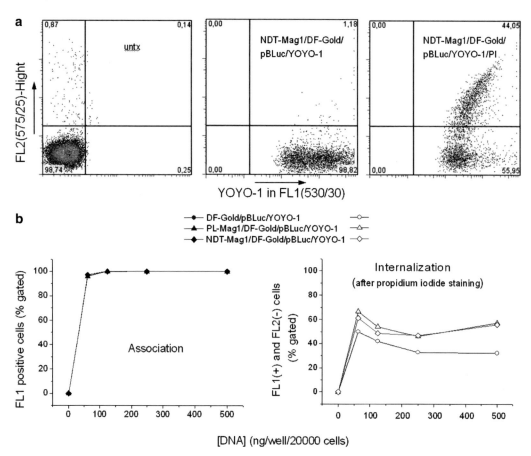

Fig. 7. Association/internalization of the plasmid DNA lipoplexes and magnetic lipoplexes in suspension-type Jurkat cells characterized by flow cytometry. Jurkat cells were transfected in a 96-well plate with plasmid DNA complexes labeled with the cell-impermeable intercalating nucleic acid stain YOYO-1 iodide (see step 1 from Subheading 3.8) as described in step 6 from Subheading 3.7, for suspension cells. Forty-eight hours post-transfection, the cells were washed with PBS and resuspended in 1% FCS in PBS as described in steps 6 and 7 from Subheading 3.8. Complex cell/association/internalization were analyzed using a FACS Vantage microflow cytometer. (a) Density plots of untransfected Jurkat cells (untx) and cells transfected with magnetic triplexes containing NDT-Mag1 magnetic nanoparticles (NDT-Mag1/DF-Gold/pBLuc/YOYO-1). For vector *association* analysis, cells were additionally incubated with propidium iodide at a final PI concentration of 1 µg ml^{-1} to stain the complexes associated with the cells, but not those internalized into cells (NDT-Mag1/DF-Gold/pBLuc/YOYO-1/PI) (see Subheading 3.8 step 11). The plasmid dose was 125 ng per well in the examples in Fig. **a**. The numbers in squares indicate the percentages of gated cells with untreated cells as a reference. (b) Percentage of YOYO-1 positive Jurkat cells (*Association*) and of Jurkat cells that have internalized complexes (*Internalization*) versus plasmid DNA concentration at transfection for plasmid lipoplexes with Dreamfect-Gold or magnetic lipoplexes made of these lipids and PL-Mag1 or NDT-Mag1 magnetic nanoparticles. The results given here clearly show that most of the cells are associated with transfection complexes. DreamFect-Gold lipoplexes exhibit less cell association/internalization compared to DreamFect-Gold magnetic triplexes

4. Observe the cells under a microscope. When the cells are completely detached, add 200 µl cell culture medium.

5. Carefully collect the cell suspension from each well using a pipette. Transfer each sample together with the pipette tip into a scintillation vial. Measure the radioactivity (CPM) in each vial using the gamma counter.

6. Calculate the amount of DNA (siRNA) associated with magnetic nanoparticles as follows:

$$\text{Internalized siRNA (\%)} = [\text{CPM}_{sample}/\text{CPM}_{ref}] \times 100,$$

where CPM_{ref} is the radioactivity measured from the reference sample.

Example of the results on nucleic acids association with siRNA and plasmid DNA transfection complexes are shown in Fig. 8.

3.10. Quantification of Luciferase Reporter Gene Expression in the Cell Lysate

1. To prepare cell lysates from adherent cells, wash transfected adherent cells from Subheading 3.7 with 150 μl per well PBS using a multichannel pipette. Add 100 μl lysis buffer per well, incubate for 10 min at RT, and then place the culture plate on ice.

2. To prepare cell lysates from suspension cells, centrifuge the cell culture plate from Subheading 3.7 at 300×g (1,200 r.p.m. on a Heraeus Megafuge 2.0) for 5 min. Place the culture plate on the magnetic plate to keep the cell pellet in place (by this time, the cells are associated with or have taken up magnetic nanoparticles). Carefully remove the supernatants with a multichannel pipette. Add 150 μl PBS to wash the cells, repeat the centrifugation and remove the supernatant. Add 150 μl lysis buffer per well. Incubate for 10 min at RT, and then place on ice.

3. To quantify luciferase reporter gene expression in cell lysates, transfer 50 μl cell lysate from each well into a 96-well black flat-bottom microplate. Add 100 μl luciferase buffer per well, and optionally mix with a pipette. Measure the chemiluminescence intensity (count time 0.20 min with background correction) using a luminometer, e.g., a Microplate Scintillation & Luminescence Counter (Canberra Packard) or a Wallac Victor 2 Multi-label Counter (PerkinElmer).

4. To construct a calibration curve to determine the amount of luciferase in transfected cell samples, add 50 μl lysis buffer per well to columns 1 and 3 of a black 96-well plate and 40 μl lysis buffer per well to columns 2 and 4. To well A1, add 30 μl lysis buffer and 20 μl luciferase standard stock (0.1 mg luciferase per ml and 1 mg BSA per ml in 0.5 M Tris-acetate buffer, pH 7.5). Pipette 50 μl from A1 to B1, mix well, and then from B1 to C1, etc. down to H1. From H1, continue the dilution series by transferring 50 μl to A3, and continue in column 3 down to G3, leaving H3 as blank. Pipette 10 μl each from column 3 to 4, and from column 1 to 2. Add 100 μl luciferase buffer to each of the wells of columns 2 and 4.

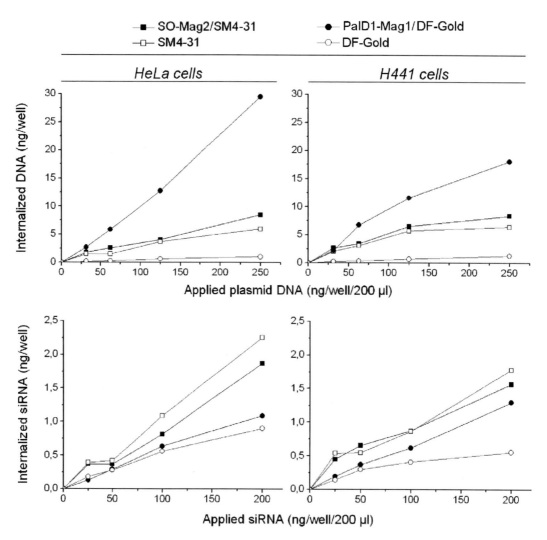

Fig. 8. Vector internalization in HeLa human cervical epithelial adenocarcinoma cells and H441 human lung epithelial cells quantified using radioactively labeled nucleic acids. HeLa and H441 cells were transfected in a 96-well plate using ^{125}I-labeled nucleic acid (plasmid DNA or siRNA) lipoplexes with DreamFect-Gold and SM4-31, respectively, or magnetic triplexes comprising DreamFect-Gold or SM4-31 and SO-Mag2 or PalD1-Mag1 magnetic particles, respectively. Iron-to-DNA and DreamFect-Gold (or SM4-31)-to-nucleic acid ratios of 0.5–1 and 4 µl/1 mg nucleic acid, respectively. At 24 h post-transfection, the cells were incubated with heparin solution in the presence of sodium azide to remove extracellularly bound complexes, washed, trypsinized and collected. Cell-associated radioactivity was measured with a gamma-counter. The applied dose of the radioactively labeled complexes was used as a reference. The results were recalculated in terms of the ng of nucleic acid internalized per well and plotted against the applied nucleic acid dosage. In both tested HeLa and H441 cells, the magnetic complexes of the DNA with DreamFect-Gold are better internalized compared to the SM4-31 magnetic lipoplexes, whereas siRNA magnetic triplexes with SM4-31 are vice versa better internalized compared to the complexes with DreamFect-Gold. Magnetofection results in overall better internalization of the transfection complexes compared to lipofection with the same vector type

Measure the chemiluminescence intensity as described above. Plot the logarithm of luciferase content in the dilution series as a function of the logarithm of measured luminescence intensity (light units). Use an approximation function (usually linear regression in this concentration range) to calculate the amount of luciferase in the transfected cell samples.

5. To be able to present the results of the luciferase expression assays as weight luciferase per weight unit total protein, determine the total protein content in the lysate as described in Subheading 3.10 steps 5–7.

6. Determine the total protein content of the samples as follows: first, add 150 μl of water to each well in a flat-bottom 96-well plate. Using a multichannel pipette, transfer 10 μl of each of the cell lysates (from steps 1 or 2 of this section) into the corresponding wells of the protein assay plate. Add 40 μl of BioRad protein assay reagent to each well and mix carefully using a plate shaker or a multichannel pipette. Measure the absorbance at 590 nm using a microplate reader (e.g. Wallac 1420 Multilabel counter; measuring time set to 0.1 s).

7. To construct a calibration curve to determine the amount of total protein in the transfected cell sample, add 25 μl of lysis buffer per well in one row (e.g. row A) of a flat-bottom 96-well plate. Add 50 μl of BSA stock solution to well 1 (e.g. A1). Mix well using a pipette. Transfer 50 μl from well 1 to well 2, mix, transfer 50 μl from well 2 to well 3 and so on to well 11, leaving well 12 as blank. Add 150 μl of water per well in another row (e.g. row B). Transfer 10 μl from row A to row B. Add 40 μl BioRad reagent to each well and mix carefully using a plate shaker or a multichannel pipette. Measure the absorbance at 590 nm (or 570 nm) using a microplate reader (e.g., a Wallac 1420 Multilabel counter; measuring time set to 0.1 s). Plot the measured absorbance versus the protein content for each well. Use linear regression to derive a calibration function, from which the protein content in the samples can be calculated. Calculate the total protein content per 10 μl cell lysate for every sample using the calibration curve.

8. Calculate weight luciferase per weight total protein (see Note 23); the results can be plotted against the applied DNA concentration or dose per well, in order to get a dose-response curve.

Sample results of luciferase expression post magnetofection versus lipofection are given in Fig. 9.

3.11. Evaluation of eGFP Gene Expression

To characterize the efficiency of eGFP plasmid delivery or anti-GFP siRNA delivery expression, prepare the transfection complexes according to steps of Subheading 3.6 and perform transfections as described in Subheading 3.7.

3.11.1. Estimation of eGFP Expression by Microscopy

After incubation at 37°C in a 5% CO_2 atmosphere, usually for 48 h, observe the plate with cells using a fluorescence microscope at 490/509 nm (green fluorescence). Take bright field and fluorescence images at a magnification of 10× to visualize the

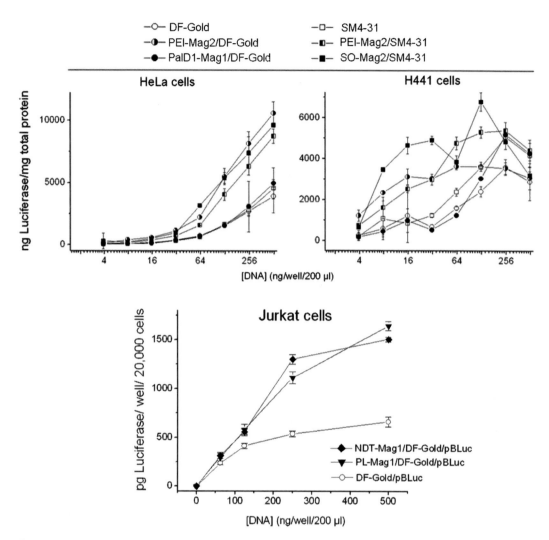

Fig. 9. Magnetofection versus lipofection efficiency characterized by luciferase reporter gene expression analysis in cell lysates. HeLa cervical epithelial adenocarcinoma cells, H441 human lung epithelial cells, and Jurkat cells were transfected with luciferase plasmid lipoplexes DreamFect-Gold/pBLuc or SM4-31/pBLuc or magnetic triplexes containing magnetic nanoparticles, according to Subheading 3.7. The figures show Luciferase expression (ng luciferase per mg total protein for adherent cells and pg luciferase per well for suspension-type Jurkat cells) versus applied plasmid concentration measured 48 h post-transfection. The iron-to-DNA ratio was 0.5 to 1, and the DreamFect-Gold(or SM4-31)-to-DNA ratio was 4 µl/1 µg plasmid. Magnetofection with selected magnetic lipoplexes results in considerable improvement of transfection efficiency compared to lipofection

cells expressing the eGFP reporter gene (see Fig. 10 for plasmid lipofection and Fig. 12 for GFP-siRNA delivery in eGFP-stably transfected cells).

3.11.2. Determination of the Percentage of eGFP-expressing Cells using Flow Cytometry

1. To characterize the transfection efficiency in terms of the percentage of transfected cells, prepare the transfection complexes with a eGFP reporter plasmid according to 3.6 and perform transfections according to 3.7. After incubation at 37°C in a

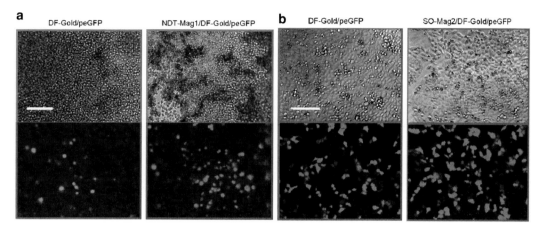

Fig. 10. Enhanced GFP (eGFP) reporter gene expression in suspension-type Jurkat T cell leukemia cells and adherent HeLa human cervical epithelial adenocarcinoma detected by microscopy. (**a**) Jurkat cells and (**b**) HeLa cells were incubated for 30 min at the magnetic plate with lipoplexes DF-Gold/peGFP or magnetic lipoplexes NDT-Mag1/DF-Gold/peGFP at a DNA concentration of 250 ng/20,000 cells/200 μl (Jurkat cells) and magnetic lipoplexes SO-Mag2/DF-Gold/peGFP at a DNA concentration of 16 ng/25,000 cells/0.33 cm^2 (HeLa cells) and observed 48 h post-transfection with a fluorescence microscope (Subheading 3.11 step 1). The iron-to-DNA wt/wt ratio was 0.5, with 4 μl DF-Gold/1 μg DNA. The figure shows bright field (*at the top*) and fluorescence images taken at 490/509 nm for eGFP fluorescence (*at the bottom*). Images were obtained at an original magnification of 10×, scale bar = 200 μm. To quantify the percentage of cells that express the eGFP reporter gene, we performed FACS analysis (see Subheading 3.11 step 2 and Fig. 11) on cells transfected as described in Subheading 3.7 with eGFP-plasmid complexes prepared according to steps from Subheading 3.6. The microscopy images show higher percentage of the eGFP expressing cells and higher eGFP fluorescence intensity post-magnetofection compared to lipofection both in adherent HeLa and suspension-type Jurkat cells

5% CO_2 atmosphere, usually over 48 h, prepare the cells for FACS analysis as described in Subheading 3.8 steps 6 and 7.

2. Analyze the cells on a flow cytometer: excite enhanced GFP (eGFP) fluorescence with an argon laser >488 nm and detect fluorescence using a 530/30 nm bandpass filter. Analyze a minimum of 10,000 events per sample.

3. The percentages of cells expressing eGFP are determined as a percentage of gated fluorescent events. To avoid overestimation of the percentage of eGFP-expressing cells, analyze untransfected cells and cells transfected with a luciferase reporter plasmid, using duplexes of magnetic nanoparticles and luciferase plasmid and magnetic lipoplexes of the luciferase plasmid as controls. This is required, because some transfection reagents such as Dreamfect-Gold, cause some apparent fluorescence, which is either an artifact or autofluorescence. Using only untreated cells as a control, may lead to an overestimation of the percentage of transfected cells. Sample results for suspension-type Jurkat cells are given in Fig. 11.

3.11.3. Quantification of eGFP Expression in Cell Lysates

1. To quantify eGFP expression in cell lysates prepared as described in Subheading 3.10. Step 1 for adherent or in Subheading 3.10 step 2 for suspension cells, transfer 50 μl cell

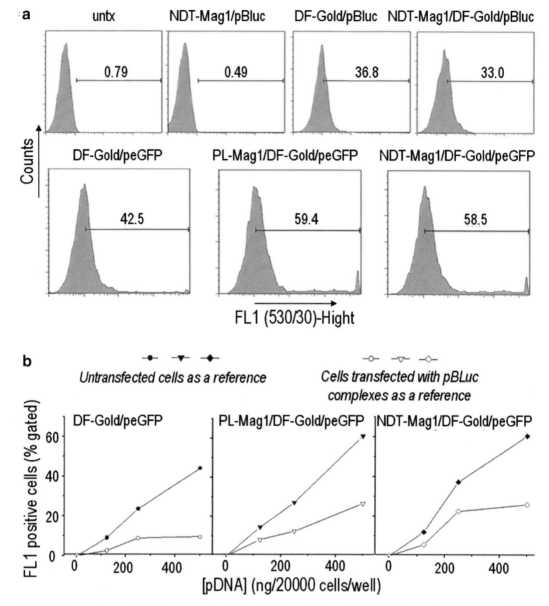

Fig. 11. Liposomal magnetofection versus lipofection efficiency. Enhanced GFP (eGFP) reporter gene expression in Jurkat cells characterized by flow cytometry. Jurkat cells were transfected in a 96-well plate as described in Subheading 3.7 step 6, for suspension cells. Forty-eight hours post-transfection, the cells were washed and resuspended in 1% FCS in PBS according to Subheading 3.8 step 7. eGFP reporter gene expression was analyzed using a FACS Vantage microflow cytometer, according to Subheading 3.11 step 2. (**a**) Histogram plots of untransfected Jurkat cells (untx); cells transfected with duplexes comprising luciferase plasmid and NDT-Mag1 magnetic nanoparticles (NDT-Mag1/pBLuc) or liposomal transfection reagent alone (DF-Gold) or triplexes comprising magnetic nanoparticles, DreamFect-Gold and pBluc (NDT-Mag1/DF-Gold/pBLuc); lipoplexes DremFect-Gold with eGFP plasmid (DF-Gold/peGFP) and magnetic triplexes comprising PL-Mag1 or NDT-Mag1 magnetic nanoparticles (PL-Mag1/DF-Gold//peGFP, NDT-Mag1/DF-Gold/peGFP) for eGFP gene expression analysis. The DNA dose was 250 ng per well. (**b**) Percentage of eGFP-expressing Jurkat cells with respect to the DNA dose per well calculated for the Jurkat cells transfected with lipoplexes DreamFect-Gold with eGFP plasmid (DF-Gold/peGFP) and magnetic triplexes comprising PL-Mag1 or NDT-Mag1 magnetic nanoparticles (PL-Mag1/DF-Gold//peGFP, NDT-Mag1/DF-Gold/peGFP) calculated as shown in **a** using untransfected cells as the reference (compact symbols) and cells transfected with similar luciferase plasmid triplexes (open symbols). Apparently, with untransfected cells as a reference, the percentage of eGFP-expressing cells is overestimated in the case of high autofluorescence of the lipid transfection reagent (as for DreamFect) and underestimated with luciferase plasmid triplexes as a reference. In both cases, the transfection efficiency in terms of the percentage of cells expressing the eGFP protein is higher for optimized magnetofection with optimized magnetic complexes compared to the results of lipofection with the same lipid transfection reagent

lysate from each well into a black 96-well plate with a transparent bottom (e.g. clear bottom black-walled plate, Greiner). Add 100 μl PBS per well and mix with the pipette. Measure the fluorescence intensity (485/535 nm, 1.0 s) using a microplate fluorescence reader, for example, a Wallac 1420 Multilabel counter. For blanks, measure wells with lysates of nontransfected cells.

2. To construct a calibration curve to determine the absolute amount of eGFP in transfected cell samples, add 3 μl of eGFP stock solution to 147 μl of lysis buffer in well A1 of a 96-well clear bottom black-walled plate and mix well. Add 50 μl of lysis buffer to each of wells A2-A12. Transfer 100 μl from A1 to A2, mix well, transfer 100 μl from A2 to A3, mix, and so on to A11. Discard the surplus 100 μl from well A11, leaving A12 as blank. Add 100 μl PBS to each well of row A and mix well. Using a microplate fluorescence reader (e.g., Wallac 1420 Multilabel counter), measure the fluorescence intensity of eGFP (excitation 485 nm, emission 535 nm, measuring time 1 s per well). Plot the measured fluorescence intensity as a function, of eGFP content per well. Use linear regression to derive a calibration function from which the eGFP content in the samples can be calculated. Use a calibration curve, constructed as described, to calculate the amount of eGFP in the transfected cell samples (see Note 24).

3. To present the results of the reporter gene expression assays as weight eGFP per weight total protein, determine the total protein content of the sample as described in Subheading 3.10 steps 5–8.

4. Calculate weight eGFP per weight total protein (see Note 25); to estimate siRNA delivery efficiency, normalize the results to the reference data determined for untransfected cells. The results can be plotted against time post-transfection to evaluate the time course of the silencing effect and to define the optimum exposure time for screening experiments (see Note 26) or against the siRNA concentration or dose per well in order to get a dose-response curve. (Sample results are given in Fig. 12).

3.12. MTT-based Test for Toxicity of Transfection Complexes

1. Wash transfected adherent cells with 150 μl PBS per well using a multichannel pipette and discard wash solutions. For transfected suspension cells, centrifuge at $300\,g$ (1,200 r.p.m. on a Heraeus Megafuge 2.0) for 5 min, remove the supernatants, carefully add 150 μl PBS, centrifuge again at $300\,g$ for 5 min, and discard the supernatants.

2. Add 100 μl per well of MTT solution and incubate in a cell culture incubator for 1.5–2 h.

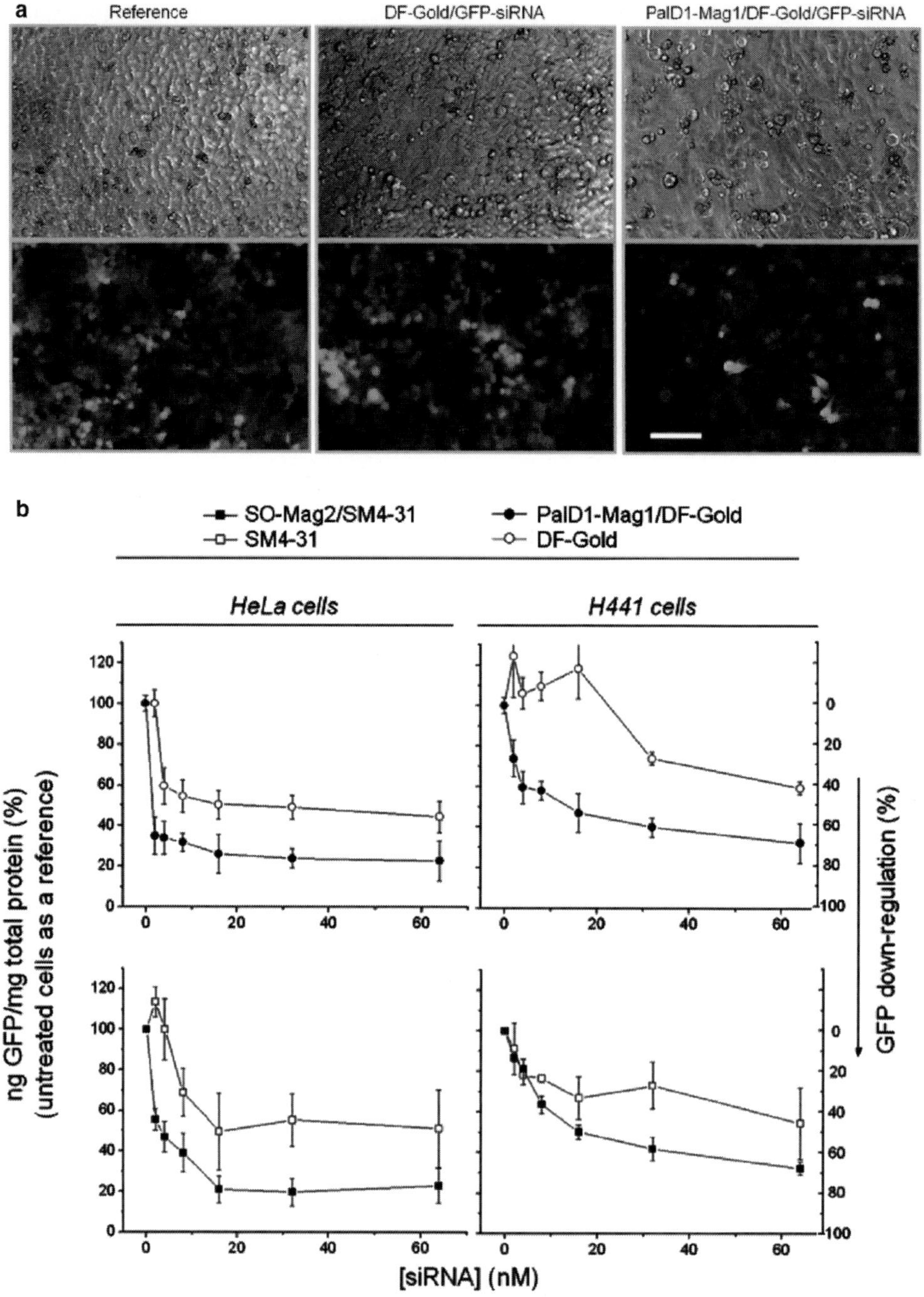

Fig. 12. Down-regulation of enhanced GFP (eGFP) reporter gene expression post-lipofection and magnetofection of short interfering RNA by microscopy and eGFP quantification in cell lysate. Stably transfected eGFP expressing HeLa human cervical epithelial adenocarcinoma (HeLa cells) and H441 human lung epithelial (H441 cells) cells were seeded in a 96-well plate, and 24 h later were transfected with 200 μL transfection volume of the magnetic and nonmagnetic antiGFP siRNA (siRNA) complexes according to steps of Subheading 3.7. (a) eGFP expression was monitored in HeLa cells 60 h post-transfection with DF-Gold/siRNA lipoplexes and PalD1/DF-Gold/siRNA magnetic lipoplexes containing 8 nM siRNA by fluorescence microscopy. The pictures show fluorescence images taken at 490/509 nm (eGFP fluorescence). Untreated cells were used as a reference. Scale bar = 100 μm. (b) eGFP expression was monitored in cell lysates according to Subheading 3.11 step 3 60 h post-transfection of the HeLa and H441 cells with SM4-31/siRNA or DF-Gold/siRNA lipoplexes and SO-Mag2/SM4-31/siRNA and PalD1/DF-Gold/siRNA magnetic triplexes (iron-to-siRNA ratio of 0.5–1, DF-to-siRNA vol/wt ratio of 4). The results show that magnetofection results in more efficient target gene down-regulation (i.e., significantly lower GFP expression levels) compared to lipofection with the same vector type

3. Observe the accumulation of insoluble violet formazan crystals. When necessary, continue the incubation to obtain an optical density of ~0.3–1.0 at 550–590 nm for untreated cells (as a reference) after product solubilization.

4. Add 100 μl MTT solubilization solution to dissolve formazan.

5. Seal the plate with parafilm or an adhesive film to avoid liquid evaporation and incubate overnight at RT until the formazan crystals completely dissolve.

6. Measure the optical density D of the MTT-formazan solution after solubilization in the range of the wide absorption spectrum maximum (550–590 nm), for example, at 590 nm, using a microplate reader (e.g. Wallac Multilabel Counter; measuring time 0.1 s). Use untransfected cells as a reference. Register the absorbance for one or several wells with a mixture of 100 μl MTT solution and 100 ml solubilization solution as a blank.

7. Cell viability in terms of cell respiration activity (18, 19) normalized to the reference data (%) is expressed as:

$$\text{Cell viability (\%)} = (D_{sample} - D_{blank})/(D_{reference} - D_{blank}) \times 100$$

where D_{sample}, D_{blank} and $D_{reference}$ are the optical densities at the maxima of the MTT-formazan absorption spectrum registered for a sample, blank and reference sample, respectively. Sample results are given in Fig. 13.

4. Notes

1. Unless stated otherwise, all solutions should be prepared in water that has a resistivity of 18.2 MΩ cm and a total organic content less than five parts per billion. This standard is referred to as "water" in this text.

2. For details on the synthesis of palmitoyl dextran (PalD1), see refs. (20, 21).

3. Any nucleic acid can be labeled with radioactive iodide isotopes.

4. The particles synthesized according to step of Subheading 3.1 are not superparamagnetic in the strict sense. To demonstrate superparamagnetic behavior, small particles (with a grain size less than approximately 10 nm) have to be stabilized with a layer of coating agents to reduce magnetic dipole-dipole interactions. Briefly, the superparamagnetic behavior of magnetic nanoparticles is not only a size effect, but may also depend on surface modifications (22). Therefore, do not magnetize magnetic nanoparticles or prepared magnetic complexes before

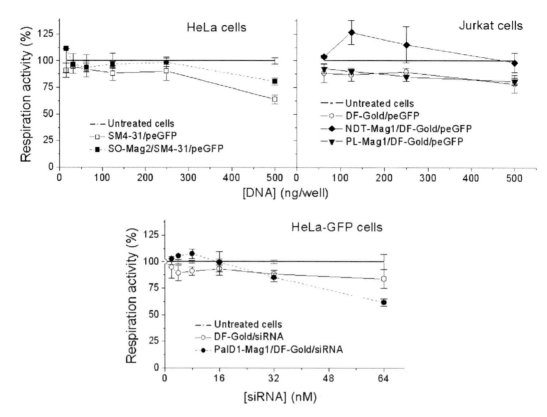

Fig. 13. MTT-based toxicity test post-lipofection and magnetofection. HeLa cells were transfected with SM4-31/peGFP lipoplexes or SO-Mag2/SM4-31/peGFP magnetic triplexes, Jurkat cells were transfected with DreamFect-Gold/peGFP lipoplexes or NDT-Mag1/DF-Gold/peGFP magnetic triplexes and HeLa-GFP cells were transfected with DF-Gold/siRNA lipoplexes and PalD1-Mag1/DF-Gold/siRNA magnetic triplexes, as described in Subheading 3.7. Respiration activity was measured 48 h post-transfection according to steps from Subheading 3.12. Iron-to-DNA and DF-Gold (SM4-31)-to-nucleic acid ratios are as shown in Figs. 10–12. The results of the MTT assay performed according to the protocol suggest that there is no additional toxicity associated with the particles compared to the lipoplexes within the tested concentration range

transfection in order to avoid aggregation. Do not freeze magnetic nanoparticle suspensions. Before use, always vortex magnetic nanoparticle suspensions very thoroughly. Optionally, sonicate magnetic nanoparticle suspensions after longer storage periods (waterbath sonicator). The measured hydrodynamic diameter of the magnetic nanomaterial suspension is a "dynamic" value and often depends on the prehistory of the sample, including storage time and conditions.

5. Excess coating compound not bound to magnetic nanoparticles can compete with binding between magnetic nanoparticles and nucleic acids. Therefore, the sonication and dialysis steps are essential.

6. Typically, core/shell magnetic nanomaterials synthesized according to steps of Subheading 3.1 consist predominantly of a magnetite core with a mean crystallite size $<d>$ of 9–11 nm that can be calculated from the broadening of the X-ray

diffraction peak (an example of the XRD-pattern is shown in Figure 1c) using the Scherer formula. The single domain size for magnetite is about 100 nm (http://www.irm.umn.edu/hg2m/hg2m_d/hg2m_d.html, 23), hence the particles have only one domain and <d> is a good approximation of the average core size of the particles. This value is useful for evaluating the weight of the insulated particle in terms of iron weight per particle. In combination with the magnetization value (an example of the magnetization curve is given in Figure 1b), the magnetic moment of the insulated particle can be calculated for any applied magnetic field value. A simple method based on measurements of the time course of the turbidity of a magnetic vector suspension when subjected to inhomogeneous magnetic fields can be used to evaluate the magnetic responsiveness as the average velocity of the complexes in defined magnetic fields, as described in ref. (21). This data can be further useful to evaluate an average magnetic moment and the composition of the magnetic transfection complexes prepared according to Subheading 3.6, in terms of the number of magnetic particles in a complex, as described in detail in references (21, 24). The mean hydrodynamic diameter of the particles/labile aggregates and ζ-potential of the magnetic nanoparticles, determined by photon correlation spectroscopy using a Malvern Zetasizer 3000 (UK), varied from 55 to 430 nm and highly positive (+55 mV) to negative (−48 mV) electrokinetic (ζ)-potential, depending on the coating composition used (Table 1). A large variety of coating compounds can be useful in magnetofection (5, 13). Particles with negative ξ-potential are not suitable to bind nucleic acids on their own. For this purpose, the particles must be combined either with enhancers (positively charged lipid or polymer) or divalent cations. We have found that particles with a magnetite crystallite size of 9–11 nm are superior to smaller particles with 3–4 nm crystallite size as components of the magnetic transfection vectors for magnetofection.

7. Appropriate dilutions for measurement have concentrations between 0.5 and 6 µg iron per ml. The suggested final dilution for measurement of the original 20 µl magnetic nanoparticle with iron concentration of 10–90 mg iron per ml is 1:15,000.

8. The iron content of the magnetic nanoparticles synthesized according to 3.1 varies from 0.41 to 0.75 g iron per g dry weight (see Table 1); aqueous suspensions after dialysis of the material usually contain ~10 mg iron per ml.

9. This protocol must be performed by authorized personnel and according to the rules and regulations for work with radioactive substances. Use pipette tips provided with an aerosol filter to avoid radioactive contamination of the pipette. This procedure can be accomplished in 2 h.

10. Preferably, use the DNA concentration during the labeling reaction as specified in steps 2.3 and 3.3. If this is not possible, increase the incubation time.
11. For complete dissolution of thallium chloride just before DNA labeling, heat the solution to 70°C using a water bath. Caution: Thallium chloride is highly toxic.
12. Self-assembly of charged colloidal particles like the magnetic nanoparticles described here is dependent on the ionic strength and ion composition of the solvent. We suggest preparing magnetic gene vectors in serum and additive-free cell culture medium or in 0.9% sodium chloride.
13. There are no established rules on whether the mixing order of components (i.e. magnetic particles, nucleic acids, and enhancers) plays a major role in terms of transfection efficiency. But order of mixing, concentrations of components and medium composition can influence the association of the nucleic aids with magnetic nanoparticles.
14. Cell culture and plating should be performed under sterile conditions. Timing: 30-min cell plating plus 24-h cell growth before transfection.
15. Cell seeding in a clear bottom black-walled plate enables eGFP expression measurements in living cells. This could be used to study the kinetics of eGFP expression post-transfection.
16. Prepare transfection complexes just before transfection; all steps should be performed under sterile conditions. Timing: 60 min.
17. A magnetic nanoparticle-to nucleic acid w/w ratio of 0.5 to 1 has proven useful for both DNA and siRNA lipoplexes and polyplexes with a variety of magnetic nanoparticle types. To determine the optimal weight ratio for an unknown particle type, it is useful to also carry out this protocol with magnetic nanoparticle stock suspensions resulting in w/w ratios of 0.25, 0.75, 1 and 1.25 or higher.
18. The order of reagent mixing and the medium for reagent dilution can be critical for the sizes, charges and compositions of the complexes, and thus, for final transfection efficiencies. To optimize the conditions for a given cell line, magnetic nanoparticle type and enhancer reagent, the mixing orders should be tested as described above.
19. Overexposure to the magnetic field may lead to toxic effects, which might negatively influence the transfection results.
20. Centrifuging the plates, before adding the transfection complexes to ensure that the cells are at the bottom of the plate. This is important to enable the transport of the transfection complexes to the cell membrane, under the influence of the magnetic force during incubation at the magnetic plate.

21. The optimal incubation conditions and the optimal exposure time at the magnetic plate to sediment the transfection complexes at the cell membrane may differ from one cell type (or transfection complex) to another and must be determined experimentally.

22. Make sure to have enough cells. Cell density $(50–100) \times 10^3$ cells per ml is sufficient to perform FACS analysis. Perform FACS analysis as quickly as possible to avoid cell aggregation and aging in the FACS buffer. Cells are magnetically labeled post-magnetofection due to association with magnetic transfection complexes; vortex cells before FACS analysis.

23. Bear in mind that the luciferase and eGFP assays are carried out with 50 μl cell lysate, while the protein assay is carried out with only 10 μl. Correspondingly, the measured values for luciferase (or eGFP) must be divided by 5 to obtain correct results when normalizing per total protein determined in 10 μl cell lysate.

24. Use the same type of 96-well clear bottom black-walled plate for both eGFP calibration curve measurements and experimental sample measurements. Make sure to measure equal volumes for the calibration curve and the experimental samples.

25. Bear in mind that the luciferase and eGFP assays are carried out with 50 μl cell lysate, while the protein assay is carried out with only 10 μl. Correspondingly, the measured values for eGFP must be divided by 5 to obtain correct results when normalizing to total protein determined in 10 μl cell lysate.

26. To evaluate the results, time point(s) post-transfection must be chosen, taking into account the target protein half-life. Proteins with longer half-lives will show a slower initial response. The duration of gene silencing is dependent on the cell doubling time and the intrinsic stability of siRNA within the cell. In vitro, luciferase protein levels recover to pre-treatment values within less than a week in rapidly dividing cell lines, but take longer than three weeks to return to steady-state levels in non-dividing fibroblasts (25).

Acknowledgments

The authors would like to thank Dr. Bob Scholte for transduction of the H441cells with eGFP and luciferase using lentiviral vectors. This work was supported by the European Union through the Project FP6-LSHB-CT-2006-019038 "Magselectofection," as well as by the German Ministry of Education and Research,

Nanobiotechnology grants 13N8186 and 13N8538. Financial support of the German Research Foundation through the project PL 281/3-1 Nanoguide and German Excellence Initiative via the "Nanosystems Initiative Munich" are gratefully acknowledged.

References

1. Plank C, Scherer F, Schillinger U, Anton M (2000) Magnetofection: enhancement and localization of gene delivery with magnetic particles under the influence of a magnetic field. J Gene Med 2:24
2. Mah CEA (2000) Microsphere-mediated delivery of recombinant AAV vectors in vitro and in vivo. Mol Ther 1:S239
3. Hughes C, Galea-Lauri J, Farzaneh F, Darling D (2001) Streptavidin paramagnetic particles provide a choice of three affinity-based capture and magnetic concentration strategies for retroviral vectors. Mol Ther 3:623–630
4. Bhattarai SR, Kim SY, Jang KY, Lee KC, Yi HK, Lee DY, Kim HY, Hwang PH (2008) N-Hexanoyl chitosan-stabilized magnetic nanoparticles: enhancement of adenoviral-mediated gene expression both in vitro and in vivo. Nanomedicine 4:146–154
5. Plank C, Anton M, Rudolph C, Rosenecker J, Krotz F (2003) Enhancing and targeting nucleic acid delivery by magnetic force. Expert Opin Biol Ther 3:745–758
6. Schillinger U, Brill T, Rudolph C, Huth S, Gersting S, Krotz F, Hirschberger J, Bergemann C, Plank C (2005) Advances in magnetofection – magnetically guided nucleic acid delivery. J Magn Magn Mater 293:501–508
7. Huth S, Lausier J, Gersting SW, Rudolph C, Plank C, Welsch U, Rosenecker J (2004) Insights into the mechanism of magnetofection using PEI-based magnetofectins for gene transfer. J Gene Med 6:923–936
8. Krotz F, de Wit C, Sohn HY, Zahler S, Gloe T, Pohl U, Plank C (2003) Magnetofection – a highly efficient tool for antisense oligonucleotide delivery in vitro and in vivo. Mol Ther 7:700–710
9. Plank C, Schillinger U, Scherer F, Bergemann C, Remy JS, Krotz F, Anton M, Lausier J, Rosenecker J (2003) The magnetofection method: using magnetic force to enhance gene delivery. Biol Chem 384:737–747
10. Isalan M, Santori MI, Gonzalez C, Serrano L (2005) Localized transfection on arrays of magnetic beads coated with PCR products. Nat Meth 2:113–118
11. Azzam T, Domb AJ (2004) Current developments in gene transfection agents. Curr Drug Deliv 1:165–193
12. Dobson J (2006) Gene therapy progress and prospects: magnetic nanoparticle-based gene delivery. Gene Ther 13:283–287
13. Mykhaylyk O, Vlaskou D, Tresilwised N, Pithayanukul P, Moller W, Plank C (2007) Magnetic nanoparticle formulations for DNA and siRNA delivery. J Magn Magn Mater 311:275–281
14. Mykhaylyk O, Antequera YS, Vlaskou D, Plank C (2007) Generation of magnetic non-viral gene transfer agents and magnetofection in vitro. Nat Protoc 2:2391–2411
15. Felgner PL, Barenholz Y, Behr JP, Cheng SH, Cullis P, Huang L, Jessee JA, Seymour L, Szoka F, Thierry AR et al (1997) Nomenclature for synthetic gene delivery systems. Hum Gene Ther 8:511–512
16. Terebesi J, Kwok KY, Rice KG (1998) Iodinated plasmid DNA as a tool for studying gene delivery. Anal Biochem 263:120–123
17. Gersdorff von K (2006) PEG-Shielded and EGF receptor-targeted DNA polyplexes: Cellular mechanisms. PhD Thesis, Ludwig Maximilian University, Munich, Germany, Supervisor Prof. Dr. E. Wagner. p 125 http://edoc.ub.uni-muenchen.de/5485/
18. Berridge MV, Herst PM, Tan AS (2005) Tetrazolium dyes as tools in cell biology: new insights into their cellular reduction. Biotechnol Annu Rev 11:127–152
19. Berridge MV, Tan AS, Hilton CJ (1993) Cyclic adenosine monophosphate promotes cell survival and retards apoptosis in a factor-dependent bone marrow-derived cell line. Exp Hematol 21:269–276
20. Suzuki M, Mikami T, Matsumoto T, Suzuki S (1977) Preparation and antitumor activity of O-palmitoyldextran phosphates, O-palmitoyldextrans, and dextran phosphate. Carbohydrate Res 53:223–229
21. Mykhaylyk O, Zelphati O, Hammerschmid E, Anton M, Rosenecker J, Plank C (2008) Recent advances in magnetofection and its potential to deliver siRNA in vitro. Meth Mol Biol Mouldy. Sioud (ed)

22. Mikhaylova M, Kim DK, Bobrysheva N, Osmolowsky M, Semenov V, Tsakalakos T, Muhammed M (2004) Superparamagnetism of magnetite nanoparticles: Dependence on surface modification. Langmuir 20:2472–2477
23. Butler RF, Banerjee SK (1975) Theoretical single-domain grain size range in magnetite and titanomagnetite. J Geophys Res 80(B29):4049–4058
24. Wilhelm C, Gazeau F, Bacri JC (2002) Magnetophoresis and ferromagnetic resonance of magnetically labeled cells. Eur Biophys J 31:118–125
25. Bartlett DW, Davis ME (2006) Insights into the kinetics of siRNA-mediated gene silencing from live-cell and live-animal bioluminescent imaging. Nucleic Acids Res 34:322–333

Chapter 35

Long-Circulating, pH-Sensitive Liposomes

Denitsa Momekova, Stanislav Rangelov, and Nikolay Lambov

Abstract

A major limiting factor for the wide application of pH-sensitive liposomes is their recognition and sequestration by the phagocytes of the reticulo-endothelial system, which conditions a very short circulation half-life. Typically prolonged circulation of liposomes is achieved by grafting their membranes with pegylated phospholipids (PEG–lipids), which have been shown, however, to deteriorate membrane integrity on one hand and to hamper the pH-responsiveness on the other. Hence, the need for novel alternative surface modifying agents to ensure effective half-life prolongation of pH-sensitive liposomes is a subject of intensive research. A series of copolymers having short blocks of lipid-mimetic units has been shown to sterically stabilize conventional liposomes based on different phospholipids. This has prompted us to broaden their utilization to pH-sensitive liposomes, too. The present contribution gives thorough account on the chemical synthesis of these copolymers their incorporation in DOPE:CHEMs pH-sensitive liposomes and detailed explanation on the battery of techniques for the biopharmaceutical characterization of the prepared formulations in terms of pH-responsiveness, cellular internalization, in vivo pharmacokinetics and biodistribution.

Key words: pH-sensitive liposomes, Steric stabilization, PEG–lipids, Block copolymers

1. Introduction

Drug discovery, at present, is highly facilitated by proteomics, genomics and high throughput screening (1, 2). The application of these powerful tools enabled the generation of large libraries of bioactive compounds, and eventually the elaboration of numerous valuable therapeutic agents of clinical significance.

Unfortunately, in many cases, the active compound cannot be applied as it stands; due to unfavorable physicochemical properties it may not reach its target site or because it is unstable after administration, or may show a high toxicity in nontarget tissues (3).

To overcome these difficulties, much research efforts have been focused on developing nanosized particulate systems that are able to deliver the active compounds to target cells or even cell organelles (4). Ever since their discovery (5) and the recognition of their structure and basic properties, liposomes comprise the most extensively studied and presumably the most successful example of the above mentioned nanoparticles for targeted drug delivery (6). The progress in the transition of liposome-mediated drug delivery from the laboratory to the clinic has been greatly facilitated by the major breakthroughs made in liposomology. Circulation lifetime of liposomes has been significantly prolonged by steric stabilization (7, 8). Advances in drug loading technologies allowed high encapsulating efficiency of some therapeutic agents using pH or other gradient methods (9–11). These have resulted in the development of liposomal drug products, which are either commercialized or are at present in advanced clinical trials (6).

However, while significant progress has been achieved in overcoming many of the setbacks associated with liposomal drug delivery, an elusive problem that still hampers the full realization of the potential of liposomes in clinical practice is their unfavorable subcellular trafficking and disposition. Liposomes have the propensity to accumulate in certain subcellular compartments, mainly lisosomes, where the encapsulating material is often degraded, thus limiting its availability at the cytosolic target site (12–14). The latter problem is a serious hurdle in the development of liposome-based carriers for intracellular delivery of peptides, proteins, nucleic acids, and certain polar anticancer drugs, which are characterized by both low cellular permeation ability and chemical/enzymatic instability. An attractive approach to avoid lysosomal sequestration and degradation of entrapped materials is the use of pH-sensitive liposomes.

This class of liposomal carriers are composed of specific lipid components such as unsaturated phosphatidylethanolamines (e.g., dioleylphosphatidylethanolamine, DOPE) and protonatable ampiphiles, e.g., Cholesteryl hemisuccinate, (CHEMs), which condition membrane phase transition in the mildly acidic environments of endosomes (15), whereby liposomes become highly fusogenic, their membrane merges with that of the endosome and eventually the cargo is effectively released in the cytosole (16–18). Hence unlike the conventional liposomes, the pH-sensitive systems escape endosomal sequestration upon entering cells. Thus pH-sensitive liposomes are considered as one of the most promising carrier systems to provide effective intracellular accumulation of genes, antisense-oligonucleotides, proteins or polar small molecules, whose inability to bypass biological barriers, in vivo instability and/or systemic toxicity hampers the fulfillment of their therapeutic potential (16).

Like the conventional liposomes, pH-sensitive liposomes are unstable in the circulation and are rapidly sequestrated in the reticulo-endothelial system (RES) which compromises their application for systemic drug delivery. The most important approach to overcoming this problem is the inclusion of pegylated lipids [poly(ethylene glycol)-derived lipids, commonly known as PEG–lipids] which hampers the opsonization and phagocytosis of liposomes in RES and dramatically increases their circulation half-lives by creating a repulsive PEG layer around the liposomes (19, 20). Unfortunately, in the case of pH-sensitive liposomes, the PEG–lipids are by far not innocent excipients as they have been reported; when incorporated at a certain concentration, the so-called saturation limit, they deteriorate membrane integrity, which significantly compromises the reservoir function of the carrier (21–23). More importantly, PEG–lipids have been reported to greatly inhibit the acidity-driven responsiveness of pH-sensitive liposomes, thus diminishing the efficient intracellular delivery of entrapped materials (24, 25).

One of the strategies to overcome this drawback is, the development of a novel class of pH-cleavable PEG-based copolymers, which can assure the loss of steric stabilization of the liposomes in an acid environment, e.g., in the endosomes or in the intratumoral microenvironment (26–29).

Another approach to achieve simultaneously steric stabilization and pH-sensitivity is, the combination of a PEG–lipid and a pH-sensitive copolymer such as terminally-alkylated copolymer of N-isopropylacrylamide and methacrylic acid (30, 31).

The impact of the conventional PEG–lipids on the acid-induced destabilization of pH-sensitive liposomes is possibly due to two main reasons: on one hand as the protonation of DOPE head group is the basis of acid-driven transition, it might be hindered by ionization of phosphate or carbamate linkages introduced by the incorporation of the conventional PEG–lipids (25). On the other hand, as the DOPE is a cone-shaped lipid, the incorporation of PEG–lipid conjugates that have a complementary inverted cone shape that can stabilize the lamellar phase even at low pH of the medium (32, 33).

Recently, a series of amphiphilic diblock copolymers based on PEG has been prepared (34–36). A common feature is the hydrophobic residue that mimics the lipid anchors of the naturally occurring phospholipids and commercially available PEG–lipids. In strong contrast to the latter, the alkyl chains and the PEG moiety are linked to the glycerol skeleton via nonionizable ether linkages. In addition, the hydrophobic anchors may be linked together, thus forming short blocks of repeating lipid-mimetic monomer units, conditioning more cylindrical form of their macromolecules. The copolymers have been shown to provide steric stabilization of liposomes based on different phospholipids and to afford higher saturation limit as compared to conventional PEG–lipids (37). On this ground, we focused our work on the development of pH-sensitive DOPE:CHEMs liposomes sterically stabilized by copolymers bearing short blocks of lipid-mimetic units (38). In the present contribution, an integral program allowing a rational characterization of the physicochemical and biopharmaceutical properties of copolymer-stabilized pH- sensitive liposomes is outlined.

2. Materials

2.1. Synthesis of Copolymers

1. 1-Dodecanol.
2. $SnCl_4$.
3. Dodecyl glycidyl ether.
4. Epibromohydrin.
5. Ethylene oxide (Clariant).
6. Monohydroxy poly (ethylene glycol) 5000 (Sigma).

2.2. Preparation of Liposomes

1. Dioleoylphosphatidylethanolamine (DOPE) (MW 744) chloroform stock solution 10 mg/ml (0.013 mM) (Sigma). Store at –20°C.
2. Cholesteryl hemisuccinate (CHEMs) (MW 486.73) (Sigma) stored at 4°C. Prepare a chloroform stock solution at a concentration of 5 mg/ml. Dissolve 25 mg CHEMs in 5 ml of chloroform. Store at –20°.
3. Copolymers $DDP(EO)_{92}$, $(DDGG)_2(EO)_{115}$ and $(DDGG)_4(EO)_{114}$ (MW 4474, 6028, and 6808, respectively).
4. Poly(ethylene glycol)(2000)-distearoyl phosphatidylethanolamine (PEG(2000)-DSPE) (MW 2748) (Lipoid GmbH, Germany). Store at –20°C.
5. 10 mM HEPES-buffered saline (HBS): Weigh 2.38 g Hepes, 8.0 g sodium chloride and 0.37 g Idranal (Sigma chemical Co.). Dissolve in 1 L of distilled water and adjust the pH to 7.4 with 1N Sodium hydroxide. Store at 4°C.

2.3. Evaluation of pH-Sensitivity of Liposomes (Leakage Assay)

1. Phosphate-buffered saline (PBS) at pH 7.4 to 5.5.

   Solution A: Weigh 9.073 g of potassium dihydrogen phosphate (KH_2PO_4) (0.066 M) and 5.1 g NaCl. Dissolve in 1 L of distilled water.

   Solution B: Weigh 11.87 g of disodium hydrogen phosphate ($Na_2HPO_4 \cdot 2\,H_2O$) (0.066 M) and 5.5 g NaCl and dissolve in 1 L of distilled water.

**Table 1
Amounts of solutions A and B for preparation of phosphate buffer at different pH**

pH	Solution A (ml)	Solution B (ml)
7.4	19.7	80.3
6.8	53.4	46.6
6	98.9	11.1
5.5	97.3	2.7

To obtain 100 ml buffer solutions with desired pH mix in a 100-ml volumetric flasks with appropriate amounts of solution A and solution B as shown in Table 1.

2. Citrate buffer pH 4.5. Weigh 24.8 g of disodium citrate (0.1 M) and dissolve in 1 L distilled water. Mix in a 100-ml volumetric flask of 66.4 ml of the prepared solution with 33.6 ml 0.1 N hydrochloric acid.

3. Calcein solution (80 mM, 5 ml) (MW 622.53). Weigh 250 mg of calcein and add 3 ml of isotonic phosphate buffer (pH 7.4). Add 500 µl of a 4 M solution of sodium hydroxide (see Note 1). After complete dissolution of calcein, adjust the solution to pH 7.4, and add phosphate buffer to final volume of 5 ml. Store at dark place until use.

4. Solution of Triton X-100 at a concentration of 10 per cent (v/v). Weigh 1 ml of Triton X-100 and dissolve in 10 ml of distillated water.

2.4. Liposome–Cell interaction

1. RPMI-1640 medium supplemented with 10% fetal bovine serum and 0.2 mM L-glutamine.
2. RPMI-1640 medium without phenol red (see Note 2)
3. Solution of trypsin (0.25%) in PBS (pH 7.4). Store at 4°C.
4. Calcein-loaded liposomes prepared immediately before treatment.

2.5. Evaluation of the Pharmacokinetic Behavior of Liposomes

1. ^{3}H-cholesteryl oleoyl ether (Amersham, Roosendaal, The Netherlands). Store at 4°C.
2. Ultima gold liquid scintillation cocktail (Perkin Elmer BioScience B.V., The Netherlands).
3. Solvable tissue solubilizer (Perkin Elmer BioScience B.V., The Netherlands).
4. Male Wistar rats (body weight approximately 250 g). House in groups of four under standard laboratory conditions and free access to food (rat chow) and water.

3. Methods

3.1. Synthesis of Copolymers

The copolymers were synthesized by anionic polymerization technique.

3.1.1. Synthesis of 1,3-Didodecyloxy-propane-2-ol

1. 20 g (0.108 mmol) of dry 1-dodecanol is placed in a two-necked, round-bottom flask flushed with nitrogen.
2. 26.1 g (0.108 mmol) of dodecyl glycidyl ether and 0.135 ml of $SnCl_4$ are added under stirring. The stirring is maintained at 110–120°C for 24 h.
3. Another portion of 0.135 ml of $SnCl_4$ is added and the mixture is stirred for another 70 h.
4. The product is recrystallized twice from hexane.

3.1.2. Synthesis of 1,3-Didodecyloxy-propane-2-glycidyl-glycerol

1. 3.9 g (9.1 mmol) of 1,3-didodecyloxy-propane-2-ol (DDP) is dissolved in 50 ml of freshly distilled terahydrofuran in a nitrogen-flushed two-necked, round-bottom flask equipped with a reflux condenser.
2. 0.35 g (14.6 mmol) of solid NaH is added to the solution under stirring. The mixture is heated to 50°C and stirred for 24 h.
3. 2.0 g (14.6 mmol) of epibromohydrin is added dropwise and the reaction mixture is stirred at 50°C for 48 h.
4. The solid material is removed by centrifugation. Tetrahydrofuran is removed by rotary evaporation. The excess of epibromohydrin is removed under vacuum.

3.1.3. Synthesis of $DDP(EO)_{92}$

1. 0.709 g (1.66 mmol) of DDP and 0.056 g (1.0 mmol) of KOH are placed in a 50 ml three-necked, round-bottom flask fitted with a reflux condenser and inlets for nitrogen and ethylene oxide (EO).
2. The mixture is heated to 110°C and is purged with 6.75 g (153 mmol) EO for 120 min. (see Note 3) for the synthesis of similar copolymers.
3. The residue is cooled to room temperature and dissolved in methylene chloride.
4. The copolymer is isolated by precipitation in hexane and washed with portions of 20–30 ml of hexane until no more hexane-soluble fraction is extracted.
5. The copolymer is dried in a vacuum up to a constant weight.

3.1.4. Synthesis of $(DDGG)_2(EO)_{115}$

1. 6 ml of toluene, 2 ml of EO, and 0.03 g (0.261 mmol) of t-BuOK are placed in a precooled to −4°C reactor flushed with nitrogen.
2. The temperature is gradually increased and the mixture is stirred at 40°C for 2 h, at 50°C for 2 h, and at 60°C overnight.

3. The system is cooled and the unreacted EO is removed under reduced pressure.
4. 1 ml from the reaction mixture is taken out and precipitated in dry diethyl ether to determine the degree of polymerization of the resulting poly(ethylene glycol).
5. 0.9 g (1.91 mmol) of 1,3-didodecyloxy-propane-2-glycidyl-glycerol (DDGG) dissolved in 4 ml of toluene is added. The reaction mixture is stirred at reflux for 28 h.
6. Toluene is removed under reduced pressure.
7. The residue is dissolved in methylene chloride and the copolymer is isolated by precipitation in hexane.
8. The copolymer is washed with portions of 20–30 ml of hexane until no more hexane-soluble fraction is extracted.
9. The copolymer is dried in a vacuum up to a constant weight.

3.1.5. Synthesis of $(DDGG)_4(EO)_{114}$

1. 2.95 g (0.59 mmol) of monohydroxy poly(ethylene glycol) of molecular weight 5,000 previously metaled by potassium naphthalide (see Note 4) is dissolved in 25 ml of toluene and placed in a 50 ml three-necked, round-bottom flask fitted with a nitrogen inlet, a reflux condenser, and a rubber septum.
2. 2.03 g (4.19 mmol) of DDGG dissolved in 5 ml of toluene is introduced.
3. The reaction mixture is heated; stirring is maintained at reflux for 48 h.
4. Toluene is removed under reduced pressure.
5. The residue is dissolved in methylene chloride and the copolymer is isolated by precipitation in hexane.
6. The copolymer is washed with portions of 20 ml of hexane until no more hexane-soluble fraction is extracted.
7. The copolymer is dried under vacuum up to a constant weight.
8. The chemical structures of the copolymer are shown in Fig. 1.

3.2. Preparation of Liposomes

One of the most popular methods for preparation of liposomes is the hydration of thin lipid films, combined with freeze/thaw and extrusion cycles.

3.2.1. Preparation of Liposomes for Leakage Assay and Liposome–Cell Interaction

1. Take lipid stocks solutions from the freezer and let them attain room temperature before use.
2. Weight calculation is based on molar ratio DOPE:CHEMs (3:2) (24). For the preparation of 2 ml of sterically stabilized liposomal suspensions with total phospholipids concentration 3 mM and copolymers at various contents from 2.5 to 10 mol % (with respect to total lipid content) mix in glass tubes 0.0036 mmol DOPE (268 μl of DOPE stock solution (10 mg/ml)), 0.0024 mmol CHEMs (233 μl of CHEMs

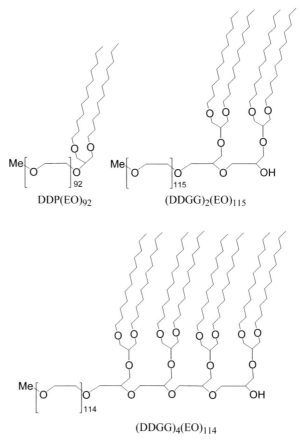

Fig. 1. Chemical structures of the copolymers

Table 2
The needed amounts of phospholipid and copolymers stock solutions for preparation of liposomes for leakage assay and fluorescence microscopy

Copolymer stock solution (10 mg/ml)	2.5 mol%	5 mol%	7.5 mol%	10 mol%
$DDP(EO)_{92}$	67 µl	134 µl	201 µl	268 µl
$(DDGG)_2(EO)_{115}$	90 µl	180 µl	270 µl	360 µl
$(DDGG)_4(EO)_{114}$	100 µl	200 µl	300 µl	400 µl

stock solution (5 mg/ml)) and 0.00015 to 0.0006 mmol copolymers. The needed amounts of copolymers stock solutions are listed in Table 2.

3. Mix well by vortexing the tubes.

4. Create a lipid film by evaporating the chloroform under a gentle stream of argon.

5. Remove any traces of the solvent by leaving under vacuum overnight.

6. Transfer 2 ml calcein solution (80 mM) to the tubes with lipid film and vortex the tubes.

7. The hydration step must be carried out at temperatures slightly above the phase transition temperature of DOPE, which is 41°C. Close the tubes with parafilm and transfer them in a water bath at 50°C for 15 min to facilitate hydration.

8. For better hydration of the lipid film and improving the entrapment efficacy of calcein, freeze the liposomal suspensions in liquid nitrogen (−140°C) and then thaw in water bath (50°C). Repeat this step eight times.

9. In order to obtain a liposomal population with uniform size distribution, extrude liposomes through polycarbonate filters with pore size of 100 nm using a handle liposomal extruder (Avestin Inc., Canada). Usually 31 cycles of extrusion through the filter is sufficient for obtaining homogenous populations of liposomes with polydispersity index below 0.1 and mean diameter around 120 nm.

3.2.2. Removing of Nonencapsulated Calcein

An easy method for removing of nonencapsulated materials is gel filtration through Sephadex G50 PD10 columns (Pharmacia, Uppsala, Sweden).

1. Wash the columns with 20 ml HBS (pH 7.4). (see Note 5).

2. When the top of the column runs dry, pipet carefully 300 μl liposome suspension onto the column; wait until the top of the column runs dry and fill the column with HBS.

3. Place a vial to collect the liposomes (first red-colored fraction which runs from the column). Store the vials with liposomes in the dark until use.

3.2.3. Preparation of Liposomes for In vivo Evaluation

For in vivo experiments, a higher lipid concentration is needed (20 μmol/ml total lipid). The copolymer content is kept at 5 mol% with regard to total lipid. At least 2.5 ml liposomal suspension is necessary to treat one animal group of four rats.

1. Weigh 30 μmol (22.32 mg) DOPE, 20 μmol (9.47 mg) CHEMs and 2.5 μmol steric stabilizing polymer (17.02 mg $(DDGG)_4(EO)_{114}$ or 7 mg PEG(2000)-DSPE). Always allow lipids to acquire room temperature before opening storage containers.

2. Transfer the lipids and polymers in round-bottom flask (50 ml) and dissolve in an appropriate volume of chloroform (3 ml).

3. Add to each formulation 10 μl of the nonexchangeable and nondegradable lipid marker ^{3}H-cholesteryl oleylether (approximately 370 kBq).

4. Prepare a dry lipid film by evaporating the chloroform using a rotation type vacuum evaporator.
5. Remove any traces of the solvent under vacuum overnight.
6. Hydrate the lipid film with 2.5 ml HBS (pH 7.4) by following steps 7–9 as described in Subheading 3.2.1. (see Note 6).

3.3. Evaluation of pH-Sensitivity of Liposomes (Leakage Assay)

1. Transfer 20 µl of calcein-loaded liposomes in quartz fluorimetric cuvettes and add 1,980 µl of buffers with varying pH (4.5–7.4) (see Subheadings 2.3.1 and 2.3.2) (dilution 1:100).
2. Measure the fluorescence intensity (I_0) of the probes immediately after dilution of the liposome suspensions (spectrofluorimeter Perkin Elmer-MPF-44B, USA, $\lambda_{ext} = 490$ nm, $\lambda_{em} = 520$ nm).
3. Incubate the cuvettes at 37°C for 10 min and measure the calcein fluorescence intensity again. Subtract I_0 from the intensity value evaluated after 10 min incubation of liposomes to obtain the fluorescence of the released calcein at different pH (I_{pH}).
4. Add to each sample, 50 µl of Triton X-100 solution (10%), mix well and measure the fluorescence intensity (I_t).
5. As the calcein fluorescence is pH-dependent to obtain the real intensity, multiply the intensity measured at pH 6, pH 5.5, and pH 4.5 by a factor of 1.2, 1.4, and 1.6, respectively.
6. Calculate the percentage of calcein leakage using the equation:

$$\text{leakage}(\%) = \frac{I_{pH} - I_{7.4}}{I_t - I_{7.4}} \times 100$$

I_{pH} = corrected intensity at acidic pH before destruction of liposomes, $I_{7.4}$ = fluorescence intensity at pH 7.4 and I_t = total fluorescence intensity after destruction of liposomes. Typical leakage profiles are shown on Fig. 2.

3.4. Liposome–Cell Interactions with EJ Carcinoma Cells

One facile method for investigation of cellular interactions of liposomes, and more precisely, their ability to accumulate within cells, is the fluorescent microscopy. To meet this objective, liposomes are loaded with fluorescent marker (calcein or other suitable dye) at high concentration, whereby its fluorescence is self-quenched. Upon cellular fusion and internalization the dye is released from the carrier, diluted in the environment so the self-quenching effect is lost and the increased fluorescence is detected by fluorescent microscopy. In the case of effective cellular internalization of pH-sensitive liposomes, that is, without endosomal sequestration, calcein would have been diluted several-hundred-fold and the cells will display uniform cytosolic fluorescence. If the liposomes have been taken up by cells by endocytosis, punctuate fluorescence will be restricted to the secondary lysosomal and endocytic vacuoles. In contrast, adsorbed liposomes should

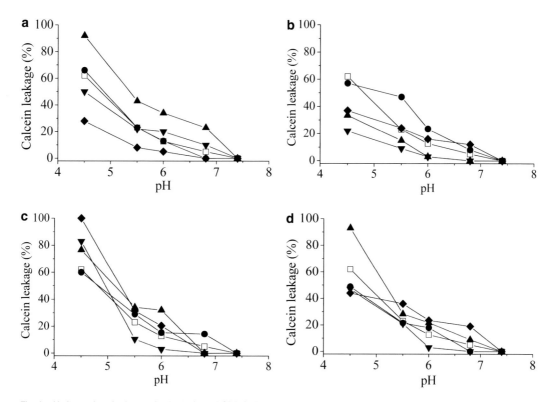

Fig. 2. pH-dependent leakage of calcein from DOPE:CHEMs (3:2) noncoated (*open squares*) and stabilized liposomes with DDP(EO)$_{52}$ (**a**), DDP(EO)$_{92}$ (**b**), (DDGG)$_2$(EO)$_{115}$ (**c**), and (DDGG)$_4$(EO)$_{114}$ (**d**), at contents of 2.5 (*circles*), 5 (*triangles*), 7.5 (*inverted triangles*), and 10 (*diamonds*) molar %. Calcein containing liposomes are incubated for 10 min at 37°C in medium at different pH. Data points represent the mean values of four independent experiments. (Reproduced from ref. 38 with permission from Elsevier)

not fluoresce at all due to the self-quenching effect unless the marker concentration has been reduced by extensive leakage. In such a case the cells appear dully fluorescent with a bright rim.

1. Adherent EJ urinary bladder carcinoma cells (see Note 7) are cultured routinely in RPMI-1640 medium supplemented with 10% FBS and L-glutamine and reset by hot trypsinization two or three times per week. For the experiments, only recently established cultures can be used advisably before the seventh passage.

2. The experimental cultures are established in 35 mm sterile petri dishes. It is advisable that the experiments are run before the monolayers reach confluence as the analysis of the photomicrographs would be troublesome. If the apparatus allows, cell cultures could be analyzed directly onto the petri dishes. Alternatively, sterile covering slides are to be placed onto cultivation dishes before the transfer of cellular suspension for subcultivation. Thereafter the cell will attach onto covering slides, which will allow them to be easily analyzed after treatment.

3. Prepare calcein-loaded (at 80 mM concentration of the dye) liposomal dispersions (3 mM lipid) shortly before the treatment procedure.

4. Treat the experimental cultures with 200 µl of liposomal suspension and incubate at 37°C for 1 h.

5. Remove the liposome-containing medium by gentle aspiration.

6. Rinse the cultures thrice with ice-cold PBS to remove residual liposome-containing medium. If the cells are cultured onto sterile cover slips, the later are extracted from the cultivation dish and rinsed with caution.

7. Fix cells with paraformaldehyde solution (3%) for 10 min at room temperature.

8. Discard paraformaldehyde (into a hazardous waste container) and wash the samples twice with PBS.

9. If samples are on a coverslip, then it is gently inverted into a drop of mounting medium on a microscope slide. Alternatively, if they are to be analyzed directly into the dishes, a cover-slip is placed onto the cells.

10. Analyze the experimental cultures by fluorescent microscopy to investigate the cellular interactions of liposomes. Typical images are displayed on Fig. 3.

3.5. In vivo Biodistribution and Pharmacokinetic Studies of Liposomes

The tissue distribution of liposomes throughout the whole body in experimental systems can be clearly determined by measuring the concentration of markers (preferably radioactive) in the blood or in each individual organ.

The aim of the following experiment is to test the circulation kinetics and biodistribution behavior of DOPE:CHEMs liposomes, either noncoated, or coated with conventional PEG–lipid PEG(2000)-DSPE or with $(DDGG)_4(EO)_{114}$.

3.5.1. Animal Experiment

All animal experiments should be performed according to national regulations and approved by the local animal experiments ethical committee.

1. House rats in groups of four (three groups are needed) under standard laboratory conditions.

2. Weigh each rat before treatment to calculate the group dose. Injection volume (~250 µl) is calculated with the average group body weight and the liposome concentration of 20 µmol total lipid/kg body weight.

3. Put the rats under light isoflurane anesthesia and take a 100 µl blood sample in duplicate (from one rat) as a blank before injection.

4. Inject the needed amount of liposomes (noncoated, coated with 5 mol% conventional PEG–lipid or $(DDGG)_4(EO)_{114}$) in the tail vein of the rats.

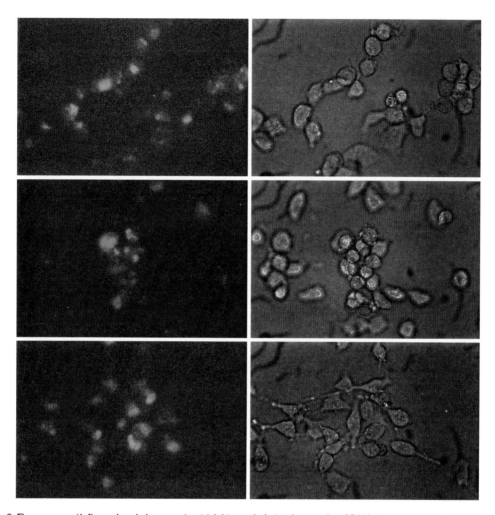

Fig. 3. Fluorescence (*left panel*) and phase contrast (*right panel*) photomicrographs of EJ bladder carcinoma cells, following exposure to pH-sensitive liposomes (100 µg/ml total lipid) at 37°C for 1 h. Noncoated DOPE:CHEMs liposomes (*top*), DDP(EO)$_{92}$-coated liposomes (*middle*), (DDGG)$_4$(EO)$_{114}$-coated liposomes (*bottom*). Copolymer content in the formulations is 5 mol%. The homogenous cytosol fluorescence observed in all images shows that the grafted copolymers do not compromise with the cellular internalization of liposomes. (Reproduced from ref. 38 with permission from Elsevier)

5. Take a 150 µl blood sample from each rat from the opposite tail vein immediately after injection of liposomes.

6. Blood sampling (150 µl) at different time points post injection (e.g., at $t=3$ min, 1, 4, 8, 24, and 48 h).

7. At the end of the treatment period, sacrifice the rats by cervical dislocation and dissect liver, spleen, lungs, and kidneys.

3.5.2. Treatment of Blood Samples

1. Fill scintillation containers with 100 µl of distilled water to prevent drying of blood; then add 100 µl of blood samples and mix.

2. Add 100 µl of a solvable tissue solubilizer and incubate for at least 1 h at room temperature.

3. To bleach the samples add 200 μl H_2O_2.
4. After the samples get decolorized, remove the excess of H_2O_2 by heating the samples at 50°C overnight.
5. Add 10 ml Ultima Gold scintillation cocktail, mix well and store the samples in dark until counting.
6. Count the activity of the samples in a Philips PW 4700 LSC for 5 min.
7. Count the activity of the 10 μl liposome samples again in the same run with the blood samples to calculate the injected dose.

3.5.3. Treatment of Organ Samples

1. Add 25 ml of water to liver and 5 ml of water to lung, spleen, and kidneys and homogenize the samples with the Ultra Turrax homogenizer (9,500 rpm).
2. Pipet 0.5 ml of the liver homogenate and 1 ml of spleen, lungs and kidneys homogenates into scintillation vials, add 200 μl of a solvable tissue solubilizer and incubate at room temperature until the tissue is dissolved.
3. Add 200 μl of 35% hydrogen peroxide and incubate for at least 24 h. (see Note 8)
4. Add 10 ml of Ultima Gold scintillation cocktail, mix well and allow to stand in the dark at room temperature for at least 24 h.
5. Count the samples activity in a Philips PW 4700 LSC.

3.5.4. Data Analysis

1. Blood concentration of liposomes at different time points can be calculated as a percentage of injected dose (ID) using the equation:

$$\%ID = \left(\frac{A_s \times \frac{V_{tot}}{V_s}}{A_{id}} \right) \times 100$$

where: A_s = activity of the blood sample, V_{tot} = rat blood volume was set as 10% of rat average group weight, V_s = volume of the blood sample, A_{id} = activity of the injected dose.

Activity of injected dose (A_{id}) is calculated as follows:

$$A_{id} = \frac{A_{10} \times V_{id}}{10}$$

where: A_{10} is average activity of 10 μl liposomal suspensions and V_{id} is the volume of the injected liposomal suspensions.

2. Activity in organs (%ID_{org}) is expressed as activity per whole organ

$$\%ID_{org} = \left(\frac{A_{org} \times F}{A_{id}} \right) \times 100$$

where: A_{org} is the activity of organ sample, F is a factor of dilution of organ samples and A_{id} is the activity of injected dose. A typical time/concentration curves and organ distribution of liposomes are presented on Figs. 4 and 5, respectively.

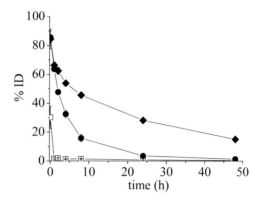

Fig. 4. Blood concentration (percentage of injected dose) versus time curves of DOPE:CHEMs liposomal formulations following intravenous injection in rats: noncoated liposomes (*open squares*), liposomes containing 5 mol% of PEG(2000)-DSPE (*circles*) and (DDGG)$_4$(EO)$_{114}$ (*diamonds*). Each data point represents the arithmetic mean ± standard deviation ($n=3$). As the results show the most important property of the liposomes stabilized with 5 mol % (DDGG)$_4$(EO)$_{114}$ is its excellent blood circulation versus both plain and DSPE-PEG(2000) stabilized vesicles. (Reproduced from ref. 38 with permission from Elsevier)

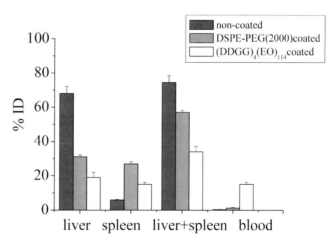

Fig. 5. Biodistribution of DOPE:CHEMs liposomal formulations 48 h after an intravenous injection in rats (20 μmol total lipid/kg body weight): noncoated liposomes, liposomes containing 5 mol% of PEG(2000)-DSPE and (DDGG)$_4$(EO)$_{114}$. Each data point represents the arithmetic mean ± standard deviation ($n=3$). The liposomal formulation sterically stabilized with 5 mol% (DDGG)$_4$(EO)$_{114}$ is obviously able to avoid the reticulo-endothelial system, localized in the liver and spleen, to a larger extent as compared to nonstabilized and liposomes stabilized by conventional PEG–lipid

4. Notes

1. For complete dissolution of calcein, it may be necessary to add 500 μl of a 4 M solution of sodium hydroxide more than once.
2. Regular RPMI-1640 contains phenol red, which has strong fluorescent properties that could interfere in the investigation. On this ground, it is advisable to establish the experimental culture in phenol red-free modification of the medium instead.
3. For the syntheses of $DDP(EO)_{30}$, $DDP(EO)_{44}$, and $DDP(EO)_{52}$ the quantities of EO are 2.20 g (50.0 mmol), 3.21 g (73.0 mmol), and 3.83 g (87.0 mmol) purged for 30, 60, and 90 min, respectively.
4. A 160 ml solution of potassium naphthalide (67 mg naphthalene, 20 mg potassium) is titrated under stirring in an inert atmosphere and room temperature with a solution of monohydroxy poly(ethylene glycol) of molecular weight 5,000 (1 g, 0.2 mmol) in freshly distilled tetrahydrofuran (5 ml) until the typical dark green colour disappears.
5. Allow the columns to acquire room temperature before running the separation procedure.
6. One can encounter difficulties regarding the extrusion of liposomes at this high lipid concentrations due to phospholipid aggregates clogging the filters. To avoid this, the extruder needs to be heated to attain a temperature of at least 40°C during the extrusion.
7. The procedure assumes the use of EJ urinary bladder carcinoma, but it is fully applicable for other adherent cell lines too, given the specific cell culture and maintenance requirements are met.
8. Repeat this step until the samples are decolorized (usually three times adding of hydrogen peroxide will suffice).

References

1. Beesley J, Roush C, Baker L (2004) High-throughput molecular pathology in human tissues as a method for driving drug discovery. Drug Discover Today 9:182–189
2. Hoever M, Zbinden P (2004) The evolution of microarrayed compound screening. Drug Discover Today 9:358–365
3. Torchilin VP (2000) Drug targeting. Eur J Pharm Sci 11(S-2):S81–S91
4. Torchilin VP (ed) (2006) Nanoparticulates as pharmaceutical carriers. Imperial College Press, London
5. Bangham AD, Standish MM, Watkins JC (1965) Diffusion of univalent ions across the lamellae of swollen phospholipids. J Mol Biol 13:238–245
6. Torchilin VP (2005) Recent advances with liposomes as pharmaceutical carriers. Nature Rev Drug Discovery 4:145–160
7. Woodle MC (1995) Sterically stabilized liposome therapeutics. Adv Drug Deliver Rev 16:249–265
8. Lasic DD, Vallner JJ, Working PK (1999) Sterically stabilized liposomes in cancer therapy and gene delivery. Curr Opin Mol Ther 1:177–185
9. Mayer LD, Tai LC, Balli MB, Mitilenes GN, Ginsberg RC, Cullis PR (1990) Characterization

of liposomal systems containing doxorubicin entrapped in response to pH gradients. Biophys Biochim Acta 1025:143–151

10. Harrigan PR, Wong KF, Redelmeier TE, Wheeler JJ, Cullis PR (1993) Accumulation of doxorubicin and other lipophilic amines into large unilamellar vesicles in response to transmembrane pH gradients. Biochim Biophys Acta 1149:329–338

11. Maurer-Spurej E, Wong KF, Maurer N, Fenske DB, Cullis PR (1999) Factors influencing uptake and retention of amion-containing drugs in large unilamellar vesicles exhibiting trans-membrane pH gradients. Biophys Biochim Acta 1416:1–10

12. Straubinger RM, Hong K, Friend DS (1983) Endocytosis of liposomes and intracellular fate of encapsulated molecules: encounter with a low pH compartment after internalization of coated vesicles. Cell 32:1096–1079

13. Huakg A, Kennel SJ, Huang L (1983) Interaction of immunoliposomes with target cells. J Biol Chem 258:14034–14040

14. Dijkstra J, Van Galen M, Scherphof GL (1984) Effects of ammonium chloride and chloroquine on endocytic uptake of liposomes by Kupffer cells in vitro. Biochim Biophys Acta 804:58–97

15. Asokan A, Cho MJ (2002) Exploitation of intracellular pH gradients in the cellular delivery of macromolecules. J Pharm Sci 91:903–913

16. Drummond DC, Zignani M, Leroux J-Ch (2000) Current status of pH-sensitive liposomes in drug delivery. Prog Lipid Res 39:409–460

17. Simões S, Slepushkin VA, Düsgüne N, Pedroso de Lima MC (2001) On the mechanism of internalization and intracellular delivery mediated by ph-sensitive liposomes. Biochim Biophys Acta 1515:23–37

18. Venugapalan P, Jian S, Sankar S, Singh P, Vyas SP (2002) pH-sensitive liposomes: mechanism of triggered release to drug and gene delivery prospects. Pharmacie 56:659–671

19. Klibanov AL, Maruyama K, Torchilin VP, Huang L (1990) Amphipatic polyethyleneglycols effectively prolong the circulation time of liposomes. FEBS Lett 268:235–238

20. Torchilin VP, Omelyanenko VG, Papisov MI, Bogdanov AAJ, Trubetskoy VS, Herron JN, Gentry CA (1994) Poly(ethylene glycol) on the liposomes surface: on the mechanism of polymer-coated liposome longevity. Biochim Biophys Acta 1195:11–20

21. Hristova K, Kenworthy A, McIntosh TJ (1995) Effect of bilayer composition on the phase behavior of liposomal suspension containing poly(ethylene glycol)-lipids. Macromolecules 28:7693–7699

22. Belisto S, Bartucci R, Montesano G, Marsh D, Sportelli L (2000) Molecular and mesoscopic properties of hydrophilic polymer-grafted phospholipids mixed with phosphatidylcholine in aqueous dispersion: interaction of dipalmitoyl N-poly(ethylene glycol) phosphatidylethanolamine with dipalmitoylphosphatidylcholine studied by spectrophotometry and spin-label electron spin. Biophys J 78:1420–1430

23. Marsh D, Bartucci R, Sportelli L (2003) Lipid membranes with grafted polymers: physicochemical aspects. Biochim Biophys Acta 1615:35–59

24. Slepushkin VA, Simões S, Dazin P, Newman MS, Guo LS, Pedroso de Lima MC, Düsgüneş N (1997) Sterically stabilized pH sensitive liposomes. Intracellular delivery of aqueous contents and prolonged circulation in vivo. J Biol Chem 272:2382–2388

25. Hong M-S, Lim S-J, Oh Y-K, Kim Ch-K (2002) pH-sensitive, serum stable and long-circulating liposomes as a new drug delivery system. J Pharm Pharmacol 54:51–58

26. Guo X, Szoka FC Jr (2001) Steric stabilization of fusogenic liposomes by a low-pH sensitive PEG-diortho ester-lipid conjugate. Bioconj Chem 12:291–300

27. Guo X, MacKay JA, Szoka FC Jr (2003) Mechanism of pH-triggered collapse of phosphatidylethanolamine liposomes stabilized by an ortho ester polyethyleneglycol lipid. Biphys J 84:1784–1795

28. Boomer A, Inerowiz HD, Zhang Z, Bergstrad N, Edwards K, Kim JM, Thompson DH (2003) Acid-triggered release from sterically stabilized fusogenic liposomes via a hydrolytic dePEGylation strategy. Langmuir 19:6408–6415

29. Masson C, Garinot M, Mignet N, Wetzer B, Mailhe P, Scherman D, Bessodes M (2004) pH-sensitive PEG lipids containing orthoester linkers:new potential tools for nonviral gene delivery. J Control Rel 99:423–434

30. Roux E, Stomp R, Pezolet M, Moreau P, Leroux J-C (2002) Steric stabilization of liposomes by pH-responsive N-isopopylacrylamide copolymer. J Pharm Sci 91:1795–1802

31. Roux E, Passirani C, Scheffold S, Benoit J-P, Leroux J-C (2004) Serum stable and long-circulating, PEGylated, pH-sensitive liposomes. J Control Release 94:447–451

32. Holland JW, Hui C, Cullis PR, Madden TD (1996) Poly(ethylene glycol)-lipid conjugates regulate the calcium-induced fusion of liposomes composed of phosphatidylethanolamine

and phosphatidylserine. Biochemistry 35: 2618–2624

33. Johnsson M, Edwards K (2001) Phase behavior and aggregate structure in mixtures of dioleoylphosphathidylethanolamine and poly(ethylene glycol)-lipids. Biophys J 80: 313–323

34. Rangelov S, Almgren M, Tsvetanov Ch, Edwards K (2002) Synthesis, characterization and aggregation behavior of block copolymers bearing blocks of lipid-mimetic aliphatic double chain units. Macromolecules 35:4770–4778

35. Rangelov S, Almgren M, Tsvetanov Ch, Edwards K (2002) Shear-induced rearrangement of self-assembled PEG-lipid structures in water. Macromolecules 35:7074–7081

36. Rangelov S, Almgren M, Edwards K, Tsvetanov Ch (2004) Formation of normal and reverse bilayer structures by self-assembly of nonionic block copolymers bearing lipid-mimetic units. J Phys Chem B 108: 7542–7552

37. Rangelov S, Edwards K, Almgren M, Karlsson G (2003) Steric stabilization of egg-phosphathidyl choline liposomes by copolymers bearing short blocks of lipid-mimetic units. Langmuir 19:172–181

38. Momekova D, Rangelov S, Yanev S, Nikolova E, Konstantinov S, Romberg B, Storm G, Lambov N (2007) Long-circulating, pH-sensitive liposomes sterically stabilized by copolymers bearing short blocks of lipid-mimetic units. Eur J Pharm Sci 32:308–317

Chapter 36

Serum-Stable, Long-Circulating, pH-Sensitive PEGylated Liposomes

Nicolas Bertrand, Pierre Simard, and Jean-Christophe Leroux

Abstract

pH-sensitive liposomes have been designed to deliver active compounds specifically to acidic intracellular organelles and to augment their cytoplasmic concentrations. These systems combine the protective effects of other liposomal formulations with specific environment-controlled drug release. They are stable at physiological pH, but abruptly discharge their contents when endocytosed into acidic compartments, allowing the drug to be released before it is exposed to the harsh environment of lysosomes.

Serum-stable formulations with minimal leakage at physiological pH and rapid drug release at pH 5.0–5.5 can be easily prepared by inserting a hydrophobically modified N-isopropylacrylamide/methacrylic acid copolymer (poly(NIPAM-co-MAA)) in the lipid bilayer of sterically stabilized liposomes. The present chapter describes polymer synthesis, as well as the preparation, and characterization of large unilamelar pH-sensitive vesicles.

Key words: pH-sensitive liposomes, N-isopropylacrylamide copolymer, Triggered release

1. Introduction

The acidification of endosomal compartments, as they evolve toward lysosomes is a well-described phenomenon (1) that can be exploited to design drug delivery systems capable of releasing their contents after endocytosis. Enhanced cytoplasmic drug concentrations can therefore be achieved with "smart" formulations, which are sensitive to acidic pHs. For this purpose, liposomal formulations are attractive, because their deformable phospholipid bilayers can be rapidly disrupted to trigger drug release. In this section, ionizable copolymers of N-isopropylacrylamide (NIPAM) are anchored in the phospholipid membrane and used to destabilize the bilayer upon acidification of the environment.

These polymers were selected owing to their ability to promptly transit from hydrated coil to dehydrated globule conformation when temperature increases above their lower critical solution temperature (LCST). Homopolymers of NIPAM have a LCST in aqueous media around 32°C, which is not compatible with physiological conditions. Thus, to be exploitable for *in vivo* drug delivery, this temperature must be increased by copolymerization with other hydrophilic monomers. When these monomers have ionizable moieties; such as methacrylic acid (MAA, pKa = 5.4), the increase in LCST is dependent on ionization, and therefore the polymers acquire pH-dependent physiological solubility (2).

Poly(NIPAM-*co*-MAA) with MAA content above 5 mol% is soluble under both physiological pH (7.4) and temperature (37°C), when its carboxylic functions are ionized. However, when the pH is decreased to the pH of late endosomes and lysosomes (~pH = 5–5.5) (1), MAA protonation triggers dehydration of the polymer chains. If the polymer is anchored to a phospholipid membrane through randomly or terminally incorporated alkyl chains, transition, to a globule conformation induces reorganization of the bilayer that leads to massive content leakage (3).

This chapter thoroughly describes the preparation of poly(NIPAM-*co*-MAA)-based liposomes that can enhance the cytoplasmic bioavailability of drugs by triggering drug release, specifically, in acidic compartments. The optimization and characterization of these systems have been described in numerous publications (3–9).

2. Materials

2.1. Polymer Synthesis

2.1.1. Alkylated Initiator (DODA-501) Synthesis

1. Anhydrous tetrahydrofuran (THF), distil freshly on an alumina column and maintain under inert atmosphere until use.
2. Anhydrous chloroform ($CHCl_3$), distil freshly on an alumina column and maintain under inert atmosphere until use.
3. Anhydrous acetonitrile (ACN), commercially available, store in a ventilated cabinet at room temperature, under inert atmosphere.
4. Acetone, reagent grade, store in a ventilated cabinet at room temperature.
5. Hexane, reagent grade, store in a ventilated cabinet at room temperature.
6. Ethyl acetate, reagent grade, store in a ventilated cabinet at room temperature.
7. 4,4′-azobis(4-cyanovaleric acid) (V-501, Sigma), store at room temperature.
8. *N*-hydroxysuccinimide (NHS, Sigma), store at room temperature.

9. 1-ethyl-3-[3-dimethylaminopropyl]carbodiimide hydrochloride (EDC, Sigma), store at –20°C and protect from humidity.

10. Dioctadecylamine (DODA, Sigma). To purify, solubilize in boiling acetone/$CHCl_3$ (4:1 v/v), cool to room temperature, and filter under vacuum with a Buchner funnel. Store the purified product at room temperature.

11. Silica thin layer chromatography (TLC) sheets, store at room temperature.

12. Silica powder, store at room temperature.

13. Potassium permanganate ($KMnO_4$) solution. To prepare, dissolve 1.5 g (10 mmol) of $KMnO_4$ and 10 g (72 mmol) of potassium carbonate (K_2CO_3) in 1.25 mL of 2.5 N sodium hydroxide (NaOH) solution and add 200 mL of deionized water. Store at room temperature for a maximum of 3 months.

2.1.2. Poly(NIPAM-co-MAA) Synthesis

1. Previously synthesized alkylated initiator (DODA-501), store at room temperature under inert atmosphere.

2. NIPAM (Sigma). Upon arrival, purify the commercially available product through crystallization by boiling heptanes/acetone (4:1 v/v) and filter under vacuum with a Buchner funnel. Store the purified product under argon, at room temperature.

3. MAA (Sigma). Upon arrival, purify the commercially available product from its polymerization inhibitor by elution on an inhibitor-removing disposable column (Sigma). Store the purified product at –20°C.

4. Anhydrous 1,4-dioxane, store in a ventilated cabinet at room temperature, under inert atmosphere.

5. THF, reagent grade, store in a ventilated cabinet at room temperature.

6. Diethylether, reagent grade, store in a ventilated cabinet at room temperature.

7. Regenerated cellulose dialysis bags, molecular *cut-off* 6–8 kDa (e.g., Spectra/Por®, Spectrum). Before use, soak the membrane in water for at least 30 min and rinse with distilled water. Standard polypropylene clips (e.g., Spectra/Por® Closures, Spectrum) can be employed to seal the tubes.

2.2. Polymer Characterization

1. *d*-Chloroform ($CDCl_3$) for proton nuclear magnetic resonance (^1H-NMR) analysis, store at room temperature.

2. THF, High Performance Liquid Chromatography-grade, for gel permeation chromatography (GPC) analysis, store at room temperature.

3. 66 mM isotonic phosphate buffer saline (53 mM Na2HPO4, 13 mM NaH_2PO4, 75 mM NaCl, PBS) pH 7.4. Sterilize by filtration on a 0.22-μm nylon filter and store at 4°C.

4. 5 N hydrochloric acid (HCl) solution. Store in a ventilated cabinet at room temperature.
5. Quartz 10 mm × 10 mm cuvettes.

2.3. Liposome Preparation

1. 1% (w/v) poly(NIPAM-co-MAA) solution in $CHCl_3$. Store at −80°C until use.
2. 4% (w/v) egg phosphatidylcholine (EPC) solution in $CHCl_3$. Store at −80°C until use (see Note 1).
3. 4% (w/v) cholesterol solution in $CHCl_3$. Store at −80°C until use.
4. 2% (w/v) 1,2-distearoyl-sn-glycero-3-phosphoethanolamine-N-(methoxy polyethylene glycol)$_{2000}$ (DSPE-PEG, NOF) solution in $CHCl_3$. Store at −80°C until use.
5. 20 mM isotonic 4-(2-hydroxyethyl) piperazine-1-ethanesulfonic acid buffer saline (HEPES) pH 7.4, containing 35 mM of trisodium 8-hydroxypyrene-1,3,6-trisulfonate (HPTS) and 50 mM of p-xylene-bis-pyridinium bromide (DPX). Store protected from light at 4°C (see Note 2).
6. Manual liposome extruder and polycarbonate filters (pore diameters of 400, 200 and 100 nm).
7. 20 mM HEPES isotonic buffer, pH 7.4. Sterilize by filtration on a 0.22-μm nylon filter and store at 4°C.
8. Agarose (e.g. Sepharose CL-4B(r), Sigma) column of adequate length and width 30 cm length and 1 cm width) for size exclusion chromatography (SEC). Store hydrated with isotonic HEPES buffer at 4°C.

2.4. Liposome Characterization

2.4.1. pH-Sensitivity Experiments

1. Quartz 10 mm × 10 mm cuvettes.
2. 10% (w/w) Triton X-100 aqueous solution, store at 4°C.
3. 20 mM HEPES, 144 mM NaCl isotonic buffer, pH 7.4. Sterilize by filtration on a 0.22-μm nylon filter and store at 4°C.
4. 20 mM isotonic 2-N-(morpholino)ethanesulfonic acid (MES) buffers, pH 5.0–6.5. Sterilize by filtration on a 0.22-μm nylon filter and store at 4°C.

2.4.2. Serum Stability Experiments

1. Quartz 10 mm × 10 mm cuvettes.
2. 10% (w/w) Triton X-100 aqueous solution, store at 4°C.
3. 20 mM HEPES, 144 mM NaCl isotonic buffer, pH 7.4. Sterilize by filtration on a 0.22-μm nylon filter and store at 4°C.
4. Rat nonsterile serum. Store at −20°C.

2.4.3. Preservation of pH-Sensitivity

1. Quartz 10 mm × 10 mm cuvettes.
2. 10% (w/w) Triton X-100 aqueous solution, store at 4°C.
3. 20 mM HEPES, 144 mM NaCl isotonic buffer, pH 7.4. Sterilize by filtration on a 0.22-μm nylon filter and store at 4°C.

4. Rat nonsterile serum. Store at −20°C.

5. Agarose (e.g. Sepharose CL-4B(r), Sigma) column of adequate length and width (30 cm length and 1 cm width) for SEC. Store hydrated with isotonic HEPES buffer at 4°C.

3. Methods

The following section thoroughly describes polymer synthesis and purification, as well as liposome preparation, purification, and basic characterization. To provide detailed protocols, the number of experiments was narrowed down to the most general procedures. However, creative investigators will find plenty of ways to modify and adjust the protocols to achieve their drug delivery needs and objectives.

In previous publications, different probes and drugs have been encapsulated successfully in the pH-sensitive liposomes described in this chapter (3–9). The vesicles alone have been shown to be stable for prolonged periods of time (over months), when stored at 4°C. However, depending on the nature of the encapsulated compounds, content may degrade or leak upon storage, reducing the overall shelf-life of the formulation.

3.1. Polymer Synthesis

3.1.1. Alkylated Radical Initiator (DODA-501) Synthesis
(see Fig. 1a) (10)

3.1.1.1. Synthesis of Disuccinimide 4,4′-azobis-(4-cyanovalerate) (A-501)

1. Flame-dry a round bottom flask containing a magnetic stirrer, sealed with a rubber septum and under a flux of argon. Cool down under a flux of argon.

2. To obtain around 1 g of purified product, weigh 2.65 g (9.5 mmol) of V-501, 2.08 g (18 mmol) of NHS and 2.06 g (10.8 mmol) of EDC. Rapidly transfer the reagents to the reaction flask, without allowing air to disturb the inert atmosphere.

3. Using anhydrous techniques, transfer 10 mL of THF and 10 mL of ACN to the reaction flask (V-501 concentration ~0.5 mmol/mL). Stir overnight (>12 h) at room temperature, under argon atmosphere.

3.1.1.2. Purification of A-501

1. Evaporate the solvent under reduced pressure and mild heating (40–50°C).

2. Solubilize the content of the reaction flask with ~30 mL acetone. Transfer the mixture to a large beaker and precipitate the reaction product by adding ~70 mL of ice-cold water. Filter under vacuum on a Buchner funnel.

3. Dry the white crust on the filter paper overnight under vacuum at room temperature. The yield is around 25%.

3.1.1.3. Synthesis of DODA-501

1. Flame-dry a round bottom flask containing a magnetic stirrer, sealed with a rubber septum. Cool down under a flux of argon.

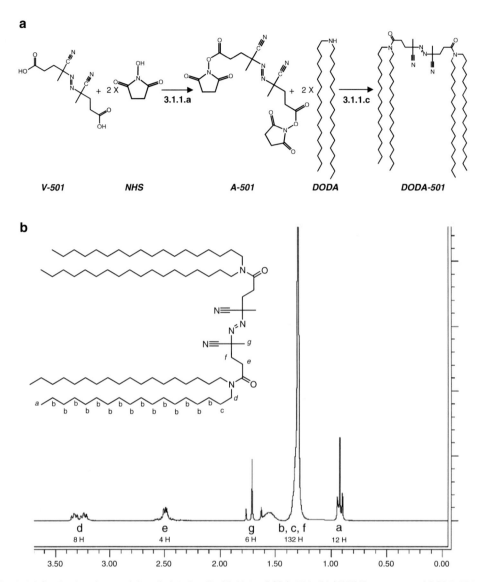

Fig. 1. (a) Synthesis scheme of the alkylated radical initiator DODA-501. (b) ^1H-NMR spectrum of DODA-501 with its characteristic protons

2. To obtain around 800 mg of alkylated initiator, weigh 1.00 g (2.1 mmol) of A-501 and 2.20 g (4.2 mmol) of purified DODA. Rapidly transfer both reagents to the round bottom flask, without allowing air to disturb the inert atmosphere, and wrap the flask with aluminum foil to protect the content from light.

3. Employing anhydrous techniques, transfer 15 mL of THF and 7.5 mL of $CHCl_3$ to the reaction flask (A-501 concentration ~0.1 mmol/mL). Stir overnight (>12 h) at room temperature, under argon atmosphere.

4. Confirm completion of the reaction by TLC with elution of a few drops of the reaction mixture with hexane/ethyl acetate (2:1 v/v). Reveal with $KMnO_4$ solution. Consider the reaction completed when the spot for A-501 is no longer visible on TLC or when no intensity changes in the spots are detectable (see Note 3).

3.1.1.4. Purification of DODA-501

1. When the reaction is completed, evaporate the solvent under reduced pressure and mild heating (40–50°C).

2. Prepare silica gel for chromatography by solvating silica powder in hexane. Deposit the gel in a chromatographic column (30 cm length, 2 cm width), and allow to settle.

3. Solubilize the reaction products in a small quantity of hexane/ethyl acetate (2:1 v/v) (see Note 4). Deposit the solution with a glass pipette on top of the silica gel column and start eluting with solvent (hexane/ethyl acetate 2:1 v/v). Collect fractions of 10 mL and identify DODA-501 by TLC. The product should elute with early fractions.

4. Pool together the fractions containing DODA-501 and evaporate the solvent under reduced pressure to obtain a brownish viscous liquid. Add a small volume of $CHCl_3$ and precipitate in ACN. Filter under vacuum with a Buchner funnel and collect the white powder. The yield is around 25%.

5. Confirm structure by ^1H-NMR in $CDCl_3$ (see Fig. 1b): δ 0.9 ppm (12 H, t, $-CH_3$), 1.1–1.6 (132 H, m, $-CH_2-$), 1.7 (6 H, t, $-C(-CH_3)-$), 2.5 (4 H, t, $-CH_2C(=O)-$), 3.1–3.5 (8 H, m, $-NCH_2-$).

3.1.2. Poly(NIPAM-co-MAA) Synthesis

3.1.2.1. Polymer Synthesis

1. Flame-dry a round bottom flask containing a magnetic stirrer, sealed with a rubber septum. Cool down under a flux of argon.

2. To obtain around 400 mg of polymer, weigh 0.025 g (0.019 mmol) of the radical initiator DODA-501, 0.425 g (3.8 mmol) of NIPAM, and 0.017 g (0.20 mmol) of MAA. Rapidly transfer the reagents to the round bottom flask, and restore inert atmosphere by purging with argon for a few minutes.

3. Using anhydrous techniques, transfer 5 mL of 1,4-dioxane into the reaction flask (monomer concentration ~0.8 mmol/mL). Degas the reaction flask by bubbling argon through the solution for 10 min. Stir overnight (>12 h) at 70°C, under argon atmosphere.

3.1.2.2 Polymer Purification

1. Evaporate the remaining solvent under reduced pressure and strong heating (80–100°C).

2. Solubilize the reaction product in a small quantity of THF and precipitate the polymer in cold diethylether. Filter under vacuum on a Buchner funnel and collect the polymer on the filter.

3.2. Polymer Characterization

3.2.1. Basic Characterization

3. Solubilize the polymer in water, transfer the aqueous solution to a dialysis bag and dialyse against deionized water under stirring at room temperature for 5–7 days, changing the water daily. Collect the contents of the dialysis bag and freeze-dry to obtain an amorphous solid. The yield is around 80%.

1. Basic polymer characterization is done by $1H$-NMR, in $CDCl_3$, and GPC relative to polystyrene standards, in THF at 1 mL/min flow. The presence of the terminal alkyl chains of the polymer is ascertained by the methyl $1H$-NMR peak at 0.9 ppm (10, 11). The expected number-average molecular weight (M_n) under these conditions should range between 10 and 15 kDa with a polydispersity comprised between 1.2 and 1.8.

3.2.2. Measuring the pH of Precipitation of the Polymer

1. The following experiment is conducted to find out the pH at which the polymer will precipitate, and induce drug release from the liposomes, once anchored on the membrane. The phase transition pH is determined at 37°C by measuring the turbidity of the polymer solutions at different pHs. As the polymer undergoes coil-to-globule transition and aggregates, the turbidity of the polymer solution increases.

2. Weigh 10 mg of polymer and dissolve in 200 mL of PBS, pH 7.4. Decrease the pH of the solution step by step (0.2 unit/step) to a final pH of 4.6, with a concentrated HCl solution. Sample aliquots of 4 mL, after each pH decrement.

3. Measure the turbidity of each aliquot (amount of light scattered at 90°) in a 10-mm × 10-mm quartz cuvette at 37°C and under magnetic stirring. Before reading, leave each sample in the cuvette for at least 5 min to reach equilibrium. Determine the turbidity of the solution with a fluorometer by setting the excitation and emission wavelengths at 480 nm.

4. Establish the fluorometer sensitivity and offset by using the aliquot of maximum turbidity and a blank buffer solution, respectively.

3.3. Liposome Preparation and Purification

3.3.1. Liposome Extrusion

1. In a clean, round bottom flask, add 693 µL (36.5 µmol) of the EPC solution, 234 µL (24.2 µmol) of the cholesterol solution, 286 µL (2.1 µmol) of the DSPE-PEG solution, and 950 µL (0.625 µmol) of the poly(NIPAM-co-MAA) solution.

2. Evaporate the $CHCl_3$ by rotary evaporation under reduced pressure, at room temperature for at least 1 h, to form a dry, uniform film on the flask wall.

3. Hydrate the polymer/lipid film with 0.9 mL of HPTS/DPX-containing HEPES buffer (see Note 2). Vortex and leave the mixture to rest overnight protected from light, at 4°C, to ensure complete lipid hydration (see Note 5).

4. Extrude the lipid preparation on a tightly closed manual extruder through a 400-nm polycarbonate membrane. Force liposome dispersion through the membrane for an *even* number of times (six or eight times). Change the membrane for one with 200-nm pore-size and repeat the extrusion process six or eight times. Finally, substitute the membrane for a 100-nm membrane and extrude an *odd* number of times, normally 7 to 21 times. Collect liposomes on the extruder side opposite to the starting point (see Note 6).

5. Confirm that the liposomes have an adequate size and size distribution, by DLS. Repeat extrusion with the smallest pore-sized membrane, if the results are not satisfactory (see Note 7).

3.3.2. Liposome Purification

1. Separate liposomes from nonencapsulated dyes and free polymer by SEC with 20-mM HEPES isotonic buffer as the mobile phase (see Note 8).

2. Pool fractions together if a high amount of liposomes is needed.

3.4. Liposome Characterization

The basic characterization of the liposomes revolves around size determination, usually by DLS, and sample liposome concentration by phosphorous assay, neither of which will be described here. Indeed, the procedures of the DLS experiments highly depend on the type of apparatus used, while the Bartlett colorimetric assay of phosphorous content has fully been described elsewhere (12–14).

3.4.1. Liposome pH-Sensitivity

1. Liposome pH-sensitivity is assessed by comparing the release kinetics of encapsulated HPTS/DPX between liposomes dispersed in neutral HEPES and acidic MES buffers (pH between 5.0 and 6.5). The method relies on monitoring the dequenching of HPTS fluorescence as it is released ($\lambda_{excitation}$ 413 nm and $\lambda_{emission}$ 512 nm) (see Note 9).

2. The following experiments can be conducted on any fluorometer. Release experiments are performed in a 10-mm × 10-mm quartz cuvette containing 1 mL of buffer maintained at 37°C and under constant magnetic stirring.

3. After the calibration of the apparatus, put 10 µL of purified liposomes in 1 mL of HEPES and monitor fluorescence emission for 30 min. At the end of the experiment, disrupt the liposomes with 10 µL of Triton X-100 solution and record this fluorescence intensity as the maximum value (100%). Plot the fraction of HPTS released *versus* time.

4. Repeat the experiment under acidic conditions (MES buffer) and plot the release kinetics at different pHs from 5.0 to 6.5. Efficient pH-sensitive liposomes should show minimal HPTS leakage at pH 7.4 and near complete (>80%) release within the first 10 min at acidic pH (see Fig. 2).

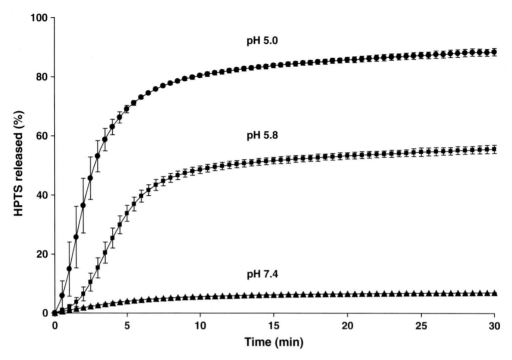

Fig. 2. Release rates of encapsulated HPTS at 37°C for liposomes prepared with the pH-sensitive polymer poly(NIPAM-co-MAA) (Mn = 11,000; M_w/M_n = 2.1) at pH 5.0 (*filled circles*), 5.8 (*filled squares*), and 7.4 (*filled triangles*). The extent of content release was calculated from HPTS fluorescence intensity (λ_{ex} = 413 nm, λ_{em} = 512 nm) relative to measurement after vesicle disruption with 10% (*v/v*) Triton X-100

3.4.2. Serum Stability

3.4.2.1. Content Leakage in Serum

1. To verify that the content of the liposome does not leak when the formulation is exposed to blood components, the kinetics of HPTS release in serum can be monitored for different periods of time.

2. Dilute a given quantity of liposomes in HEPES buffer with a given quantity of serum to give a final serum concentration of 50% (*v/v*). Incubate for a prolonged period of time (up to a few hours).

3. At given time intervals, sample 1-mL aliquots of the liposome suspension and measure fluorescence as described in subheading 3.4.1. Total fluorescence of each sample can be determined by disruption of the liposome with 10 μL of Triton X-100 solution. Serum-stable liposomes should exhibit minimal release of HPTS in serum.

3.4.2.2. Preservation of pH-Sensitivity

1. To verify that the liposomes maintain their pH-sensitivity after exposure to blood proteins, the pH-sensitivity experiment can be conducted after incubation of the liposomes with serum or plasma.

2. Incubate the liposome formulation for 1 h with preheated serum, stirring at 37°C. Separate the liposomes from excess serum by SEC with a 20-mM HEPES mobile phase.

3. Conduct the pH-sensitivity experiment on serum-exposed liposomes as described in Subheading 3.4.1.

3.5. Summary

From the bases presented in this chapter, the imaginative scientist will find many ways to use the pH-sensitive liposomes to its advantage. Now that serum compatibility and pH-sensitivity have been established with the *in vitro* experiments presented above, the next major step is to inject the formulation *in vivo* and study their biological fate.

As protocols for *in vivo* experiments highly depend on the desired purposes of the investigator and on local animal welfare legislation and guidelines, detailed procedures for animal studies will not be presented here. However, a few helpful suggestions will be proposed to guide the design of pH-sensitive PEGylated liposomes from a preclinical *in vivo* perspective. Although these recommendations could probably be adapted to any animal model, they were mainly drawn from experiments on rodents.

The PEG corona provided by DSPE-PEG in the preparation of liposomes prolongs their circulation time, but the vesicles are still considered foreign to the body and are eventually caught up by the mononuclear phagocytosis system. Therefore, PEGylated liposomes will likely be cleared from the circulation within a few hours to a few days after injection, depending on their size, charge and composition. For the same reasons, although long-circulating liposomes may accumulate at tumor and inflammation sites when present, the vesicles will mostly distribute in the liver and spleen. The expected plasmatic circulation half-life for these liposomes is usually between 3 and 12 h, contingent on PEG chain length, the doses injected and the tracking marker deployed (7, 8).

Although tracing the content of the liposomes can be easily achieved by developing analytical methods for each type of payload, following the vesicles themselves can offer interesting complementary information. To this end, radioactive nonexchangeable probes inserted in the phospholipid membranes, such as ^{14}C-cholesteryl oleate or ^{3}H-cholesterol hexadecyl ether, are commonly used. The amount of radioactivity injected depends on the isotope, purpose of the experiment and design of the formulation. However, doses from 5 to 10 µCi/kg can be used to accurately depict liposome circulation profiles. Also, although these markers are known to be retained in the bilayer as long as the liposomes remain intact, it is worth mentioning that redistribution or biotransformation of the probes can occur once the vesicles are destroyed. Hence, data must be critically considered if these probes are to be monitored over time frames compatible with possible liposome disruption and metabolism.

At last, the molecular weight of the NIPAM copolymers must also be carefully chosen. Indeed, these polymers are not expected to be biodegradable because of their chemical structure.

In order to avoid accumulation in the body after injection, it is of primary importance that all polymer chains be of a size shorter than the glomerular filtration *cut-off* for the polymer. In rats, this maximum value is in the range of 40,000 (15).

4. Notes

1. Although EPC has been the most commonly used phospholipid for the preparation of poly(NIPAM-*co*-MAA)-based liposomes, other lipids can be chosen if desired. The successful utilization of 1,2-dioleoyl-*sn*-glycero-3-phosphatidylcholine (DOPC) (5) and 1-palmitoyl-2-oleoyl-*sn*-glycero-3-phosphatidylcholine (POPC) (6) has been reported. Use of lipids with high transition temperature (T_m) complicates the preparation procedure, because the formulation must be heated to reach the fluid phase during the extrusion step. When the lipid T_m is above the polymer's LCST, the two-step film hydration/extrusion procedure described in Subheading 3.3.1 cannot be performed for a longer period. The fixation of the polymer in the bilayer is achieved by incubating the preformed liposomes with the polymer solution overnight (>12 h) at 4°C (5, 6, 9). However, this post incorporation method is not as efficient for the production of pH-sensitive liposomes, because the release triggered at acidic pH is lower (6).

2. Any isotonic buffer can be selected. This buffer will form the internal compartment of the liposomes, so it must be chosen according to the experiments that will be carried out. To be able to quantify the phospholipids after liposome preparation, the buffer must be phosphate-free. The buffer can contain either the dissolved drug or fluorescent marker, depending on the intended use of the liposomes.

3. This step is not essential, but rapidly confirms the success of the reaction before starting silica gel chromatography. The $KMnO_4$ solution reveals the cyanide groups present on A-501 and DODA-501. With these specific elution solvents, the retention factor of DODA-501 is much higher ($R_f \sim 0.80$) than that of the initial product, A-501 ($R_f \sim 0$). Thus, a fast eluting spot on the TLC sheet confirms that the reaction product is present.

4. Utilize the smallest amount of solvent possible to solubilize the products to ensure homogenous migration of the product.

5. Although leaving the film to hydrate overnight facilitates incorporation of the polymer in the bilayer and extrusion through the polycarbonate membranes, it is also possible to hydrate for only a couple of hours. Likewise, higher lipid concentrations could also be used, but this may further impede extrusion.

6. Employing an odd number of extrusions with the last set of membranes ensures that any large aggregates or contaminants will be kept off the final preparation. Assuming that all glassware is sterilized and handled according to sterile techniques, it is noteworthy that extrusion through the last membrane safeguards sterility of the formulation.

7. Theoretically, to achieve long-circulating properties and avoid extensive uptake by the liver and spleen, injected colloids should have a size between 70 nm and 200 nm (16). A typical diameter of 100–200 nm is obtained for the pH-sensitive PEGylated liposomes presented here. The number of times the formulation is extruded through the membrane regulates size distribution, the latter becoming more uniform as the number increases. In DLS, the polydispersity index (PdI) is calculated from the square of the normalized standard deviation ($PdI = (\sigma/Z_{avg})^2$). Acceptable PdI values will depend on the desired properties of the vesicles, but values under 0.1 can be usually attained.

8. Once again, HEPES is utilized here as the mobile phase, because it is a phosphate-free buffer, but other types can be used if desired. The elution buffer will constitute the external phase of the final formulation. In cases where the liposomes are loaded with a fluorescent probe, fractions which contain the colloids can be identified by absorbance or fluorescence. Alternatively, turbidity can also be used for nonfluorescent liposomes, assuming that the initial lipid quantities are large enough. Finally, it is essential to verify that the column used can properly separate free poly(NIPAM-co-MAA) and liposomes. Separating properties of chromatographic conditions can be established through the use of fluorescently- (9) or radiolabelled (15) copolymers.

9. Although this experiment is the simpler way to ascertain the pH-sensitivity of the liposomes, similar experiments can be conducted with encapsulated drugs. The same procedure can be used if the drug is fluorescent, and self-quenched when encapsulated (e.g., doxorubicin). However, in the case of nonfluorescent compounds, the drug released must be separated from the liposome by SEC or dialysis at each time-point and then assayed by the appropriate analytical method.

Acknowledgements

Financial support from the CIHR and the FRSQ (scholarships to NB and PS) is acknowledged.

References

1. Mukherjee S, Ghosh RN, Maxfield FR (1997) Endocytosis. Physiol Rev 77:759–803
2. Chen G, Hoffman AS (1995) Graft copolymers that exhibit temperature-induced phase transitions over a wide range of pH. Nature 373:49–52
3. Meyer O, Papahadjopoulos D, Leroux J-C (1998) Copolymers of N-isopropylacrylamide can trigger pH sensitivity to stable liposomes. FEBS Lett 421:61–64
4. Leroux J-C (2004) pH-responsive carriers for enhancing the cytoplasmic delivery of macromolecular drugs. Adv Drug Deliv Rev 56:925–926
5. Leroux J-C, Roux E, Le Garrec D, Hong K, zcopolymers for the preparation of pH-sensitive liposomes and polymeric micelles. J Control Release 72:71–84
6. Roux E, Francis M, Winnik FM, Leroux J-C (2002) Polymer based pH-sensitive carriers as a means to improve the cytoplasmic delivery of drugs. Int J Pharm 242:25–36
7. Roux E, Passirani C, Scheffold S, Benoit J-P, Leroux J-C (2004) Serum-stable and long-circulating, PEGylated pH-sensitive liposomes. J Control Release 94:447–451
8. Roux E, Stomp R, Giasson S, Pézolet M, Moreau P, Leroux J-C (2002) Steric stabilization of liposomes by pH-responsive N-isopropylacrylamide copolymer. J Pharm Sci 91:1795–1802
9. Zignani M, Drummond DC, Meyer O, Hong K, Leroux J-C (2000) *In vitro* characterization of a novel polymeric-based pH-sensitive liposome system. Biochim Biophys Acta 1463:383–394
10. Kitano H, Akatsuka Y, Ise N (1991) pH-responsive liposomes which contain amphiphiles prepared by using lipophilic radical initiator. Macromolecules 24:42–46
11. Winnik FM, Davidson AR, Hamer GK, Kitano H (1992) Amphiphilic poly(N-isopropylacrylamide) prepared by using a lipophilic radical initiator: Synthesis and solution properties in water. Macromolecules 25:1876–1880
12. Bartlett GR (1959) Phosphorous assay in column chromatography. J Biol Chem 234:466–468
13. Satish PR, Surolia A (2002) Preparation and characterization of glycolipid-bearing multilamellar and unilamellar liposomes. In: Basu SC, Basu M (eds) Liposomes: methods and protocols, vol 199. Humana Press, Totowa, NJ, pp 193–202
14. Zuidam NJ, de Vrueh R, Crommelin DJ (2003) Chapter 2: Characterization of liposomes. In: Torchilin VP, Weissig V (eds) Liposomes: a practical approach. Oxford University Press, Oxford, pp 31–78
15. Bertrand N, Fleischer JG, Wasan KM, Leroux J-C (2009) Pharmacokinetics and biodistribution of N-isopropylacrylamide copolymers for the design of pH-sensitive liposomes. Biomaterials 30(13):2598–2605.
16. Stolnik S, Ilum L, Davis SS (1995) Long circulating microparticulate drug carriers. Adv Drug Deliv Rev 16:194–214

Index

A

Acetal 215, 216
Acid-labile liposome
    degradation 418, 420
    lipoplex formation 407
Acute myocardial infarction 10, 306,
    309–310, 361, 362
Adjuvant
    lipid A 17
    mannosylated liposomes 185–187
Adriamycin 19, 215, 342, 343
Alza Corporation 8
Ambisome® 2
Amphotericin B 2, 307, 308
Angiogenic vessels 335–346
Anionic lipoplex 436, 442
Annamycin 4
Anthracycline 139–145
Antifungal 2
Antigenic peptide 167
Antigen-presenting cells (APC)
    dendritic cells 164
    macrophages 164, 178
    receptors 164, 178
Antimyosin antibody 307–309
Antiparasitic activity 152
Antisense oligonucleotides 11, 12, 215, 447, 487, 528
Antitrypanocidal activity 151
Anti-viral cytokine 169
Apoptosis
    ceramide 295, 300
    mitochondria 295, 300
Arabinocytidylyl-N⁴-octadecyl-1-β-D-arabinofuranosyl-
    cytosine 131
Archae bacteria
    lipids 87–95
    thermophilic 87
Arsenic trioxide 53
Arsonolipids 148–152, 154
Arsonoliposomes 147–161
Asialofetuin 425–434
Asialoglycoprotein 178, 426
Asialoglycoprotein receptor (ASGPr) 426
Atomic force microscopy 43, 89, 93–94
ATP-loaded liposomes 361–374

Autoantibodies 323
Autoclavation 91–93
Avanti Polar Lipids 34, 37, 58, 59,
    99, 151, 153, 165, 218, 219, 247, 280, 281, 284,
    296, 297, 307, 308, 323, 324, 350, 363, 364, 378,
    394, 406, 427, 437, 453
Avestin Inc 34, 37, 535
Avidin–biotin affinity
    xchromatography 229–230, 234, 308
Azidothymidine 132

B

B-cell lymphoma 4, 9
Benzoporphyrin 17
Bioconjugation 268, 274, 275
Biodistribution
    liposomes 2, 7, 15, 18, 139,
    267, 280, 300, 301, 363, 538–541
Biopanning 338, 346
Block copolymer 215, 217, 218
Bolus release 114
Bradford assay 245–246, 249
Bubble liposomes
    imaging 473, 475, 476, 478, 482, 483
    transfection 474, 478–480, 482
Bupivacaine 53
Butyl vinyl ether (BVE) 217

C

Caelyx® 8, 140, 279, 280, 282, 288, 290
Calcein 45, 53, 58, 63, 64,
    70, 71, 116, 118–121, 124, 125, 149, 153, 155–160,
    211, 247, 251–253, 283, 531, 535–538, 542
Carboxyfluorescein-5,6 (CF)
    liposomal encapsulation 60–62, 72, 157
    membrane integrity 156, 157
    self-quenching 274
Cardiocytes 306, 309
Cardiomyocytes 230, 362, 382–384
Cardiomyopathy(ies) 305
Cardiotoxicity 6, 139
Cationic liposomes 10, 185, 195,
    393, 405, 426, 438–440, 461–471
C. elegans 449
Cell penetrating peptide (CPP) 5, 13, 164,

238, 309, 349, 350, 368, 369
Ceramide
  apoptosis 295, 300
  liposome 14, 295, 296, 299–302
  mitochondria 295, 300
C-ethynylcytidylyl-(5′-5′)-N⁴-octadecyl-1-β-D-
  arabinofuranosylcytosine 131
Cholesterol 2, 4, 32–35, 41, 42,
  46, 47, 56, 58, 59, 119, 120, 130, 132, 135, 136, 141,
  142, 144, 145, 149, 153, 164, 166, 167, 173, 178,
  180, 195–198, 206, 207, 214, 219, 247, 269, 271,
  281, 283, 296, 297, 308, 324, 327, 337, 342, 343,
  345, 352, 363–365, 394, 427, 438, 448, 450, 452,
  462, 471, 548, 552, 555
Cholesteryl hemisuccinate (CHEM) 437, 528, 530
Choriomeningitis virus 165
Cisplatin 8, 10, 11
Click chemistry
  bioconjugation 268, 274, 275
  cycloaddition 274, 275
  preformed liposomes 267–275
Clodronate 189–200
Colloids 5, 77, 97, 108, 110, 420
Computed tomography (CT) 14, 15,
  259, 260, 269, 321, 475
Concanavalin A (Con-A) 178, 183, 268, 275
Contrast agent 15, 16, 322, 323, 474
Copolymer 13, 214, 215, 217, 218,
  529, 530, 532–535, 539, 545, 546, 555
Cosmetics 30, 77
Cremophor EL formulation 214
Cryofixation 231
Cryoprotectant 53, 57, 66, 115, 166
Cryotransmission electron microscopy 395, 396, 400
Cycloaddition 274, 275
Cyclodextrin 53, 56, 58, 65
Cytarabine 4
Cytosine-arabinoside 130
Cytoskeletal-antigen specific immunoliposomes
  (CSIL) 305–316
Cytotoxicity 10, 11, 42, 149, 151, 169,
  214, 217, 282, 289–291, 299, 476, 479
Cytotoxic T cell response 170

D

Daunoxome® 2
DC-Chol See 3β-[N-(N,N-dimethylaminoethane)
  carbamoyl]cholesterol
Dehydration-rehydration liposomes See DRV liposomes
Dendritic cells 16, 164, 166, 172,
  178, 179, 183–186, 194
Deoxy-5-fluorouridylyl-N⁴-octadecyl-1-β-D-
  arabinofuranosylcytosine 131
DepoCyt® 4

Dicetylphosphate 32
Dideoxycytidine 132
Dideoxyinosine 132
Didodecyldimethyl ammoniumbromide
  (DDAB) 165, 167, 173
Dihexanoyl phosphatidylcholine (DHPC) 125, 126
1,2-Dilauroyl-sn-Glycero-3-
  Phosphoethanolamine 450
3β-[N-(N,N-Dimethylaminoethane)carbamoyl]
  cholesterol (DC-Chol) 394
1,2-Dimyristoyl-sn-Glycero-3-
  Phosphoethanolamine 450
1,2-Dioleoyl-3-trimethylammoniumpropane 59, 78,
  219, 363, 364, 378, 379, 427, 450, 451
Dipalmitoyl phosphatidylcholine (DPPC) 33, 45, 58,
  60, 136, 151, 161, 269, 271, 281, 283, 345, 437, 440
Diractin 78, 84
DLPE. See 1,2-Dilauroyl-sn-Glycero-3-
  Phosphoethanolamine
DMPE. See 1,2-Dimyristoyl-sn-Glycero-3-
  Phosphoethanolamine
Dock and lock liposomes
  adapter protein 258
  targeted liposome 257–266
  targeting protein 258, 260, 263
DOTAP. See 1,2-Dioleoyl-3-trimethylammoniumpropane
DOTMA. See N-[1-(2,3-Dioleyloxy) propyl]-N,N,N-
  triethylammonium
Doxorubicin 4, 6, 8–11, 18, 19, 33, 139–142,
  144, 145, 214, 216, 260, 279–292, 337, 557
Dragendorff spray reagent 219, 225
DreamFect™ 488
Dried reconstituted vesicles. See DRV liposomes
Dried rehydrated vesicles. See DRV liposomes
Drug delivery system 7, 214, 231,
  238, 243, 279, 335, 545
Drug release 7, 8, 79, 81–82, 210,
  215, 234, 243, 545, 546, 552
DRV liposomes 51–73
Dry-film method 394, 397
Duplex drugs 131, 132, 134

E

Echogenicity measurement 121–123
Echogenic liposomes, drug and gene delivery 114
Elan Pharmaceuticals 2
Elasticity 80–81
Elastic liposomes 77–85
Electrophoretic mobility 158, 439, 441
Encapsulation
  hydrophilic drugs 53, 116, 126, 206
  hydrophobic drugs 129
  protein 53, 57
Encapsulation efficiency 44–46, 53,

63, 65, 71, 72, 116, 121, 126, 139, 144, 174, 182–183, 206, 210, 286, 287
Endosomal release
　triggered release 123–124, 243–252
Endosomes 10, 12, 13, 214, 215, 349, 405, 406, 421, 422, 436, 438, 450, 528, 529, 546
Enhanced permeability and retention
　(EPR) effect 6, 279, 323, 336, 362
Entrapment efficiency 43, 52, 53, 56, 81, 153–154, 156–158, 452, 535
Environment-responsive liposomes 213–239
Enzyme linked immunosorbent assay (ELISA) 166, 186, 188, 308, 364–365, 367–368
Epaxal® 17
Epidermis 78, 82, 83, 384
Epithelial cells 191, 500, 504, 506, 512, 514
Estradiol 79–82, 85
Ethanol injection method 394–396
Ethinylcytidine 131
Ethosomes 78
Ethylene glycol vinyl ether (EGVE) 217
Extravasation 336
Extrusion 34, 37–38, 71, 80, 81, 92, 132–134, 141, 142, 166, 167, 206, 284, 343, 352, 433, 452–456, 533, 535, 542, 552–553

F

FACS. *See* Fluorescence-activated cell sorting
FITC-dextran 350, 352, 353, 355, 363, 364, 366
FITC-labeled liposomes 118
Flow cytometry 298, 351, 353–354, 514–515
Fluorescence-activated cell sorting
　(FACS) 184–186, 282, 287, 288, 291, 298, 353, 354, 493, 504–510, 515, 516, 523
Fluoro-deoxyuridine 131
Folate-modified liposomes 10
Folate receptor 10, 215, 280, 282, 287–292
Folic acid 215, 268, 280, 282, 283, 287, 288, 290–292
Freeze-drying 48, 52, 53, 61–63, 115, 116, 119–120, 206, 207, 327, 366
Fusogenic lipid 217, 436
Fusogenic liposomes 377–389
Fusogenic peptide 10, 436

G

Gadolinium-loaded liposomes
　imaging 7
　magnetic resonance imaging 7
　preparation 7
Galactosylated liposomes 11
Gamma scintigraphy (GS) 15, 321
Gas-carrying liposomes 114
Gene delivery

non-viral delivery 435, 449, 484
viral gene delivery 173, 177, 178, 426, 448, 487
Gene transfection 306, 426, 474, 484
Gentamycin 53
Giant liposomes 59, 69–70
Gilead Sciences 2
Green fluorescence protein (GFP) 13, 237, 491–494, 506, 508, 513–516, 518, 520

H

Hepatitis C 165, 168
Hepatocyte 216, 425, 426, 449
Hepatoma cells 217
High-pressure filter extrusion 133–134, 166–168
High-pressure homogenization 206, 208
High throughput screening 527
Human RNase I 258, 260, 266
Humectants 77
Humoral immunity 177
Hyaluronic acid 77
Hydrazone 8, 215, 217, 218, 220–229, 231–232, 234, 235, 238, 268
Hydrophobic drugs 114, 129–136, 216
Hypoxia 381–383

I

Idea AG 84
Immunization 166, 168–173, 178, 185–186
Immunoconjugate 216
Immunodominant epitope 165
Immunogenicity 17
Immunoliposomes 5, 6, 9–11, 18, 305–319, 329, 330, 364, 365, 367–368
Immunostimulatory CpG
　oligonucleotides 164, 167–170, 173
Inflexal® 17
Influenza virus 16, 449
Intra-articular injection 194
Intracellular delivery 10, 13, 16, 186, 380, 528, 529
Intracellular distribution 215, 350, 351, 355
Intranasal administration 193
Intraperitoneal administration 193
Intratracheal administration 193
Intravenous administration 2, 4, 19, 192
Intraventricular administration 193–194
Iontophoresis 17
Irinotecan 4
Ischemia 18, 306, 309–318, 361–374, 377, 389
Ischemia-reperfusion injury 361, 362

K

Ketoprofen 78, 84
Kupffer cells 2, 178, 192

L

Lamellarity 43, 44, 470
Langendorff............................309, 315–317, 365, 368–369
Large unilamellar vesicles (LUV)............................ 37, 470
Lecithine 34, 35
Lipidated ligands.. 268
Lipid dispersion...71, 123, 400
Lipid nanobubbles.. 484
Lipid packing 2
Lipopeptides.. 164
Lipophilic drugs 53, 56, 65, 88, 129, 130, 133–136
Lipoplatin™ 3, 8
Lipoplex
    serum stability.. 448
    transfection 430–432
Liposomal drug 5, 10, 126, 127, 243, 528
Liposomal product 3, 143
Liposomal topotecan 4
Liposomal vaccines 53, 57, 60, 164
Liposomal vinorelbine 4
Liver.. 2, 5–7, 15, 18, 192, 193, 301, 317, 363, 425, 426, 456, 539–541, 555, 557
Long-circulating liposomes 5–12, 195, 555
Luciferase assay 395, 400, 428, 431, 442, 476, 483
Lurtotecan (OSI-211)... 4, 129
Lymph nodes..7, 172, 191–193

M

Macrophage
    depletion... 189–201
    suicide.. 190
Magnetic liposomes... 19, 279–292
Magnetic nanoparticles 97, 280, 281, 283–287, 292, 487–494, 496–502, 505–508, 510, 511, 514–516, 519, 521, 522
Magnetic resonance imaging (MRI) 7, 14, 15, 280, 321–323, 331
Magnetite (Fe_3O_4)...............................19, 98, 106, 108, 109, 496, 520, 521
Magnetofection 487–523
Magnetoliposomes 97–110
Magnetophoresis 102–103
Mannan 178, 180, 181, 183, 185
Mannitol...34, 115–120, 124, 126, 127, 135, 174
Mannosylated liposomes 177–188
Marqibo® 4
Maximal tolerated dose (MTD) 300
Mechanosensitive channel.. 244
Melting temperature (T_m).. 33, 47
Membrane channel protein 2437

Membrane protein reconstitution.................................. 2437
Methotrexate 7, 130
Methylthiosulfonate (MTS)....................................246, 249, 282, 289, 296, 299, 302
Micelles 134–136, 214, 215, 217, 218, 223, 229, 235, 238, 253, 327, 362, 367, 420, 484
Microfluidization.. 69, 73
Minimate™ 142, 143
miRNA (microRNA) 449
Mitochondria
    apoptosis 10, 191, 295, 305
    ceramide 14, 2950
    drug delivery 295–303
Mitochondriotropics... 14, 295
Mitoxantrone.. 4, 9
Mozafari method.. 42
Mucoadhesivity 178
Multifunctional liposomes...................................... 213–239
Multilamellar vesicles (MLV) 32, 36, 37, 43–45, 53, 60, 61, 65, 71, 155, 428
Myelosupression 139
Myocardial infarction 10, 306, 309–310, 361–363, 365, 369, 372
Myocet® 2

N

Nanoliposomes 29–48
N-[1-(2,3-Dioleyloxy) propyl]-N,N, N-triethylammonium 59, 67
Necrosis 18, 305
NeoPharm Inc. 4
Neutraceuticals 30
N-glutarylphosphatidyl ethanolamine
    (NGPE)....................................307, 308, 323, 326
Niosomes 78
Non-phagocytic cells 189, 194
Nonviral vector 449
N/P ratio
    DNA delivery 464–466
    transfection 464
Nucleosome-specific antibodies 10
Nucleosomes... 10, 323

O

Octafluorocyclobutane.. 118
Oligomannose-coated l
    iposomes 12
Oligopeptides 335–346
Oncosis 305, 306
O-palmitoylmannan (OPM).................180–183, 185–187
Opsonin 6, 336
Osteoarthritis 78, 84
Osteoblasts 190
Osteoclasts.. 190

P

Paclitaxel, liposomal encapsulation 53, 139
Parathymic lymphnodes 193
Particulate carrier 166
Passive accumulation 323
Peptides 9, 11, 13, 52, 53, 57, 59,
    163–165, 167, 169, 173, 268, 339, 349, 352, 436
Perfluoropropane 16, 58, 474, 477, 483, 484
Pharmaceutical carrier 1
Pharmacokinetics 2, 7, 33, 83–84,
    132, 267, 448, 531, 538
Phase transition temperature 33, 84, 101,
    103, 108, 126, 208, 211, 454, 535
pH-degradable hydrazone bond 218, 238
pH-gradient 140
Phosphatidic acid (PA) 57, 59, 67, 264
Phosphatidylcholine 32, 33, 57, 59, 78–80,
    149, 173, 195–198, 205, 206, 208, 215, 308, 363
Phospholipid anchor 15, 130, 214, 268–272
Phospholipon 89–91, 94
Phospholipon GmbH 34
Photodynamic therapy (PDT)
    liposomes 10, 17, 18, 214
    phthalocyanine 214
Photofrin 18
Photolabile drugs 66
Photosensitive drugs 54, 56
pH sensitive lipoplex 435–443
pH-sensitive liposomes 12, 13, 235,
    527–542, 549, 553, 555, 556
Physical stability assessment 81
Picogreen™ 438
Picogreen® 437, 441
Pirarubicin 53
Polychelating polymer 321–333
Polydispersity index (PdI) 37, 94, 182, 352, 535, 557
Polyethylene glycol (PEG)
    PEGylated lipid 259, 263
    PEGylated lipoplex 438
    PEGylation of liposomes 336, 345
    stealth liposomes 166
Polyhistidine 215 7
Polyketals 215
Polyplex
    DNA delivery 522
    transfection 488, 522
Prednisolone (PRE) 56, 58, 64–66, 73
Pressure-freezing 117
Prodrugs 129, 130
Propamidine 190, 191
Proteoliposome 251
Pulmonary drug delivery 88

R

Radiolabelling
    biodistribution 301
    imaging 2807
    liposomes 11
Reconstitution 57, 63, 247, 249–254, 261
Regulon Inc. 8
Rehydration 52, 53, 63, 67, 381, 451, 452
Relaxation time
    gadolinium 322
    magnetic resonance imaging 321–323, 331
Remote loading 139–145, 342, 396
Reperfusion 18, 306, 309–316,
    318, 361–363, 369, 370, 372, 389
Reticulo-endothelial system
    (RES) 2, 5, 6, 336, 345, 529, 541
Retinol 77
RGD peptides 11
Riboflavin 54, 56, 62
RISC. See RNA-induced silencing complex
RNA-induced silencing complex 446, 447
RNA interference (RNAi) 173, 445–449, 461

S

Serum-resistant lipoplex 425–433
SibTech, Inc. 259, 260
siRNA. See Small interfering RNA
Skin. See also Epidermis; Stratum corneum
    human 82, 83
    penetration enhancement 78
    permeability 82–83
Small interfering RNA 17, 173, 217,
    218, 346, 445–456, 461–471, 487–489, 491–494,
    497–504, 506–509, 511–514, 517, 518
Sonication technique, small unilamellar
    liposomes 35
Sonopore 476, 478, 479, 481, 485
Soy phosphatidylcholine (SPC) 4, 132,
    141, 164, 167, 378, 379
Sphingomyelin (SM) 2, 57
Spleen 2, 15, 168, 169, 171,
    186, 190–195, 301, 539–541, 555, 557
Stealth liposomes
    biodistribution 166
    circulation time 363, 555
Stearyl triphenylphosphonium (STPP) cation 296–301
Stearyl triphenylphosphonium (STPP)
    liposomes 295–301
Stereotaxical injection 193
Steric stabilization 2, 528, 529
Stewart assay reagent 59, 154
Stimuli-sensitive liposomes 12–13

STPP cation. *See* Stearyl triphenylphosphonium cation
Stratum corneum 78, 82, 84
Subcutaneous administration 7, 192–193
Subtilisin 259–262, 265
Superparamagnetic iron oxide 98
Surface charge 2, 67, 158, 178, 462
Surfactant 45, 78, 80, 81, 101
Synovial cavity 194

T

Tangential flow filtration 142
Targeted drug delivery 11, 13, 18, 238, 528
Targeting protein 258, 260, 263
TAT-peptide 13, 14, 349–358
Taxol® 4
T cell induction 168–171
Testes 194
Tetraetherlipid
    extraction 88–90
    thermostability 92, 93
Thermostability testing 92
Thioguanin 53, 54
Thiol-reactive functions 268
Thiopropionate 217, 218
Tocopherol 132, 133, 164, 167
Transdermal drug delivery 77–85
Transferosomes 17
Transferrin (Tf) 10, 178
Transmission electron microscopy (TEM)
    bubble liposomes 475, 477–478
    surface morphology 80
    vesicular shape 80
Transphosphatidylation 343
Trauma 377
Triggered liposomal release 244
Triolein (TO) 59
Trityls 215, 216
Tumor
    growth inhibition 19, 301–302
    imaging 16
    targeting 323

U

Ultrasette™ 142, 143
Ultra-sonography (US) 14, 321
Ultrasound
    ultrasound-responsive liposomes 113–127
    ultrasound-triggered release 123–125
Uranylacetate 100, 310, 475, 478

V

Vancomycin 53
Vasoactive intestinal peptide (VIP) 11
Vesicular phospholipid gels (VPGs) 205–211
Vincristine 2, 129
Viral vector 448
Virosomes 13–14, 16, 17

W

Wound care 384–385
Wound healing 385, 386

Z

Zeta potential 43, 44, 79, 80, 91,
    92, 94, 149, 150, 154, 158, 182, 297, 429, 433, 438,
    439, 441, 464, 469
Zetasizer 79, 80, 158, 179,
    182, 342, 345, 451, 521
Zwitterionic phospholipids 97